Quantum Mechanics in Curved Space-Time

NATO ASI Series

Advanced Science Institutes Series

A series presenting the results of activities sponsored by the NATO Science Committee, which aims at the dissemination of advanced scientific and technological knowledge, with a view to strengthening links between scientific communities.

The series is published by an international board of publishers in conjunction with the NATO Scientific Affairs Division

A	**Life Sciences**	Plenum Publishing Corporation
B	**Physics**	New York and London
C	**Mathematical and Physical Sciences**	Kluwer Academic Publishers Dordrecht, Boston, and London
D	**Behavioral and Social Sciences**	
E	**Applied Sciences**	
F	**Computer and Systems Sciences**	Springer-Verlag
G	**Ecological Sciences**	Berlin, Heidelberg, New York, London,
H	**Cell Biology**	Paris, and Tokyo

Recent Volumes in this Series

Volume 227—Dynamics of Polyatomic Van der Waals Complexes
edited by Nadine Halberstadt and Kenneth C. Janda

Volume 228—Hadrons and Hadronic Matter
edited by D. Vautherin, F. Lenz, and J. W. Negele

Volume 229—Geometry and Thermodynamics: Common Problems of Quasi-Crystals, Liquid Crystals, and Incommensurate Systems
edited by J.-C. Tolédano

Volume 230—Quantum Mechanics in Curved Space-Time
edited by Jürgen Audretsch and Venzo de Sabbata

Volume 231—Electronic Properties of Multilayers and Low-Dimensional Semiconductor Structures
edited by J. M. Chamberlain, L. Eaves, and J.-C. Portal

Volume 232—New Frontiers in Quantum Electrodynamics and Quantum Optics
edited by A. O. Barut

Volume 233—Radiative Corrections: Results and Perspectives
edited by N. Dombey and F. Boudjema

Volume 234—Constructive Quantum Field Theory II
edited by G. Velo and A. S. Wightman

Series B: Physics

Quantum Mechanics in Curved Space-Time

Edited by

Jürgen Audretsch

Konstanz University
Konstanz, Federal Republic of Germany

and

Venzo de Sabbata

Bologna University
Bologna, Italy

Plenum Press
New York and London
Published in cooperation with NATO Scientific Affairs Division

Proceedings of a NATO Advanced Research Workshop
on Quantum Mechanics in Curved Space-Time,
held May 2-12, 1989,
in Erice, Sicily, Italy
at Ettore Majorana Centre for Scientific Culture

Library of Congress Cataloging in Publication Data

NATO Advanced Research Workshop on Quantum Mechanics in Curved Space-Time (1989: Erice, Italy)
Quantum mechanics in curved space-time / edited by Jürgen Audretsch and Venzo De Sabbata.
p. cm.—(NATO ASI series. Series B, Physics; vol. 230)
"Proceedings of a NATO Advanced Research Workshop on Quantum Mechanics in Curved Space-Time, held May 2-12, 1989, in Erice, Sicily, Italy"—T.p. verso.
"Published in cooperation with NATO Scientific Affairs Division.
Includes bibliographical references and index.
ISBN 0-306-43661-2
1. Space and time—Congresses. 2. Quantum field theory—Congresses. 3. Quantum gravity—Congresses. 4. Astrophysics—Congresses. I. Audretsch, Jürgen, date. II. De Sabbata, Venzo. III. North Atlantic Treaty Organization. Scientific Affairs Division. IV. Title. V. Title: Curved space-time. VI. Series: NATO ASI series. Series B, Physics; v. 230.
QC173.59.S65N37 1989 90-14188
530.1'4—dc20 CIP

A Division of Plenum Publishing Corporation
233 Spring Street, New York, N.Y. 10013

Printed in the United States of America

PREFACE

Quantum mechanics and quantum field theory on one hand and Gravity as a theory of curved space-time on the other are the two great conceptual schemes of modern theoretical physics. For many decades they have lived peacefully together for a simple reason: it was a coexistence without much interaction. There has been the family of relativists and the other family of elementary particle physicists and both sides have been convinced that their problems have not very much to do with the problems of the respective other side. This was a situation which could not last forever, because the two theoretical schemes have a particular structural trait in common: their claim for totality and universality. Namely on one hand all physical theories have to be formulated in a quantum mechanical manner, and on the other hand gravity as curved space-time influences all processes and vice versa. It was therefore only a question of time that physically relevant domains of application would attract a general interest, which demand a combined application of both theoretical schemes. But it is immediately obvious that such an application of both schemes is impossible if the schemes are taken as they are. Something new is needed which reconciles gravity and quantum mechanics. During the last two decades we are now doing the first steps towards this more general theory and we are confronted with fundamental difficulties.

This field of research was subject of the NATO Advanced Study Institute on "Quantum Mechanics in Curved Space-Time" which was held at the Ettore Majorana Center for Scientific Culture in Erice (Sicily, Italy) from May 2nd through May 12th, 1989. It was at the same time the 11th course of the International School of Cosmology and Gravitation of the Center.

The fundamental difficulties encountered in reconciling quantum mechanics with gravity are the reason that there are today still considerable uncertainties as to what a quantum theory of gravity should be.

Therefore a reasonable approximation to the final version of a full quantum theory of gravity is of great importance, because it i.) will help to clarify from a physical point of view the underlying conceptual problems causing the intrinsic difficulties mentioned above and at the same time ii.) will lead to physically important results which may be applicable in astrophysically relevant situations.

Such an approximation which has successfully been studied over the last fifteen years, is the semi-classical approximation in which the gravitational field is retained as a classical background field, interacting with quantized matter. This is today generally regarded as an intermediate step leading to conceptual solutions and to characteristic effects, the physical structure of which will survive in a more complete theory, in which these effects will show up among others or as limiting cases.

In the very early universe, matter must be described by quantum fields. Effects of curvature, topology and non-zero temperature have been of considerable importance. This applies as well to astrophysical situations in which black holes play a role. In both cases the application for the semi-classical approximation seems to be justified. The Advanced Study Institute considered a variety of aspects of this theory and put also emphasis on the following fields which precede or succeed the semi-classical approximation in the hierarchy of theories: Experimental neutron interferometry provides us with devices to investigate fundamental quantum mechanical effects related to topology and the influence of weak gravitational fields. An axiomatic of space-time geometry can be based on quantum mechanical fields in first quantization. There is a path integral approach to quantum mechanics in curved space-time. On the other end of the spectrum of subjects the semi-classical approximation was exceeded by discussions of quantum strings, superspace cosmology and the nonlinear sigma model.

With regard to the central subject the following topics have been studied in detail: Accelerated particle detectors and the operational meaning of the particle concept in curved space-time. Equivalence principle in the quantum domain. Back reaction and renormalization problems. Finite temperature effects and the laws of black hole thermodynamics in a very general case. Mutual interaction between several quantum fields. Self-consistent cosmologies and realizations of the inflationary paradigm. Thus a survey of the empirical aspects, the conceptual basis and the related fundamental problems as well as the astrophysical and cosmological applications has been obtained.

The editors wish to conclude this preface by thanking the NATO Scientific Affairs Division which provided the basic funding of the School, and the Ettore Majorana Center for Scientific Culture which made a considerable financial contribution. We should like to express our strong appreciation to Dr. Gabriele and his staff at the Center in Erice who provided excellent administrative services and continuous assistence. Finally we wish to thank the lecturers and seminar speakers, who did so much to make this School successful, and last but not least the students and other participants for contributing to the very stimulating scientific and human atmosphere.

Jürgen Audretsch
Konstanz, F.R. Germany

Venzo de Sabbata
Bologna, Italy

December, 1989

CONTENTS

LECTURES

SEMINARS

NEUTRON INTERFEROMETRY - MACROSCOPIC MANIFESTATIONS OF QUANTUM MECHANICS

Samuel A. Werner and Helmut Kaiser

Physics Department and Research Reactor

University of Missouri, Columbia, Missouri, U.S.A.

ABSTRACT

The perfect-silicon-crystal neutron interferometer has provided us with a very sophisticated didactic device to investigate fundamental quantum mechanical phenomena occurring over macroscopic distances. In these lectures, we review the history and progress in this field over the past 15 years. We describe in some detail the following three experiments: (a) Gravitationally-induced quantum interference; (b) the effect of the Earth's rotation upon the quantum mechanical phase of the neutron (Sagnac Effect); (c) the quantum phase shift of a neutron diffracting around a charged-electrode (Aharonov-Casher effect).

INTRODUCTION

A Mach-Zehnder interferometer is topologically equivalent to a ring. At some point A on the ring an incident wave is brought in and split coherently into two parts, one propagates clockwise and the other propagates counterclockwise around the ring. At some point B on the opposite side of the ring, these two coherent waves are recombined and allowed to leak out to be detected by an observer, who interprets the results in terms of constructive and destructive interference of waves. This idea is not new. In fact, it goes back more than 2000 years to the time of Alexander the Great, which is illustrated in Fig. 1. Clearly, this device generates many interference harmonics.

Lectures at: International School of Cosmology and Gravitation. Erice, Italy. May 2-12, 1989.

Neutron interferometry, based upon the Bonse-Hart[1] perfect-silicon-crystal x-ray interferometer, was first demonstrated by Rauch, Treimer and Bonse[2] in Vienna in 1974. Since that time, this single-crystal device has proven to be a marvelous didactic laboratory for probing and elucidating the fundamental quantum mechanical principles of nature.

Fig. 1. The tuba of Alexander the Great.

A list of general references is given in Table I. Four years after the original experiment of Rauch et al., a conference on neutron interferometry was held in Grenoble, France. The conference proceedings were published in a book by Oxford University Press. In 1980 one of us (S.A. Werner) wrote an article for Physics Today summarizing the various neutron interferometry experiments up to that time. A.G. Kein and S.A. Werner wrote two review articles on neutron optics which took a broad view toward both wave-front division and amplitude splitting interference phenomena with thermal and cold neutrons. The one appeared in Reports on Progress in Physics and the other as a chapter in a three-volume book on Neutron

TABLE I

NEUTRON INTERFEROMETRY

GENERAL REFERENCES:

(1) Neutron Interferometry, Proceedings of an Internat. workshop held June 5-7, 1978 in Grenoble, France. At the ILL, Ed. U. Bonse & H. Rauch. Oxford Univ. Press (1979).

(2) "Neutron Interferometry," S.A. Werner Physics Today, Dec. 1980.

(3) "Neutron Optics." A.G. Klein & S.A. Werner, Rep. Prog. Phys. **46**, pp. 259-335, 1983.

(4) "Neutron Optics." S.A. Werner & A.G. Klein. Chap. 4 of Neutron Scattering. Methods of Experimental Physics, Vol. 23A, Ed. K. Skold & D.L. Price. Academic Press, (1986).

(5) New Techniques and Ideas in Quantum Measurement Theory. Conf. in honor of Eugene Wigner in NYC, Jan. 21-24, 1986. Ed D.M. Greenberger, Annals of N.Y. Acad. of Sciences. Vol. 480, Dec. (1986).

(6) Matter Wave Interferometry, Proceedings of an Internat. Workshop in Vienna, Austria, Sept. 14-16, 1987. Ed. G. Badurek, H. Rauch, A. Zeilinger. North-Holland (1988).

(7) Neutron Optics, Varley F. Sears. Oxford University Press. (1989).

TABLE II a

NEUTRON INTERFEROMETRY (1974-1989)

1. **First Test of Si-crystal Interferometer.**
   Vienna (1974)

---

2. **Sign Change of Fermion ψ During 2π Precession.**
   ILL, MURR (1975, 1976).

---

3. **Gravitationally - Induced Quantum Interference.**
   Ann Arbor, MURR (1975, 1980, 1985, 1988)

---

4. **Neutron Sagnac Effect.**
   - **Earth's rotation. MURR (1979).**
   - **Turntable. MIT (1984).**

---

5. **Neutron Fizeau Effect.**
   - **Moving boundaries, ILL (1981, 1985)**
   - **Stationary boundaries, MURR (1985, 1988)**

---

6. **Search for Non-linear Terms in the Schrodinger Equation.**
   - **LL- 2-crystal interferometer. MIT (1981).**
   - **Long-λ Fresnel diffraction, ILL (1981).**

---

7. **Search for an Aharonov-Bohm Effect for Neutrons.**
   MIT (1981).

---

8. **Measurement of Longitudinal Coherence Length of a Neutron Beam.**
   MURR (1983).

TABLE II b

NEUTRON INTERFEROMETRY (1974-1989)

(9) **Coherent Superposition of Spin States ("Wigner Phenomena") ILL (1983).**

---

(10) **Search for Quaternions in Quantum Mechanics.**
MURR (1984).

---

(11) **Quantum Interference in Accelerated Frames.**
ILL (1983).

---

(12) **Search for New Guage Fields - Rotating U Rod.**
MIT (1983).

---

(13) **Precision Measurement of Scattering Lengths ^{149}Sm, ^{235}U, ^{3}He, ^{3}H (Four-Body Nuclear Interaction).**
ILL, MURR (1975-1985).

---

(14) **Observation of the Aharonov-Casher Effect**
MURR (1989).

---

(15) **Stochastic vs. Deterministic Attenuation of a Neutron Beam.**
ILL (1987).

---

(16) **Neutron Spin-Pendellosung Resonance.**
MIT (1988).

Scattering. In 1986, a wonderful conference was held in New York City in honor of Eugene Wigner on New Techniques and Ideas on Quantum Measurement Theory. The proceedings were published in a book by the New York Academy of Sciences. About one-half of the papers were theoretical, and the others dealt with subjects ranging from neutron interferometry to Bell's inequalities. In the fall of 1987 a conference was held in Vienna, in connection with the Schrödinger centennial, on Matter Wave Interferometry. There were many papers on neutron interferometry, and others on electron and atom beam interference phenomena. The proceedings are published in a North-Holland book. Recently, a beautiful book by Varley Sears on Neutron Optics has appeared which traces the development of this field from the early experiments of Fermi up through fairly recent neutron interferometry experiments.

We show in Table II a list of the important neutron interferometry experiments to date. The labels MURR means University of Missouri, ILL means Institut Laue Langevin, and MIT means Massachusetts Institute of Technology. We would like to focus on three of these experiments:

(1) Gravitationally-induced quantum interference.
(2) Neutron Sagnac Effect.
(3) Aharonov-Casher Effect.

But, before doing that we will describe the Bonse-Hart perfect-silicon-crystal interferometer, and remind you of the basic idea of a quantum mechanical phase shift caused by a potential.

PERFECT CRYSTAL INTERFEROMETERS AND QUANTUM PHASE SHIFTS

An important breakthrough in interferometry in the angstrom wavelength range of wavelengths was achieved by Bonse and Hart in 1964. Their x-ray interferometer is based upon two important ideas. First, Bragg reflection is used for splitting and recombining the two interfering beams, and second beam splitters and mirrors which make up the interferometer are machined from a monolithic piece of a large, highly perfect silicon single crystal.

A schematic diagram of the original, three-crystal, LLL-type device is shown in Fig. 2. A photograph of three of our interferometers at the University of Missouri is shown in Fig. 3. It consists of three perfect crystal slabs cut perpendicular to a set of strongly reflecting Bragg planes, typically (220). The distances between the slabs are usually 3 to 5 centimeters and must be equal to within approximately 1μm. A collimated, nominally monochromatic beam is directed along the line SA and is coherently split by Bragg reflection in the first crystal slab. These two coherent beams are split again in the second crystal slab near points B and C; two of these beams are brought back together to overlap and interfere in the third crystal slab near point D. This device is topoligically identical to the Mach-Zehnder interferometer of classical optics. The label LLL for this interferometer type signifies that it involves three Laue transmission-geometry crystals.

If one wants to fully understand the operation of this interferometer, it is necessary to describe the Bragg reflection process within the crystal slabs using the dynamical theory of diffraction. However, these details are not really necessary in order to understand the conceptual operational features. If the beam traversing path I is phase-shifted relative to the beam traversing path II, by increasing the "optical" path length (say, by introducing a slab of material in this path), the count rates in detectors C_2 and C_3 will change. It can be shown on very general grounds that the expected intensities in these detectors, as a function of the phase shift $\Delta\beta$, are of the form

$$I_2 = a_2 - b_2 \cos(\Delta\beta) \quad , \tag{1}$$

and

$$I_3 = a_3 + b_3 \cos(\Delta\beta) \quad , \tag{2}$$

where the constants a_2, b_2, a_3 and b_3 depend upon the incident beam intensity, the structure of Si and its nuclear scattering amplitude. Furthermore $b_2 = b_3$, and $a_2/a_3 \approx 2.7$. The sum $(I_2 + I_3)$ of the intensities of the outgoing interfering beams is a constant. Thus, one sees that as the phase shift $\Delta\beta$ is varied, the intensity is swapped back and forth between C_2 and C_3.

When one realizes that there are of order 10^9 oscillations of the deBroglie wave on each path, one must admit that it is a miracle that this device has the stability to work as an interferometer. There are very stringent requirements on the microphonic and thermal stability of the device. In our set-up the interferometer

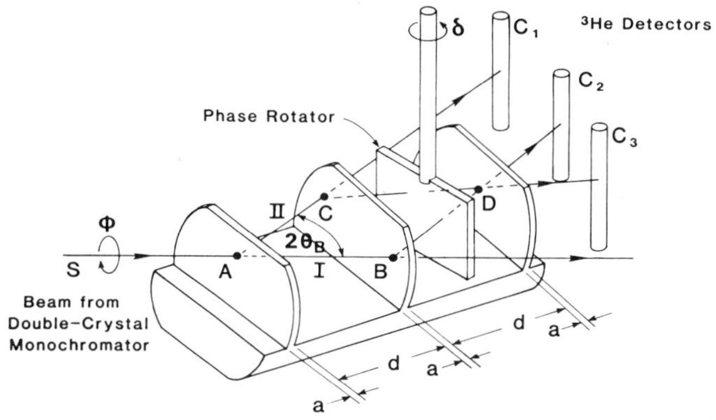

Fig. 2. Schematic diagram of our LLL Bonse-Hart Si-crystal interferometer. The dimensions are $d = 34.518 \pm 0.002$mm, $a = 2.464 \pm 0.002$mm, for the interferometer used in the gravity experiments.

Fig. 3. Photograph of three LLL interferometers

is positioned inside of a metallic, isothermal enclosure, which is mounted on a vibration isolation pad which restricts the floor vibrations (normally severe in a reactor hall) from reaching the interferometer.

The incident beam need not be precisely monochromatic. Typically, experiments are carried out with $\delta\lambda/\lambda \approx 0.01$. The important aspect of this interferometer is that it utilizes Bragg reflections in perfect crystals, where the Darwin width is about 1/10 arcsec. Thus, the Bragg reflection process defines the wavelength along a given trajectory (ray) to within about 1 part in 10^6. This precise wavelength definition is therefore accomplished by the interferometer itself, and not by the incident beam preparation.

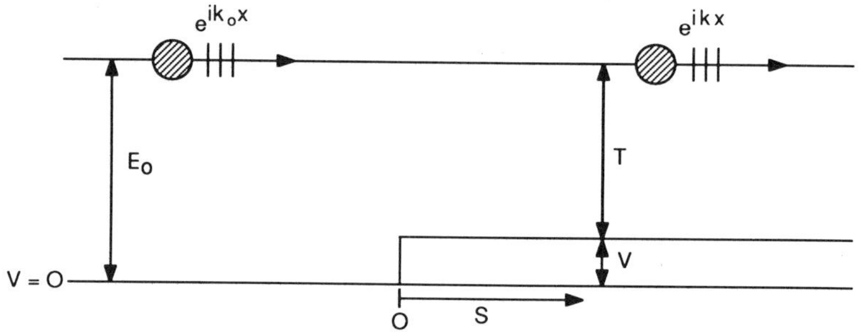

Fig. 4. Diagram used in deriving the formula, Eq. (6), for the phase shift caused by a potential.

Let us turn now to the most elementary description of a quantum phase shift caused by a potential. Consider a thermal neutron (wavelength $\lambda \approx 1$Å, energy $E_0 \approx 82$ meV) moving in the x-direction as shown in Fig. 4. At a point $S = 0$ it encounters a potential step of small magnitude V; conservation of energy requires

$$E_0 = T + V \quad , \tag{3}$$

where T is the kinetic energy in the region $S > 0$. For non-relativistic neutrons this means that the neutron's momentum p is related to its initial momentum by

$$\frac{p_0^2}{2m} = \frac{p^2}{2m} + V \tag{4}$$

Writing $p = \hbar k$, where $k = 2\pi/\lambda$, we see that the shift in wavevector due to the potential is

$$\Delta k = -\frac{1}{2}\frac{V}{E_0}k_0 \quad , \tag{5}$$

and the phase shift is

$$\Delta\beta = \Delta k \cdot S = -\frac{1}{2}\frac{V}{E_0}k_0 S \quad . \tag{6}$$

This is a simple but fundamental formula which we will now utilize.

GRAVITATIONALLY-INDUCED QUANTUM INTERFERENCE

Gravity and quantum mechanics do not simultaneously play an important role in most experimentally accessible phenomena in physics. This fact poses a real challenge to fundamental physics. However, a neutron interferometer experiment, for which the outcome necessarily depends upon both the gravitational constant G and Planck's constant h was first carried out by Colella, Overhauser and Werner (COW)[3] and subsequently a series of increasingly precise experiments have been carried out by our group.[4,5]

A schematic diagram of the overall experimental setup is shown in Fig. 5. A thermal neutron beam is brought out of the University of Missouri Research Reactor (MURR) through a helium gas-filled beam tube, and monochromated by a double-crystal monochromator assembly (using either pyrolytic graphite crystals or copper crystals). The monochromatic beam then passes through a series of collimating slits onto the interferometer. The double-crystal monochromator provides a variable wavelength incident beam directed along the local North-South axis of the Earth; a fact which we will see is important in these experiments.

The experimental procedure involves turning the interferometer, including the entrance slit and the three detectors C_1, C_2 and C_3 about the incident beam line SAB as shown in Fig. 2. Neutrons are counted at each angular setting ϕ for a preset length of time. This procedure allows the path CD to be somewhat higher above the Earth's surface than the beam path AB. The difference in the Earth's gravitational potential between these two levels results in a quantum mechanical phase shift of the neutron wave on trajectory ACD relative to the trajectory ABD. The phase accumulated on the rising path AC is equal to the phase accumulated on the opposite rising path BD. If we assume that the gravitational potential is Newtonian, then

$$V(\phi) = mgH_0 \sin\phi \quad , \tag{7}$$

where m is the neutron mass, g is the local acceleration due to gravity, and H_0 is the perpendicular distance between the lines AB and CD. We note that $mg = 1.023 \times 10^{-9}$ eV/cm at Columbia, Missouri. The incident neutron kinetic energy is

$$E_0 = \frac{\hbar^2 k_0^2}{2m} \quad ,(= 0.0406\text{eV for } \lambda = 1.419\text{Å}). \tag{8}$$

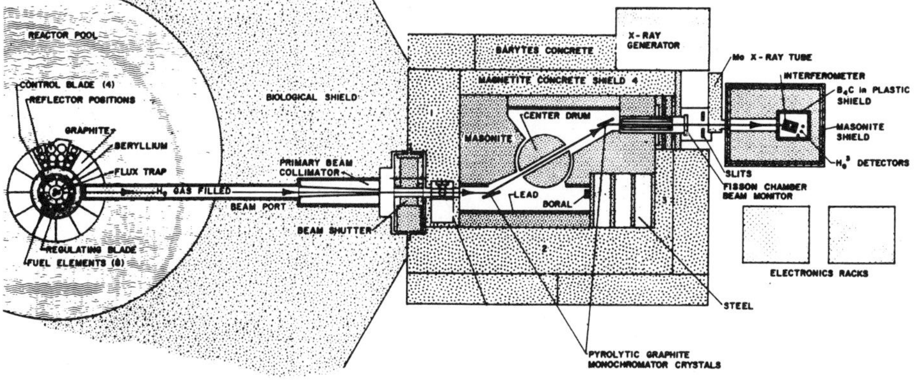

Fig. 5. Schematic diagram of the double-crystal monochromator assembly and the interferometry set-up at beam port B at MURR.

Using Eq. (6) we have the gravitationally-induced phase shift given as

$$\Delta\beta_{grav} = -2\pi m^2 \frac{g}{h^2} \lambda A_0 \sin\phi = q_{grav} \sin\phi \tag{9}$$

where A_0 is the area enclosed by the beam trajectories in the interferometer, and λ is the neutron wavelength. We note that this formula contains both the gravitational constant and Planck's constant. As the interferometer is turned through various "tilt angles" ϕ, always maintaining the Bragg condition, we expect to observe quantum oscillations in the count rates for the two detectors C_2 and C_3, induced by the gravitational field of the Earth.

The results of an experiment carried out with incident neutrons of wavelength $\lambda = 1.419$ Å is shown in Fig. 6.(4) The loss of contrast was originally interpreted as being due to the fact that the interferometer bends and warps under its own weight (on the scale of angstroms) as it is rotated about the incident beam axis, which is not an axis of elastic symmetry of the Si-crystal interferometer. We have studied these effects in detail with *in situ* x-ray experiments. X-rays were directed along the same incident beam path and the interfering x-ray beams were monitored as a function of rotation angle. The effect of gravity on x-rays (gravitational red shift) over the distances involved in the interferometer is negligible. The frequency of oscillation of the interferogram due to bending was therefore measured directly with x-rays and subtracted from the frequency of oscillation measured with neutrons, leaving only the effects of gravitationally-induced quantum interference. Recently, Horne(6) has reanalyzed these COW-type experiments, pointing out that the three-crystal LLL interferometer is not a simple two-path device, but really an eight-path interferometer. This has the effect that the single interferometer area A_0 appearing in the COW phase shift formula, Eq. (9), should be replaced by a dynamical diffraction intensity-weighted average over three areas, $A_0, A_0 + \delta A$ and $A_0 - \delta A$, where $\delta A/A_0 = \Gamma a/d$. Here a is the thickness of each of the three slabs, d is the distance between them and Γ is a factor (less than unity) dependent upon the misset angle $\Delta\theta$ of a given incident ray from the exact Bragg condition. The eight paths are shown in Fig. 7. Using Horne's theory, we find that a correction of 4.8% to the experimental frequency of the gravity interferogram should be made before comparison with the COW phase shift formula. The fact that the experiment actually involves more than one interferometer area also explains the loss of contrast with increasing $|\phi|$. Similar conclusions, using a complete spherical-wave dynamical diffraction treatment, were reached by Bonse and Wrablewski(7,8) in analyzing their acceleration-induced interferometry experiment. We have shown

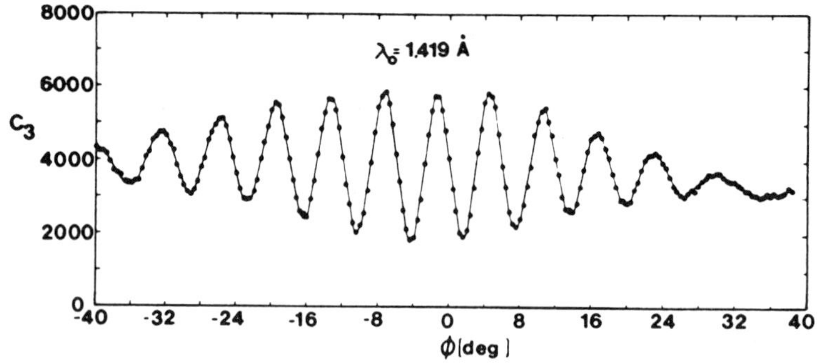

Fig. 6. Gravitationally-induced quantum interferogram taken with the interferometer of Fig. 2 with 1.419 Å neutrons. (Staudenmann et al. Ref. 4).

that this geometrical correction to the COW formula for the frequency of oscillation is in fact given by[5]

$$q_{grav} = \left(1 + \frac{2}{3}\frac{a}{d}\right) q_{cow} \quad . \tag{10}$$

For our interferometer $a/d = 0.0714$, so that

$$q_{grav} = 1.0476 q_{cow} \quad . \tag{11}$$

We now turn to comparing this predicted frequency of the gravitationally-induced quantum interferogram with our most recent experiments. We have followed a somewhat different procedure than in the original COW experiments. The phase shift $\Delta\beta_{grav}(\phi) = q_{grav} \sin\phi$ is measured directly by first setting $\phi = 0$, and rotating the phase rotator through successive angle δ. Due to the neutron-nuclear optical potential of the phase rotator, this results in a sinusoidal interferogram. Tilting the interferometer through an angle ϕ, and repeating the rotation of the phase rotator, gives rise to another sinusoidal interferogram. The difference in phase between these two interferograms is $\Delta\beta_{grav}(\phi)$. The results of a series of very accurate measurements, using 1.417 Å neutrons are summarized in Fig. 8. The abscissa is $\sin(\phi - \phi_0)$, where ϕ_0 is a correction due to the Sagnac Effect ($\phi_0 = 1.41°$), which will be discussed below. The slope of $\Delta\beta_{grav}$ vs. $\sin(\phi - \phi_0)$ is the experimental frequency q_{exp} of the gravity interferogram. To compare this frequency with q_{grav}, we must make a correction for bending and for the Sagnac effect, namely

$$q_{grav} = (q_{exp}^2 - q_{Sagnac}^2)^{1/2} - q_{bend} \tag{12}$$

Theory gives $q_{Sagnac} = 1.45$ rad, and with x-rays we measure $q_{bend} = 1.41$ rad. Thus, our current experimental result is

$$q_{grav}(observed) = 58.72 \pm 0.03\, rad \quad . \tag{13}$$

Eq. (11) predicts

$$q_{grav}(theory) = 59.19\, rad. \tag{14}$$

Thus, the observed frequency due to gravity is 0.8% lower than theory predicts. Recently, Greene and Layer[9] have suggested that this discrepancy might be due to the fact that x-rays interrogate a slightly different region of the *Si* crystal slabs than the neutron beams. We intend to pursue this suggestion experimentally.

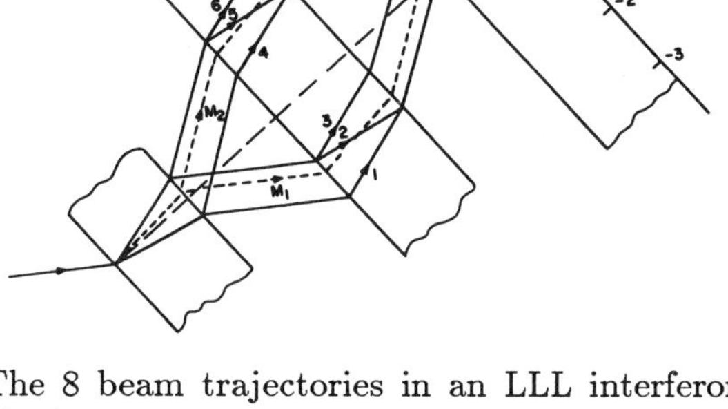

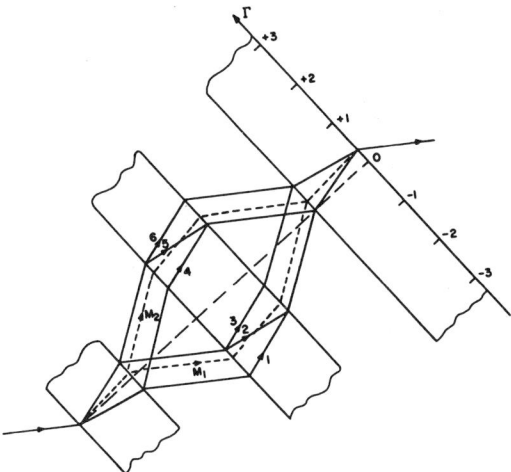

Fig. 7. The 8 beam trajectories in an LLL interferometer (M. Horne. Ref. 6).

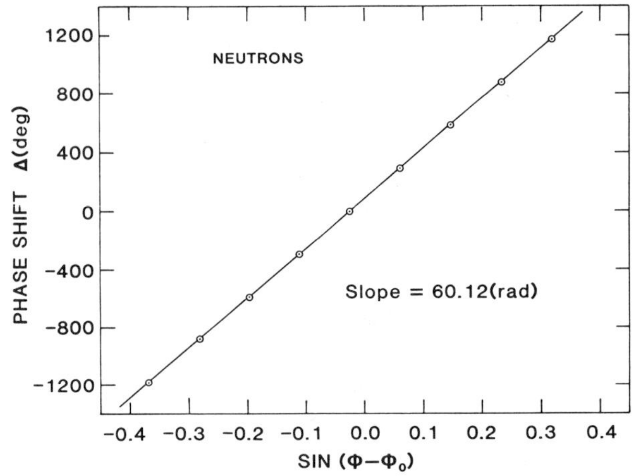

Fig. 8. Gravity-induced phase shift as a function of the sine of the interferometer tilt angle ϕ. (S.A. Werner et al. Ref. 5).

THE NEUTRON SAGNAC EFFECT

In 1913 the French scientist M.G. Sagnac demonstrated that angular rotation can be detected by means of optical interferometry.(10) This effect is the physical basis for the navigational Sagnac gyroscope. Subsequently, in 1925 Michelson, Gale and Pearson carried out an heroic experiment in which they constructed an interferometer in the form of a rectangle of size 2010 ft × 1113 ft and were able to detect the retardation of light due to the Earth's rotation, corresponding to about 1/4 of a fringe, in agreement with theory.(11) The experiment which we will now describe is the quantum mechanical analog of the MGP experiment.

Since the gravitationally-induced quantum interference experiments are carried out on the surfaces of our rotating Earth, a non-inertial frame, the Hamiltonian governing the neutron's motion is

$$\mathcal{H} = \frac{p^2}{2m_i} - m_g\vec{g}\cdot\vec{r} - \vec{\omega}\cdot\vec{L} \quad . \tag{15}$$

Here, the angular momentum of the neutron's motion about the center of the Earth ($\vec{r} = 0$) is

$$\vec{L} = \vec{r}\times\vec{p} \quad , \tag{16}$$

where the canonical momentum $\vec{p}$ in terms of the neutron's velocity $\vec{v}$ and the angular rotation frequency $\vec{\omega}$ of the Earth is

$$\vec{p} = m_i\vec{v} + m_i\vec{\omega}\times\vec{r} \quad . \tag{17}$$

We are more careful now to distinguish between the neutron's inertial mass m_i and its gravitational mass m_g.

The third term in Eq. (15) gives rise to the Coriolis force and the centrifugal force. In general, the phase shift in an interferometer experiment can be calculated by evaluating the line integral of the canonical momentum around the interferometer loop, namely

$$\Delta\beta = \frac{1}{\hbar}\oint\vec{p}\cdot d\vec{r} \quad . \tag{18}$$

It is a straight forward matter to see that $\Delta\beta$ is given by two terms

$$\Delta\beta = \Delta\beta_{grav} + \Delta\beta_{Sagnac} \quad , \tag{19}$$

where $\Delta\beta_{grav}$ is given by Eq. (9), with $m^2 \to m_i m_g$, and the Sagnac phase shift is[12]

$$\Delta\beta_{Sagnac} = \frac{2m_i}{\hbar}\vec{\omega} \cdot \vec{A}_0 \tag{20}$$

Here $\vec{A}_0$ is the normal area vector of the interferometer loop. For an incident beam directed along the local (horizontal) north-south axis of the Earth, Eq. (20) gives

$$\Delta\beta_{Sagnac} = \frac{4\pi m_i \omega A_0}{h} \cos\Theta_L \cos\phi = q_{Sagnac} \cos\phi \quad . \tag{21}$$

where $\Theta_L = 51.37°$ is the colatitude angle at Columbia, Missouri. The fact that this phase shift depends on the cosine of the interferometer tilt angle ϕ, explains why its contribution q_{Sagnac} to the interferogram frequency must be added in quadrature to the effect of gravity, q_{grav}. Numerically q_{Sagnac} is only about 2% of q_{grav}.

However, in an experiment carried out by Werner, Staudenmann and Colella[13] a vertically directed beam was used as shown in Fig. 9.[14] The Sagnac phase shift was measured (using the phase-rotator technique described earlier for the gravity experiments) as a function of the interferometer orientation angle ϕ about the vertical axis. From symmetry, it is clear that there is no ϕ-dependent gravity-induced phase shift for this geometry. In terms of the colatitude angle Θ_L at the point on the Earth's surface where the experiment was done, and the angle ϕ, the Sagnac phase shift (from Eq. (20)) is given by

$$\Delta\beta_{Sagnac} = \frac{2m_i}{\hbar}\omega A_0 \sin\Theta_L \sin\phi \quad . \tag{22}$$

The experimenal results are shown in Fig. 10. When the normal area vector $\vec{A}_0$ points West or East, the phase shift is zero; while when $\vec{A}_0$ is directed North or South it is + 95° and −95° respectively. This experimental result is in reasonable

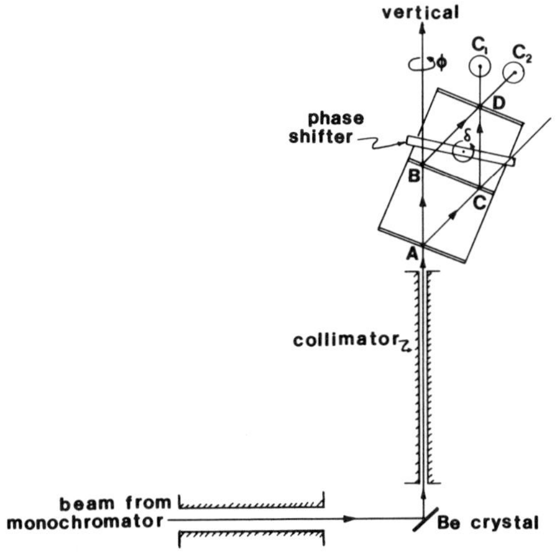

Fig. 9. Schematic diagram of the vertical beam geometry used in observing the neutron Sagnac Effect. (S.A. Werner et al. Ref. 13).

agreement with Eq. (22) which predicts $\Delta\beta_{sagnac}$ should be $+92°$ and $-92°$ for the North and South orientations, respectively.

It is interesting to note that the results of this experiment depend only upon the inertial neutron mass m_i, while the results of the gravity experiment depend upon the product of the gravitational mass m_g with the inertial mass m_i. Thus, one can interpret the combination of the two experiments as a quantum mechanical interference measurement of the inertial and gravitational masses.

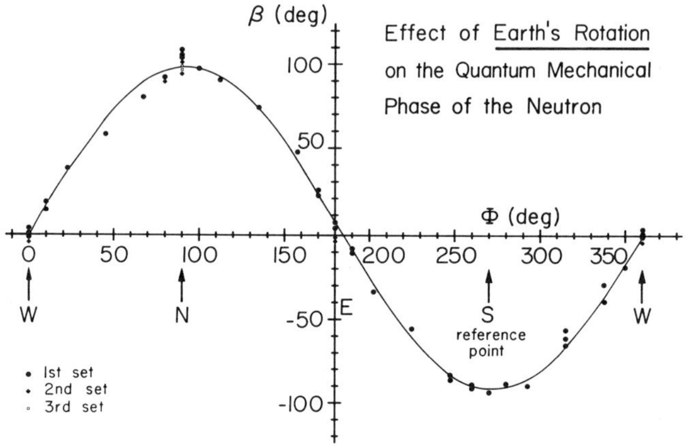

Fig. 10. Phase shift due to the Earth's rotation. The angle ϕ specifies the orientation of the interferometer normal area vector with respect to the local N-S axes of the Earth. (S.A. Werner et al. Ref. 13).

THE AHARONOV-CASHER EFFECT

In 1984 Aharonov and Casher[15] proposed that a beam of neutral particles with a magnetic dipole moment passing around opposite sides of a line charge will undergo a relative quantum phase shift. This AC effect is considered to be an electrodynamic and quantum mechanical dual of the Aharonov-Bohm effect for charged particles, as can be understood by the schematic diagram shown in Fig. 11. The AB flux tube has been replaced by a line of electric charge of lineal density Λ, and the electron beam of charge e^- has been replaced by a beam of neutrons with magnetic moment $\vec{\mu}$. If one views the AB flux tube as a line of magnetic dipoles, one sees that the role of charge and magnetic dipole have been interchanged between the AB effect and the AC effect. According to AC, the Lagrangian which describes this duality is

$$\mathcal{L} = \frac{1}{2}mv^2 + \frac{1}{2}MV^2 + \frac{e}{c}\vec{A}(\vec{r} - \vec{R}) \cdot (\vec{v} - \vec{V}) \quad . \tag{23}$$

Here $m, \vec{v}$, and $\vec{r}$ are the mass, velocity and position vector of the charge particle (system), while $M, \vec{V}$ and $\vec{R}$ are the mass, velocity and position vector of the neutral particle or system (solenoid in the AB effect), and e is the charge of the charged particle (system). $\vec{A}(\vec{r} - \vec{R})$ is the magnetic vector potential due to the neutral particle or system (solenoid). $\mathcal{L}$ is Galilean and translationally invariant at low velocity.

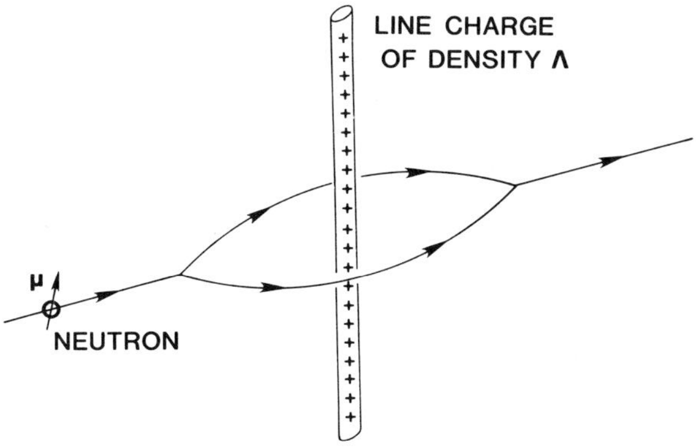

Fig. 11. Schematic diagram showing a neutron of magnetic moment $\vec{\mu}$ diffracting around a line charge. (AC effect).

For an electron moving in the region external to a long solenoid, it experiences no Lorentz force, and its canonical momentum is

$$\vec{p} = m\vec{v} + \frac{e}{c}\vec{A} \quad . \tag{24}$$

The well-known AB phase shift is readily obtained from Eq. (24) as

$$\Delta\beta_{AB} = \frac{e}{\hbar c}\oint \vec{A}\cdot d\vec{r} = \frac{e}{\hbar c}\Phi \quad , \tag{25}$$

where Φ is the magnetic flux inside the solenoid.

For a neutron of mass m and magnetic moment $\vec{\mu}$, moving in the region of an electric field $\vec{E}$, the Hamiltonian is[16]

$$\mathcal{H} = \frac{p^2}{2m} - \frac{1}{mc}\vec{\mu}\cdot(\vec{E}\times\vec{p}) \tag{26}$$

Thus, using Hamilton's equations, we see that the canonical momentum is

$$\vec{p} = m\vec{v} + \frac{\vec{\mu}}{c}\times\vec{E} \quad . \tag{27}$$

It is a straightforward matter to show that there is no force on the neutron. For a neutron diffracting around a line charge of lineal density Λ, one obtains the AC phase shift by evaluating the line integral of $\vec{p}$, namely

$$\Delta\beta_{AC} = \frac{1}{\hbar}\oint \vec{p}\cdot d\vec{r} = \sigma\frac{4\pi\mu\Lambda}{\hbar c} \quad , \tag{28}$$

where $\sigma = \pm 1$ depending upon whether the neutron spin is up or down with respect to the plane of the neutron's motion. As pointed out by AC, this phase shift depends only upon the lineal charge density Λ, enclosed by the beam paths, but not on any details of the geometrical shape. In this sense the effect is topological. Thus, instead of a line charge, a prism-shaped electrode was placed between the splitter (S) and mirror plate (M) of the interferometer as shown in Fig. 12, thereby enabling a much higher lineal charge density to be obtained. The experiment is a collaborative University of Melbourne-University of Missouri project carried out in Columbia, Missouri.[17]

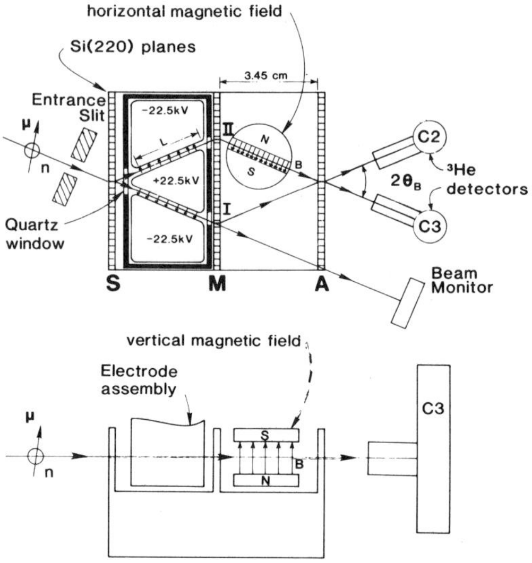

Fig. 12. Schematic diagram of the neutron interferometer set-up used in observing the Aharonov-Casher phase shift at MURR. (A. Cimmino et al. Ref. 17).

For an electrode, Gauss' law allows us to replace Λ by $2VL/4\pi D$. Here V is the potential difference between the electrodes, D is their separation, and L is the effective path length as shown in Fig. 12. In terms of these parameters, for $D = 0.154$ cm, $L = 2.53$ cm and $V = 45$ kV (= 150 statvolts), we find

$$\Delta\beta_{AC} = 1.50\sigma \text{ milliradians} \quad . \tag{29}$$

It was assumed in the derivation of Eq. (28) that the neutrons are polarized along an axis parallel to the line charge. However, it is not necessary to use polarized neutrons if an additional spin-independent phase shift is judiciously introduced and fine-tuned. In this experiment, gravity was used for this purpose. The introduction of a further spin-dependent phase shift $\Delta\beta_M$, by means of a magnetic bias field, enabled maximum sensitivity to the AC effect to be accomplished.

For spin-up neutrons the count rate in detector C_3 is

$$I_3^{\uparrow} = \frac{1}{2}[a_3 + b_3\cos(\Delta\alpha + \Delta\beta)] \quad , \tag{30}$$

whereas, for spin-down neutrons, it is

$$I_3^{\downarrow} = \frac{1}{2}[a_3 + b_3\cos(\Delta\alpha - \Delta\beta)] \quad . \tag{31}$$

The spin-independent phase shift is called $\Delta\alpha$ here, and the spin-dependent phase shift is called $\Delta\beta$. Thus for unpolarized incident neutrons we have

$$I_3 = I_3^{\uparrow} + I_3^{\downarrow} = a_3 + b_3\cos\Delta\alpha\cos\Delta\beta \tag{32}$$

We adjust $\Delta\alpha$ by gravitationally-induced quantum interference to be zero (mod 2π), and we adjust $\Delta\beta_M$ to be $\pi/2$ or $3\pi/2$, where

$$\Delta\beta = \Delta\beta_M + \Delta\beta_{AC} \ . \tag{33}$$

Thus, since $\Delta\beta_{AC}$ is such a small phase angle, we have

$$I_3(\pm) = a_3 \pm b_3 \mid \Delta\beta_{AC} \mid \ , \tag{34}$$

where the + sign is for negative center electrode polarity, and the − sign is for positive center electrode polarity. A similar expression applies to the count rate in detector C_2, namely

$$I_2(\pm) = a_2 \mp b_2 \mid \Delta\beta_{AC} \mid \ . \tag{35}$$

Thus, the count rates are linearly proportional to $\Delta\beta_{AC}$.

The above analysis was carried out for vertically directed (z-direction) magnetic bias field. For a horizontal magnetic bias field, the analysis is a bit more subtle, since the neutron first precesses about the z-axis (parallel to $\frac{\vec{v}}{c} \times \vec{E}$) in the electrostatic cell, and then about the x-axis (horizontal) in the magnet. In this case, the leading term is quadratic in $\Delta\beta_{AC}$, and clearly unobservable, since $\Delta\beta_{AC}$ is only of order 1 milliradian. We have carried out an extensive run under these conditions as a check on the stability of, and spurious contributions to the experiment. We refer to these results as the "null experiment."

The magnetic bias field was varied by changing the reluctance of a magnetic circuit (see Ref. 18). To find the correct operating point for $\Delta\alpha$ and $\Delta\beta_M$, the following procedure was followed: With the magnetic field B set equal to zero, the interferometer was tilted in steps. The first maximum near zero tilt angle, sets $\Delta\alpha = 0$ (mod 2π). Leaving the tilt angle fixed at this maximizing value, operating points of negative slope ($\Delta\beta_M = \pi/2$) and positive slope ($\Delta\beta_M = 3\pi/2$) were selected by scanning the magnetic field as shown in Fig. 13. The magnetic field was then changed in small increments around the value giving $\Delta\beta_M \approx \pi/2$ or $3\pi/2$. Gravity scans were then carried out for each value of the magnetic field and the optimum was chosen as that giving minimum oscillation. A complete series of gravity scans is shown in Fig. 14 which graphically display the oscillation amplitude as a function of magnetic field.

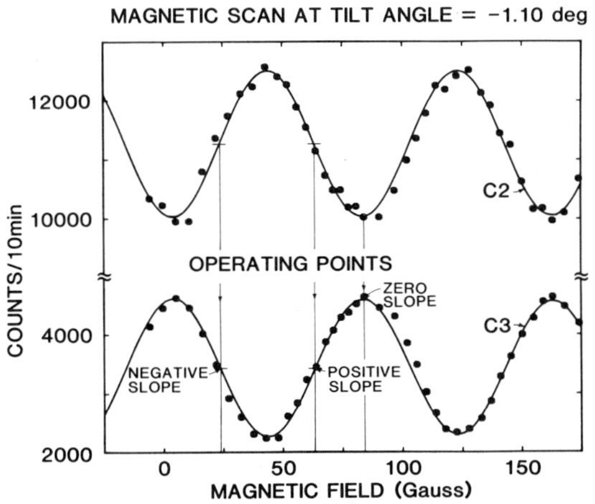

Fig. 13. Magnetic scan interferogram (A. Cimmino et al. Ref. 17).

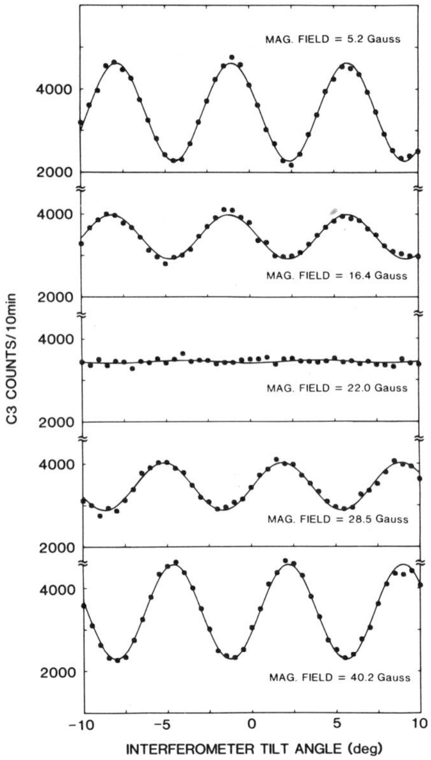

Fig. 14. Gravity scans at various magnetic bias fields. the scan at 22.0 gauss shows no oscillations, thus guaranteeing that the magnetic phase $\Delta\beta_M = 3\pi/2$.

Once the optimum operating conditions were established, the 45 kV high voltage was switched on and periodically reversed in polarity. Each switching cycle (positive, zero, negative) was controlled by a preset-count beam monitor, taking approximately 30 minutes. A summary of the data is given in Table III. $C_2(+)$ and $C_3(+)$ are the total counts accumulated in these detectors for a positive center electrode polarity, while $C_2(-)$ and $C_3(-)$ are for a negative center electrode potential polarity. There were four separate experiments, each taking about 35 days ($\approx$ 1700 voltage cycles), namely:

(a) null experiment	(bias field horizontal)
(b) positive slope	($\Delta\beta_M = 3\pi/2$)
(c) negative slope	($\Delta\beta_M = \pi/2$)
(d) zero slope	($\Delta\beta_M = \pi$).

Statistically combining the data for negative and positive slopes we found

$$\overline{\Delta C} = w_2 \mid C_2(+)-C_2(-) \mid +w_3 \mid C_3(+)-C_3(-) \mid = 9010 \pm 2130 \text{ counts/1700 cycles}, \tag{36}$$

where w_2 and w_3 are statistical weighting factors (inversely proportional to the square of the standard deviations).

Table III

This table summarizes the total accumulated neutron counts for positive (+)and negative (-) voltage polarity for the four operating conditions of the bias magnetic field. There were approximately 1700 voltage cycles for each operating condition, taking approximately 35 days each.

Experimental Condition	C_2 (+)	C_3 (+)	C_2 (-)	C_3 (-)	$C_2(+) + C_3(+)$	$C_2(-) + C_3(-)$	$C_2(+) - C_2(-)$	$C_3(+) - C_3(-)$
(1) Null	20,660,214 ±4,545	6,016,395 ±2,453	20,665,350 ±4,546	6,015,752 ±2,453	26,676,609 ±5165	26,681,102 ±5,165	-5136 ±6,428	+643 ±3,469
(2) Positive Slope	19,367,164 ±4,401	6,098,373 ±2,469	19,353,702 ±4,399	6,106,392 ±2,471	25,465,538 ±5,046	25,460,094 ±5,045	+13,462 ±6,223	-8,019 ±3,494
(3) Negative Slope	18,584,314 ±4,311	5,830,053 ±2,415	18,594,300 ±4,312	5,821,810 ±2,413	24,414,368 ±4,941	24,416,110 ±4,941	-9,986 ±6,097	+8,243 ±3,413
(4) Zero Slope	18,119,184 ±4,257	7,912,480 ±2,813	18,117,466 ±4,256	7,916,111 ±2,814	26,031,664 ±5,102	26,033,578 ±5,102	+1,718 ±6,020	-3,631 ±3,979

A fit to the magnetic scan of Fig. 13 gives us the slope of the interferogram at the operating points, namely

$$b = b_2 = b_3 = 2,057,000 \pm 57,000 \text{ counts/1700 cycles.} \tag{37}$$

Thus, using Eqs. (34) and (35) we have

$$\mid \Delta\beta_{AC} \mid = \frac{\overline{\Delta C}}{2b} = 2.19 \pm 0.52 \text{ milliradians.} \tag{38}$$

This result is to be compared to the theoretical prediction of 1.50 milliradians. The accuracy of the experiment is limited by available neutron intensity and long-term apparatus stability. Obviously, the much larger interferometers currently being designed and fabricated will be important in the next generation A-C effect experiments.

CONCLUDING REMARKS

Our future plans in this field include the following experiments:

(1) Neutron Michelson-Morley experiment.
(2) Neutron coherence in the time domain.
(3) Wheeler delayed choice experiment.
(4) Mashhoon effect - phase shift due to the neutron spin angular momentum in a rotating frame.[19]

Acknowledgements: This work is supported by the Physics Division of the U.S. National Science Foundation, Grant No. NSF-PHY8813253. We have benefitted from the theoretical advice of Bahram Mashhoon. We thank Alberto Cimmino for the drawing of the circular tuba (Fig. 1).

REFERENCES

1. U. Bonse and M. Hart, Appl. Phys. Lett **6**, 155(1965).
2. H. Rauch, W. Treimer and U. Bonse, Phys. Lett. **47A**, 425 (1974).
3. R. Colella, A.W. Overhauser and S.A. Werner, Phys. Rev. Lett. **34**, 1472 (1975).
4. J.-L. Staudenmann, S.A. Werner, R. Colella and A.W. Overhauser, Phys. Rev. **A21**, 1419 (1980).
5. S.A. Werner, H. Kaiser, M. Arif and R. Clothier, Physica **B151**, 22 (1988).
6. M.A. Horne, Physica **B137**, 260 (1986).
7. U. Bonse and T. Wroblewski, Phys. Rev. Lett. **51**, 1401 (1983).
8. U. Bonse and T. Wroblewski, Phys. Rev. **D30**, 1214 (1984).
9. G.L. Greene and H. Layer, submitted to Phys. Rev. Lett (1989).
10. M.G. Sagnac, C.R. Acad. Sci. Paris **157**, 708 (1913).
11. A.A. Michelson, H.G. Gale and F. Pearson, Astrophys. J. **61**, 140 (1925).

12. This formula was first given by L. Page. Phys. Rev. Lett. **35**, 543 (1975). An interesting derivation of it, based upon a Doppler shift by each mirror in the interferometer was given by M. Dresden and C.N. Yang, Phys. Rev. **D20**, 1846 (1979).

13. S.A. Werner, J.-L. Staudenmann and R. Colella, Phys. Rev. Lett. **42**, 1102 (1979).

14. This vertical geometry was suggested by J. Anandan, Phys. Rev. **D15**, 1448 (1977).

15. Y. Aharonov and A. Casher, Phys. Rev. Lett. **53**, 319 (1984).

16. L.L. Foldy, Rev. Mod. Phys. **30**, 471 (1958).

17. A. Cimmino, G.I. Opat, A.G. Klein, H. Kaiser, S.A. Werner, M. Arif and R. Clothier, submitted to Phys. Rev. Lett. April 26, 1989.

18. S.A. Werner, R. Colella, A.W. Overhauser and C.F. Eagen, Phys. Rev. Lett. **35**, 1053 (1975).

19. B. Mashhoon, Phys. Rev. Lett. **61**, 2639 (1988).

THE GEOMETRY OF MATTER FIELDS

Claus Lämmerzahl

Fakultät für Physik der Universität Konstanz

D-7750 Konstanz, Fed. Rep. Germany

Abstract: At first we derive a field equation by means of essentially geometry-free demands, that is, by means of a Cauchy-problem, superposition principle, finite propagation speed and a conservation law. These requirements characterise a symmetric hyperbolic system of partial differential equations of first order.

In what follows, the field equation will define the geometry of space-time (in the same way as light rays and point particles define 'their' geometry, cf. Ehlers, Pirani and Schild (1972)). A general discussion of the derived field equation leads to a metrical structure of Finslerian type, to the propagation of helicity states, to a path structure and to the propagation of spin states. We thereby investigate the concept of a generalised Clifford-algebra.

Specialisation to two light cones, two helicity and spin states and the requirement that there is at least one time-like group velocity, then leads to the usual Dirac equation which defines as its geometry a Riemann-Cartan space-time with axial torsion only and interaction with the Maxwell field.

1. Introduction

1.1. Description of the problem

The purpose of what follows is to show that on physical grounds at least in the classical region space-time geometry when defined by means of matter fields turns out to be a Riemann-Cartan geometry with axial torsion. Moreover, the Maxwell field seems to be the only additional field which can interact with the classical limit of these matter fields.

Usually space-time geometry is characterised by the motion of point particles. By means of physically reasonable assumptions this motion is described by a geodesic equation in a Weylian structure as was shown in the constructive axiomatics of Ehlers, Pirani and Schild (1972) (see also Castagnino (1971)). The point particles are thereby *test particles* which react on but do not influence geometry.

However, we know that all physical matter must be described quantum mechanically. Quantum mechanics is the theory of matter superior to classical mechanics. Therefore the above classically defined point particles must be replaced by quantum mechanical objects. And one way of describing them is by means of fields obeying some field equation.

Quantum Mechanics in Curved Space-Time
Edited by J. Audretsch and V. de Sabbata
Plenum Press, New York, 1990

The present approach to the determination of space-time geometry therefore makes use of matter fields as physical objects which 'feel' the geometry. That is, matter fields play the role of a *test field*. (For the measurement of fields compare the lectures of W. Unruh in this volume.) It is clear that the notion of geometry will thereby change. On the level of the field equations there will be neither any metric nor any connection; some generalised metric and connection will be introduced in discussing the propagation of the singularities of these fields and our usual Riemannian metric and linear connection will enter the consideration at a late stage only.

The program of our approach may be subdivided into three steps:

(i) First of all it is necessary to determine the dynamics of the test fields by means of *basic experiences* which will be formulated by means of *geometry-free axioms*, i.e. exclusively by means of differential topological demands on the test fields. In our case this dynamics results in the evolution equation for fields, respectively in the field equation.

(ii) The second step consists in an exploration of the geometry of the field equation involving a formulation of the geometry in physical terms. In this context 'physical terms' mean the development of field theoretical terms of geometry, for example, field theoretical analogues of metric, connection, curvature * and so on. Up to now there seem to be just two possible terms of geometry: first, as a characterisation of distinguished objects z_0 by a relation $A(z_0) = 0$, where the function A will be regarded as geometry. Secondly, as a (possibly universal) transportation law of some physical object z (with $z^{(k)}$ the kth derivative with respect to some parameter): $z^{(k)} = B(x, z, z', z'', \ldots, z^{(k-1)})$, where now the function B will be considered as geometry. The special cases of the conformal and projective structures are obvious. In this sense field equations can be discussed with respect to the characterisation and propagation of singularities and of the WKB-limit. The sharpness of wave fronts may be of interest, too.

(iii) Demanding special properties from the matter fields of course specialises these fields. So one may try to characterise the Dirac-, the Schrödinger- or the Petiau-Duffin-Kemmer-equation. And the restriction to some special field equation also restricts the possible geometry which can be defined by means of these matter fields.

The aim of these steps is to obtain a description of the geometry of fields. According to the remarks in (ii) two kinds of geometry thereby arise: the geometry of the tangent space, that is the *kinematical geometry* (e.g. the Minkowski space as tangent space of a space with (pseudo-)Riemannian metric) and the geometry of the space-time manifold, that is the geometry of dynamical evolution, the *dynamical geometry* (e.g. a Riemann-Cartan geometry). Since the dynamics of the fields has been postulated in (i) this procedure provides us with a *dynamical definition* of both kinds of geometry.

Within this approach to field equations possible "reasonable" generalisations of the Dirac equation may also arise. In addition, hints towards or justification for the minimal coupling procedure in field equations can possibly be given. A better understanding of the usual Dirac equation in curved space-time may also be achieved firstly by recognising which ideas related to this equation will also survive in these more general cases and secondly by uniquely characterising the usual Dirac equation.

To emphasise, no coupling of any given field equation (for example a special relativistic one, for which already a special kinematical geometry has to be assumed) to some prescribed geometry will be carried out. But precisely the *fields and their dynamics will determine the geometry of space-time*. What we want is to proceed by way of analysis to geometry.

We emphasise in addition that all geometric fields, e.g. metric, torsion etc., will be considered as external fields; geometric fields are *not dynamical*.

There is a rather small body of literature dealing with this and related problems. To begin with, the following remark made by Ehlers (1973) may be regarded as the

* For example, for point particles curvature appears in describing geodesic deviation. Is there a notion of curvature in field theory?

initial motivation for dealing with field theory in connection with space-time axiomatics: "It is quite possible that such an analysis (maybe in terms of possible wave equations for massive particles) would show that a 'reasonable' description of matter only if the Weyl space is, in fact, Riemannian, ..." . Then Audretsch (1983) (followed by Audretsch *et al.* (1984), Loinger (1985) and Kasper (1986)) was the first to apply field theoretical concepts in this context by showing that the classical limit of the Dirac equation in a Weyl-space, which EPS arrived at, reduces that space indeed to a Riemann-space. Subsequently Liebscher (1985) derived the metric of space-time from a 4-component generalised Dirac equation alone. Together with Bleyer he discussed this approach in more detail in the following series of papers: Bleyer and Liebscher (1986) and (1987) and Bleyer (1988). The dynamical generation perspective of geometry has already been spelled out by Andersson (1987). In the present work even the dynamics of the fields will be given a foundation and, on this basis, such ideas will be consequently pursued to include the determination of some affine structures.

1.2. Description of the Work

The present work attempts to determine the space-time geometry according to the scheme described above.

First we assume that the space-time is a four-dimensional differentiable manifold which admits a (3+1)-slicing. Then the time evolution of a field, i.e. the possibility of preparing a field, will lead to an abstract Cauchy-problem. In combination with other first principles like a superposition principle, finite velocity of propagation and a conservation law, we get as an equation of motion for the test fields a wave equation which is a symmetric hyperbolic linear homogeneous system of partial differential equations of first order.

The geometrical content of this equation is given by the characteristics and the propagation of singularities of this system and leads to a conformal Finslerian structure of space-time. In addition we obtain the propagation of the helicity states which can test more than the Finslerian part of space-time structure, for example torsion. By

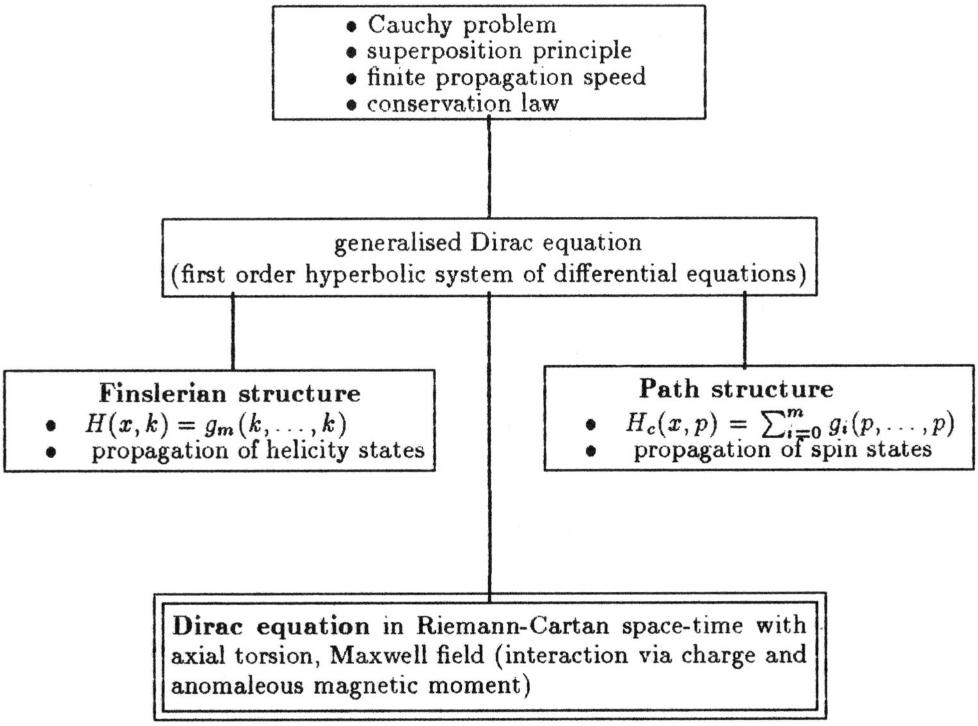

introducing a classical limit we get a path structure and also the propagation of spin states. The way of calculating the propagation of the singularities and the spin states, especially the notion of a generalised Clifford-algebra, seems to be new in literature.

In the end the special case of the Dirac-equation is characterised by means of the number of light cones and mass shells and their respective number of inner degrees of freedom. In this case we can derive the usual Clifford algebra and get as structure of space-time a Riemann-Cartan geometry with axial torsion together with an electromagnetic field with which the classical particles interact via charge and anomalous magnetic moment, as it is determined by the classical particle paths and the spin motion. The characterisation of the Dirac equation and the derivation of its geometrical content may be regarded as the main result of this work.

This result may be compared with the constructive axiomatics of Ehlers, Pirani and Schild (1972) who use the concepts of light rays and freely falling point particles to arrive at a Weylian structure with projective parameters and torsion of space-time remaining unspecified. Also only a conformal class of metrics can be defined. When the concept of a spin-vector according to Audretsch and Lämmerzahl (1988) is added to the light rays and point particles, torsion may be introduced to arrive at a Weyl-Cartan structure.

2. The field equation

To begin with, we want to set up the field equation on the basis of a number of experimentally motivated first principles. The geometrical contents of this field equation will be determined later in chapters 3 and 4. To this end a 4-dimensional differentiable manifold $\mathcal{M}$ will be assumed. Every point of this manifold is realised by an apparatus measuring quantities from which we assume the considered field φ can be calculated; thereby the fields φ are vector valued complex functions: $\varphi : \mathcal{M} \to \mathbf{C}^m : x \mapsto \varphi(x)$.

In order to set up the field equation we start with the most fundamental principle of physics, that is, the possibility of determining some measurable quantities out of a set of given quantities. The determined quantities are assumed to depend linearly on the given one. Next, finite propagation velocity and a conservation law restrict the determination equation to a first order symmetric hyperbolic system of partial differential equations.

In other words, in this chapter our strategy will be the following: We regard the given fields as solutions of some physical equation of motion (in fact, they *are* solutions because nothing else can be observed). Then from these solutions the feature of certain properties, which give information about its equation of motion, are axiomatically demanded . This is just the other way round than is usually found in mathematics where assumptions are made about the differential operator (e.g. Lipshitz-conditions) and information (existence, uniqueness) is derived about possible solutions.

2.1. The dynamical system

The dynamics of a field should have an evolutional structure: there should be at least one class of 3-dimensional non-intersecting hypersurfaces Σ_t consisting of measuring apparatus filling up all 'space' at a 'time' t. The Σ_t are parametrised by t, and all Σ_t are diffeomorphic to one another *. With reference to this parameter the evolution of the field, i.e. the progression of the field from Σ_t to $\Sigma_{t+\delta t}$ and so on, takes place. Before stating the corresponding axiom in ch. 2.2, we first want to express the above description by means of a mathematical definition (cf. Fisher and Marsden (1979)) and describe some related notions:

---

* The diffeomorphicity of all hypersurfaces Σ_t means that there will be no change in their topology. Such pecularities must be considered seperately. But as we are interested in local aspects of the space-time geometry only, we limit ourselves to an interval $[T_0, T]$ of the parameter range where no change in topology will occur.

Definition: *A* $(3+1)$*-slicing is a 3-dimensional differentiable manifold* Σ *and a class of embeddings* $e_t : \Sigma \to \mathcal{M}$, $t \in \mathrm{I} = [T_0, T] \subset \mathrm{R}$, *so that* $e : \mathrm{I} \times \Sigma \to \mathcal{M} : (t, \mathrm{x}) \mapsto x = e(t, \mathrm{x}) = e_t(\mathrm{x})$ *is a* C^∞ *-diffeomorphism.*

Therefore we can define

$$\Sigma_t := \{x \in \mathcal{M} \mid x = e_t(\mathrm{x}), \mathrm{x} \in \Sigma\}$$

$$\varphi_t := \varphi \mid_{\Sigma_t} : \Sigma_t \to \mathbf{C}^m$$

$$\widehat{\varphi}_t := e_t^*(\varphi) : \Sigma \to \mathbf{C}^m : \mathrm{x} \mapsto \widehat{\varphi}_t(\mathrm{x}) = \varphi(e_t(\mathrm{x})). \tag{2.1}$$

To put it the other way around: there should be a mapping e_t so that $\widehat{\varphi}_t$ via $\varphi(e_t(\mathrm{x}))$ fits to a field φ possessing no additional non-smoothness arising from e_t. That is, if $\widehat{\varphi}_t$ is a $\mathrm{C}^\infty(\Sigma)$-function then φ_t should be $\mathrm{C}^\infty(\Sigma_t)$ too. We can think of $\widehat{\varphi}_t$ as the field φ, as being measured by apparatus moving along the lines $v_\Sigma := \dfrac{de_t(\mathrm{x})}{dt}$.

The $\mathrm{C}^\infty(\Sigma)$-functions form a locally convex vector space, i.e. a Fréchet-space (Dieudonné (1971)). In addition, since Σ is a differentiable manifold, there is a partition of unitiy inducing a definite Riemannian metric $g^{(\Sigma)}$ on Σ so that integration can be performed. However, this metric is by no means physically characterised. Physical metrics will be encountered in ch. 3.

Now we want to demand that the fields obey an evolutional structure with respect to the slicing e_t. This means that if a field φ_{t_0} is prepared on Σ_{t_0} then the field φ_t on Σ_t for $t \geq t_0$ will be uniquely determinable. This usually is called Hadamard's *principle of scientific determinism* (cf. Hille and Phillips (1957)) *. Thereby it should be possible to pose arbitrary initial data φ_{t_0}. Especially all $\mathrm{C}^\infty(\Sigma_t)$-functions could be posed. **

Mathematically the initial dates play the role of labels of the set of solutions. If the manifold $\mathcal{M}$ possesses some symmetries, e.g. a 'time'-like Killing vector, then the respective conserved quantities can serve as labels as well. While the latter labels are appropriate in connection with quantum mechanics (these labels are measurable) the former are of no physical importance in quantum mechanics. However, in the case of symmetries, one label can be transformed to the other one, therefore being mathematically equivalent. Therefore, also in the context of a realisation of quantum mechanics by means of classical field theory there exists labels in the form of initial dates although they are not measurable. In the following, the existence of this kind of label will be assumed.

With respect to the slicing e_t this means that posing an arbitrary $\widehat{\varphi}_{t_0} =: \widehat{\varphi}_0 \in \mathrm{C}^\infty(\Sigma)$ will uniquely determine the $\widehat{\varphi}_t$ for $t \geq t_0$. We fix the regularity of $\widehat{\varphi}_t$ by demanding to be in $\mathrm{C}^\infty(\Sigma)$ too; in addition the determined $\widehat{\varphi}_t$ should be twice differentiable with respect to t:

Axiom 1: (Evolution)

(i) $\exists(3+1)$*-slicing so that* $\forall t, t_0 \in \mathbf{I} \subset \mathbf{R}, t \geq t_0$ *posing* $\widehat{\varphi}_{t_0} \in \mathrm{C}^\infty(\Sigma)$ *uniquely determines* $\widehat{\varphi}_t \in \mathrm{C}^\infty(\Sigma)$.

(ii) *The mapping* $\mathrm{I} \to \mathrm{C}^\infty(\Sigma) : t \mapsto \widehat{\varphi}_t$ *whereby* $\widehat{\varphi}_t$ *is determined according to* (i) *is in* $\mathrm{C}^2(\mathrm{I}, \mathrm{C}^\infty(\Sigma))$. ***

---

* We do not consider equations of motion with constraints.

** Indeed, more general functions, e.g. distributions, may be taken. However, since we want to derive necessary conditions on the evolution only, limitation to smaller function spaces is appropriate. Beside this, it may seem to be necessary to limit ourself on functions or distributions with compact support because only such functions can be experimentally prepared. However, if solutions propagate with infinite velocity then $\mathrm{supp}\varphi_t$ which then acts as initial data for $\varphi_{t'}$ with $t' \geq t$ may not be compact; therefore it is necessary to choose as initial dates functions with noncompact support.

*** A mapping $t \mapsto \widehat{\varphi}_t$ is called *differentiable* if there exists a $\chi \in \mathrm{C}^\infty(\Sigma)$ so that

$$\lim_{h \to 0} p_{K,j}\left(\frac{1}{h}(\widehat{\varphi}_{t+h} - \widehat{\varphi}_t) - \chi\right) = 0$$

The first part of the axiom implies that there is a mapping

$$U : \mathbf{I} \times \mathbf{I} \times C^{\infty}(\Sigma) \to C^{\infty}(\Sigma) : (t, t_0, \widehat{\varphi}_{t_0}) \mapsto \widehat{\varphi}_t = U(t, t_0, \widehat{\varphi}_0) \tag{2.2}$$

in the triangle $t \geq t_0$. Since $\widehat{\varphi}_t$ again acts as an initial date we get the composition law

$$U(t', t, U(t, t_0, \widehat{\varphi}_0)) = U(t', t_0, \widehat{\varphi}_0) \qquad \forall t' \geq t \geq t_0. \tag{2.3}$$

Of course

$$U(t, t, \widehat{\varphi}_t) = \widehat{\varphi}_t \qquad \forall t \in \mathbf{I}. \tag{2.4}$$

The second part of the axiom implies

$$\lim_{s \to t'} U(s, t, \widehat{\varphi}_t) = U(t', t, \widehat{\varphi}_t). \tag{2.5}$$

The properties (2.2-5) define U to be a *dynamical system*.

2.2. The superposition principle

All interference experiments with quantum matter show that it obeys the superposition principle, see for example Rauch (1983). We will formalise this experience in demanding the linearity of the determination of $\varphi_{t'}$ from the functions φ_t. With respect to the slicing e_t we therefore demand:

Axiom 2:: (Superposition).
$\forall \widehat{\varphi}_t, \widehat{\psi}_t \in C^{\infty}(\Sigma), \forall \alpha, \beta \in \mathbf{C} : \widehat{\varphi}_{t'} = U(t', t, \widehat{\varphi}_t) \wedge \widehat{\psi}_{t'} = U(t', t, \widehat{\psi}_t) \Rightarrow \alpha\widehat{\varphi}_{t'} + \beta\widehat{\psi}_{t'} = U(t', t, \alpha\widehat{\varphi}_t + \beta\widehat{\psi}_t)$.

Therefore U will be linear: $U(t', t, \widehat{\varphi}_t) =: U(t', t)\widehat{\varphi}_t$ and (2.3.) changes to

$$U(t'', t')U(t', t) = U(t'', t).$$

The set of all U constitutes a *semigroup* and each U is called *propagator* or *evolution operator* in the locally convex vector spaces $\mathcal{E}(\Sigma)$.

2.3. The abstract Cauchy-problem

For the above semigroup we can define the *infinitesimal generator* $\widehat{G}_t$ which depends on t (e.g. Komura (1968), Yoshida (1971)): from the definition of the derivative of $\widehat{\varphi}_t$ which evolved from a prepared $\widehat{\varphi}_0$ we have in $\mathcal{E}$

$$\begin{aligned} \frac{d\widehat{\varphi}_t}{dt} &= \lim_{h \to 0+} \frac{1}{h}(\widehat{\varphi}_{t+h} - \widehat{\varphi}_t) \\ &= \lim_{h \to 0+} \frac{1}{h}(U(t+h, t_0) - U(t, t_0))\widehat{\varphi}_{t_0} \\ &= \lim_{h \to 0} \frac{1}{h}(U(t+h, t) - \mathrm{Id})\widehat{\varphi}_t \\ &=: \widehat{G}_t\widehat{\varphi}_t \end{aligned} \tag{2.6}$$

with the domain of $\widehat{G}_t$, $D(\widehat{G}_t) \subset C^{\infty}(\Sigma)$ consisting of those $\widehat{\varphi}_t$ for which this limit exists. With this definition we get the *abstract Cauchy-problem* or the *evolution equation*

$$\frac{d\widehat{\varphi}_t}{dt} = G_t\widehat{\varphi}_t \qquad \text{with} \qquad \widehat{\varphi}_{t_0} = \widehat{\varphi}_0. \tag{2.7}$$

---

where $p_{K,j}$ is the set of semi-norms in $\mathcal{E}(\Sigma)$. The function χ is then called $\frac{d}{dt}\widehat{\varphi}_t$.

of which $U(t, t_0)\varphi_0$ is a solution. According to axiom 1(ii) $D(\widehat{G}_t)$ is nonempty for all $t \geq T_0$ and indeed equals $\mathcal{E}(\Sigma)$.

If $\widehat{G}_t = \widehat{G}$ does not depend on time, then the assumption of strong continuity of the mapping $t \to \widehat{\varphi}_t$ is enough to show that the abstract Cauchy-problem is well posed (Yosida (1971), Komura (1968)). However, for our general case there seems to be no corresponding theorem so that we need a much stronger assumption.

2.4. Finite propagation speed

It is a fundamental experience in physics that all phenomena do not propagate instantaneously but always need a finite time to reach a separated point. To formalise this experience we first define the speed of propagation.

Let $\widehat{\varphi}_{n,t}$ be a sequence in $C^\infty(\Sigma)$ with $\mathrm{supp}\widehat{\varphi}_{n,t} \to \{\mathrm{x}\}$ for $n \to \infty$ then the x- and t-dependent propagation speed may be defined as

$$c(\mathrm{x}, t) := \lim_{t' \to t} \lim_{n \to \infty} \sup\left\{ \frac{\sigma(\mathrm{x}', \mathrm{x})}{t' - t} \;\Big|\; \mathrm{x}' \in \mathrm{supp}\widehat{\varphi}_{n,t'} \right\}.$$

$\sigma(\mathrm{x}', \mathrm{x})$ is the geodesic length within Σ between x and x$'$ and $\widehat{\varphi}_{n,t'} = U(t', t)\widehat{\varphi}_{n,t}$. The definition is independent of the chosen sequence. * For some bounded region $\Omega \subset \Sigma$ and for some finite time-interval I we can define the maximum speed $c(\Omega, \mathrm{I}) := \sup\{c(\mathrm{x}, t) \mid \mathrm{x} \in \Omega, t \in \mathrm{I}\}$.

The basic experience is described by our

Axiom 3: *All fields propagate with finite velocity.*

First this means that compact $\mathrm{supp}\varphi_t$ implies compact $\mathrm{supp}\varphi_{t'}$. Since an embedding maps compact sets into relatively compact sets we get with respect to the slicing e_t

$$U(t', t) : C_0^\infty(\Sigma) \to C_0^\infty(\Sigma).$$

Following an idea of Svendsen (1982) we can show

Theorem: *$\widehat{G}_t$ is a differential operator.*

Proof. We show first that $\mathrm{x} \notin \mathrm{supp}\widehat{\varphi}_t \Rightarrow \mathrm{x} \notin \mathrm{supp}\left(\widehat{G}_t\widehat{\varphi}_t\right)$. If $\mathrm{x} \notin \mathrm{supp}\widehat{\varphi}_t$ then $\exists U \subset \Sigma, U$ compact, with $\mathrm{supp}\widehat{\varphi}_t \subset U, \mathrm{supp}\widehat{\varphi}_t \neq U$ and $\mathrm{x} \notin U$. Since $\widehat{\varphi}_t$ propagates with finite velocity there is a $\delta t > 0$ so that $\mathrm{supp}\widehat{\varphi}_{t+\delta t} \subset U$. Then $\mathrm{x} \notin \mathrm{supp}\widehat{\varphi}_{t+\delta t}$ and therefore $\mathrm{x} \notin \mathrm{supp}\frac{1}{\delta t}(\widehat{\varphi}_{t+\delta t} - \widehat{\varphi}_t)$ which implies by definition (2.6) that $\mathrm{x} \notin \mathrm{supp}\left(\widehat{G}_t\widehat{\varphi}_t\right)$.

Therefore $\widehat{G}_t$ does not increase support, i.e. $\widehat{G}_t$ is a local operator. In addition, $\widehat{G}_t$ is a linear operator with $D(\widehat{G}_t) = C_0^\infty(\Sigma)$. According to Dieudonné (1971) then it must be a differential operator of finite order with coefficients which are $C^\infty(\Sigma)$-functions.■ **

For the evolution eqtn. we then get (in the multi-index notation; tensors on Σ are written with latin indices ranging from 1 to 3)

$$\frac{d\widehat{\varphi}_t}{dt} = \sum_{|a|=0}^{N} {}^{|a|}\widehat{G}_t^{(a)} \widehat{\partial}_{(a)} \widehat{\varphi}_t \tag{2.8}$$

* According to (2.5) $U(t', t)\widehat{\varphi}_t \to \widehat{\varphi}_t$ for $t' \to t$. Therefore we can conclude that $\mathrm{supp}\widehat{\varphi}_t = \mathrm{supp}\widehat{\psi}_t \Rightarrow \mathrm{supp}(U(t', t)\widehat{\varphi}_t) = \mathrm{supp}(U(t', t)\widehat{\psi}_t)$ for $t' \to t$.

** Since the cited theorem is true only if the premises concerning the domain of $\widehat{G}_t$ hold, we had to make the strong assumption of axiom 1(ii).

where ${}^{|a|}\widehat{G}_t^{(a)}$ are functions of x and $\widehat{\partial}$ acts on x only. Therefore finite velocity forces $\widehat{G}_t$ to be a ***local operator*** and the evolution eqtn. to be a system of partial differential equations.

2.5. A conservation law

Next we want to explore the basic experience that during evolution no matter will be created nor annihilated ($d^3\Sigma$ is the invariant volume element $\sqrt{g^{(\Sigma)}}d^3\mathrm{x}$):

Axiom 4: *For all solutions of* (2.8): $\int_\Sigma \widehat{\varphi}_t^+ \widehat{\varphi}_t d^3\Sigma = const.$

For the propagator U this means that

$$\int_\Sigma (U(t',t)\widehat{\varphi}_t)^+ U(t',t)\widehat{\varphi}_t d^3\Sigma = const. \tag{2.9}$$

From this eqtn. we shall deduce that $\widehat{G}_t$ is of order one only * and has Hermitian coefficients (Svendson (1984)).

At first, from differentiating (2.9) and using (2.7) we see that $\widehat{G}_t$ is formally skew-adjoint, that is $\int_\Sigma ((\widehat{G}_t\varphi)^+\varphi + \varphi^+\widehat{G}_t\varphi)d^3\Sigma = 0 \quad \forall\varphi \in C_0^\infty(\Sigma)$ which implies

$$(\widehat{G}_t\widehat{\varphi}_t)^+\widehat{\varphi}_t + \widehat{\varphi}_t^+\widehat{G}_t\widehat{\varphi}_t = \widehat{\partial}_b\left(\mathrm{Re}\sum_{|a|=0}^{N-1} \widehat{\varphi}_t^+ A^{b(a)}\widehat{\partial}_{(a)}\widehat{\varphi}_t\right) \tag{2.10}$$

for some coefficients $A^{b(a)}$. On the right hand side inside the bracket the order of the derivative is one less than in $\widehat{G}_t$. Especially the highest order coefficients obey $\left({}^N\widehat{G}_t\right)^+ = (-1)^N\left({}^N\widehat{G}_t\right)$. The skew-adjointness of $\widehat{G}_t$ implies the time-independence of $\int_\Sigma \widehat{\varphi}_t^+\widehat{\psi}_t d^3\Sigma$ and the reality of $\int_\Sigma \widehat{\varphi}_t^+\left(i\frac{d}{dt}\widehat{\varphi}_t\right)d^3\Sigma$ for all $\widehat{\varphi}_t, \widehat{\psi}_t \in C_0^\infty(\Sigma)$.

For an appropriate small neighbourhood $U(\mathrm{x}_0)$ we define closed balls $B(\mathrm{x}_0, R) \subset U(\mathrm{x}_0) \subset \Sigma$ with center x_0 and geodesic radius R. With these notions and ($c := c(U(\mathrm{x}_0), \mathrm{I})$)

$$I(t',t) := \int_{B(\mathrm{x}_0, r+c|t'-t|)} (U(t',t)\widehat{\varphi}_t)^+ U(t',t)\widehat{\varphi}_t d^3\Sigma$$

it is easy to prove (Svendsen (1982))

Lemma: $\quad I(t',t) \geq I(t,t)$.

Interpreting $\widehat{\varphi}_t^+\widehat{\varphi}_t$ as probability density this lemma means that no probability can leave the truncated cone (with supp$\widehat{\varphi}_t$ as one base and supp$\widehat{\varphi}_{t'}$ as the other) through its mantle.

This lemma shows that

$$\lim_{t'\to t+} \frac{dI(t',t)}{dt'} \geq 0.$$

---

* Hörmander (1984), §23, proved that if for a *scalar* partial differential eqtn. the Cauchy problem is valid and solutions propagate with finite velocity only, then the hypersurface Σ_t cannot be characteristic. If this theorem is true for systems also, finite velocity is enough to reduce the order of $\widehat{G}_t$ to one. However, up to now a proof is lacking.

Using the definition of $I(t',t)$ and $\widehat{G}_t$ we get

$$-\int\limits_{B(x_0,r)} \left((\widehat{G}_t\widehat{\varphi}_t)^+\widehat{\varphi}_t + \widehat{\varphi}_t^+\widehat{G}_t\widehat{\varphi}_t\right)d^3\Sigma \quad \leq \quad c \int\limits_{\partial B(x_0,r)} \widehat{\varphi}_t^+\widehat{\varphi}_t d^2\Sigma. \tag{2.11}$$

This is a necessary condition which the differential operator has to fulfill.

For later purpose we show that the abstract time-derivative can be transcribed to a derivative within $\mathcal{M}$. For $\widehat{\varphi}_t \in C^\infty(\Sigma)$ we also have $\frac{d}{dt}\widehat{\varphi}_t \in C^\infty(\Sigma)$ and show

Lemma: $\left(\frac{d\widehat{\varphi}_t}{dt}\right)(\mathrm{x}) = \frac{d}{dt}(\widehat{\varphi}_t(\mathrm{x}))$.

Proof: Defining $\widehat{\varphi}(t,\mathrm{x}) := \widehat{\varphi}_t(\mathrm{x})$ we have $\left(\frac{d}{dt}\widehat{\varphi}_t\right)(\mathrm{x}) = \frac{\partial}{\partial t}(\widehat{\varphi}(t,\mathrm{x}))$ since

$$\lim_{h\to 0} p_{K,j}\left(\frac{1}{h}(\widehat{\varphi}(t+h,\mathrm{x}) - \widehat{\varphi}(t,\mathrm{x})) - (\frac{d\widehat{\varphi}_t}{dt})(\mathrm{x})\right) = 0$$

for all $j = 1,2,\ldots$ ($p_{K,j}$ is the set of seminorms in $\mathcal{E}(\Sigma)$). Regarding x in $\frac{\partial}{\partial t}\widehat{\varphi}(t,\mathrm{x})$ as parameter we can write $\frac{\partial}{\partial t}\widehat{\varphi}(t,\mathrm{x}) = \frac{d}{dt}(\widehat{\varphi}(t,\mathrm{x})) = \frac{d}{dt}(\widehat{\varphi}_t(\mathrm{x}))$.■

Therefore this lemma especially means that $\widehat{\varphi}_t \in C^1(\mathbf{I}, C_0^\infty(\Sigma))$ implies $\widehat{\varphi} \in C^1(\mathbf{I}\times\Sigma)$. Since in addition according to axiom 1(ii) $\widehat{\varphi}_t \in C^2(\mathbf{I}, C_0^\infty(\Sigma))$ we have $\widehat{\varphi} \in C^2(\mathbf{I}\times\Sigma)$, so that $\varphi \in C^2(\mathcal{M})$.

Now one can show (slightly generalising Svendsen (1982))

Theorem: *The evolution eqtn. is a first order symmetric hyperbolic system of partial differential eqtns. on* $\mathcal{M}$

Proof: Using Stoke's theorem we get from the inequality (2.11) and eqtn. (2.10)

$$-\int\limits_{\partial B(x_0,r)} 2\mathrm{Re}\sum_{|a|=0}^{N-1}\widehat{\varphi}_t^+ A^{b(a)}\widehat{\partial}_{(a)}\widehat{\varphi}_t n_b d^2\Sigma \quad\leq\quad c\int\limits_{\partial B(x_0,r)}\widehat{\varphi}_t^+\widehat{\varphi}_t d^2\Sigma, \quad \forall\widehat{\varphi}_t \in C_0^\infty(\Sigma)$$

where n with $g^{(\Sigma)}(n,n) = -1$ is the normal to $\partial B(\mathrm{x}_0,r)$. For $N > 1$ on the left hand side there appear derivations while on the right hand side there are no. Therefore we can construct a function $\widehat{\varphi}_t$ contradicting this inequality so that necessarily $N \leq 1$, that is $\widehat{G}_t = {}^1\widehat{G}_t^a\widehat{\partial}_a + {}^0\widehat{G}_t$ reducing (2.7) to a first order system:

$$\frac{d}{dt}\widehat{\varphi}_t = {}^1\widehat{G}_t^a\widehat{\partial}_a\widehat{\varphi}_t + {}^0\widehat{G}_t\widehat{\varphi}_t. \tag{2.12}$$

From (2.10) we conclude that $\widehat{G}_t^a$ is Hermitian, so that we arrive at a first order symmetric hyperbolic system.

This equation can be written on the manifold $\mathcal{M}$: with the help of the above lemma we have

$$\left(\frac{d\widehat{\varphi}_t}{dt}\right)(\mathrm{x}) = \frac{d}{dt}(\widehat{\varphi}_t(\mathrm{x})) = \frac{d}{dt}((e_t^*(\varphi))(\mathrm{x})) = \frac{d}{dt}(\varphi(e_t(\mathrm{x}))) = (v_\Sigma(\varphi))(x) = (\partial_0\varphi)(x)$$

at $x = e_t(\mathrm{x})$ in a chart with $v_\Sigma^\mu = \delta_0^\mu$. With this result and the definitions $de_t({}^1\widehat{G}_t) =: {}^1G \in T\Sigma_t \subset T\mathcal{M}$, $e_t({}^0\widehat{G}_t) =: {}^0G$ (both coefficients are $C^1(\mathcal{M})$) we arrive at $\partial_0\varphi = {}^1G^{\hat{\mu}}\partial_{\hat{\mu}}\varphi + {}^0G\varphi$ (vectors lying within Σ_t are written with hatted indices) which after multiplication with an arbitrary matrix $i\tilde{\gamma}^0$ with $\det\tilde{\gamma}^0 \neq 0$ can be rewritten as ($\tilde{\gamma}^{\hat{\mu}} := -\tilde{\gamma}^0\,({}^1G^{\hat{\mu}})\,, M := -i\tilde{\gamma}^0\,({}^0G)$)

$$0 = i\tilde{\gamma}^\mu \partial_\mu \varphi - M\varphi \tag{2.13}$$

$$\exists A, \det A \neq 0: \quad (A\tilde{\gamma}^\mu)^+ = A\tilde{\gamma}^\mu. \tag{2.14}$$

The fact that $\varphi \in C^2(\mathcal{M})$ especially implies that also the coefficients $\tilde{\gamma}^\mu, M \in C^1(\mathcal{M})$.

In order that (2.13) is a hyperbolic system we have to show that the polynomial $H(x,k) := \det(\tilde{\gamma}^\mu k_\mu)$ is hyperbolic. This is easy to see: Since $H(x,k) = 0 \Leftrightarrow \det({}^1G^{\hat{\mu}} k_{\hat{\mu}} - k_0) = 0$ with Hermitian ${}^1G^{\hat{\mu}}$ the eigenvalues k_0 are real, so that for given real $k_{\hat{\mu}}$ the equation $H(x,k) = 0$ has only real k_0 as solutions. In addition, $H(x, \partial\Sigma_t)$ cannot vanish: if it vanishes then the surface Σ_t is characteristic contradicting axiom 1(i) which states that one can pose arbitrary initial data on Σ_t (cf. Courant and Hilbert (1962)).

Therefore (2.13) is a symmetric hyperbolic system of first order. ■.

(2.13) is the differential equation, the *generalised Dirac equation*, we looked for. (For a survey of partial differential equations and especially the notion of hyperbolicity see e.g. Gårding (1985)) The main principles we used to arrive at this equation are the Cauchy-Problem, superposition-principle, finite propagation speed and a conservation law.

It should be noted that according to our approach the matrices $\tilde{\gamma}^\mu$ and M are not uniquely fixed by the solutions φ but are free to be multiplied by a common arbitrary nonsingular matrix. Therefore only the equivalence class defined by $(\tilde{\gamma}^\mu, M) \sim (\tilde{\gamma}'^\mu, M') := (A\tilde{\gamma}^\mu, AM)$, A arbitrary, nonsingular, is physically meaningful. These requirements leave the Hermiticity condition invariant too.

The generalised Dirac equation above does *not* provide us with the notion of any connection (as for example the path of a particle does). However, demanding invariance under $Gl(m, \mathbf{C})$-transformations $\varphi \rightarrow \varphi' := S\varphi$ the coefficients in (2.12) have to transform according to $\tilde{\gamma}'^\mu = S\tilde{\gamma}^\mu S^{-1}, M' := SMS^{-1} - iS\tilde{\gamma}^\mu\partial_\mu S^{-1}$. Therefore M transforms rather similar to a connection.

We note that from axiom 4 a conservation law can be inferred: starting from (2.10) with $N = 1$ we get $\frac{d}{dt}(\widehat{\varphi}_t^+ \widehat{\varphi}_t) = \widehat{\partial}_b(\widehat{\varphi}_t^+ {}^1\widehat{G}_t^b \widehat{\varphi}_t)$. With the above lemma we get for the left hand side $\partial_0(\varphi^+\varphi)$ and similarly for the right hand side $\partial_{\hat{\mu}}(\varphi^+ {}^1G^{\hat{\mu}}\varphi)$, that is, with the above definitions

$$\partial_\mu(\varphi A\tilde{\gamma}^\mu \varphi) = 0. \tag{2.15}$$

Of course, for symmetric hyperbolic systems a uniqueness result holds (Courant and Hilbert (1962)) recovering the uniqueness of axiom 1.

Another observation is that with the above results it is possible to show that one can choose other slicings of the manifold $\mathcal{M}$ for which the axioms 1 to 4 are true.

3. The propagation of singularities: introduction of a metrical structure and propagation of helicity states

Singularities are discontinuities in the solutions of partial differential equations or in one of their derivatives. The occurence of singularities is one of their definite mathematical and physical features and so can be regarded as a physical phenomenon impressing some characteristic structure onto the differential manifold $\mathcal{M}$.

As one main result of this chapter these characteristics will define a conformal Finslerian stucture, endowing the manifold with a (quite general) metric.

Up to now, methods of Finslerian geometry found physical applications only in the domain of solid state physics describing the geometry of sound waves in acoustical anisotropic and inhomogeneous media (cf. Meyer and Schroeter (1981)). It is also be conceivable that some possible optical anisotropic and inhomogeneous media effects (i) the splitting of the light cone inside this medium into two cones due to the interaction of the helicity of light with the electrons (compare Nicol's prism) and (ii) different speeds of light in different directions (see the letter of W. Barthel in Matsumoto (1970)). Other possibilities of physical realisations of Finsler geometry are purely speculative in the sense that there might be (beyond today's measuring possibilities) in nature a more general than a Riemannian measuring of lengths (cf. Riemann (1854), Reichenbach (1928) and the letter of Finsler in Matsumoto (1970)).

In our approach the possibility of a Finslerian structure is traced back to the discussion of some field equation, in this way providing some theoretical reasons for taking this concept into account. If we do not observe any Finslerian effects, then we have to reduce the field equation so as to lead to a Riemann space only.

Another main result of this chapter concerns the idea of a characteristic Clifford-algebra that forms part of a generalised Clifford-algebra, which will be derived in section 4.3. The main feature of such an algebraic structure consists in connecting the 'inner' space $\mathbf{C}^m$ represented by the γ-matrices and derived quantities, with the tangent space $T\mathcal{M}$.

There is a large amount of physical and mathematical examinations of the Clifford algebra connected with the usual Dirac equation. On the contrary, up to now generalised Clifford algebras are used in a very formal way only to describe the linearising of n-forms (Roby (1969)), which can be connected with the linearising of some special hypothetical n^{th}-order scalar wave equations (Nono (1971)). A short review of present concepts of generalised Clifford algebras has been given by Childs (1978). All these are purely mathematical considerations. A special case based on the Lorentz-group has been introduced on physical grounds by Duffin (1938) Kemmer (1939), Bhabba (1949) and Harish-Chandra (1947) (see also Wightman (1970)). The generalised Clifford-algebra obtained by us is even more general than those considered in the literature.

In our case this concept - which can be used to discuss systems of partial differential equations, especially to calculate the propagation of helicity states along the light rays - will be given more mathematical and physical motivation and meaning. In this work the calculation of the propagation of helicity states is done independently of the work of Dencker (1982) who takes a broader class of differential operators - pseudodifferential operators of real principal type - into account but limits himself to a Hamiltonian formulation of the equation of motion of the polarisation sets, the distribution formulation of helicity states.

Another result of this chapter is that all the notions used in connection with a Riemannian metric, e.g. measuring of length, invariant volume element, orthogonality, Lorentz-group, tetrads, an isomorphism between covariant and contravariant vectors etc., can also be transcribed to the Finslerian case without changing its principles; 'only' technical complications will come along with it.

3.1. The singularities

We assume that a jump, i.e. a singularity *, in the function φ or in its derivatives may occur at some surface $\Phi(x) = 0$, called jump-surface. For describing a function φ having jumps up to order N (that means up to the Nth derivative) let us begin with the ansatz

$$\varphi(x) = \sum_{i=0}^{N} \left(S^{(i)}(\Phi)\right)(x) a^{(i)}(x) + R(x)$$

with $a^{(i)}, R \in C^{\infty}(M)$ and $S^{(n)}(\Phi) := \frac{1}{n!}\eta(\Phi)\Phi^n$ (η is the Heaviside function). Since φ should be a (generalised) solution of the field equation we insert this series into (2.13) and perform the differentiation taking $\partial_\mu \left(S^{(n)}(\Phi)\right)(x) = (S^{(n-1)}(\Phi))(x)(\partial_\mu \Phi)(x)$

* Singularities of solutions of differential equations are here defined along the lines of Courant and Hilbert(1962). Mathematically more sound may be the notion of the wave front set (Hörmander (1984)) resp. of the polarisation set (Dencker (1984)).

($S^{(-1)}$ is the δ-function) into account:

$$0 = i\check{\gamma}^\mu(\partial_\mu\Phi)a^{(0)}S^{(-1)}(\Phi) + \sum_{i=0}^{N-1}\left[i\check{\gamma}^\mu(\partial_\mu\Phi)a^{(i+1)} + i\check{\gamma}^\mu\partial_\mu a^{(i)} - Ma^{(i)}\right]S^{(i)}(\Phi)$$
$$+\left(i\check{\gamma}^\mu\partial_\mu a^{(N)} - Ma^{(N)}\right)S^{(N)}(\Phi) + i\check{\gamma}^\mu\partial_\mu R - MR$$

All coefficients to the $S^{(i)}(\Phi)$ have to vanish independently so that the following series of equations relating the hypersurface Φ to the functions $a^{(i)}$ arise:

$$0 = i\check{\gamma}^\mu(\partial_\mu\Phi)a^{(0)} \tag{3.1}$$
$$0 = i\check{\gamma}^\mu(\partial_\mu\Phi)a^{(i+1)} + i\check{\gamma}^\mu\partial_\mu a^{(i)} - Ma^{(i)}, \quad i = 1, ..., N-1 \tag{3.2}$$
$$0 = (i\check{\gamma}^\mu\partial_\mu a^{(N)} - Ma^{(N)})S^{(N)}(\Phi) + i\check{\gamma}^\mu\partial_\mu R - MR$$

These $N+1$ equations describing the jumps $a^{(i)}$ along a hypersurface $\Phi = 0$ are now the basis for the remainder of this chapter. If the function φ has no jump of order 0 then it is clear that (3.1) is always valid for the jump of lowest order.

3.2. The geometry of the jump-surfaces

If the field possesses a jump, say of order j, then $a^{(j)}$ cannot be zero implying that the coefficients in (3.1) fulfill the characteristic equation as solvability condition (after multiplication with A from (2.14)) with $\tilde{\gamma}^\mu := A\check{\gamma}^\mu$:

$$0 = \det(\tilde{\gamma}^\mu\partial_\mu\Phi). \tag{3.3}$$

Because of (2.14) this is just one real equation relating the hypersurfaces $\Phi = 0$ to the coefficients $\check{\gamma}^\mu$. Defining the normal $k_\mu := -\partial_\mu\Phi$ (3.3) gives $0 = \det\left((\check{\gamma}^0)^{-1}\check{\gamma}^{\hat{\mu}}k_{\hat{\mu}} + k_0\right)$ which is a determination equation for the eigenvalues k_0 of $(\check{\gamma}^0)^{-1}\check{\gamma}^{\hat{\mu}}k_{\hat{\mu}}a = -k_0 a$. The set of eigenvectors a (we omit the index (j)) is complete and all k_0 are real. Introducing $H(x,k) := \det(\tilde{\gamma}^\mu k_{\mu'})$ (3.3) reads

$$0 = H(x,k) = \sum_{|\alpha|=m} g_m^{(\alpha)}k_{(\alpha)} =: g_m(k, \ldots, k)$$

defining a totally symmetric covariant tensorfield g_m of mth rank. g_m is determined by the $m \times m$-matrices γ^μ and is therefore $C^\infty(\mathcal{M})$.

In general $H(x,k)$ consists of a product of several irreducible polynomials:

$$H(x,k) = \prod_{r=1}^{R}\left(H^{(r)}(x,k)\right)^{\alpha_r} \quad \text{with} \quad \sum_{r=1}^{R} m_r\alpha_r = m. \tag{3.4}$$

m_r is the degree of $H^{(r)}$ and all the $H^{(r)}$ are irreducible. Each $H^{(r)}$ must possess m_r solutions because H possesses m of it (some of them may be equal). The multiplicity of the zeros of the characteristic equation becomes important for the determination of the jumps a (see ch. 3.4). For the rest of this chapter we will drop the index $^{(r)}$; all statements are with respect to one of the irreducible polynomials.

If we determine the coefficients γ^μ by measuring the field φ, then we also know the g_m only up to multiplication with a scalar leading to the conformal structure

$$[g_m]_c := \{g'_m \mid g'_m = e^{\lambda(x)}g_m, \lambda(x) \in \mathbf{R}\}. \tag{3.5}$$

g_m can also be determined by measuring the normals k_μ and using (3.4).

(3.4) represents a differential equation of Hamilton-Jacobi-type from which the possible jump-hypersurfaces may be determined. This equation always possesses a solution if $\frac{\partial}{\partial k_\mu} H(x,k)\big|_{H(x,k)=0} \neq 0$ for at least one $\mu = 0, \ldots, 3$ (Carathéodory (1956)). However, especially $\frac{\partial}{\partial k_0} H(x,k)$ cannot vanish since, as will be seen in ch. 3.4, $H(x,k) = 0$ can possess simple zeros only. Consequently, in our case $H(x,k) = 0$ always possesses a solution $\Phi(x)$.

Jump-surface can also be interpreted as characteristic surfaces (Courant and Hilbert (1962)).

The *normal cone* at $x \in \mathcal{M}$ is $N_x := \{k \in T_x^*\mathcal{M} \mid H(x,k) = 0\}$ and describes a conic algebraic surface in the cotangent space. N_x is a cone without isolated k's and there cannot exist two closed parts of two cones which neither intersect one another nor lay inside one another. Therefore there is an open cone $\Gamma(\gamma, \Sigma_t)$ consisting of all $q \in T_x^*\mathcal{M}$ so that for each $k \neq q$ the equation $H(x, k + \lambda q) = 0$ has m solutions. The *hyperbolicity cone* $\Gamma(\gamma, \Sigma_t)$ is convex and $\partial_\mu \Sigma_t \in \Gamma(\gamma^\mu, \Sigma_t)$.

For obtaining information about the smoothness of the normal cone we recall that $\frac{\partial}{\partial k_0} H(x,k)$ never vanishes. According to lemma A.1.1. of Hörmander (1984) solutions $k_0 = k_0(x, k_{\hat\mu})$ of $H(x,k) = 0$ are therefore locally analytical which implies that $\frac{\partial}{\partial k_{\hat\mu}} k_0(x, k_{\hat\mu})$ always exists.

The *propagation cone* $K(\gamma, \Sigma_t)$ dual to $\Gamma(\gamma, \Sigma_t)$ is defined through $K(\gamma, \Sigma_t) := \{v \in T_x\mathcal{M} \mid k(v) \geq 0 \quad k \in T_x^*\mathcal{M}\} \subset T_x\mathcal{M}$. The boundary of K consists in those v for which $k(v) = 0$ for at least one $k \in \partial\Gamma$. Therefore for this k and v we have $k(v) = 0$ and $H(x,k) = 0$ from which, because of the smoothness of the normal cone, the relation

$$v^\mu = \frac{1}{m} \frac{\partial H(x,k)}{\partial k_\mu}\bigg|_{k \in \partial\Gamma(\gamma, \Sigma_t)} \tag{3.6}$$

giving the *fastest signals* can be derived. Since $\frac{\partial}{\partial k_0} H(x,k) \neq 0$ the measured velocity $\dot{x}^{\hat\mu} = v^{\hat\mu}/v^0$ of the fastest signals is always finite. (3.5) is valid also for all $k \in N_x$ giving rise to the *bicharacteristics* resp. the *light rays*.

Here it will only be mentioned that with the above formalism we can define the notions of length, invariant volume-element, orthogonality, light-, time- and space-likeness and generalised Lorentz-transformations.

3.3. The geometry of the light rays

We now want to derive the equation of motion for the light rays starting from (3.4). Since $H(x,k)$ is homogeneous of degree m in the variable k we may get a Finslerian metric function (cf. Rund (1959)) by taking the mth root. However, we cannot proceed this way, because the indefiniteness of $H(x,k)$ makes it impossible to perform derivatives of the metric function. * We must therefore develop a generalised Finslerian formalism which starts from a function of arbitrary degree of homogeneity.

Taking the total derivative of (3.4) we get

$$0 = m v^\nu \partial_\nu k_\mu + \partial_\mu H(x,k). \tag{3.7}$$

* This corresponds to the fact that $0 = \delta \int ds$ in Riemannian geometry is defined only for $ds \neq 0$. However in the equation of motion the limit $ds \to 0$ can be carried out. For light rays one uses $0 = \delta \int (ds)^2$.

The difficulty now consists in expressing k through v which is possible only if (3.6) is solvable for k. To this end we define a (generalised) *Finslerian metric* in k homogeneous of degree $m-2$

$$g^{\mu\nu}(x,k) := \frac{1}{m(m-1)} \frac{\partial^2 H(x,k)}{\partial k_\mu \partial k_\nu} \tag{3.8}$$

and demand

Axiom 5: $\det\left(g^{\mu\nu}(x,k)\big|_{H(x,k)=0}\right) \neq 0.$

By means of the implicit function theorem this condition implies that on the normal cone the mapping (3.6) between k and v is one-to-one leading to a canonical isomorphism between the cotangent and the tangent space: $T_x^*(\mathcal{M}) \to T_x(\mathcal{M}) : k \mapsto v$. There is an inverse mapping $f : T_x(\mathcal{M}) \to T_x^*(\mathcal{M}) : v \mapsto k := f(x,v)$. The homogeneity of $H(x,k)$ implies $g^{\mu\nu}(x,k)k_\mu k_\nu = H(x,k)$ and $v^\mu = g^{\mu\nu}(x,k)k_\nu$. Defining the inverse metric $g_{\mu\nu}(x,v) := (m-1)\frac{\partial}{\partial v^\nu} f_\mu(x,v)$ it is not difficult to see that $\delta^\nu_\mu = g_{\mu\sigma}(x,v)g^{\sigma\nu}(x,k) = g_{\sigma\mu}(x,v)g^{\sigma\nu}(x,k)$ which implies that $g_{\mu\nu}(x,v)$ is also symmetric. Additionally, $\det(g_{\mu\nu}(x,v)) \neq 0$ and $k_\mu = g_{\mu\nu}(x,v)v^\nu$.

Geometrically the above condition leads to restrictions on the normal cone: one or two different components of the normal cone cannot intersect and the cone cannot be non-convex.

With these results it is now possible to extract from (3.7) the equation of motion for the light rays v:

$$0 = v^\nu \partial_\nu v^\mu + \left\{ {\mu \atop \nu\sigma} \right\}(x,v) v^\nu v^\sigma \tag{3.9}$$

with the generalised Finslerian transport coefficient

$$\left\{ {\mu \atop \nu\sigma} \right\}(x,v) := g^{\mu\rho}(x,f(x,v)) \left((m-1)\partial_{(\nu} g_{\sigma)\rho}(x,v) - \frac{(m-1)^2}{m} \partial_\rho g_{\sigma\nu}(x,v) \right)$$

which is homogeneous of degree zero in v. For $m=2$ it reduces to the usual Finslerian transport coefficient and for velocity independent metrics to the Christoffel-connection. Therefore, via (3.5) and the above equation of motion the light rays define a (generalised) *conformal Finslerian structure*. In general, for light rays the strong equivalence principle * is not valid.

Although $\left\{ {\mu \atop \nu\sigma} \right\}(x,v)$ constitutes no connection, we can obtain one by means of

$$\overset{\{\}}{\Gamma}{}^\sigma_{\mu\nu}(x,v) := \frac{1}{2} \frac{\partial^2}{\partial v^\mu \partial v^\nu} \left(\left\{ {\sigma \atop \tau\rho} \right\}(x,v) v^\tau v^\rho \right) \tag{3.10}$$

We remark that it is possible to set up a Lagrangian formulation with the Lagrange function $L(x,v) := H(x,f(x,v))$.

3.4. The characteristic Clifford-algebra

In this chapter we treat some general properties of the γ-matrices. They fulfill a characteristic Clifford-algebra first showing that the notion of a Clifford-algebra is intimately connected with the analysis of systems of partial differential equations, and second giving us a tool at hand to calculate the propagation of helicity states along the light-rays. Later (in ch. 4.3.) we shall derive the more specific generalised Clifford-algebra.

---

* A set of paths obeys the *strong equivalence principle* if there is a coordinate system and there are parametrisations so that for all paths the inertial law $\ddot{x} = 0$ holds.

For determining the characteristic Clifford-algebra the determinant of a matrix by means of multiplication can be represented by its minor B':

$$\prod_{r=1}^{R}\left(H^{(r)}(x,k)\right)^{\alpha_r} = H(x,k) = \det(\tilde{\gamma}^\mu k_\mu) = B'(x,k)\tilde{\gamma}^\mu k_\mu = \tilde{\gamma}^\mu k_\mu B'(x,k).$$

Since $\tilde{\gamma}^\mu k_\mu$ is Hermitian it is diagonalisable. Therefore there exists another matrix $B(x,k)$ so that

$$H_0(x,k) := \prod_{r=1}^{R} H^{(r)}(x,k) = B(x,k)\tilde{\gamma}^\mu k_\mu \tag{3.11}$$

with $H_0(x,k)$ being homogeneous of degree $m' := \sum m_r$ and possessing simple zeros only. Therefore no $H^{(r)}$ can possess a common zero with some other $H^{(r')}$. This shows that all normal cones, and therefore also all light cones, can intersect neither themselves nor one another.

$B(x,k)$ is homogeneous of degree $m'-1$ and has the form $B(x,k) = B^{\mu_1\cdots\mu_{m'-1}} k_{\mu_1}\cdots k_{\mu_{m'-1}}$. Already equation (3.11) shows one of the most important features of a Clifford-algebra: that the product of matrices is proportional to the unit matrix. We obtain further features by differentiating this equation with respect to k

$$\left(\frac{\partial B}{\partial k_\mu}\right)\tilde{\gamma}^\nu k_\nu + B\tilde{\gamma}^\mu = \tilde{\gamma}^\nu k_\nu\left(\frac{\partial B}{\partial k_\mu}\right) + \tilde{\gamma}^\mu B = \frac{\partial H_0}{\partial k_\mu} =: m'V^\mu \sim v^\mu \tag{3.12}$$

$$\begin{aligned}\left(\frac{\partial^2 B}{\partial k_\mu \partial k_\nu}\right)\tilde{\gamma}^\sigma k_\sigma + 2\left(\frac{\partial B}{\partial k_{(\mu}}\right)\tilde{\gamma}^{\nu)} &= 2\tilde{\gamma}^{(\mu}\left(\frac{\partial B}{\partial k_{\nu)}}\right) + \tilde{\gamma}^\sigma k_\sigma\left(\frac{\partial^2 B}{\partial k_\mu \partial k_\nu}\right) \\ &= \frac{\partial^2 H_0}{\partial k_\mu \partial k_\nu} =: m'(m'-1)G^{\mu\nu}\end{aligned} \tag{3.13}$$

....

$$m'\left(\frac{\partial^{m'-1}B}{\partial k_{(\mu_1}\cdots\partial k_{\mu_{m'-1}}}\right)\tilde{\gamma}^{\mu_{m'})} = m'\tilde{\gamma}^{(\mu_1}\left(\frac{\partial^{m'-1}B}{\partial k_{\mu_2}\cdots\partial k_{\mu_{m'})}}\right) = m'!G^{\mu_1\cdots\mu_{m'}} \tag{3.14}$$

The above equations (3.12 to 14) show most clearly the main feature of a Clifford algebra: it relates products of $\mathbf{C}^m$-valued matrices to space-time tensors. Equation (3.14) may be considered superior to all the other equations because the other equations may be derived from it. This equation, together with the explicit form of B, may be regarded as a Clifford algebra, which we will call a *characteristic Clifford algebra* because it is defined by means of the light cone:

$$B^{(\mu_1\cdots\mu_{m'-1}}\tilde{\gamma}^{\mu_{m'})} = \tilde{\gamma}^{(\mu_1}B^{\mu_2\cdots\mu_{m'})} = g_{m'}^{\mu_1\cdots\mu_{m'}}.$$

However, for our purposes the most important equations are (3.12) and (3.14) which connect the light-velocity and the metric with matrices acting on $\mathbf{C}^m$.

3.5. Propagation of helicity states

The singularities a are solutions of (3.1) whereby k is determined by (3.3), i.e. is a zero of one of the irreducible polynomials. The power α_r is the number of solutions a corresponding to one solution k. The a's are called *helicity states*. Since $a^+\tilde{\gamma}^\mu a \sim v^\mu$ these helicity states are also carrying information about the k's resp. the v's.

It is now our task to set up an equation of motion, a propagation law, for the a's, i.e. an equation of the form $v^\mu\partial_\mu a = v(a) \sim f(x,v,a)$ up to proportionality ((3.1) defines a only up to a factor). Thereby the function f shall depend on the values of v and a along the path of the light rays only. This can easily be done with the help of the characteristic Clifford-algebra. Using (3.10) we get with some appropriate proportionality factor α ($\widetilde{M} := AM$):

$$\begin{aligned} \alpha v(a) &= \alpha \frac{1}{m'} \left(\frac{\partial H_0}{\partial k_\mu}\right) \partial_\mu^{\text{total}} a \\ &= \left(\frac{\partial B}{\partial k_\mu}\right) \tilde{\gamma}^\nu k_\nu \partial_\mu^{\text{total}} a + B \tilde{\gamma}^\mu \partial_\mu^{\text{total}} a \\ &= \left(\frac{\partial B}{\partial k_\mu}\right) \left[\partial_\mu^{\text{total}} (\tilde{\gamma}^\nu k_\nu a) - \partial_\mu^{\text{total}} (\tilde{\gamma}^\nu k_\nu) a\right] - B \left(\tilde{\gamma}^\nu k_\nu a^{(1)} + i\widetilde{M} a\right). \end{aligned}$$

The first term in the square bracket vanishes by virtue of (3.1) and the first term in the round bracket vanishes because of (3.11) so that, as far as a is concerned, we arrive at a propagation law

$$\alpha v(a) = -\left(\frac{\partial B}{\partial k_\mu}\right) \left[(\partial_\mu \tilde{\gamma}^\nu) k_\nu + \tilde{\gamma}^\nu \partial_\mu k_\nu\right] a - iB\widetilde{M} a.$$

However, the term $\partial_\mu k_\nu$ is not in order: it depends on k close by the path. It can be eliminated by means of (3.13) and the symmetry of $\partial_\mu k_\nu$: we insert $\frac{\partial B}{\partial k_{(\mu}} \tilde{\gamma}^{\nu)}$ and get

$$v(a) = -\frac{m'(m'-1)}{2\alpha} G^{\mu\nu} \partial_\mu k_\nu a - \frac{1}{\alpha} \left[\left(\frac{\partial B}{\partial k_\mu}\right) (\partial_\mu \tilde{\gamma}^\nu) k_\nu + iB\widetilde{M}\right] a \qquad (3.14).$$

In principle we arrived at our aim. Note that the 'mass' $\widetilde{M}$ enters the propagation law and - besides the metrical interaction - causes an additional rotation of the helicity states.

We can go further in showing that this equation can be written in a manifest covariant manner: to this end, we define a 'spinorial' connection through

$$0 = \left(\overset{\{\}}{D}_\mu \tilde{\gamma}^\nu\right) k_\nu := (\partial_\mu \tilde{\gamma}^\nu) k_\nu + \overset{\{\}}{\Gamma}{}^\sigma_{\mu\nu}(x,k) \tilde{\gamma}^\nu k_\sigma + \left[\overset{\{\}}{\Gamma}_\mu(x,k), \tilde{\gamma}^\nu k_\nu\right] \qquad (3.15)$$

which are $4m^2$ equations for the $4m^2$ unknown $\overset{\{\}}{\Gamma}_\mu(x,k)$. The vector connection is obtained from (3.9) and v is replaced by k. Inserting (3.15) into (3.14) and using (3.11) and (3.12) we finally get

$$v^\mu \overset{\{\}}{D}_\mu a = \frac{1}{\alpha} \left[\frac{m'(m'-1)}{2} G^{\mu\nu} \overset{\{\}}{D}_\mu k_\nu + B \left(\tilde{\gamma}^\mu \overset{\{\}}{\Gamma}_\mu - iM\right)\right] a \qquad (3.16)$$

which represents a manifest covariant transportation law for the direction of a. Here we defined $\overset{\{\}}{D}_\mu(x,v)a := \partial_\mu a + \overset{\{\}}{\Gamma}_\mu(x,v)a$ and $\overset{\{\}}{D}_\mu k_\nu := \partial_\mu k_\nu - \overset{\{\}}{\Gamma}{}^\sigma_{\mu\nu}(x,k)k_\sigma$.

By means of the above transportation law the underlying geometry may be described. This will be undertaken in ch. 5 where a special case will be discussed.

4. The classical limit: path structure and spin-propagation

In this chapter a new structure, the classical limit of the field equation (2.13), will be introduced. This amounts to a theory bearing some similarities to that outlined in the preceding chapter, so that some of the already developed ideas can now be taken over and represented in shorter form. While the former theory was a definite mathematical feature of partial differential equations, the classical limit which postulates that plane waves are approximate solutions of (2.13), is valid in certain physical situations only.

The geometrical structure we arrive at will be even more general than the Finslerian: the basic metric function, resp. the Hamilton function, will be some non-homogeneous polynomial in the momenta. With this Hamilton function we can describe mass shells and the paths of the 'classical' particles as well as the propagation of spin states. Again, the paths are related to the metrical structure of space-time while the propagation of the WKB-states gives some affine structure. In the high-energy-limit the particle velocity approaches the light velocity, while the paths of the classical particles may yet be different from the light rays - thereby showing that kinematically though not necessarily dynamically the geometry induced by the singularities is contained in the classical limit. Since the classical limit shows more features than the singularities it is now possible to derive the full generalised Clifford-algebra which (i) again is essential in deriving the propagation of WKB-states, (ii) also connects space-time tensors with the γ-matrices and (iii) gives the principle underlying the linearisation of higher order scalar partial differential equations.

4.1. The classical limit

The classical limit, or WKB-limit, describes situations where there exist solutions of (2.13) which are approximately, that is within a certain sufficently small region, plane waves.

Definition: *$\varphi : \mathcal{M} \to \mathbf{C}^m$ is a local plane wave if within an appropriate neighbourhood of $x \in \mathcal{M}$ there is a field of $\mathbf{C}^m$-bases and functions $S \in C^2(\mathcal{M}, \mathbf{R})$ and $a \in C^1(\mathcal{M}, \mathbf{C}^m)$ so that it can be represented as*

$$\varphi(x) = a(x)e^{iS(x)} \tag{4.1}$$

*with * $\|i\gamma^\mu \partial_\mu a\| \ll \|a\|$ if a is represented with respect to the field of $\mathbf{C}^m$-bases.*

A transformation $\varphi \to \varphi' := S\varphi$ which leaves $\|a\|$ invariant changes the above estimate to

$$\|i\tilde{\gamma}^\mu \partial_\mu a' - M^{(1)\prime} a'\| \ll \|a'\| \tag{4.2}$$

with $(M^{(1)})' := i\tilde{\gamma}'^\mu S\partial_\mu S^{-1}$. This means that the existence of local plane waves distinguishes a special class of base fields and therefore induces the existence of a matrix $M^{(1)}$ transforming according to $M^{(1)} \to (M^{(1)})' = SM^{(1)}S^{-1} - i\tilde{\gamma}'^\mu S\partial_\mu S^{-1}$, that is, like M. Of course, $M^{(1)}$ like M is no connection, but shares some features of it. $M^{(1)}$ is defined only up to matrices annihilating a. With the help of this matrix the covariance of the estimate (4.2) can be maintained.

A constant *active* transformation of a leaving $\|a\|$ invariant gives a new local plane wave solution.

Note that our classical limit is no expansion with respect to $\hbar$, because a field equation like (2.13) by itself does not define any such quantity.

---

* $\|a\| := a^+\tilde{\gamma}^\mu n_\mu a$ with $n_\mu = \partial_\mu \Sigma_t$.

Now we are looking for equations which must be valid if (2.13) will admit local plane waves. Inserting (4.1) into the field equation we get with $M^{(0)} := M - M^{(1)}$ which transforms homogeneously, and the *momentum* $p_\mu := -\partial_\mu S$

$$0 = \left(\check{\gamma}^\mu p_\mu - M^{(0)}\right) a + i\check{\gamma}^\mu \partial_\mu a - M^{(1)} a.$$

If (4.1) is a local plane wave solution, then the second term can be neglected so that the first term, the term of lowest order, vanishes. Therefore the other term, determining the derivative of those solutions, also has to vanish. Therefore, as equations describing local plane waves, the WKB-equations we get

$$0 = \left(\check{\gamma}^\mu p_\mu - M^{(0)}\right) a \tag{4.3}$$

$$0 = i\check{\gamma}^\mu \partial_\mu a - M^{(1)} a. \tag{4.4}$$

We show now that the 'mass' $M^{(0)}$ is Hermitian: The current conservation (2.15) together with (2.13) implies $0 = \partial_\mu(\varphi^+\tilde{\gamma}^\mu\varphi) = \varphi^+\left(i(AM)^+ - iAM + \partial_\mu\tilde{\gamma}^\mu\right)\varphi$. Since this is true for all $\varphi(x)$ (axiom 1(i)) we conclude $0 = i(AM)^+ - iAM + \partial_\mu\tilde{\gamma}^\mu$. On the other hand, $\varphi^+\tilde{\gamma}^\mu\varphi = a^+\tilde{\gamma}^\mu a$ so that with the help of (4.4) we get $0 = \partial_\mu(a^+\tilde{\gamma}^\mu a) = a^+\left(i(AM^{(1)})^+ - iAM^{(1)} + \partial_\mu\tilde{\gamma}^\mu\right)a$. Comparison gives $a^+\left((AM^{(0)})^+ - AM^{(0)}\right)a = 0$ for all a so that $\left(AM^{(0)}\right)^+ = AM^{(0)}$.

It is clear that (4.3) can be brought to the form

$$0 = \left(\bar{\gamma}^\mu p_\mu - \bar{E}\right) a \tag{4.5}$$

where $\bar{E}$ is a unit matrix up to some diagonal zeros according to the rank of $M^{(0)}$ and that there is another Hermitian matrix $\bar{\beta}$ which hermitises $\bar{\gamma}$: $(\bar{\gamma}^\mu)^+ = \bar{\beta}\bar{\gamma}^\mu\bar{\beta}^{-1}$.

4.2. The geometry of the mass shell

For WKB-solutions to exist, equation (4.3 resp. 5) must have nontrivial solutions. This is possible only if the coefficient matrix fulfills as solvability condition the *Hamilton-Jacobi equation*

$$0 = \det\left(\bar{\gamma}^\mu p_\mu - \bar{E}\right) =: \sum_{i=1}^{m} \bar{g}_i(p,\ldots,p) - \bar{\epsilon} =: H_c(x,p), \qquad \bar{\epsilon} := \begin{cases} 0 & \text{if } \det M^{(0)} = 0 \\ 1 & \text{if } \det M^{(0)} = 1. \end{cases} \tag{4.6}$$

Again this is one real equation defining m totally symmetric $C^\infty(\mathcal{M})$-tensor fields $\bar{g}_i$. For each given $p_{\hat{\mu}}$ there are m solutions $p_0 = p_0(x, p_{\hat{\mu}})$ of (4.6). Of course $H_c = H + Q$ for some polynomial Q of degree lower than m. From the Laplace expansion theorem for determinants it is clear that

$$\bar{g}_1^\mu = (-1)^m \operatorname{tr}\bar{\gamma}^\mu, \qquad \bar{g}_2^{\mu\nu} = \tfrac{1}{2}(-1)^{m-1}\left(\operatorname{tr}\bar{\gamma}^\mu \operatorname{tr}\bar{\gamma}^\nu - \operatorname{tr}(\bar{\gamma}^\mu\bar{\gamma}^\nu)\right).$$

In general $H_c(x,p)$ consists of a product of irreducible polynomials $H_c(x,p) = \prod_{r=1}^{R}\left(H_c^{(r)}(x,p)\right)^{\alpha_r}$ with $\sum_{r=1}^{R} m_r\alpha_r = m$. Each $H_c^{(r)}(x,p)$ possesses m_r solutions. The *mass shell* at $x \in \mathcal{M}$ is $S := \{p \in T_x^*\mathcal{M} \mid H_c(x,p) = 0\}$. In the following we will consider one equation $H^{(r)}(x,p) = 0$ only and drop the index (r).

(4.6) represents a Hamilton-Jacobi differential equation for the phase $S(x)$. It always possesses a solution because $\dfrac{\partial}{\partial p_0}H(x,p) \neq 0$ which again results from the fact that $H(x,p) = 0$ possesses simple zeros only. This can be established with the same argument as in the case of the normal cone. Also, since the above derivative does not vanish, $p_0(x, p_{\hat{\mu}})$ is locally analytical. Therefore the derivative $\frac{\partial}{\partial p_{\hat{\mu}}} p_0(x, p_{\hat{\mu}})$ exists so that the vector

$$\bar{v}_c^\mu(x,p) := \frac{1}{m} \frac{\partial H_c(x,p)}{\partial p_\mu} \Big|_{H_c(x,p)=0}$$

can be interpreted as *group velocity* of a wave packet. Of course $\lim_{p_{\hat{\mu}} \to \infty} \bar{v}_c = v$.

4.3. The path structure

We now want to derive the equation of motion for the group velocity starting from (4.6). Taking the total derivative we again get $0 = m\bar{v}_c^\nu \partial_\nu p_\mu + \partial_\mu H_c(x,p)$. Again we define $\bar{g}_c^{\mu\nu}(x,p) := \frac{1}{m(m-1)} \frac{\partial^2 H_c(x,p)}{\partial p_\mu \partial p_\nu}$ and have to demand

Axiom 6: $\det\left(\bar{g}_c^{\mu\nu}(x,k)\big|_{H_c(x,k)=0}\right) \neq 0$.

Therefore the mapping $p \to \bar{v}_c$ is an isomorphism and there is an inverse mapping $\bar{f}_c : \bar{v}_c \to p$. We also define $\bar{g}_{c\mu\nu}(x,\bar{v}_c) := (m-1)\frac{\partial}{\partial \bar{v}_c^\nu} \bar{f}_{c\mu}(x,\bar{v}_c)$ and can show that $\bar{g}_{c\mu\nu}(x,\bar{v}_c)$ is symmetric too. *

Defining $L_c(x,\bar{v}) := H_c(x, \bar{f}(x,\bar{v}))$ we get after some calculation the equation of motion for the group velocity

$$\bar{v}_c^\nu \partial_\nu \bar{v}_c^\mu + \overset{\{\}}{\bar{G}^\mu}(x,\bar{v}_c) = 0 \tag{4.7}$$

with $\overset{\{\}}{\bar{G}^\mu}(x,\bar{v}_c) := (m-1)\bar{g}_c^{\mu\nu}(x,\bar{f}_c(x,\bar{v}_c))\left(2\bar{v}_c^\sigma \partial_{[\sigma} \bar{f}_{c\nu]}(x,\bar{v}_c) + \frac{1}{m}\partial_\nu L_c(x,\bar{v}_c)\right)$. For $m = 2$ and $\bar{g}_1 = 0$ it reduces to the usual Riemannian geodesic equation and for $\bar{g}_i = 0, i = 1,\ldots,m-1$ to (3.9). For this path structure the strong equivalence principle in general does not hold.

In general for $p_{\hat{\mu}} \to \infty$ (4.7) does not approach (3.9) showing that null geodesics do not necessarily fill out the whole light cone.

For a later purpose it is convenient to perform a transformation on the momenta in such a way that in the resulting function $\mathcal{H}$ the linear term vanishes. To this end we define the *kinematical momentum* $P := p - A$ with $A(x) := \bar{f}_c(x,0)$. Then we have

$$'\mathcal{H}_c(x,P) := H_c(x,P+A) = \sum_{i=0}^{m} {'g_i}(P,\ldots,P) \tag{4.8}$$

with $'g_1 = 0$. In the case of $\mathcal{H}_c(x,A) \neq 0$ (4.8) may be divided through $-'\mathcal{H}_c(x,A)$ to get

$$\mathcal{H}_c(x,P) = \sum_{i=2}^{m} g_i(P,\ldots,P) - \epsilon \quad \text{with} \quad \epsilon = \begin{cases} 1 & \text{if } '\mathcal{H}_c(x,A) \neq 0 \\ 0 & \text{if } '\mathcal{H}_c(x,A) = 0 \end{cases} \tag{4.9}$$

with renormalised coefficients g_i of which g_1 vanishes. We can again define the velocity v_c and the metric g_c in analogy to $\bar{v}_c$ and $\bar{g}_c$. It is $v_c = \bar{v}_c$ and $g_c = \bar{g}_c$. These equalities and the fact that $v_c = 0$ for $P = 0$ justifies calling P the kinematical momentum.

---

* The relations for the light rays arising from the homogeneity of the function H are in general not valid in this case.

For this Hamilton function the above formalism is valid mutatis mutandis: From (4.9) the equation of motion can also be derived:

$$0 = v_c^\nu \partial_\nu v_c^\mu + \overset{\{\}}{G^\mu}(x, v_c) + g_c^{\mu\nu}(x, f(x, v_c))F_{\nu\sigma}v_c^\sigma .$$

with $F_{\nu\sigma} := -\partial_\mu A_\sigma + \partial_\sigma A_\mu$. Later we interpret $F_{\nu\sigma}$ as Maxwell field. We will call the formalism resting on $\mathcal{H}(x, P)$ the *kinematical formalism*.

4.4. The generalised Clifford algebra

In this section we want to examine the properties of the γ-matrices in their totality. This will lead to the notion of a generalised Clifford algebra which is useful for the same purposes as the characteristic Clifford algebra, that is, for the calculation of a propagation law of the WKB-states. It additionally allows discussion of the linearisation of scalar wave equations of higher order.

In analogy to ch. 3.3. we are here dealing with the algebra corresponding to the solvability condition (4.6). The formalism is absolutely the same except that the matrix $\bar{B}(x, p)$ is now not homogeneous:

$$\bar{B}(x,p) = \sum_{i=1}^{m'-1} \bar{B}^{\mu_1,\mu_2,\dots,\mu_i}(x) p_{\mu_1} p_{\mu_2} \cdots p_{\mu_i} .$$

Again we have

$$\bar{B}(x,p)\left(\bar{\gamma}^\mu p_\mu - \bar{E}\right) = \left(\bar{\gamma}^\mu p_\mu - \bar{E}\right)\bar{B}(x,p) = H_{c0}(x,p) = \sum_{i=1}^{m'} \bar{g}_i(p,\dots,p) - \bar{\epsilon} \quad (4.10)$$

with a H_{c0} possessing simple zeros only.

Differentiation of (4.10) with respect to p gives

$$\left(\frac{\partial \bar{B}}{\partial p_\mu}\right)(\bar{\gamma}^\nu p_\nu - \bar{E}) + \bar{B}\bar{\gamma}^\mu = (\bar{\gamma}^\nu p_\nu - \bar{E})\left(\frac{\partial \bar{B}}{\partial p_\mu}\right) + \bar{\gamma}^\mu \bar{B} = \frac{\partial H_{c0}}{\partial p_\mu} =: m' V^\mu \sim \bar{v}_c^\mu \quad (4.11)$$

$$\left(\frac{\partial^2 \bar{B}}{\partial p_\mu \partial p_\nu}\right)(\bar{\gamma}^\sigma k_\sigma - \bar{E}) + 2\left(\frac{\partial \bar{B}}{\partial p_{(\mu}}\right)\bar{\gamma}^{\nu)} = 2\bar{\gamma}^{(\mu}\left(\frac{\partial \bar{B}}{\partial p_{\nu)}}\right) + (\bar{\gamma}^\sigma k_\sigma - \bar{E})\left(\frac{\partial^2 \bar{B}}{\partial p_\mu \partial p_\nu}\right)$$
$$= \frac{\partial^2 H_{c0}}{\partial p_\mu \partial p_\nu} =: m'(m'-1)\bar{G}^{\mu\nu} \quad (4.12)$$

....

$$m'\left(\frac{\partial^{m'-1}\bar{B}}{\partial p_{(\mu_1} \cdots \partial p_{\mu_{m'-1}}}\right)\bar{\gamma}^{\mu_{m'})} = m'\bar{\gamma}^{(\mu_1}\left(\frac{\partial^{m'-1}\bar{B}}{\partial p_{\mu_2} \cdots \partial p_{\mu_{m'})}}\right) = m'! g_{m'}^{\mu_1\dots\mu_{m'}} . \quad (4.13)$$

The last equation was also arrived at by dealing with the characteristic Clifford-algebra. Beside this series of equations, we can insert (4.10) into (4.9). In order that the resulting equation is fulfilled for all p each coefficient has to vanish. We get

$$-\bar{B}^0\bar{E} = -\bar{E}\bar{B}^0 = -1$$
$$\bar{B}^0\bar{\gamma}^\mu - \bar{B}^\mu\bar{E} = \bar{\gamma}^\mu\bar{B}^0 - \bar{E}\bar{B}^\mu = \bar{g}_c^\mu$$
$$\bar{B}^{(\mu}\bar{\gamma}^{\nu)} - \bar{B}^{\mu\nu}\bar{E} = \bar{\gamma}^{(\mu}\bar{B}^{\nu)} - \bar{E}\bar{B}^{\mu\nu} = \bar{g}_2^{\mu\nu}$$

....

$$\bar{B}^{(\mu_1,\mu_2,\dots,\mu_{m'-1}}\bar{\gamma}^{\mu_{m'})} = \bar{\gamma}^{(\mu_1}\bar{B}^{\mu_2,\dots,\mu_{m'-1},\mu_{m'})} = \bar{g}_{m'}^{\mu_1,\dots,\mu_{m'}} . \quad (4.14)$$

For $E = 1$ this system of equations can now be successively solved for $\bar{B}^{\mu_1,\mu_2,\ldots,\mu_{m'}}$:

$$\begin{aligned}
\bar{B}^0 &= 1 \\
\bar{B}^\mu &= \bar{\gamma}^\mu - \bar{g}_1^\mu \\
\bar{B}^{\mu\nu} &= \bar{\gamma}^{(\mu}\bar{\gamma}^{\nu)} - \bar{\gamma}^{(\mu}\bar{g}_1^{\nu)} - \bar{g}_2^{\mu\nu} \\
&\ldots\ldots\ldots\ldots\ldots\ldots \\
\bar{B}^{(\mu_1,\mu_2,\ldots,\mu_{m'-1})} &= \sum_{j=0}^{m'} \bar{\gamma}^{(\mu_1} \ldots \bar{\gamma}^{\mu_j} \left(-\bar{g}_{m'-j}^{\mu_{j+1}\ldots\mu_{m'-1})}\right)
\end{aligned}$$

However $\bar{B}^{(\mu_1,\mu_2,\ldots,\mu_{m'-1})}$ has to fulfill (4.14) giving a relation which we will call a *generalised Clifford-algebra*:

$$0 = \sum_{i=0}^{m'} \bar{\gamma}^{(\mu_1} \ldots \bar{\gamma}^{\mu_i} \left(-\bar{g}_{m'-j}^{\mu_{i+1}\ldots\mu_{m'})}\right). \tag{4.15}$$

This equation relates the metrical coefficients $\bar{g}_i$ to the $\bar{\gamma}$-matrices. Assuming the special case $\bar{g}_i = 0$ for $i = 1, \ldots, m'-1$ we get $\bar{g}_{m'}^{\mu_1,\ldots,\mu_{m'}} = \bar{\gamma}^{(\mu_1} \ldots \bar{\gamma}^{\mu_{m'})}$ which already was considered by Nono (1971) in a different context.

The above relations are valid mutatis mutandis in the kinematical formalism: replacing in (4.3) p by $P + A$ we get $0 = \left(\bar{\gamma}^\mu P_\mu - M_A^{(0)}\right) a$ with the new 'mass' $M_A^{(0)} := M^{(0)} - \bar{\gamma}^\mu A_\mu$. With the same reasoning as below (4.4) we arrive at an equation $0 = (\gamma^\mu P_\mu - E)\, a$ together with an Hermitising matrix β: $(\gamma^\mu)^+ = \beta\gamma^\mu\beta^{-1}$. Of course $\mathrm{tr}\gamma^\mu = 0$.

4.5. Propagation of the spin states

The *spin states* a are solutions of (4.5) whereby for one given p as solution of (4.6) many a's may occure according to the multiplicity of the zero. For each two solutions belonging to one p we get for an arbitrary matrix Q

$$\begin{aligned}
0 &= \bar{a}Q\left(\bar{\gamma}^\mu p_\mu - \bar{E}\right) a \\
0 &= \bar{a}\left(\bar{\gamma}^\mu p_\mu - \bar{E}\right) Qa
\end{aligned}$$

with $\bar{a} := a^+\bar{\beta}$. Addition and subtraction of these equations gives

$$(\bar{a}\{Q, \bar{\gamma}^\mu\}a)\, p_\mu = 2\bar{a}Qa \tag{4.16}$$

$$(\bar{a}[Q, \bar{\gamma}^\mu]a)\, p_\mu = 0 \tag{4.17}$$

The remainder of this chapter is completely analogous to ch. 3.5. It is a matter of replacing H_0, (3.4), helicity-states, (3.11 and 12) by H_{c0}, (4.6), spin-states, (4.11 and 12). For the equation of motion of the spin states we get

$$v^\mu \overset{\bar{\Gamma}}{D}_\mu a = \frac{1}{\alpha}\left[-\frac{m'(m'-1)}{2}\bar{G}^{\mu\nu}\overset{\bar{\Gamma}}{D}_\mu p_\nu + B\left(\bar{\gamma}^\mu\bar{\Gamma}_\mu - iM^{(1)}\right)\right] a \tag{4.18}$$

with the definition

$$0 = \left(\overset{\bar{\Gamma}}{D}_\mu \bar{\gamma}^\nu\right) p_\nu = (\partial_\mu \bar{\gamma}^\nu)\, p_\nu + \bar{\Gamma}^\nu_{\mu\sigma}(x,p)\bar{\gamma}^\sigma p_\nu + \left[\bar{\Gamma}_\mu(x,p), \bar{\gamma}^\nu p_\nu\right]$$

whereby the $\bar{\Gamma}^{\nu}_{\mu\sigma}(x,p)$ was obtained from $\overset{\{\}}{\bar{G}}(x,\bar{v}_c)$ according to (3.10).

In the kinematical formalism (4.18) remains unchanged but the bars are omitted and an extra term $\frac{1}{2}F_{\mu\nu}\frac{\partial B}{\partial P_{[\mu}}\gamma^{\nu]}$ appears.

After these general considerations we can now further specify the field equation to arrive at the usual Dirac equation and discover its geometry.

5. The Dirac equation

In this chapter we want to find some minimal requirements in order to arrive at, resp. to characterise the usual Dirac equation. These demands essentially consist in two internal degrees of freedom for the helicity and spin states and two light cones and one time-like group velocity. Subsequently the most general Hamilton-Jacobi equation for the classical limit meeting these demands can be set up and discussed. The equation of motion for the paths turns out to be the Lorentz force equation in a Riemannian space-time. With the help of the propagation of the WKB-states the bilinear forms can be studied further. Since we are led to the usual Clifford-algebra more specific results can be concluded than in the general case. As the main result of this work, the propagation of the bilinear forms will define a Riemann-Cartan space-time with axial torsion only and a Maxwell field with which they interact via the usual coupling terms and a non-minimal Pauli term as well.

5.1. The propagation of singularities

Up to now, in nature never more than two light cones (one future and one past cone) have been observed to exist in 'empty space' (for a careful analysis of experimental data cf. Bleyer (1988).) In addition, there should be no more than two helicity states:

Axiom 7:

(i) *There are two light cones only.*
(ii) *There are two helicity states.*

As there are two helicity states on the light cone, the multiplicity of the zeros of the characteristic polynomial is two, so that the characteristic polynomial is the square of another one: $H(x,k) = (H_0(x,k))^2$. Furthermore, H_0 shall lead to two light cones only, being therefore a polynomial of order two: $H_0(x,k) = g^{\mu\nu}(x)k_\mu k_\nu$. As the characteristic polynomial is of order four, the γ's and the M are complex-valued 4×4-matrices.

It is clear that the metrical structure is given by the tensor $g^{\mu\nu}(x)$ with $\det g^{\mu\nu}(x) \neq 0$ so that there is an inverse $g_{\mu\nu}(x)$. These metrical tensors define a *conformal structure*: The light rays are given by $v^\mu = g^{\mu\nu}k_\nu$. The equation of motion for these light rays is the usual geodesic equation with the Christoffel symbols made up of the $g^{\mu\nu}$. It is clear from the hyperbolicity of H that the metric $g^{\mu\nu}$ has the correct signature.

The characteristic Clifford-algebra is given by a matrix $B = B^\mu p_\mu$ obeying $B^{(\mu}\tilde{\gamma}^{\nu)} = g^{\mu\nu}$. The helicity states obey the following propagation equation $v^\mu \overset{\{\}}{D}_\mu a = -\frac{1}{2}g^{\mu\nu}\overset{\{\}}{D}_\mu k_\nu a + \frac{1}{2}B\left(\tilde{\gamma}^\nu \overset{\{\}}{\Gamma}_\nu - M^{(1)}\right)a.$

5.2. The classical limit

Since in the Hamilton-Jacobi-equation the square of $H(x,k)$ appears as highest order part, the Hamilton-Jacobi-equation must be of the form $H_c(x,p) = (g^{\mu\nu}(x)k_\mu k_\nu)^2 +$

$\bar{g}_3^{\mu\nu\sigma}(x)k_\mu k_\nu k_\sigma + \bar{g}_2^{\mu\nu}(x)k_\mu k_\nu + \bar{g}_1^\mu(x)k_\mu - \bar{\epsilon}$. We take as a further basic experience and demand towards a characterisation of the Dirac-equation:

Axiom 8:
(i) *There are two spin states.*
(ii) *There is a time-like group velocity.*

The existence of two spin-states means that $H_c(x,p)$ must be the square of another polynomial. Since the degree of $H_c(x,p)$ is four we get up to sign $H_c(x,p) = (H_{c0}(x,p))^2$ with $H_{c0}(x,p) := \bar{g}^{\mu\nu}(x)p_\mu p_\nu + \bar{g}^\mu(x)p_\mu - \bar{\epsilon}$. Again, the metric is $\bar{g}_c^{\mu\nu}(x) = \bar{g}^{\mu\nu}(x)(= g^{\mu\nu}(x))$, and the group-velocity is $\bar{v}_c^\mu = \bar{g}^{\mu\nu}p_\nu + \frac{1}{2}\bar{g}^\mu$ which can be solved for $p : p_\mu = \bar{f}_\mu(x,\bar{v}_c) = \bar{g}_{\mu\nu}\left(\bar{v}_c^\nu - \frac{1}{2}\bar{g}^\nu\right)$.

Choosing an A according to (4.3), i.e. in our case $A_\mu = -\frac{1}{2}\bar{g}_{\mu\nu}\bar{g}^\nu$, we get

$$'\mathcal{H}_c(x,P) = \bar{g}^{\mu\nu}(x)P_\mu P_\nu - \tfrac{1}{4}\bar{g}_{\mu\nu}(x)\bar{g}^\mu(x)\bar{g}^\nu(x) - \bar{\epsilon} = \bar{g}^{\mu\nu}(x)P_\mu P_\nu - m^2(x)$$

with $m^2(x) := -H_{c0}(x,A)$. For $'\mathcal{H}(x,P) = 0$ to possess two solutions P_0 for arbitrarily given $P_{\hat{\mu}}$ forces $m^2(x) \geq 0$ an inequality which has to be satisfied by the coefficients $\bar{g}^\mu$, resp. by the A_μ. The requirement that there is at least one v_c within the light cone forces $m^2 > 0$ because $0 < g_{c\mu\nu}v_c^\mu v_c^\nu = g_c^{\mu\nu}P_\mu P_\nu = m^2$. Therefore $'\mathcal{H}_c(x,P)$ can be divided through m^2. It is easily veryfiable that $v_c^\mu = \dfrac{\partial}{\partial p_\mu}\mathcal{H}_c = g_c^{\mu\nu}P_\nu = g^{\mu\nu}\left(p_\nu + \frac{1}{2}g_{\nu\sigma}g^\sigma\right) = \dfrac{1}{2m^2}\dfrac{\partial}{\partial p_\mu}H_{c0} = \dfrac{1}{m^2}\bar{v}_c^\mu$. Because of $v_c^\mu = g^{\mu\nu}P_\nu$, P is indeed the kinematical momentum. The fact that $m^2 > 0$ especially means that $\epsilon = 1$ in $\mathcal{H}_c(x,P)$ and that $E = 1$.

Starting from the Hamilton-function $\mathcal{H}_c(x,P)$ we get as equation of motion

$$0 = v_c^\nu \partial_\nu v_c^\mu + \left\{{\mu \atop \nu\sigma}\right\} v_c^\nu v_c^\sigma + g^{\mu\nu}F_{\nu\tau}v_c^\tau. \tag{5.1}$$

with $\left\{{\mu \atop \nu\sigma}\right\}$ as the Christoffel-symbol made up of the $g^{\mu\nu}$. Therefore the considered geometry of the paths turns out to be a *Riemannian structure with a Lorentz-force.* The reduction of the Weylian structure as defined by the singularities to a Riemannian one was possible only because of the inhomogeneity of the Hamilton-Jacobi equation, i.e. because of $m^2 \neq 0$.

5.3. The Clifford-algebra and propagation of spin-states

In this special case the Clifford algebra in the kinematical formalism turns out to be the usual one

$$\gamma^{(\mu}\gamma^{\nu)} = g^{\mu\nu}$$

with $\mathrm{tr}\gamma^\mu = 0$ from which all the well known results of the usual γ-matrices can be derived, especially that there is a complete set of matrices $\mathbf{1}, i\gamma_5, \gamma^\mu, \gamma_5\gamma^\mu, \frac{i}{4}[\gamma^\mu,\gamma^\nu]$ called the *Dirac algebra.*

Defining the bilinear forms

$$S := \bar{a}a, \quad P := \bar{a}i\gamma_5 a, \quad J^\mu := \bar{a}\gamma^\mu a, \quad S^\mu := \bar{a}\gamma_5\gamma^\mu a, \quad T^{\mu\nu} := \bar{a}i\gamma^{[\mu},\gamma^{\nu]}a.$$

with $\bar{a} := a^+\beta$ we infer from (3.16) and (3.17) by taking as Q successively the elements of the Dirac-algebra, that the only independent equations are

$$P = 0, \quad Sv_c^\mu = J^\mu, \quad T^{\mu\nu} = \epsilon^{\mu\nu\rho}{}_\sigma P_\rho S^\sigma. \tag{5.2}$$

Therefore $T^{\mu\nu}$ can be derived from P and S^σ. We can show especially that $S \neq 0$ for WKB-states.

As the equation of motion for the spin-states we get

$$v_c^\mu \overset{\{\}}{D}_\mu a = -\tfrac{1}{2}g^{\mu\nu}\left(\overset{\{\}}{D}_\mu P_\nu\right)a + \tfrac{1}{2}\left(\gamma^\nu P_\nu + 1\right)Ka - \tfrac{1}{2}F_{\mu\nu}\gamma^\mu\gamma^\nu$$

with $K := \frac{1}{2}\left(\gamma^\mu\Gamma_\mu - iM^{(1)}\right)$ which is an arbitrary matrix.

A tedious calculation now gives the propagation laws for the bilinear forms whereby each derivative of one bilinear form depends on all the other bilinear forms. Since in all propagation laws there is an open proportionality factor, and because of (5.2) we can restrict ourselves to the normalised current vector, which however is identical to the particle velocity, and to the normalised *spin-vector* $\widehat{S}^\mu := \dfrac{1}{S}S^\mu$, so that the only independent entities are now v_c and $\widehat{S}$. Of course, the propagation equation for the current is equivalent to the previously derived equation (5.1).

More calculation then implies that

$$v_c^\nu \overset{\{\}}{D}_\nu \widehat{S}^\mu = -g^{\mu\nu}\left(F_{\nu\sigma} + K_{[\nu\sigma]}\right)\widehat{S}^\sigma + v_c^\nu \epsilon_{\nu\rho}{}^{\mu\sigma}K_\sigma \widehat{S}^\rho + v_c^\mu K_{[\nu\sigma]}v_c^\nu \widehat{S}^\sigma . \qquad (5.3)$$

The tensor $K_{[\mu\nu]}$ and axial vector K_μ result from the arbitrary matrix K. However, not all degrees of freedom of the matrix K survive.

The first term on the right hand side of (5.3) is the spin precession in the electromagnetic field. The term with K_σ can be identified with a connection-term which has the form of an axial torsion (For a survey of space-time theories with torsion see Baekler, Hehl and Mielke (1986)).

However, beside the Maxwell field no other field has been observed to cause the precession of the spin so that we state the last axiom:

Axiom 9: *The spin precession is always proportional to the Maxwell field.*

This requirement implies $K_{[\mu\nu]} = \alpha F_{[\mu\nu]}$ for some coefficient α. Defining the anomalous gyromagnetic ratio $g := 2(1+\alpha)$ we get as final propagation equation

$$v_c^\nu \overset{*}{D}_\nu \widehat{S}^\mu = -\frac{g}{2}g^{\mu\nu}F_{\nu\sigma}\widehat{S}^\sigma + \left(\frac{g}{2} - 1\right)v_c^\mu F_{[\nu\sigma]}v_c^\nu \widehat{S}^\sigma \qquad (5.4)$$

whereby $\overset{*}{D}$ is the covariant derivative with the axial torsion. This equation is the BMT-equation (see Bargmann, Michel and Telegdi (1959), Rubinov and Keller (1963) and Rafanelli and Schiller (1964)) in a Riemann-Cartan space (cf. Audretsch (1981), (1981a)).

Because of the last equation we can state the following result; that the Dirac equation which has been characterised by means of the axioms 1 to 9 indeed *defines as geometry a Riemann-Cartan geometry with axial torsion.* In addition, a *Maxwell field couples to the Dirac particles in the classical limit via charge and anamalous magnetic moment.*

Acknowledgement

This work forms part of the author's thesis written under the direction of Prof. J. Audretsch to whom I am grateful for encouragement and support. For discussions I thank Prof. H. Dehnen, Dipl.-Phys. D. Kovacs and especially Prof. W. Watzlawek, who was most helpful and referred me to some useful literature.

References

Andersson, S.I. (1987): Quantitative measures of geometric and topological structure as generated by dynamics, *Physica Scripta* **35** 225

Audretsch, J. (1981): Trajectories and Spin Motion of Massive Spin $\frac{1}{2}$ Particles in Gravitational Fields, *J. Phys.* A: *Math.Gen.* 14 411.
Audretsch, J. (1981a): Dirac Electron in Space-Time with Torsion: Spinor Propagation, Spin Precession and Non-Geodesic Motion, *Phys. Rev.* **D 24** 1470.
Audretsch, J. (1983): The Riemannian Structure of Space-Time as a Consequence of Quantum Mechanics, *Phys. Rev.* **D 27**, 2872.
Audretsch, J.; Gähler, F.; Straumann, N. (1984): Wave Fields in Weyl-Space and Conditions for the Existence of a Preferred Pseudo-Riemannian Structure, *Comm. Math. Phys.* **95** 41
Audretsch, J.; Lämmerzahl, C. (1988): Constructive Axiomatic Approach to Space-Time Torsion, *Class. Quantum Grav.* **5** 1285
Baekler, P.; Hehl, F.W.; Mielke, E. (1986): Nonmetricity and Torsion: Facts and Fancies in Gauge Approaches to Gravity, in Ruffini, R. (Hrsgb.): *Proceedings of the* 4th *Marcel Grossmann Meeting on General Relativity*, Elsèvier, Amsterdam.
Bargmann, V.; Michel, L.; Telegdi, V.L. (1959): Precession of the Polarisation of Particles Moving in a Homogeneous Electromagnetic Field, *Phys. Rev. Lett.* **2** 435
Bleyer, U. (1988): *Eine nicht-Lorentz-invariante Verallgemeinerung der Dirac-Gleichung: Begründungen und Konsequenzen*, Dissertation B, Akad. d. Wiss. DDR, Potsdam.
Bleyer, U.; Liebscher, D.-E. (1986): Induced causality, *Astron. Nachr.* **307** 267
Bleyer, U.; Liebscher, D.-E. (1987): Quantum mechanical consequences of pregeometry, *Preprint, Potsdam (will appear in: Proceedings of the* IVth *Seminar on Quantum Gravity*, Moscow 1987, World Scientific).
Bhabha, H.J. (1949): On the Postulational Basis of the Theory of Elemtary Particles, *Rev. Mod. Phys.* **21** 451.
Carathéodory, C. (1956): *Variationsrechnung und Partielle Differentialgleichungen erster Ordnung*, B.G. Teubner, Leipzig
Castagnino, M. (1971): The Riemannian Structure of Space-Time as a Consequence of a Measurement Method, *J. Math. Phys.* **12** 2203.
Childs, L. N. (1978): Linearising of n-ic Forms and Generalized Clifford Algebras, *Linear and Multilinear Algebra* **5** 267.
Courant, R; Hilbert, D. (1962): *Methods of Mathematical Physics*, Vol. II, Interscience Publishers, New York.
Dencker, N. (1982): On the Propagation of Polarization Sets for Systems of Real Principal Type, *J. Functional Analysis*, 46 351.
Dieudonné, J. (1971): *Èlements d'Analyse*, Gauthier-Villars, Paris.
Duffin, R.J. (1938): On the Characteristic Matrices of Covariant Systems, *Phys. Rev.* **54** 1114.
Ehlers, J.; Pirani, F.A.E.; Schild, A. (1972): The Geometry of Free Fall and Light Propagation, in: L. O'Raifeartaigh (ed.): *General Relativity, Papers in Honour of J.L. Synge*, Clarendon Press, Oxford.
Ehlers, J. (1973): Survey of General Relativity Theory, in Israel, W. (ed.): *Relativity, Astrophysics and Cosmology*, D. Reidel, Dordrecht.
Fisher, A.E.; Marsden, J.F. (1979): The Initial Value Problem and the Dynamical Formulation of General Relativity, in S.W. Hawking, W. Israel (ed.): *General Relativity, an Einstein centennary survey*, Cambridge Univ. Press.
Gårding, L. (1985): Hyperbolic Differential Operators, in Jäger, W.; Moser, J.; Remmert, R. (Hrsgb.): *Perspectives in Mathematics, Anniversary of Oberwolfach* 1984, Birkhäuser Verlag, Basel.
Harish-Chandra (1947): On Relativistic Wave Equations, *Phys. Rev.* **71** 793.
Hille, E.; Phillips, R.S. (1957): *Functional Analysis and Semi-Groups*, American Mathematical Society.
Hörmander, L. (1984): *The Analysis of Linear Partial Differential Operators*, Volume I - IV, Springer-Verlag, Berlin.
Kasper, U. (1986): On the importance of quantum mechanics for the axiomatic approach to the theory of gravitation, in *Proceedings of the Conference on Differential Geometry and its Aopplications*, in August 1986 in Brno, CSSR.
Kemmer, N. (1939): The particle aspect of meson theory, *Proc. Roy. Soc.* **A173** 91.

Komura, T. (1968): Semigroups of Operators in Locally Convex Spaces, *J. Functional Analysis* **2** 258
Liebscher, D.-E. (1985a): The Geometry of the Dirac Equation, *Ann. Physik* **42** 35
Loinger, A. (1985): Weylian geometry and first order wave equations, *Nouvo Cim.* **88B** 9
Matsumoto, M. (1970): *The Theory of Finsler Connections*, Publications of the Study Group of Geometry.
Meyer, R.; Schroeter, G. (1981): The Application of Differential Geometry to Ray Acoustics in Inhomogeneous and Moving Media, *Acustica* **47** 105.
Nono, T. (1971): Generalised Clifford algebras and linearisations of a partial differential equation. in Ramakrishnan, A. (ed.): *Proceedings of the Conference on "Clifford-Algebra, its Generalisations and Applications"*, Matscience, The Institute of Mathematical Sciences, Madras.
Rafanelli, K.; Schiller, R. (1964): Classical Motions of Spin $\frac{1}{2}$ Particles, *Phys. Rev.* **135** B279.
Rauch, H. (1983): in S. Kamefuchi (ed.): *Proceedings of the International Symposium on Foundations of Quantum Mechanics*, Tokyo.
Reichenbach, H. (1928): *Philosophie der Raum-Zeit-Lehre*; deGruyter, Berlin
Riemann, B. (1854): Ueber die Hypothesen, die der Geometrie zugrunde liegen, Habilitationsvortrag, in Weber, H. (Hersg.): *Bernhard Riemanns gesammelte mathematische Werke und wissenschaftlicher Nachlass*, B.G. Teubner, Leipzig 1876.
Roby, N. (1969): Algèbres de Clifford des formes polynomes, *C.R. Acad. Sc. Paris* **A268** 484.
Rubinow, S.J.; Keller, J.B. (1963): Asymptotic Solution of the Dirac Equation, *Phys. Rev.* **131** 2789.
Rund, H. (1959): *The Differential Geometry of Finsler Spaces*, Grundlehren der Mathematischen Wissenschaften in Einzeldarstellungen, Springer-Verlag, Berlin
Svendsen, E.C. (1982): Unitary one-parameter groups with finite speed of propagation, *Proc. Amer. Math. Soc.* **84**, 357.
Wightman, A.S. (1970): Relativistic Wave Equations as Singular Hyperbolic Systems, in Spencer, D.C. (ed.): *Partial Differential Equations*, Proceedings of the Symposium in Pure Mathematics, American Mathematical Society.
Yoshida, K. (1971): *Functional Analysis*, Springer-Verlag.

QUANTUM MECHANICS IN CURVED SPACETIMES

STOCHASTIC PROCESSES ON FRAME BUNDLES

Cécile DeWitt-Morette

Center for Relativity and Department of Physics
The University of Texas at Austin
Austin, Texas 78712-1081

I. FORMALISM FOR QUANTUM MECHANICS: PATH INTEGRATION

It is customary, nowadays, to pay homage to path integration at the beginning of a course in Quantum Physics—but later on to use only its most obvious properties. So much so that it is often said that,

> "Feynman path integrals are mathematically meaningless."
>
> "We only know how to integrate gaussian path integrals."
>
> "A path integral is only the solution of a parabolic partial differential equation, satisfying some initial conditions."

I hope that my brief presentation will show that none of the above is true. I shall refute these three statements while summarizing works where

Feynman path integrals are defined;
path integrals are computed;
path integrals appear in varied situations.

I will then have in hand an excellent tool to study quantum mechanics in curved spacetimes—a tool to solve problems and an "instrument de pensée".

A. Defining Feynman path integrals (as opposed to Wiener integrals)

A path integral I is an integral over a space of paths X. A path $x \in X$ is a map

$$x : T \times \Omega \to M \qquad \text{by} \qquad x(t, \omega) \in M \ .$$

---

[1] "When dealing with less simple and concrete equations, physical intuition is less reliable and often borders on wishful thinking" (van Kampen, 1976).

Quantum Mechanics in Curved Space-Time
Edited by J. Audretsch and V. de Sabbata
Plenum Press, New York, 1990

A point $\omega \in \Omega$ labels a path, t parametrizes the path $x(\cdot, \omega)$. M can be a Riemannian manifold, a multiply connected space, a μ-sheeted Riemann surface, a group manifold, or a symmetric space. To begin with, M will be $\mathbb{R}^n$.

$$I = \int_\Omega F\big(x(t,\omega)\big)\, d\mu(\omega) \ . \tag{I.A–1}$$

μ is an integrator—possibly a measure; $F : M \to \mathbb{R}$.

The space of paths X is, in general, infinite-dimensional, and our intuition based on the properties of $\mathbb{R}^n$ can lead us astray[1] if we think of the path integral I as the limit when $n = \infty$ of an integral over $\mathbb{R}^n$ [see DeWitt-Morette et al., 1979, Appendix A, for some differences between integration over $\mathbb{R}^n$ and functional integration]. Leaving our intuition aside, we can learn functional integration from mathematicians. But most mathematicians have developed functional integration with a restriction that precludes its uses in Feynman path integrals. Namely, the integral is over a probability space $(\Omega, \mathcal{F}, \mu)$ where $\mathcal{F}$ is a σ-algebra of subsets of Ω and μ a bounded measure on $\mathcal{F}$, normalized to unity

$$\mu(\Omega) = 1 \ . \tag{I.A–2}$$

Feynman path integrals are not sums of probabilities but sums of probability amplitudes expressible in terms of unbounded measures. And already on $\mathbb{R}^n$ for $n \geq 2$, most of integration theory is cast in terms of bounded measures.

Dyson calls the mathematical definition of Feynman integrals over spaces of paths and over spaces of histories an "opportunity" not to be missed. Several approaches have been investigated for making mathematical sense of Feynman integrals. See the Appendix.

These different approaches serve different purposes. Many of them aim at defining path integrals for as large a class of physical systems as possible. I will present one of them, the prodistribution definition, because it leads to computational techniques that have effectively been used to solve problems of quantum mechanics in curved spacetimes. For example, to quote a nontrivial problem, the calculation of glory scattering of polarized waves by black holes can be expressed in terms of the solution of an associated system—and this associated system is the quantum mechanics of particles moving in curved spacetimes (DeWitt-Morette, 1984). The cross section obtained from path integrals defined and computed with prodistributions is

$$d\sigma(\Omega)/d\Omega = 4\pi^2\lambda^{-1}B^2(\theta)\frac{dB(\theta)}{d\theta}\, J^2_{2s}\big(2\pi\lambda^{-1}B(\theta)\sin\theta\big) \tag{I.A–3}$$

where Ω is the solid angle $d\Omega = 2\pi \sin\theta\, d\theta$, λ is the wave length of the scattered wave, $B(\theta)$ is the inverse of the deflection function $\Theta(B)$, giving the scattering angle θ of a ray as a function of its impact parameter B. For example, for a Schwartzschild black hole of mass M (Darwin, 1959)

$$B(\theta) = M\big(3\sqrt{3} + 3.48 \exp(-\theta)\big) \ . \tag{I.A–4}$$

J_{2s} is the Bessel function of order $2s$ with $s = 0$ for a scalar wave, $s = 1$ for an electromagnetic wave, $s = 2$ for a gravitational wave. This formula matches perfectly with the Handler-Matzner numerical calculations based on partial wave decomposition (Futterman et al., 1988).

I refer you to DeWitt-Morette (1984) for the complete detailed calculation of the glory cross section (I.A–3). Here I give only the building blocks; first the definition of a prodistribution; then, in Section I.B, a computational technique based on this definition.

A number of years ago, at the suggestion of Yvonne Choquet-Bruhat, I found in the chapter of Bourbaki on integration on topological vector spaces (Bourbaki, 1969) the means of defining an object, which could be used to build and compute some Feynman integrals. Whereas most texts on probability begin with $(\Omega, \mathcal{F}, \mu_{\mathcal{F}})$, a set Ω, a Borel σ-algebra $\mathcal{F}$, a measure $\mu_{\mathcal{F}}$ on $\mathcal{F}$, Bourbaki begins with $(\Omega, \mathcal{Q}, \mu_{\mathcal{Q}})$, a topological vector space Ω Hausdorff and locally convex, a projective family of finite-dimensional spaces $\mathcal{Q}$ related to Ω and a *pro*jective family of bounded *measures* $\mu_{\mathcal{Q}}$ on $\mathcal{Q}$, called "promeasures".

A projective family of finite-dimensional spaces is a mathematical construction, which accounts for Feynman's heuristic procedure. It states qualitatively the idea, "Let us replace a path by an arbitrary finite set of its values"; it makes it simple to ensure that the result is independent of the chosen set. But it does more because the finite-dimensional spaces are not restricted to spaces of paths defined by a finite number of their values. And it has computational power, as we shall see shortly. It is defined as follows.

Let $\mathcal{F}(\Omega)$ be the set of closed subspaces $V, W, \ldots$ of Ω of finite codimension partially ordered by the inclusion relation. The space V belongs to $\mathcal{F}(\Omega)$ if and only if it consists of points $\omega \in \Omega$ such that

$$\langle \omega_j', \omega \rangle = 0 \qquad \text{for a finite set} \qquad V_0 \stackrel{\text{def}}{=} \{\omega_j'\} \ , \tag{I.A–5}$$

where $\omega_j' \in \Omega'$, the topological dual of Ω. If $V_0 \subset W_0$, then $W \subset V$.

Let Ω/V be the space of equivalence classes $[\omega]$ defined by the equivalence $\omega_1 \sim \omega_2 \langle = \rangle (\omega_1 - \omega_2) \in V$. Let P_V be the canonical mapping from Ω into Ω/V. Let $W \subset V$ and let P_{VW} be defined by

$$P_V = P_{VW} \circ P_W \ . \tag{I.A–6}$$

The quotient spaces $\Omega/V, \Omega/W, \ldots$ together with the canonical mappings P_{VW} : $\Omega/W \to \Omega/V \cdots$ form the projective system of finite-dimensional quotient spaces of Ω indexed by $\mathcal{F}(\Omega)$.

Example: For instance, let Ω be the space of continuous paths ω with fixed origin,

$$x(t, \omega) = \omega(t) \ , \quad \omega(t_0) = 0, \qquad t \in T \ .$$

The topological dual Ω' is the space of bounded measures. The duality in Ω is

$$\langle \omega', \omega \rangle = \int_T \omega(t)\, d\omega'(t) \tag{I.A–7a}$$

or

$$\langle \omega', \omega \rangle = \int_T \omega^\alpha(t)\, d\omega_\alpha'(t) \tag{I.A–7b}$$

or

$$\langle \delta'_t, \omega \rangle \ = \omega(t) \tag{I.A–7c}$$

as the case may be. Let θ_V be a partition of T

$$\theta_V = \{t_1, \ldots, t_V\}$$

and let $V_0 = \{\delta_{t_j}\}$ be indexed by θ_V. A path x belongs to V if $\langle \delta_{t_j}, x \rangle = x(t_j) = 0$. Two paths ω_1 and ω_2 are equivalent if

$$(\omega_1 - \omega_2)(t_j) = 0 \qquad \text{for every} \qquad t_j \in \theta_V \ .$$

A point $[\omega] \in \Omega/V$ is defined by its values $\omega(t_j)$ for every $t_j \in \Omega_V$. The canonical map

$$P_V : \Omega \to \Omega/V \ \text{ by } \ \omega \mapsto [\omega] = \{u^1, \ldots, u^V\} \ \text{ with } \ u^j = \langle \delta_{t_j}, \omega \rangle \ . \tag{I.A–8}$$

If we choose V_0 to be a set of Dirac distributions $\{\delta_{t_j}\}$, we replace a path by a finite set of its values. How to choose V_0 depends on the path integral to be computed.

So much for the projective family $\mathcal{Q}$ of finite-dimensional spaces related to Ω. We can now define a promeasure. A promeasure $\mu_{\mathcal{Q}}$ is a family of bounded measures $\{\mu_V\}$ on $\{\Omega/V\}$ such that

$$\begin{cases} \mu_V(\Omega/V) \text{ is independent of } V; \\ \text{when } W \subset V, \ \ \mu_V \text{ is the image of } \mu_W \text{ under } P_{VW} \, . \end{cases} \tag{I.A–9}$$

It is an easy matter, and it will be profitable, to restate the coherence conditions satisfied by the family $\mu_{\mathcal{Q}}$ of bounded measures as coherence conditions satisfied by the family $\mathcal{F}\mu_{\mathcal{Q}}$ of their Fourier transforms. The Fourier transform of a measure μ_V on Ω/V is defined on the dual of Ω/V. The space Ω/V is finite-dimensional (in the previous example, $[\omega] = \{u^1, \ldots, u^V\} \equiv u$). The dual $(\Omega/V)'$ is also finite-dimensional, of the same dimension as Ω/V. We shall label its points $[\omega]'$ or u'. The Fourier transform of μ_V is, by definition,

$$\mathcal{F}\mu_V([\omega]') = \int_{\Omega/V} \exp(-i\langle [\omega]', [\omega] \rangle)\, d\mu_V([\omega]) \ . \tag{I.A–10}$$

We introduce the transposed mapping $\tilde{P}_V : (\Omega/V)' \to \Omega'$ in order to go back to our aim, which is to define objects on the infinite-dimensional space Ω.

The transposed mapping $\tilde{P}_V$ of the linear continuous mapping P_V is by definition

$$\langle u', P_V\,\omega \rangle_{\Omega/V} = \langle \tilde{P}_V\, u', \omega \rangle_\Omega \qquad \text{for} \qquad \omega \in \Omega \ , \tag{I.A–11}$$

where $\langle \ , \ \rangle_{\Omega/V}$ is the duality in Ω/V and $\langle \ , \ \rangle_\Omega$ is the duality in Ω. If one labels $[\omega]'$ a point in $(\Omega/V)'$, then

$$\tilde{P}_V([\omega]') = \omega' \ .$$

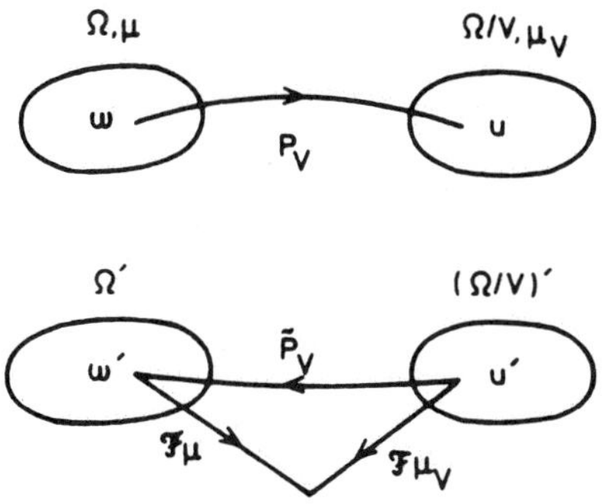

Figure 1. P_V is a linear continuous mapping, $\tilde{P}_V$ its transposed. μ_Q is a promeasure on Ω defined as a family of measures μ_V on Ω/V for all $V \in \mathcal{F}(\Omega)$. The Fourier transform $\mathcal{F}\mu_V = \mathcal{F}\mu_Q \circ \tilde{P}_V$.

Note that there is no canonical isomorphism between either of these spaces and its duals, so that the label u' $\left[\text{resp}\,[\omega]'\right]$ bears no relationship with u $\left[\text{resp}\,[\omega]\right]$. Thus, it is often better to write $\tilde{P}_V(u') = \omega'$ than the previous equation.

Let $\omega' = \tilde{P}_V\, u'$, the Fourier transform of the family μ evaluated at ω' is

$$\mathcal{F}\mu_Q(\omega') = \mathcal{F}\mu_V(u') = \int_{\Omega/V} \exp\left(-i\langle u', u\rangle\right) d\mu_V(u) \ . \tag{I.A–12}$$

Since Ω' is the union of all V_0 for all $V \in \mathcal{F}(\Omega)$, equation (I.A–12) defines $\mathcal{F}\mu_Q$ on Ω'.

The coherence conditions satisfied by $\mathcal{F}\mu_Q$ are

$$\begin{cases} \mathcal{F}\mu_V(0) \text{ is independent of } V; \\ \text{when } W \subset V, \ \ \mathcal{F}\mu_V = \mathcal{F}\mu_W \circ \tilde{P}_{VW} \,. \end{cases} \tag{I.A–13}$$

We note that the coherence conditions (I.A–13) are simpler to implement than the coherence conditions (I.A–9) because they are not set statements.

Now, measures need not be defined as set functions; they can equivalently be defined as distributions of order zero. As set functions, the distinction between bounded and unbounded measures is formidable; as distributions, the distinction is interesting but not overwhelming. Given the facts that

$$\begin{cases} \text{(I.A–13) is easier to handle than (I.A–9), and} \\ \text{unbounded measures belong more naturally to distribution theory} \\ \text{than to set theory,} \end{cases}$$

I propose to work with the *projective* family $\mathcal{F}\mu_Q$ of *distributions* $\{\mathcal{F}\mu_V; V \in \mathcal{F}(\Omega)\}$ satisfying the coherence conditions (I.A–13). Originally called "pseudomeasure," the family $\mathcal{F}\mu_Q$ was shortly afterwards called "prodistribution" by Dieudonné. The Fourier transform $\mathcal{F}\mu_V$ is not restricted to being the Fourier transform of a bounded measure, thus prodistributions are generalizations of promeasures.

Important issues remained to be solved:

i) The space of test functions for $\mathcal{F}\mu_V$ is V-dependent. The set of all test functions for all $V \in \mathcal{F}(\Omega)$ is a difficult object to handle.

ii) The following theorem in the theory of promeasures has not been generalized to prodistributions:

"The mapping $\mu_Q \to \mathcal{F}\mu_Q$ of the set of promeasures on Ω to the set of functions on Ω' is injective."

However, it is possible to bypass these issues if one develops a theory of integration based on $\mathcal{F}\mu_Q$ rather than on μ_Q. This theory is still rudimentary compared to the Lebesgue theory of integration but sufficient to compute many nontrivial physics problems.

Let $F = f \circ P_V$ for $V \in \mathcal{F}(\Omega)$ and $f \in \mathcal{S}(\Omega/V)$ where $\mathcal{S}$ is the Schwartz space whose dual is the space of tempered distributions. Then we can define

$$\int_\Omega F(\omega)\, d\mu_Q(\omega) \text{ to be the number equal to } \int_{\Omega/V} f(u)\, d\mu_V(u) \ . \tag{I.A–14}$$

with

$$\mathcal{F}\mu_V = \mathcal{F}\mu_Q \circ \tilde{P}_V \ . \tag{I.A–15}$$

This definition has computing power.

Exercise: Compute the transposed $\tilde{P}_V$ of P_V defined by (I.A–8). Answer: $\tilde{P}_V$ is defined by (I.A–11), which here reads

$$\langle u', u\rangle_{\Omega/V} = \sum_i u'_i u^i = \sum_i u'_i \langle \delta_{t_i}, \omega\rangle = \langle \sum_i u'_i \delta_{t_i}, \omega\rangle = \langle \tilde{P}_V u', \omega\rangle_\Omega \ .$$

Hence,

$$\tilde{P}_V u' = \sum_i u'_i \delta_{t_i} \ . \tag{I.A–16}$$

It is easy to give an intuitive presentation of prodistributions when μ is a gaussian integrator. Let μ be a gaussian integrator on $\mathbb{R}^n$,

$$d\mu_s(x) = (2\pi s)^{-n/2} (\det A^{-1})^{1/2} \exp\left(-\tfrac{1}{2s} (A^{-1})_{ij}\, x^i x^j\right) dx^1 \ldots dx^n \tag{I.A–17}$$

where $s = 1$ for real gaussian (a bounded measure) and $s = i$ for a complex gaussian (an unbounded measure). Its Fourier transform

$$(\mathcal{F}\mu_s)(x') = \exp\left(-\tfrac{s}{2} A^{ij} x'_i x'_j\right) \tag{I.A–18}$$

is equally well behaved whether $s = 1$ or $s = i$. Moreover, one cannot write in a compact form a gaussian μ on Ω, whereas its Fourier transform is the obvious continuum limit of (I.A–18), namely,

$$(\mathcal{F}\mu_s)(\omega') = \exp\left(-\tfrac{s}{2} \int_T \int_T G(t,s)\, d\omega'(t)\, d\omega'(s)\right) \ . \tag{I.A–19}$$

The use of Fourier transforms in functional integration in field theory has long been proved useful. There it is written formally

$$\int \mathcal{D}\varphi \exp\left(i\, S(\varphi) + i \int_x \varphi(x)\, J(x)\, dx\right) \ . \tag{I.A–20}$$

This expression can be considered as the Fourier transform of the integrator $\mathcal{D}\varphi \exp\big(iS(\varphi)\big)$ evaluated at J.

B. Computing path integrals. $I = \int_\Omega F(\omega)\, d\mu(\omega)$.

1. *Linear Methods.* It often happens that F can be decomposed into two maps

$$F = f \circ P \tag{I.B–1}$$

where P is a continuous linear map from Ω into a linear space U,

$$P : \Omega \to U \text{ by } \omega \mapsto u$$

and f is an "arbitrary" map $f : U \to \mathbb{R}$. In this case we can immediately write down (see previous section)

$$I = \int_\Omega (f \circ P)(\omega)\, d\mu(\omega) = \int_U f(u)\, d(P\mu)(u) \qquad \text{(I.B–2)}$$

where $P\mu$ is the image of μ under P and the Fourier transform of $P\mu$ is given in terms of the Fourier transform of μ by

$$\mathcal{F}(P\mu) = \mathcal{F}(\mu) \circ \tilde{P} \ . \qquad \text{(I.B–3)}$$

Examples of linear mappings P on vector spaces of paths X. First, two examples of vector spaces of paths: the space of continuous paths, $x : T \to \mathbb{R}^n$, and the space of continuous paths x with vanishing boundary conditions.

i) $P : X \to y$ by $x \mapsto y$ such that $y(t) = \int_0^\infty K(t,s)\, x(s)\, ds$. (I.B–4)

$K(t,s)$ can be a matrix; it can also be proportional to a step function restricting the domain of integration.

ii) If the space X admits a denumerable basis $\{\Phi_k\}$ (e.g., in a Sturm-Liouville problem), the following map is often useful

$$P : X \to \mathbb{R}^\infty \text{ by } y_k = \int_T x(s)\, \Phi_k(s)\, ds \ . \qquad \text{(I.B–5)}$$

This map is particularly useful whenever one encounters eigenvectors with vanishing eigenvalues.

iii) We can generalize to arbitrary elements of the dual X' of X the example (I.A–7, I.A–8)

$$P : X \to \mathbb{R}^p \text{ by } y^k = \langle \mu_k, x \rangle_X \text{ for } \mu_1, \ldots \mu_p \in X' \ . \qquad \text{(I.B–6)}$$

Having used (I.B–2) the linear map P to rewrite the path integral, what next? It depends if P is of the type *i*), *ii*), or *iii*). It often happens that P is of type *iii*). The integrand, upon careful examination, is not a function of $x(\cdot,\omega)$ but a function of several $\langle \mu_k, x(\cdot,\omega)\rangle$, say p of them. Then the right-hand side of (I.B–2) is an integral over $\mathbb{R}^p$. If μ is a gaussian, $P\mu$ is a gaussian, and it is straightforward to write down explicitly the right-hand side of (I.B–2) given the left-hand side of (I.B–3). See (I.A–17) and (I.A–18). The work consists of inverting a $p \times p$ matrix and computing its determinant. If p is small—say less than 3, at most 4, as is often the case—then one does it. If p is large, for instance, the path has been discretized and $\{\mu_k\}$ is the family $\{\delta_{t_k}\}$ for a large number of t_k, then usually the system is defined by an action; the matrix to be inverted comes from the second variation of the action; and the calculus of variation gives techniques fully worked out to invert the matrix of interest (DeWitt-Morette, 1976; DeWitt-Morette, 1984).

If P is of type *ii*) with $\{\Phi_k\}$ a complete set of eigenfunctions of a Sturm-Liouville problem, then usually most of the y_k defined by (I.B–5) can be trivially integrated, leaving a few y_k physically important. The integral over $\mathbb{R}^\infty$ is a trivial integral over a space of finite codimension, say p, and a nontrivial integral over $\mathbb{R}^p$.

If P is of type *i*), there often exists another linear map from Y into $\mathbb{R}^\infty$ or $\mathbb{R}^p$ of type *ii*) and *iii*).

If none of the above is true, one can resort, if justified, to an expansion of $F = \sum_k f_k \circ P_k$.

The linear methods are particularly useful when the integrator is gaussian. This does not imply that we then integrate gaussians because in (I.B–1) the mapping f is "*arbitrary*." When the system is defined by an action, the best gaussian to choose is the gaussian whose Fourier transform is

$$(\mathcal{F}\mu)(\omega') = \int_T d\omega'_\alpha(t) \int_T d\omega'_\beta(s)\, G^{\alpha\beta}(t,s) \qquad \text{(I.B–7)}$$

where $G^{\alpha\beta}(t,s)$ is the Green function of the Jacobi operator obtained from the second (functional) derivative of the action. Its boundary conditions are determined by the space of paths over which one integrates. For instance, consider the space of paths $x : [t_a, t_b] \times \Omega \to \mathbb{R}^n$ with $x(t_a, \omega) = a$, $x(t_b, \omega) = b$. Since Ω must be a vector space, we can only integrate over paths with vanishing boundary conditions: set[2]

$$x = q + \eta \qquad \text{(I.B–8)}$$

where $q(t_a) = a$, $q(t_b) = b$, and make a change of variable of integration so that the new variable of integration is η, which indeed has vanishing boundary conditions. Since

$$\int_\Omega \eta^\alpha(t,\omega)\, \eta^\beta(s,\omega)\, d\mu(\omega) = G^{\alpha\beta}(t,s)\ , \qquad \text{(I.B–9)}$$

$G^{\alpha\beta}(t,s)$ inherits the boundary conditions of η.

The interesting fact is that one does not *choose* this gaussian; one derives it from the Feynman-Kac's formula (I.B–20a) by appropriate linear mappings (DeWitt-Morette et al., 1979).

2. Stochastic Methods. These linear procedures have been used extensively in a great variety of situations. See DeWitt-Morette (1987) for references. But if nonlinear procedures are needed, we can turn to stochastic calculus for suggestions, possibly using Fourier transforms as a bridge to transfer techniques. But first a word of caution.

In addition to the profound difference between bounded and unbounded measures, there are two other easy-to-handle but noteworthy differences between Wiener and Feynman integrals.

---

[2] If $x(t,\omega)$ does not belong to $\mathbb{R}^n$, consider all the one-parameter families of paths around q and integrate over the vector fields along q defined by these families (see DeWitt-Morette et al., 1979).

i) In a Wiener integral a path has dimention $T^{1/2}$, e.g.,

$$\int_\Omega d\mu(\omega)\,(x(t,\omega) - x(t_0,\omega))^2 = t - t_0 \ . \tag{I.B–10}$$

In a Feynman integral a path has dimension L, e.g., in a WKB approximation, one writes

$$x(t,\omega) = q(t) + \eta(t,\omega) \tag{I.B–11}$$

where $q(t)$ is the classical path. We shall label z the Brownian path satisfying (I.B–10), and the corresponding path in Feynman integral will be

$$x = \mu z \tag{I.B–12}$$

where μ is a constant of dimension $LT^{-1/2}$, e.g., $\mu = \sqrt{\hbar/m}$. Then if in (I.B–11) we wish to use a variable of integration ξ with the dimension of a Brownian path, we write

$$x = q + \mu\xi \ ; \tag{I.B–13}$$

and semiclassical expansions are naturally cast as expansions in powers of $\mu = \sqrt{\hbar/m}$.

ii) In a diffusion equation one models systems such as the following. Some matter, concentrated at x_0 at time t_0, diffuses randomly. One gives a probability distribution of the matter at time t by means of a random variable, defined on a probability space $(\Omega, \mathcal{F}, \mu)$

$$x(t,\cdot) : \Omega \to \mathbb{R}^n \ , \quad \text{or} \quad x(t) : \Omega \to \mathbb{R}^n \ . \tag{I.B–14}$$

Often, but not necessarily, $x(t)$ is defined by a stochastic differential equation defined by a Brownian path $z(t)$ in $\mathbb{R}^m$

$$dx(t) = X(x(t))\,dz(t) + A(x(t))\,dt \ , \qquad x(t_0,\omega) = x_0 \tag{I.B–15}$$

where

$$\begin{aligned} X(x(t)) &: \mathbb{R}^m \to \mathbb{R}^n \ , \quad \text{i.e., } X : \mathbb{R}^n \to L(\mathbb{R}^m, \mathbb{R}^n) \ , \\ A(x(t)) &: \mathbb{R} \ \ \to \mathbb{R}^n \ , \quad \text{i.e., } A : \mathbb{R}^n \to L(\mathbb{R} \ , \mathbb{R}^n) \ , \end{aligned}$$

are given "arbitrary" maps. One is usually interested in a function φ of the diffused matter. One knows $\varphi(x(t,\omega))$ only probabilistically. A quantity of interest is its average Ψ over the whole diffused matter. The average Ψ is the function of the starting point x_0 and the time t when one looks at the diffusing matter

$$\Psi(x_0,t) = \int_\Omega d\mu(\omega)\,\varphi(x(t,\omega)) \ . \tag{I.B–16}$$

The path integral (I.B–16) satisfies the following diffusion equation

$$\frac{\partial \Psi}{\partial t} = \frac{1}{2} \sum_i X_i^\alpha(x_0) X_i^\beta(x_0) \frac{\partial^2 \Psi}{\partial x_0^\alpha \, \partial x_0^\beta} + A^\alpha(x_0) \frac{\partial \Psi}{\partial x_0^\alpha} \tag{I.B–17a}$$

$$\Psi(x_0, t_0) = \varphi(x_0) \ . \tag{I.B–17b}$$

The path integral solution $\Psi(x,t)$ of a Schrödinger equation is evaluated at the final position x of the paths: we integrate over all the paths that end at x at time t. We can map the space Ω_0 of paths on $T = [t_0, t]$ such that $x(t_0) = 0$ into the space of paths Ω_t such that $x(t) = 0$ by reparametrizing the paths in Ω_0. To $\tilde{x} \in \Omega_0$ we associate $x \in \Omega_t$ by

$$x(s) = \tilde{x}(t + t_0 - s) \ . \tag{I.B–18}$$

If $\tilde{x}$ is differentiable, then $dx(s)/ds = -d\tilde{x}(t + t_0 - s)/dt$. For more details, see DeWitt-Morette and Elworthy (1978).

It is, however, in some respects more intuitive to work with path integral solutions of diffusion equations than with path integral solutions of Schrödinger equations: We know if a Brownian path has started at 0; we do not know ahead of time if it will end at 0. On the other hand, $\Psi(x,t) = \int_{\Omega_t} d\mu(\omega)\, \varphi(x(t_0, \omega))$, and it is more intuitive to work with the initial wave function φ evaluated at $x(t_0, \omega)$ rather than at $x(t, \omega)$ as in (I.B–16). See Section III.B.2 for the difference.

For the sake of brevity, I shall present the stochastic techniques in their usual setting; to obtain (formally) the corresponding results in quantum mechanics, one has to make sure that:

i) in the path integral $\int_\Omega F(\omega)\, d\mu(\omega)$, Ω is a vector space, μ a prodistribution;
ii) the paths are scaled to be of dimension L (I.B–12);
iii) the paths are run "backwards" (I.B–18).

One can look at the path integral (I.B–16)

$$\Psi(x_0, t) = \int_\Omega d\mu(\omega)\, \varphi(x(t, \omega))$$

from different points of view; this is the very reason a path integral is a versatile tool, and many techniques can be brought to bear on its computation or its approximations or in obtaining information without computing it.

Let us look at a slightly richer system than the one presented in equations (I.B–14) to (I.B–17). Consider the system of stochastic differential equations

$$\begin{cases} dx(t) = X\big(x(t)\big)\, dz(t) \ , \quad x(t_0) = x_0 \in \mathbb{R}^n \\ dv(t) = V\big(x(t)\big)\, v(t)\, dt \ , \quad v(t_0) = 1 \, ; \text{ i.e., } v(t) = \exp \displaystyle\int_{t_0}^t V\big(x(s)\big)\, ds \, , \end{cases} \tag{I.B–19}$$

with X and V given, "arbitrary." The effect of including a term $A(x(t))\,dt$ in the first equation can be read off equation (I.B–17) and need not be included for the purpose of the present discussion. Consider the path integral

$$\begin{aligned}\Psi(x_0,t) &:= \int_\Omega \varphi(x(t,\omega))\, v(t,\omega)\, d\mu(\omega) \\ &\equiv \int_\Omega \varphi(x(t)) \exp\Big(\int_{t_0}^{t} V(x(s))\, ds\Big)\, d\mu(\omega)\ ,\end{aligned} \tag{I.B–20a}$$

also written by probabilists

$$\Psi(x_0,t) := \mathbb{E}\Big(\varphi(x(t))\, v(t)\Big)\ , \tag{I.B–20b}$$

also written by analysts

$$\Psi(x_0,t) := (P^t_{t_0}\varphi)(x_0)\ , \tag{I.B–20c}$$

also written by geometers

$$\Psi(x_0,t) := \mathbb{E}\,\varphi(F_{t_0}(x_0,t))\ , \qquad \text{where} \quad x(t,\cdot) = F_{t_0}(x_0,t,\cdot)\ . \tag{I.B–20d}$$

It can be shown by various methods that

$$\begin{cases} \dfrac{\partial\Psi}{\partial t} = \dfrac{1}{2}\displaystyle\sum_i X_i^\alpha(x_0)\, X_i^\beta(x_0)\frac{\partial^2\Psi}{\partial x_0^\alpha\, \partial x_0^\beta} + V(x_0)\Psi \equiv \mathcal{A}(x_0)\Psi\ , \\ \Psi(x_0,t_0) = \varphi(x_0)\ . \end{cases} \tag{I.B–21}$$

The different expressions (I.B–20) reflect the different points of view.

i) A physicist says, "The *path integral* (I.B–20a) solves the diffusion equation (I.B–21)," and he[3] computes the path integral by the classic methods: Feynman diagrams, semiclassical expansions, Monte Carlo numerics, or by the methods presented here (linear methods and, in Section III, stochastic methods). Ω and μ are spelled out to make use of equations such as (I.B–2).

ii) A probabilist says, "The *process* (I.B–19) solves the diffusion equation (I.B–21)," and he investigates various processes; he is usually more interested in the properties of the average (expectation value) (I.B–20b) than in its value, so that he works with a fixed probability space $(\Omega, \mathcal{F}, \mu)$ and does not spell it out in (I.B–20b).

iii) An analyst says, "The generator of the *semigroup* $\{P^t_s\}$ (I.B–20c) on the space of functions φ is the differential operator $\mathcal{A}$ (I.B–21). If φ is the characteristic function $\mathbb{1}_B$ of the set $B \subset \mathbb{R}^n$, then $P^t_s(\mathbb{1}_B)$ defines a transition probability $p(s, x_0; t, B)$, which solves various partial differential equations." He investigates the semigroups $\{P^t_s\}$ on various spaces of functions, e.g., L^∞ spaces or L^2 spaces.

iv) A geometer says, "$F_{t_0}(\cdot, t, \omega)$ (I.B–20d) is a *flow*, which maps $\mathbb{R}^n$ into $\mathbb{R}^n$." He investigates flows on manifolds other than $\mathbb{R}^n$.

---

[3] Read "he/she" where appropriate.

What do physicists learn by talking to their colleagues?

From a probabilist, a physicist learns that a change of variable of integration in (I.B–20a), say, from $x(t)$ given in Cartesian coordinates to $x(t)$ given in polar coordinates, is not done in stochastic calculus as in differential calculus because in the first case $(\Delta x)^2 \sim \Delta t$, whereas in the second case $\Delta x \sim \Delta t$. The correct change of variables in stochastic calculus introduces terms that have no counterpart in differential calculus. The complication can be removed by a path-dependent time reparametrization "adapted" to the change of coordinates. The final expression is totally different from the one obtained by change of coordinates in differential calculus, but it is far simpler than the result obtained without time reparametrization. For explicit calculations, see Young and DeWitt-Morette (1986).

From a geometer, a physicist learns to set up processes (I.B–20) on fibre bundles and to compute a path integral with integrands defined on the base space. This is obviously a suitable approach for quantum mechanics in curved spacetimes. We shall set up stochastic processes on a frame bundle over spacetime and consider quantities defined on the base space. But stochastic processes on fibre bundles do more than solving problems of quantum mechanics in curved spacetimes. They solve problems of quantum mecanics in the presence of gauge fields; they solve problems of quantum mechanics on multiply connected spaces; they solve problems whenever fibre bundles offer an appropriate description of the system under consideration.

The next chapter will be devoted to stochastic processes on fibre bundles and, in particular, to stochastic processes on frame bundles.

Stochastic processes on fibre bundles appear to be so useful that it is natural to present a quantization scheme (last lecture), which begins with a stochastic process on a fibre bundle (defined by the system under consideration); path integrals are defined by the stochastic process—they yield the Lagrangian and the Hamiltonian operator of the system. This procedure removes the ambiguities one encounters when one *begins* with the Lagrangian.

II. STOCHASTIC PROCESSES ON FIBRE BUNDLES

A. Differential Versions of Stochastic Processes on Fibre Bundles

1. Principal bundle. For concreteness we shall consider the frame bundle. We shall assume that the bundle has been trivialized and we shall work on one patch. Consider a path $q(t)$ on the base space, its horizontal lift $\rho(t)$ in the bundle, and the image $g(t)$ of the horizontal lift on the typical fibre $G = GL(n, \mathbb{R})$. Set $q(t_0) = q_0$, $\rho(t_0) = \rho_0$, $g(t_0) = g_0$. The horizontal lift is defined by a connection σ

$$\frac{d\rho(t)}{dt} = \sigma\big(\rho(t)\big)\,\frac{dq(t)}{dt}\ . \tag{II.A–1}$$

We cannot construct a stochastic version "$d\rho(t) = \sigma\big(\rho(t)\big)\,dz(t)$" of this equation because a Brownian path $z(t) \in \mathbb{R}^n$ while $dq(t)/dt \in T_{q(t)}M^n$. (Note the t-dependence of this space.) However, we can use the fact that a frame

$$\rho(t) : \mathbb{R}^n \to T_{q(t)}M^n\ , \quad \text{hence } \big(\rho(t)\big)^{-1} : T_{q(t)}M^n \to \mathbb{R}^n$$

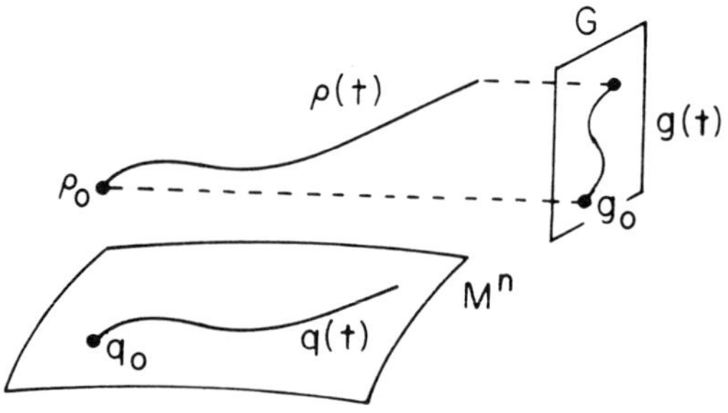

Figure 2. A frame bundle with structure group G.

to rewrite (II.A–1) in a form suitable for our purpose:

$$\frac{d\rho(t)}{dt} = \Big(\sigma\big(\rho(t)\big) \circ \rho(t)\Big) \left(\big(\rho(t)\big)^{-1} \frac{dq(t)}{dt}\right) . \tag{II.A–2a}$$

The soldering form $\big(\rho(t)\big)^{-1}(dq(t)/dt) \in \mathbb{R}^n$, and anticipating further development we call it $(dz(t)/dt)$. Set $X\big(\rho(t)\big) := \sigma\big(\rho(t)\big) \circ \rho(t)$, then[4]

$$\frac{d\rho(t)}{dt} = X\big(\rho(t)\big) \frac{dz(t)}{dt} \tag{II.A–2b}$$

has a stochastic version

$$d\rho(t) = X\big(\rho(t)\big)\, dz(t) . \tag{II.A–3}$$

Exercise: Properties of (II.A–2b). Set $x(t) = \big(\rho(t), q(t)\big)$ a path in the frame bundle; *i*) show that if (in (II.A–2b)) $dz(t)/dt$ is constant, then $q(t)$ is a geodesic; *ii*) show that (II.A–2b) implies that $q(t)$ is the Cartan development of $z(t)$.

i) $dz/dt = c$ implies $\rho^{-1}\, dq/dt = c$; i.e., the components of the tangent vector to q remain constant in the frame parallel transported along q. Hence, q is a geodesic. We can also check that the covariant derivative of dq/dt along $q(t)$ vanishes.

$$\frac{dq(t)}{dt} = \frac{d}{dt}\Big(\Pi\big(x(t)\big)\Big) \quad \text{where } x(t) = \big(\rho(t), q(t)\big) ,$$

but since here $\rho(t)$ is the horizontal lift of $q(t)$,

$$\begin{aligned} \frac{dq(t)}{dt} &= \Pi'\big(\dot\rho(t)\big) = \Pi'\Big(X\big(\rho(t)\big)\,\dot z(t)\Big) = \rho(t)\,\dot z(t) \\ \frac{D}{dt}\,\dot q(t) &= \rho(t)\,\frac{d}{dt}\big(\rho^{-1}(t)\,\dot q(t)\big) = \rho(t)\,\frac{d}{dt}\,c = 0 . \end{aligned} \tag{II.A–4}$$

[4] In principal bundles other than frame bundles (i.e., when there is no soldering form), one can use the other definition of connections; namely, a connection is a one-form on the tangent bundle of the principal bundle with values in the Lie algebra of the structure group. This definition provides the linear spaces needed to define Brownian paths.

ii) q is said to be the development of z if

$$\rho_0^{-1} \int_0^t (\text{parallel transp.})^{-1}\, \dot{q}(s)\, ds = \int_0^t \dot{z}(s)\, ds \quad \text{for every } t\ , \tag{II.A–5}$$

the parallel transport being from $q(s)$ to q_0 along q. The left-hand side of (II.A–5) is equal to

$$\begin{aligned}
\rho_0^{-1} \int_0^t \rho_0 \big(\rho(s)\big)^{-1} \dot{q}(s)\, ds &= \int_0^t \big(\rho(s)\big)^{-1}\, \Pi'\big(\dot{\rho}(s)\big)\, ds \\
&= \int_0^t \big(\rho(s)\big)^{-1}\, \Pi'\Big(X\big(\rho(s)\big)\, \dot{z}(s)\Big)\, ds \\
&= \int_0^t \big(\rho(s)\big)^{-1}\, \rho(s)\, \dot{z}(s)\, ds\ . \quad \blacksquare
\end{aligned}$$

2. Associated vector bundle. Given a principal G-bundle, and a representation (ρ, F) of G on a vector space F, the parallel transport $v(t)$ of v_0 from q_0 to $q(t)$ is

$$v(t) = \rho\big(g(t)\, g_0^{-1}\big)\, v_0 \tag{II.A–6}$$

where $g(t)$ is the image on the typical fibre G of the horizontal lift of $q(t)$. The connection stated in terms of σ in the previous example (II.A–1) can also be stated in terms of a form ω with values in the Lie algebra $\mathcal{L}(G)$. If we already know σ, then ω is the same connection as σ if

$$\omega\ (\text{horizontal vector}) = 0\ .$$

If we do not know σ, the choice of the connection ω determines the vertical component of a tangent vector to the bundle (hence indirectly its horizontal component) by

$$\omega(tg\ \text{vector at } p) = \gamma \in \mathcal{L}(G) \Leftrightarrow \text{vertical component of the } tg \text{ vector at } p = \hat{\gamma}(p)$$

where γ and $\hat{\gamma}(p)$ are related by the canonical isomorphism between $\mathcal{L}(G)$ and the tangent space at p to the fibre at $\Pi(p)$. If $\gamma = dg(s)/ds|_{s=0}$ where $g(s)$ is a one-parameter subgroup, then

$$\hat{\gamma}(p) = \frac{d}{ds}(\tilde{R}_{g(s)}p)\,|_{s=0} \tag{II.A–7}$$

where $\tilde{R}_g$ is the right action of G on the bundle. Given the curve $g(t)$, in paragraph 1), image in G of the horizontal lift defined by σ, we shall now express it in terms of ω, the one form with values in $\mathcal{L}(G)$ that vanish on the horizontal vectors defined by σ.

Let $\phi : \Pi^{-1}\big(q(t)\big) = U \times G$ be a local trivialization such that (see caption of Fig. 4) for $x_0(t)$ defined by

$$\phi\big(x_0(t)\big) = \Big(\Pi\big(x_0(t)\big),\ \overset{\vartriangle}{\phi}\ \big(x_0(t)\big)\Big) = \big(q(t), e\big)\ .$$

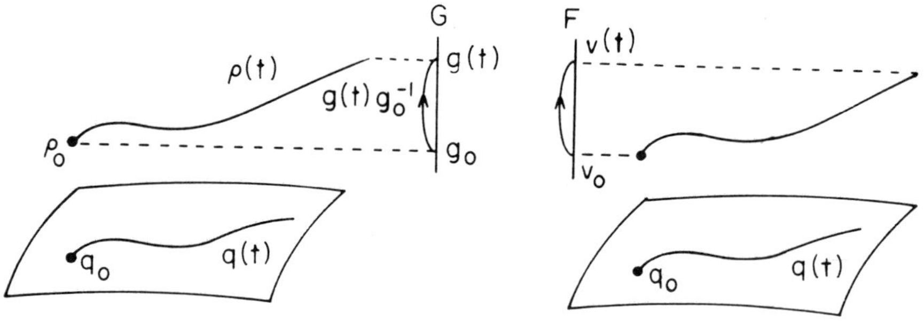

Figure 3. On the left a principal bundle, on the right an associated vector bundle.

Then

$$\phi'_{x_0(t)} : T_{x_0(t)}\left(\Pi^{-1}(q(t))\right) \to T_{q(t)}U \times T_e G$$

and

$$\phi'_{x_0(t)} \circ \sigma_{x_0(t)} = (id, \alpha) \ ;$$

where α is a one form on the base space with value in $T_e G$.

Let s be a section on the bundle such that

$$\phi \circ s : q(t) \mapsto (q(t), e) \ .$$

Then it can be proved (see, e.g., Choquet-Bruhat and DeWitt-Morette (1982), Part I, p. 367) that, for s^* the pull back mapping

$$\alpha = -s^* \omega \ .$$

If $x(t) = \tilde{R}_{g(t)}\, x_0(t)$, then $\sigma_{x(t)} = \tilde{R}'_{g(t)}\, \sigma_{x_0(t)}$ and

$$(\phi'_{x(t)} \circ \sigma_{x(t)}) \left(\frac{dq(t)}{dt} \right) = \left(\frac{dq(t)}{dt}, \frac{dg(t)}{dt} \right)$$

with

$$\begin{aligned} \frac{dg(t)}{dt} &= R'_{g(t)}\, \alpha \left(\frac{dq(t)}{dt} \right) = -R'_{g(t)}\, s^* \omega \left(\frac{dq(t)}{dt} \right) \\ &= -s^* \omega \left(\frac{dq(t)}{dt} \right) g(t) \ . \end{aligned}$$

For example, the pull back $s^*\omega$ of a connection in a $U(1)$-bundle is the electromagnetic potential, or more precisely,

$$s^* \omega \left(\frac{dq(t)}{dt} \right) = i \frac{e}{\hbar c}\, A_\alpha(q(t))\, \frac{dq^\alpha(t)}{dt} \in \mathcal{L}(U(1)) \ .$$

Then

$$dg(t) = -i \frac{e}{\hbar c} A_\alpha \big(q(t)\big)\, dq^\alpha(t)\, g(t) \ , \tag{II.A-8}$$

and since $U(1)$ is abelian,

$$g(t) = \exp\Big(- i \frac{e}{\hbar c} \int_0^t A_\alpha \big(q(t)\big)\, dq^\alpha(t)\Big) \ . \tag{II.A-9}$$

3. Weakly associated bundle. This bundle is not defined by a representation of the structure group of the principal bundle but by a representation of a group homomorphic to the structure group. For example, consider the wave function of a particle moving on a circle S^1. We are interested in the sum over all paths beginning at $q_0 \in S^1$ and ending at $q(t) \in S^1$. The space of paths on S^1 splits in an infinite number of homotopy classes labelled by the number $n \in \mathbb{Z}$ of loops a path has made in going from q_0 to $q(t)$. The universal covering of S^1 is $\mathbb{R}$.

A universal covering $\widetilde{M}$ of a manifold M is a simply connected covering of M. It is a principal bundle.

$$\Pi : \widetilde{M} \to M = \widetilde{M}/G \tag{II.A-10}$$

where G is the discrete group of automorphisms of M isomorphic to the fundamental group (first homotopy group) $\Pi_1(M)$. In the above example

$$S^1 = \mathbb{R}/\mathbb{Z} \ .$$

The universal covering has a unique connection: a horizontal lift $\tilde{q}$ of a path q in the base space is the unique lift in $\widetilde{M}$ with a given $\tilde{q}(t_0)$.

A scalar wave function over S^1 is a section of a vector bundle ξ with typical fibre $\mathbb{C}$; we are interested in the action of $U(1)$ on the wave function. Let

$$a : \mathbb{Z} \to U(1) \qquad \text{by} \qquad n \mapsto \exp in\alpha \ , \qquad \alpha \in \mathbb{R} \ . \tag{II.A-11}$$

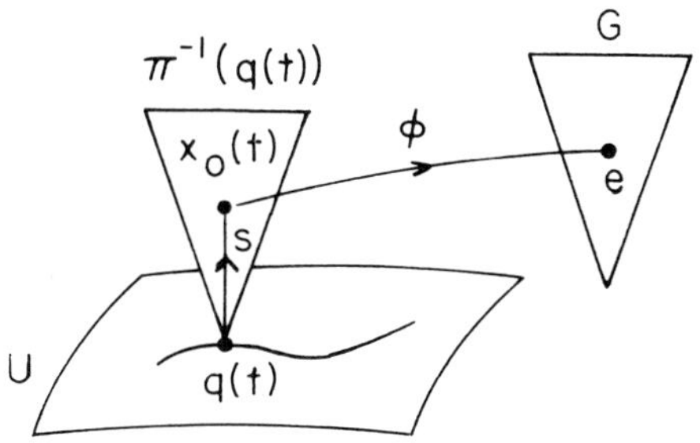

Figure 4. s is the section defined by the trivialization, i.e., $s : q(t) \mapsto x_0(t)$ such that $\overset{\triangle}{\phi}\big(x_0(t)\big) = e \in G$. The mapping $s^*\big(x_0(t)\big) : T^*_{x_0(t)}$ Bundle $\to T^*_{q(t)}$ Base.

Let $\tilde{a}$: Prin $\mathbb{Z}$-bundle $\to$ Prin $U(1)$-bundle over S^1, i.e., $\tilde{a}$ maps the universal cover of S^1 into the principal bundle associated to the vector bundle ξ:

$$\tilde{a}(\tilde{q}(t)n) = \tilde{a}(\tilde{q}(t))\, a(n) \ , \qquad n \in \mathbb{Z} \ .$$

The parallel transport of a wave function from $q(t)$ back to q_0 is

$$\tau_t^{t_0} \varphi(q(t)) = \tilde{a}(\tilde{q}(t_0)) \left[\tilde{a}(\tilde{q}(t))\right]^{-1} \varphi(q(t)) \ , \tag{II.A–12a}$$

which is often abbreviated to

$$\tau_t^{t_0} \varphi(q(t)) = \tilde{q}(t_0) \left(\tilde{q}(t)\right)^{-1} \varphi(q(t)) \ , \tag{II.A–12b}$$

although, as it stands, this last equation is incorrect.

The space of paths Ω from $\tilde{q}(t_0)$, chosen abitrarily, to any point $\tilde{q}(t)$ covering $q(t)$ decomposes into subspaces characterized by the final point $\tilde{q}(t)$

$$\Omega = \bigcup_{\mathbb{Z}} \Omega^n$$

where $\tilde{q}_n \in \Omega^n$ means that $\tilde{q}_n(t) = \tilde{q}_n$ such that $\pi(\tilde{q}_n) = q(t)$. We have

$$\tilde{q}_k(t) = \tilde{q}_n(t)\,(n)^{-1}\,(k) \tag{II.A–13}$$

where (n) is the group action of $\mathbb{Z}$ on $\mathbb{R}$: $\tilde{q}_0(t)(k) = \tilde{q}_o(t) + 2\pi\, k\,$Radius. The parallel transport (II.A–9) along a path q with lift $\tilde{q}_n$ is

$$\begin{aligned} \tau_t^{t_0} \varphi(q(t)) &= \tilde{a}\Big(\tilde{q}(t_0)\,\left(\tilde{q}_n(t)\right)^{-1}\Big)\, \varphi(q(t)) \\ &= \tilde{a}\Big(\tilde{q}(t_0)\,\left(\tilde{q}_0(t)\right)^{-1}\Big)\, \left(a(n)\right)^{-1} \varphi(q(t)) \\ &= \exp\left(-in\alpha\right) \tilde{q}(t_0)\,\left(\tilde{q}_0(t)\right)^{-1} \varphi(q(t)) \end{aligned} \tag{II.A–14}$$

in abbreviated form.

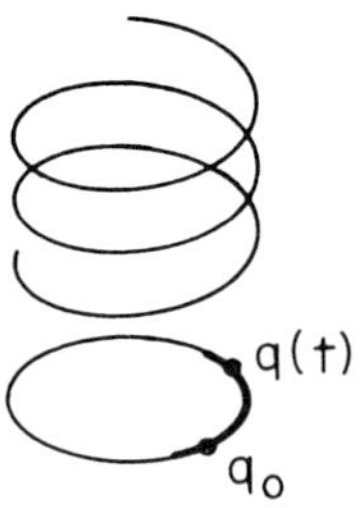

Figure 5. A circle and its universal cover.

There are simpler ways (Laidlaw and DeWitt-Morette, 1971) to establish this result; but I wanted to present the variety of problems that can be solved by stochastic processes on fibre bundles. In general, the propagator on a multiply connected space M is a linear combination of the propagators; each propagator is obtained by summing over paths in the same homotopy class; the coefficients of the linear combinations form a unitary representation of the fundamental group of M. The propagator is defined modulo on overall phase factor.

B. Stochastic differential equations on fibre bundles

We have written down three candidates for stochastic differential equations that may lead to interesting diffusion manifolds.

a) On the frame bundle, the stochastic system[5] (II.A–3)

$$\begin{cases} dx(t) = X\big(x(t)\big)\,dz(t)\,, \text{ where } x(t) = \big(q(t),\rho(t)\big)\,, & x(t_0) = x_0\,, \\ dv(t) = V\Big(\Pi\big(x(t)\big)\Big)\,v(t)\,dt\,, & v(t_0) = 1\,, \end{cases} \tag{II.B–1}$$

defining a path integral over paths taking their values in a Riemannian manifold:

$$\Psi(q_0,t) = \int_\Omega \varphi\Big(\Pi\big(x(t,\omega)\big)\,v(t,\omega)\,d\mu(\omega)\,, \qquad \varphi : M \to \mathbb{R}\,. \tag{II.B–2}$$

b) On a vector bundle associated with a $U(1)$ bundle over $\mathbb{R}^n$, the stochastic equation (II.A–8) with $dq(t) = \mu\,dz(t) \in \mathbb{R}^n$,

$$dg(t) = -i\frac{e}{\hbar c}\,A_\alpha\big(q(t)\big)\,dq^\alpha(t)\,g(t)\,, \qquad g(t_0) = g_0\,, \tag{II.B–3}$$

defining a path of a particle in an electromagnetic field

$$\begin{aligned} \Psi(q_0,t) &= \int \varphi\big(q(t,\omega)\big)\,\big(g(t,\omega)\big)^{-1}\,d\mu(\omega) \\ &= \int_\Omega \varphi\big(q(t,\omega)\big)\exp\Big(i\int_0^t \frac{e}{\hbar c}\,A_\alpha\big(q(t,\omega)\big)\,dq^\alpha(t,\omega)\Big)\,d\mu(\omega)\,. \end{aligned} \tag{II.B–4}$$

c) On a vector bundle weakly associated to a universal covering, the stochastic equation obtained from (II.A–13) when $q(t)$ is a stochastic process

$$\tilde{q}_{g_1}(t) = \tilde{q}_{g_2}(t)\,g_2^{-1}\,g_1 \tag{II.B–5}$$

values in a multiply connected space:

$$\begin{aligned} \Psi(q_0,t) &= \sum_{g\in G}\int_{\Omega^g} \tilde{q}(t_0,\omega)\big(\tilde{q}_g(t,\omega)\big)^{-1}\varphi\big(q(t,\omega)\big)\,d\mu(\omega)\,, \quad \Omega = U\Omega^g \\ &= \sum_{g\in G} a(g^{-1})\int_{\Omega^e} \tilde{q}(t_0,\omega)\,\big(\tilde{q}_e(t,\omega)\big)^{-1}\varphi\big(q(t,\omega)\big)\,d\mu(\omega)\,. \end{aligned} \tag{II.B–6}$$

---

[5] In order not to clutter the equations, we have abbreviated $\overset{\triangle}{\phi}\big(x(t)\big)$ to $x(t)$. Recall that $\overset{\triangle}{\phi}\big(x(t)\big) = \rho(t)$.

So far we have worked on one patch U assumed trivialized $\Pi^{-1}(U) = U \times G$ (principal bundle) or $\Pi^{-1}(U) = U \times F$ (associated bundle). We now have to face changes of coordinates on the overlap of two patches. A stochastic differential equation is in fact an integral, e.g.,

$$d\rho(t) = X\big(\rho(t)\big)\, dz(t), \rho(t_0) = \rho_0 \Leftrightarrow \rho(t) = \rho_0 + \int_{t_0}^{t} X\big(\rho(s)\big)\, dz(s) \ .$$

The integral, as it stands, is meaningless because $z(t)$ is not of bounded variation—intuitively, $dz(t) \simeq \sqrt{dt}$, then $\int_{t_0}^{t} (X \frac{dz}{dt})\, dt = \infty$ if defined as a Lebesque integral.

1. Stratonovich calculus. However, Itô and Stratonovich have proposed definitions of stochastic integrals that are the starting points of two very different stochastic calculi. We hope to decide whether to define the stochastic equations ((II.B–1), (II.B–3), (II.B–5)) by Itô integrals or Stratonovich integrals. The answer is "Stratonovich" because the Itô formula for change of variables is not tensorial. Indeed, an Itô integral is, by definition

$$\int X(t,\omega)\, dz(t,\omega) := \lim_{\text{in measure}} \sum_j X(\tau_j,\omega)\big(z(t_{j+1},\omega) - z(t_j,\omega)\big) \qquad \text{(II.B–7)}$$

where $\tau_j \leq t_j$. One says "the differentials stick out in the future," or "the integrand is nonanticipating." It follows from this definition that under a change of coordinates $\theta : \mathbb{R}^n \to \mathbb{R}^n$, the new coordinates $\theta\big(x(t)\big)$ are obtained from the old coordinates $x(t)$ by

$$\begin{aligned} \theta\big(x(t_b)\big) = \theta\big(x(t_a)\big) &+ \int_{t_a}^{t_b} D\theta\big(x(t)\big)\, dx(t) \\ &+ \frac{1}{2} \int_{t_a}^{t_b} D^2\theta\big(x(t)\big)\, dx(t)\, dx(t) \ . \end{aligned} \qquad \text{(II.B–8)}$$

The term

$$D^2\theta\, dx\, dx = \frac{\partial^2\theta}{\partial x^\alpha\, \partial x^\beta}\, dx^\alpha\, dx^\beta \qquad \text{(II.B–9)}$$

is not a tensor.

On the other hand, a Stratonovich integral is, by definition,

$$\begin{aligned} \int X(t,\omega)\, dz(t,\omega) := \lim_{\text{in measure}} \sum_j \frac{1}{2} \big(X(t_{j+1},\omega) \\ + X(t_j,\omega)\big)\big(z(t_{j+1},\omega) - z(t_j,\omega)\big) \ . \end{aligned}$$

The mid-point rule suggests that Stratonovich change of variable of integration may be tensorial. Indeed, the mid-point rule applied to a Taylor expansion wipes out the second differentials

$$\begin{cases} \theta(b) - \theta(a) &= \theta'(a)(b-a) + \frac{1}{2}\theta''(a)(b-a)^2 + 0\big((b-a)^3\big) \\ \theta(a) - \theta(b) &= \theta'(b)(a-b) + \frac{1}{2}\theta''(b)(a-b)^2 + 0\big((b-a)^3\big) \end{cases}$$

gives

$$\theta(b) - \theta(a) = \tfrac{1}{2}\left(\theta'(a) + \theta'(b)\right)(b-a) + 0\left((b-a)^3\right) \ . \tag{II.B–10}$$

We shall prove that the Stratonovich integral is tensorial by using the Itô formula (II.B–8) and the relationship between Itô and Stratonovich integrals.

$$\text{Strat.} \int X\, dz = \text{Itô}\Big(\int X\, dz + \frac{1}{2}\int dX\, dz\Big) \ . \tag{II.B–11}$$

We have written the Itô and the Stratonovich integrals in terms of z, the letter we have used for Brownian paths. They are in fact more general. The question is, "What are the processes Z indexed by "time" for which $\int_0^t X_s\, dZ_s$ can be defined on a reasonably large class of integrands X in such a way that the Dominated Convergence Theorem holds?" (Bichteler, 1984). This is a task we shall be happy to let mathematicians do for us. But in order to be able to use their results, I shall use their notation

$$\text{Itô} \int_0^t X_s\, dZ_s \equiv \int_0^t X_s\, dZ_s \equiv (X * Z)_t \ . \tag{II.B–12}$$

Thus, $X * Z$ is thought of as a process indexed by "time." Similarly,

$$\text{Strat.} \int_0^t X_s\, dZ_s \equiv \int_0^t X_s\, \delta Z_s \equiv (X \bullet Z)_t \ . \tag{II.B–13}$$

With this notation, (II.B–11) is written

$$X \bullet Z = X * Z + \tfrac{1}{2}\langle X|Z\rangle \ . \tag{II.B–14}$$

The bracket $\langle X|Z\rangle$ can be defined by this equation given the definition of the Itô and Stratonovich integrals. Alternatively, (II.B–14) can be used to define Stratonovich integrals and the bracket defined by either of the following two equations

$$X_t^2 = X_0^2 + 2\int_0^t X\, dX + \langle X|X\rangle_t \tag{II.B–15}$$

obtained from Itô formula (II.B–8) with $\theta : X_t \to X_t^2$

$$X_t Y_t = X_0 Y_0 + \int_0^t X_s\, dY_s + \int_0^t Y_s\, dX_s + \langle X|Y\rangle_t \ . \tag{II.B–16}$$

In the following we consider only continuous integrators X, Y, Z. Since $\langle X|Z\rangle = \int dX_s\, dZ_s$, it vanishes if one of the integrators X or Z is of finite variation. As a process $\langle X|Z\rangle_t$ is of finite variation, hence

$$\begin{aligned}\langle X|Y \bullet Z\rangle = \langle X|Y * Z\rangle &= \int dX\, d\Big(\int Y\, dZ\Big) \\ &= \int Y\, dX\, dZ = Y * \langle X|Z\rangle \ .\end{aligned} \tag{II.B–17}$$

Exercise: Show by computing both sides of the equations that

$$X \bullet (Y \bullet Z) = (XY) \bullet Z \tag{II.B–18a}$$

$$X * (Y * Z) = (XY) * Z \; . \tag{II.B-18b}$$

Exercise: Show that

$$\langle \Phi_{,i}(X) | Y \rangle = \langle \Phi_{,ij}(X) * X^j | Y \rangle \; . \tag{II.B–19}$$

Answer: By Itô formula (II.B–8)

$$\Phi_{,i}(X) = \Phi_{,i}(X_0) + \Phi_{,ij}(X) * X^j + \tfrac{1}{2}\Phi_{,ijk}(X) * \langle X^j | X^k \rangle \; , \tag{II.B–20}$$

and since $\langle X^j | X^k \rangle$ is of finite variation, equation (II.B–19) follows.

Theorem: *Stratonovich stochastic differential equations transform like ordinary differential equations.*

Proof: Let $\Phi : U \to \mathbb{R}^n$ be a twice-differentiable map defined on an open subset $U \subset \mathbb{R}^n$. The derivative map of Φ at x

$$D\Phi(x) : \mathbb{R}^n \to \mathbb{R}^n \; .$$

Let X be a continuous vector-valued integrator whose values lie in U at all times; then

$$\Phi(X_t) = \Phi(X_0) + \int_0^t D\Phi(X_s)\,\delta X_s \; , \qquad 0 \le t \le \infty \; . \tag{II.B-21}$$

Indeed,

$$\begin{aligned}
D\Phi(X) \bullet X &= \Phi_{,i}(X) \bullet X^i && \text{by definition} \\
&= \Phi_{,i}(X) * X^i + \tfrac{1}{2}\langle \Phi_{,i}(X) | X^i \rangle && \text{by (II.B–14)} \\
&= \Phi_{,i}(X) * X^i + \tfrac{1}{2}\langle \Phi_{,ij}(X) * X^j | X^i \rangle && \text{by (II.B–19)} \\
&= \Phi_{,i}(X) * X^i + \tfrac{1}{2}\Phi_{,ij} * \langle X^j | X^i \rangle && \text{by (II.B–17)} \\
&= \Phi(X) - \Phi(X_0) && \text{by (II.B–8)} \; .
\end{aligned}$$

Suppose X solves the stochastic differential equation

$$X = x + A_\alpha(X) \bullet Z^\alpha \; . \tag{II.B–22}$$

It follows from (II.B–21) that

$$\begin{aligned}
\Phi(X) &= \Phi(x) + D\Phi(X) \bullet X = \Phi(x) + D\Phi(X) \bullet \big(A_\alpha(X) \bullet Z^\alpha\big) \\
&= \Phi(x) + D\Phi(X)\, A_\alpha(X) \bullet Z^\alpha \qquad \text{by (II.B–18)} \; .
\end{aligned}$$

Set $Y = \Phi(X)$ and $y = \Phi(x)$, then Y solves

$$Y = y + B_\alpha(Y) \bullet Z^\alpha \ , \tag{II.B–23}$$

where $B_\alpha = D\Phi\, A_\alpha \circ \Phi^{-1}$ is the image of A_α under the derivative map $D\Phi$: Equation (II.B–22) transforms as in ordinary differential calculus. ■

2. Path integrals defined by stochastic processes on fibre bundles. We shall assume that the stochastic processes (II.B–1) and (II.B–3) are defined by Stratonovich integrals and investigate the properties of the path integrals (II.B–2) and (II.B–4). In Section III we shall compute pathwise linear approximations and semiclassical approximations of the Feynman-Kac formula for quantum systems in curved Riemannian manifolds (II.B–2).

a) Given the process $x(t)$ defined by the Stratonovich integral (II.A–3)

$$\begin{cases} x_t &= x_0 + (X \bullet z)_t \ ; \quad \Pi(x(t)) = q(t) \\ v_t &= \exp \displaystyle\int_0^t V\Big(\Pi\big(x(s)\big)\Big)\, ds \ . \end{cases} \tag{II.B–24}$$

We shall show that the following path integral, to be understood as (II.B–2) although the ω are not explicitly written,

$$\Psi(q_0, t) := \int_\Omega \varphi\Big(\Pi\big(x(t)\big)\Big) \exp\left(\int_0^t V\Big(\Pi\big(x(s)\big)\Big)\, ds\right) d\mu(\omega) \tag{II.B–25}$$

is a solution of the diffusion equation

$$\begin{cases} \dfrac{\partial \Psi}{\partial t} &= \Big(\dfrac{1}{2}\Delta + V\Big)\Psi \ , \\ \Psi(q_0, 0) &= \varphi(q_0) \ , \end{cases} \tag{II.B–26}$$

where Δ is the Laplacian defined by the connection implied in the process x by (II.B–24). As in $\mathbb{R}^n$, the proof rests on the Itô formula (II.B–8) applied to the function φ of the process $\Pi \circ x$ (II.B–24); it is clear from the results derived on $\mathbb{R}^n$ (I.B–19) and (I.B–21) that it is sufficient to prove the result when $V = 0$. First we express the Stratonovich equation $x(t) = x_0 + (X \bullet z)_t$ in terms of an Itô equation:

$$x(t) = x_0 + (X * z)_t + \tfrac{1}{2}\langle X|z\rangle_t \ . \tag{II.B–27}$$

The Itô formula (II.B–8) gives

$$\begin{aligned} \varphi\Big(\Pi\big(x(t)\big)\Big) &= \varphi\Big(\Pi\big(x(0)\big)\Big) + \int_0^t D(\varphi \circ \Pi)\,\big(x(s)\big)\, dx(s) \\ &\qquad + \tfrac{1}{2}\int_0^t D^2(\varphi \circ \Pi)\big(x(s)\big)\, dx(s)\, dx(s) \qquad \text{(II.B–28)} \\ &\equiv \varphi(q_0) + \big(D(\varphi \circ \Pi) * x\big)_t + \tfrac{1}{2}\big(D^2(\varphi \circ \Pi) * \langle x|x\rangle\big)_t \ . \end{aligned}$$

The explicit calculation of this equation is obviously intricate and we shall break it up into two steps: *i)* we concentrate on the contribution of the Stratonovich correction term $\frac{1}{2}\langle X|z\rangle$ and *ii)* we concentrate on the properties of the process x.

i) Set $\varphi \circ \Pi = f$; then

$$
\begin{aligned}
f\big(x(t)\big) = f\big(x(0)\big) + \int_0^t Df\big(x(s)\big)\Big(X\big(x(s)\big)\, dz(s) \\
+ \tfrac{1}{2} DX\big(x(s)\big)\, X\big(x(s)\big)\, dz(s)\, dz(s)\Big) \qquad \text{(II.B–29)} \\
+ \tfrac{1}{2}\int_0^t D^2 f\big(x(s)\big)\, X\big(x(s)\big)\, dz(s)\, X\big(x(s)\big)\, dz(s)\ .
\end{aligned}
$$

The Stratonovich correction term $\frac{1}{2}\langle X|z\rangle_t$ modifies the first integral but not the second one in (II.B–29) because $\int dz\, dz\, dz = 0$—alternatively because of (II.B–19). The integrals with respect to $dz\, dz$ can be combined:

$$
\begin{aligned}
&Df\big(x(s)\big)DX\big(x(s)\big)X\big(x(s)\big)\, dz(s)\, dz(s) + D^2 f\big(x(s)\big)\, X\big(x(s)\big)\, dz(s)\, X\big(x(s)\big)\, dz(s) \\
&= \partial_i f(\partial_j X^i_k) X^j_l\, dz^l\, dz^k + (\partial_i \partial_j f) X^i_k\, dz^k\, X^j_m\, dz^m \\
&= \sum_k \partial_i f(\partial_j X^i_k) X^j_k\, ds + \sum_k (\partial_i \partial_j f) X^i_k X^j_k\, ds \\
&= \sum_k \partial_j(\partial_i f X^i_k) X^j_k\, ds = \sum_k X_k(X_k(f))\, ds \equiv \mathcal{L}^2_X(f)\, ds\ . \qquad \text{(II.B–30)}
\end{aligned}
$$

Hence, (II.B–29) has been simplified to read

$$
\begin{aligned}
f\big(x(t)\big) = f\big(x(0)\big) + \int_0^t X\Big(f\big(x(s)\big)\Big)\, dz(s) \\
+ \tfrac{1}{2}\int_0^t \mathcal{L}^2_X f\big(x(s)\big)\, ds\ .
\end{aligned} \qquad \text{(II.B–31)}
$$

ii) Let $S(t, x_0)e$ be a solution of

$$
dx(t) = X\big(x(t)\big)e\, dt\ , \qquad \text{where } e \in \mathbb{R}^n\ . \qquad \text{(II.B–32)}
$$

We have established (II.A–4) that $\Pi\big(x(t)\big)$ is a geodesic at q_0; the components of its tangent are equal to e in the ρ_0 frame. Hence, $x(t)$ is the horizontal lift of a geodesic, at x_0 at $t = 0$. Then (II.B–31) can be rewritten

$$
\begin{aligned}
f\big(x(t)\big) = f\big(x(0)\big) + \int_0^t \frac{d}{dr} f \circ S\big(r, x(s)\big)\Big|_{r=0} dz(s) \\
+ \frac{1}{2}\int_0^t \frac{d^2}{dr^2} f \circ S\big(r, x(s)\big)e\Big|_{r=0}\big(dz(s), dz(s)\big)\ .
\end{aligned} \qquad \text{(II.B–33)}
$$

To prove that this equation is the same as (II.B–31) with $dS(t)/dt = X(S(t))$, one computes explicitly the integrands: Equation (II.B–32) says

$$\begin{cases} \frac{d}{dt}\Big(S\big(t,x(s)\big)e\Big) = X\Big(S\big(t,x(s)\big)e\Big)e\ , & e \in \mathbb{R}^n \\ S\big(0,x(s)\big)e = x(s)\ . \end{cases} \tag{II.B–34}$$

$$\frac{d}{dr} f \circ S\big(r,x(s)\big)e = Df \frac{d}{dr} S\big(r,x(s)\big)e = Df\, X\Big(S\big(r,x(s)\big)e\Big)e$$

$$\frac{d}{dr} f \circ S\big(r,x(s)\big)e\Big|_{r=0} = Df\, X\big(x(s)\big)e\ . \tag{II.B–35}$$

$$\begin{aligned} \frac{d^2}{dr^2} f \circ S\big(r,x(s)\big)e &= \frac{d}{dr} Df\, X\Big(S\big(r,x(s)\big)e\Big)e \\ &= D^2 f\, X\Big(S\big(r,x(s)\big)e\Big)e\, X\Big(S\big(r,x(s)\big)e\Big)e \\ &\quad + Df\, DX\Big(S\big(r,x(s)\big)e\Big)e\, X\Big(S\big(r,x(s)\big)e\Big)e \end{aligned}$$

$$\begin{aligned} \frac{d^2}{dr^2} f \circ S\big(r,x(s)\big)e\Big|_{r=0} &= D^2 f\, X\big(x(s)\big)e\, X\big(x(s)\big)e \\ &\quad + Df\, DX\big(x(s)\big)e\, X\big(x(s)\big)e\ . \end{aligned} \tag{II.B–36}$$

Since $e\,e(dz,dz) = e_k\, e_l(dz^k, dz^l) = ds$, the above quadratic form applied to the pair (dz,dz) is identical to (II.B–30). ∎

Finally, we can work out explicitly (II.B–28) using (II.B–33), (II.B–35) and (II.B–36) with $f = \varphi \circ \Pi$. We shall also use the fact that $S(r,x_0)e$ is the lift of a geodesic.

$$\gamma(r,x_0) = \Pi \circ S(r,x_0)\,e\ . \tag{II.B–37}$$

Then, D/dr being the covariant derivative and Δ the Laplacian defined by the connection σ,

$$\begin{aligned} \frac{d^2}{dr^2}\varphi \circ \gamma\big(r,x(s)\big)\Big|_{r=0} &= \frac{d}{dr}\Big(D\varphi \frac{d\gamma}{dr}\big(r,x(s)\big)\Big)\Big|_{r=0} \\ &= \frac{d}{dr}\Big(\partial_k\varphi \frac{d\gamma^k}{dr}\big(r,x(s)\big)\Big)\Big|_{r=0} \\ &= \Big(\frac{D}{dr}D\varphi\Big)\frac{d\gamma}{dr}\big(r,x(s)\big)\Big|_{r=0} \\ &\quad + D\varphi \frac{D}{dr}\frac{d}{dr}\gamma\big(r,x(s)\big)\Big|_{r=0} \\ &= \Delta\varphi\big(x(s)\big)\ ; \end{aligned} \tag{II.B–38}$$

the second term vanishes because γ is a geodesic.

Exercise: With the notation of Section II.A.2, use (II.B–36) and (II.B–30) to show that if the connection σ defining X (II.A–2b) is the Levi-Civita connection,

$$\big(\overset{\Delta}{\phi}{}'_{x_0(t)} \circ \sigma_{x_0(t)}\big)^k_{ij} = -(\Gamma_i)^k_j$$

with

$$\Gamma^k_{ij} = \tfrac{1}{2}\, g^{km}\,(\partial_i\, g_{jm} + \partial_j\, g_{im} - \partial_m\, g_{ij})\ , \tag{II.B–39}$$

then

$$\Delta_x = g^{ij} \frac{\partial}{\partial x^i} \frac{\partial}{\partial x^j} - g^{ij} \Gamma^k_{ij} \frac{\partial}{\partial x^k} \ . \tag{II.B–40}$$

Hint: $\sum X^i_k X^j_k = g^{ij}$, $\quad \sum(\partial_j X^i_k) X^j_k = -g^{lm} \Gamma^i_{lm}$.

Remark: To obtain (II.B-26) one proceeds as usual. In brief, insert either (II.B-31) or (II.B-33) in the path integral and note that the martingale $\int \ldots dz$ does not contribute.

b) Given the process defined by the Stratonovich integral (II.B–3)

$$g(t) = \exp\Big(- i \frac{e}{\hbar c} (A_\alpha \cdot q^\alpha)_t \Big) \ , \qquad g(0) = 1 \ , \tag{II.B–39}$$

we shall show that the following path integral, to be understood as (II.B–4) even though we do not explicitly write the ω,

$$\Psi(q_o, t) := \int_\Omega \varphi(q(t)) \exp\Big(i \frac{e}{\hbar c} (A_\alpha \cdot q^\alpha)_t \Big)\, d\mu(\omega)$$

is a solution of

$$\begin{cases} \dfrac{\partial \Psi}{\partial t} &= \dfrac{1}{2} \Big(D_\alpha + i \dfrac{e}{\hbar c} A_\alpha(x_0) \Big)^2 \Psi \\ \Psi(q_0, 0) &= \varphi(q_0) \ . \end{cases}$$

$D_\alpha + i \frac{e}{\hbar c} A_\alpha(x)$ is the covariant derivative defined by the pull back of the connection on the base space $i \frac{e}{\hbar c} A_\alpha(x)$.

This is basically the same result as in the previous paragraph (II.B–26).

In conclusion, Stratonovich equations solve diffusion equations expressed in terms of covariant derivatives.

III. COMPUTING PATH INTEGRALS FOR SYSTEMS IN CURVED SPACETIMES

I shall present only computations based on stochastic processes on frame bundles—and only a few of them. Moreover, I shall present them briefly because they have already been published and because they are far too long to be reproduced here in full.

I want to write down approximations not only for the solution (II.B–25) but also for the solution of the Schrödinger equation.

$$\begin{cases} i\hbar \dfrac{\partial \psi}{\partial t} &= -\dfrac{\hbar^2}{2m} \Delta\psi + V\psi \ , \qquad t \in [t_a, t_b] \equiv T \ ; \\ \psi(a, t_a) &= \varphi(a) \ , \qquad a \in M \ . \end{cases} \tag{III.A–1}$$

The diffusion and the Schrödinger equations can both be written

$$\begin{cases} \dfrac{\partial\psi}{\partial t} &= \dfrac{1}{2}\mu^2 s\,\Delta\psi + \dfrac{1}{\mu^2 s}\dfrac{1}{m}V\psi \\ \psi(a,t_a) &= \varphi(a) \end{cases} \tag{III.A–2}$$

with ($\mu^2 = \hbar/m,\ s = i$) for the Schrödinger equations and ($\mu^2 = 1,\ s = 1$) for the diffusion equation. The path integral solution (II.B–25) of the diffusion equation $\Psi(a, t_b)$ is over the space Ω_- of paths that vanish at t_a, and the path integral solution (III.A–4) of the Schrödinger equation $\Psi(b, t_b)$ is over the space Ω_+ of paths that vanish at t_b. The solution of the diffusion equation is computed over the probability space $(\Omega_-, \mathcal{F}, \mu_s)$

$$\Psi(a,t_b) = \int_{\Omega_-} d\mu_s(\omega)\exp\left(\frac{1}{\mu^2 s}\frac{1}{m}\int_{t_a}^{t_b} V\Big(\Pi(x(t))\Big)dt\right)\varphi\Big(\Pi(x(t_b))\Big). \tag{III.A–3}$$

It suggests that the solution of the Schrödinger equation[6]

$$\Psi(b,t_b) = \int_{\Omega_+} d\mu_s(\omega)\exp\left(\frac{1}{\mu^2 s}\frac{1}{m}\int_{t_a}^{t_b} V\Big(\Pi(x(t))\Big)dt\right)\varphi\Big(\Pi(x(t_a))\Big) \tag{III.A–4}$$

be computed on the projective system $(\Omega_+, \mathcal{Q}, \mu_s)$ defined in section I.A. The Fourier transform of μ_s is (I.A–19)

$$\mathcal{F}\mu_s(\omega') = \exp\Big(-\frac{s}{2}\int_T d\omega'_\alpha(t)\int_T d\omega'_\beta(s)\,G_\pm^{\alpha\beta}(t,s)\Big)\ . \tag{III.A–5}$$

$$\text{For } \mu_s \text{ on } \Omega_+,\quad G_+^{\alpha\beta}(t,s) = \delta^{\alpha\beta}\inf(t_b - t, t_b - s)\ ; \tag{III.A–6a}$$

$$\text{For } \mu_s \text{ on } \Omega_-,\quad G_-^{\alpha\beta}(t,s) = \delta^{\alpha\beta}\inf(t - t_a, s - t_a)\ . \tag{III.A–6b}$$

In Section III.A, we give the piecewise linear approximation of (III.A–3), and in Section III.B, we give the semiclassical approximations of (III.A–4). A little bit of tedious bookkeeping gives the other cases.

A. Piecewise Linear Approximations

1. The piecewise linear approximations of the path integrals (II.B–25) or (III.A–3). They have been obtained by replacing a Brownian path z by its piecewise linear approximation z_Π for a partition $\Pi = (t_a, t_1, \ldots, t_k = t_b)$.

$$z_\Pi : T = [t_a, t_b] \times \Omega \to \mathbb{R}^n$$

by

$$z_\Pi(r,\omega) = (\Delta_j t)^{-1}\big((t_{j+1} - r)\,z(t_j,\omega) + (r - t_j)\,z(t_{j+1},\omega)\big) \tag{III.A–7}$$

[6] If there is a term proportional to $A_\alpha \partial\psi/\partial x^\alpha$ in the equations, the corresponding terms in the solutions (III.A–3) and (III.A–4) are affected by opposite signs. For details, see DeWitt-Morette and Elworthy (1978).

for $t_j \le r \le t_{j+1}$ and $\Delta_j t = t_{j+1} - t_j$. The basic theorems for piecewise linear approximations have been developed by Elworthy (1982) and Elworthy and Truman (1981). The detailed calculation of the piecewise linear approximation I_k of (III.A–3) can be found in DeWitt-Morette and Elworthy (1978).

$$I_k = \int_{\mathbb{R}^{nk}} \overline{K}(k;k-1)\,\overline{K}(k-1;k-2)\ldots\overline{K}(1;0)\,\varphi(q_0)\,dv(q_0)\ldots dv(q_{k-1}) \tag{III.A–8}$$

where $dv(q_j)$ is the Riemannian volume element

$$dv(q_j) = \left(|g(q_j)|\right)^{1/2} dq_j^1 \ldots dq_j^n \ ,$$

g being the determinant of the metric tensor, and

$$\overline{K}(j+1;j) = \left(\frac{m}{2\pi s\hbar}\,\Delta_j t\right)^{n/2} \mathcal{D}(j+1;j)\,\exp\left(\frac{s}{\hbar}\,S(j+1;j)\right) \ . \tag{III.A–9}$$

$S(j+1;j)$ is the action function evaluated along the geodesic from (q_j, t_j) to (q_{j+1}, t_{j+1}), and $\mathcal{D}(j+1;j)$ is the absolute value of the invariant Van Vleck determinant for the action $S(j+1;j)$

$$\mathcal{D}(j+1;j) = (\Delta_j t)^{-n}\, g^{-1/2}(q_{j+1})\, g^{-1/2}(q_j) \left| \det_{\alpha\beta} \frac{\partial(\exp^{-1}_{q_{j+1}})^\alpha(q_j)}{\partial q_j^\beta} \right| \tag{III.A–10}$$

where $\exp_{q_{j+1}} : T_{q_{j+1}}M \to M$ and $D(\exp^{-1}_{q_{j+1}})(q_j) : T_{q_j}M \to T_{q_{j+1}}M$. It has been shown (eq. (6.38) in DeWitt (1957)) that

$$\mathcal{D}(j+1;j) = (\Delta_j t)^{-n}(1 + \tfrac{1}{6}\, R_{\alpha\beta}\, \Delta_j q^\alpha\, \Delta_j q^\beta + \cdots) \tag{III.A–11}$$

where $\Delta_j q = q_{j+1} - q_j$ and R is the Ricci tensor.

I had been expecting not the Van Vleck determinant but its square root. However, checking and rechecking the long calculation of (III.A–4) when z is replaced by its piecewise linear approximation (III.A–7) did confirm (III.A–10). My expectation was based on the erroneous assumption that one can use the WKB approximation of the propagator for its short-time approximation ($\overline{K}(j+1;j)$. But all we can say, *a priori,* is that if K is the exact propagator, then $\overline{K}$ such that

$$K(j+1;j) = \overline{K}(j+1;j)\,(1 + O(\Delta_j t)^2) \tag{III.A–12}$$

is a good choice.

2. A semihistorical summary of other piecewise linear approximations. In the early fifties, it was thought that the WKB approximation of the exact propagator K was a good choice for $\overline{K}$ for the following reasons:

a) It was obtained (Morette, 1951) in the flat case by requiring probability conservation (i.e., unitarity): if the L_2-norm of the initial wave function ψ_0 is unity, requiring the L_2-norm of the wave function ψ at time $t_0+\varepsilon$ to be unity and ignoring terms of order Δx^2 and $\Delta x\,\Delta t$ leads to $|\overline{K}|^2 = |K_{\mathrm{WKB}}|^2$ where

$$K_{\mathrm{WKB}}(j+1;j) = (m/2\pi i\hbar)^{n/2}\,\mathcal{D}^{1/2}(j+1;j)\exp\left(\frac{i}{\hbar}\,S(j+1;j)\right)\,.$$

b) Since $K = K_{\mathrm{WKB}}\bigl(1+0(\Delta_j t)^2\bigr)$ for many physical systems (Pauli, 1952), it was assumed that $\overline{K}$ could be taken equal to K_{WKB}.

However, when the configuration space is a Riemannian manifold, DeWitt (1957) showed that for a system with Lagrangian

$$L=\tfrac{1}{2}m|\dot{q}|^2-V(q)\,;\quad H=\langle p,q\rangle-L\,;\quad \widehat{H}=\frac{1}{2m}g^{-1/4}\,p_i\,g^{1/2}g^{ij}p_j\,g^{-1/4}+V(q)$$

if one chooses $K = K_{\mathrm{WKB}}$, the limit of I_k—if its exists—should satisfy the Schrödinger equation

$$i\hbar\partial\psi_+/\partial t = \widehat{H}_+\,\psi_+ \qquad \text{with} \qquad \widehat{H}_+ = \widehat{H} + \tfrac{1}{12}\hbar^2 R$$

where R is the Riemann curvature scalar. It is often thought that the Correspondence Principle favors this choice. The original remark by DeWitt (1957) said only, "The quantum theory that one arrives at by applying the Correspondence Principle via [choosing $\overline{K}$ to be the W. K. B. approximation] is determined not by the operator $\widehat{H}$ but by the operator $\widehat{H}_+$." Moreover, DeWitt (1957) also showed that if one chooses $\overline{K}$ equal to the simplest guess one can make, namely, $\overline{K}(j+1,j) = (m/\Delta_j t2\pi i\hbar)^{n/2}\exp\left(\frac{i}{\hbar}\,S(j+1;j)\right)$, then for the same Lagrangian, the limit of I_k should satisfy

$$i\hbar\partial\psi_{++}/\partial t = \widehat{H}_{++}\,\psi_{++} \qquad \text{with} \qquad \widehat{H}_{++} = \widehat{H} + \tfrac{1}{6}\hbar^2 R\,.$$

Both $\widehat{H}_+$ and $\widehat{H}_{++}$ agree with $\widehat{H}$ when $\hbar$ tends to zero and are self-adjoint when $\widehat{H}$ is self-adjoint, i.e., both schemes are unitary.

If one had insisted that, for a system whose classical Lagrangian and Hamiltonian are L and H, the canonical pair $(L,\widehat{H})$ was to be preferred over the pairs $(L,\widehat{H}_+)$ or $(L,\widehat{H}_{++})$, one would have said

$$\overline{K}(j+1;j) =(m/2\pi i\hbar\Delta_j t)^{n/2}\left(1+\frac{1}{6}R_{\alpha\beta}\,\Delta_j q^\alpha\,\Delta_j q^\beta\right)\exp\left(\frac{i}{\hbar}S(j+1;j)\right), \tag{III.A–13}$$

i.e., an expression which agrees with (III.A–9), (III.A–11), but one would not have thought of (III.A–9), which looks wrong to anyone expecting K to be K_{WKB}.

Formally, $\overline{K}(j+1;j)$ given by (III.A–13) can be replaced by

$$K_{\mathrm{WKB}}(j+1;j)\exp\left(\tfrac{1}{12}\,R_{\alpha\beta}\,\Delta_j q^\alpha\,\Delta_j q^\beta\right)$$

and in the Feynman integral, where $\Delta q^\alpha\,\Delta q^\beta$ can be replaced by $i(\hbar/m)\delta^{\alpha\beta}\Delta t$, by

$$K_{\mathrm{WKB}}(j+1;j)\,\exp\left(\frac{i}{12}\,\frac{\hbar}{m}\,\Delta_j t\,R\right)\,, \qquad \text{i.e., } K_{\mathrm{WKB}}\times\text{ phase}.$$

Thus, $|\overline{K}| = |K_{\mathrm{WKB}}|$ but $\overline{K}$ is not necessarily equal to K_{WKB}; this is precisely the mistake made "in the early fifties."

B. Semiclassical Approximations

1. Strict[7] WKB approximation of the path integral (III.A–4) defined by (II.A–3). The stochastic process $x(t)$. If $x(t)$ were defined by an ordinary—rather than by a stochastic differential—equation (i.e., by (II.A–2b) rather than (II.A–3)), $\Pi(x(t))$ would be the Cartan development of a path in $\mathbb{R}^n$ (see exercise following (II.A–2b)), and one could use the properties of the development map, Dev, to compute the path integral. One could proceed as follows. Let

$$\text{Dev: Space of paths on } \mathbb{R}^n \to \text{space of paths on } M$$

by $z \mapsto \text{Dev}\, z$. Set

$$X(t,z) = (\text{Dev}\, z)(t) \ , \tag{III.B–1}$$

then rewrite (III.A–4)

$$\Psi(b,t_b) = \int_{\Omega_+} d\mu_s(\omega) \exp\left(\frac{1}{\mu^2 s}\frac{1}{m}\int_{t_a}^{t_b} V\big(X(t,b+\mu z)\big)\,\varphi\big(X(t_a,b+\mu z)\big)\right) \tag{III.B–2}$$

and expand the right-hand side in powers of $\mu = (\hbar/m)^{1/2}$ (I.B–12). The result would be the semiclassical expansion of $\Psi(b,t_b)$. The WKB approximation could be computed explicitly with the linear methods presented in Section I.B.1 because the derivative of any map (in particular, the development map) is linear. To compute $\Psi(b,t_b)$ one needs to choose the initial wave function φ.

To compute the WKB approximation of Ψ, a natural choice for φ, first proposed by Truman (1976, 1977), is

$$\varphi(\cdot) = \exp\left(\frac{i}{\hbar}S_0(\cdot)\right)T(\cdot) \ , \tag{III.B–3}$$

where T is an "arbitrary" function on M whose support determines the localization of the system and S_0 is the initial value of the solution of the Hamilton-Jacobi of the system with potential V.

Embarrassingly, this illegitimate procedure (using the smooth Cartan development map rather than the measurable stochastic development map) has been used (DeWitt-Morette et al, 1979). The result is in agreement with the stochastic calculation of Elworthy and Truman (1981) and Watling (1986) (see below).

$$\Psi_{\text{WKB}}(t_b,b) = \left(\det_{\alpha\beta}\partial Z^\alpha(t_a)/\partial Z^\beta(t_b)\right)^{1/2}\exp\left(\frac{i}{\hbar}\bar{S}(t_b,b)\right)T\big(Z(t_a)\big) \tag{III.B–4}$$

[7] We call WKB approximation the first meaningful term of a semiclassical expansion, and "strict" WKB the first term when it happens to be of the form $(\det)^{1/2}\exp(\frac{i}{\hbar}$ classical action), i.e., when the critical points of the action are not degenerate (no caustics).

where $\bar{S}$ is the action function computed for the classical path Z on M

$$m g_{\alpha\beta} \ddot{Z}^{\beta}(t) + \nabla_{\alpha} V(Z(t)) = 0 \tag{III.B–5a}$$

satisfying the following boundary conditions

$$\begin{cases} Z(t_b) = b & \text{(III.B–5b)} \\ \nabla_{\alpha} S_0(Z(t_a)) = m g_{\alpha\beta}(Z(t_a)) \dot{Z}^{\beta}(t_a) \, . & \text{(III.B–5c)} \end{cases}$$

The determinant in (III.B–4) gives the rate at which a flow of classical paths originating in a neighborhood of $S_0(Z(t_a))$ diverges or converges. It reflects both the choice of the initial wave function (III.B–3) and the dynamical properties of the system (III.B–5a). The WKB approximation (III.B–4) is a "strict" WKB approximation, i.e., valid only if the flow up to time t_b of classical paths originating in a neighborhood of $S_0(Z(t_a))$ has an inverse: The flow of classical paths on M is a map

$$\Phi_t : M \to M \qquad \text{by} \qquad a \mapsto Z(t, a) \tag{III.B–6}$$

where $Z(\cdot, a)$ is a classical path originating at a. The result (III.B–4) is very plausible—indeed, so plausible that some physicists would think the long calculation leading to it unnecessary. On the other hand, mathematicians[8] would think it unreliable.

I shall now present briefly[9] a legitimate calculation of the WKB approximation of (III.A–3). Minor modifications would give the approximation (III.B–4) of (III.A–3). For a discussion of flows of stochastic dynamical systems, see Carverhill and Elworthy (1983) and Elworthy (1982).

Elworthy (1982) has shown that the stochastic equation (II.A–3) defines a "stochastic development" from the space of Brownian paths defined on a fixed tangent space $T_a M$ (or $T_b M$ if preferred) to the space of paths on M—a path and its development beginning at a (respectively ending at b). The stochastic development map is measurable (not differentiable), and one can apply the Girsanov-Cameron formula to (III.A–3). The Girsanov-Cameron-Martin formula gives the flexibility of writing two versions of a path integral, one in terms of a process x, the other in terms of a process $\bar{x}$. I shall present the procedure on $\mathbb{R}^n$ in a form easy to generalize and choose $\bar{x}$ such that one can easily write down semiclassical expansions. Let

$$\begin{cases} dx_t = \mu \, dz_t \, , \; \mu = \sqrt{\hbar/m} \text{ (not to be confused with the integrator } \mu\,!) \\ v_t = \exp \int_0^t V(x_s) \, ds \end{cases} \tag{III.B–7}$$

[8] Having in a first draft of this calculation excused my use of (II.A–3) as if it were (II.A–2b) with the conventional "By a convenient abuse of language," my mentor (K.D.E.) wrote back, "That 'convenient abuse of language' phrase is almost as strong an abuse of the word 'convenient' as the Mafia could ever have made, since it knocks out one of the major mathematical difficulties!"

[9] This presentation is based on unpublished lectures given by Elworthy at Les Houches in 1983.

be the flat space version of (II.B–24); let Y be an arbitrary map: $\mathbb{R}^n \to \mathbb{R}$, which will be chosen later so that the path integral expressed in terms of the potential V may be restated as a path integral in terms of the action S; and let

$$\begin{cases} d\bar{x}_t &= \mu\, dz_t + \nabla Y(\bar{x}_t)\, dt \equiv \mu\, dz_t + \nabla Y_t\, dt \ , \qquad \bar{x}_0 = x_0 \ ; \\ dv_t &= \exp \int_0^t V(\bar{x}_s)\, ds \ . \end{cases} \tag{III.B–8}$$

Then,

$$\Psi(x_0, t) = \int_\Omega \Big(M_t \exp\Big(\int_0^t V(y_s)\, ds\Big)\, \varphi(y_t)\Big)\, d\mu(\omega) \ , \qquad M_b = 1 \ , \tag{III.B–9}$$

solves (II.B–26) when $M = \mathbb{R}^n$, with

$$M_t = 1 \ , \qquad y_t = x_t \text{ given by (III.B–7)}\,, \tag{III.B–10}$$

or

$$\begin{cases} M_t &= \exp\Big(\frac{1}{\mu^2}\big(Y(x_0) - Y(\bar{x}_t)\big) \\ &\quad + \frac{1}{\mu^2}\int_0^t \Big(\frac{\partial Y}{\partial s}(\bar{x}_s) + \frac{1}{2}\|\nabla Y(\bar{x}_s)\|^2 + \frac{1}{2}\nabla Y_s(\bar{x}_s)\Big)\, ds\Big) \\ y_t &= \bar{x}_t \text{ given by (III.B–8)}\,. \end{cases} \tag{III.B–11}$$

We choose the apparently more intractable version (III.B–11) because we can choose Y to yield computable semiclassical approximations. Elworthy made the following choice suggested by classical mechanics. Recall the equations of classical mechanics. Let $\Phi : M \to M$ be defined

$$\begin{cases} \ddot{\Phi}_s(a) &= -\nabla V(\Phi_s(a)) \ , \\ \Phi_0(a) &= a \ , \quad \dot{\Phi}_0(a) = \nabla S_0(a) \ . \end{cases} \tag{III.B–12}$$

Assume Φ is a diffeomorphism for $0 \le s \le \tau$. Let $S : [0, \tau) \times M \to \mathbb{R}$ be the solution of the Hamilton-Jacobi equation

$$\frac{1}{2}\|\nabla S_t(a)\|^2 + V(a) + \frac{\partial S_t}{\partial t}(a) = 0 \tag{III.B–13}$$

such that S_0 is precisely the function S_0 in (III.B–12). Then

$$\dot{\Phi}_t(a) = \nabla S_t(\Phi_t(a)) \ .$$

For $t < \tau$, set[10]

$$Y_s(a) := -S_{t-s}(a) \ , \quad 0 \le s \le t \ . \tag{III.B–14}$$

[10] This and (III.B–17) are the complications announced earlier (I.B–18) because the Brownian path z in (III.B–7) and (III.B–8) vanish at time $t = 0$; the reversal (III.B–14) would not be necessary if the paths z were to vanish at the final time. In Carverhill and Elworthy (1983) and Elworthy (1988), much of the work is done with z vanishing at the final time.

Then, with the choice (III.B–11) and the initial distribution (III.B–3), the limit of (III.B–9) is

$$\Psi_{\mathrm{WKB}}(x_0,t) = \exp\Big(-\frac{1}{\mu^2}\,S_t(x_0)\Big)\int_\Omega \exp\Big(-\frac{1}{2}\int_0^t \Delta S_{t-s}(\bar{x}_s)ds\Big)T(\bar{x}_t)d\mu(\omega)\,, \tag{III.B–15}$$

which is the same, *mutatis mutandis,* as (III.B–4) because

$$\Delta S_{t-s}(\Theta_s(a)) = \frac{\partial}{\partial s}(\log\det T_{\Theta_s(a)}\,\Phi_{t-s}^{-1}) \tag{III.B–16}$$

where Φ is defined by (III.B–12),

$$\Theta_s(a) = \Phi_{t-s}(\Phi_t^{-1}(a))\ , \quad \dot{\Theta}_s(a) = -\nabla S_{t-s}(\Theta(a))\ . \tag{III.B–17}$$

To obtain the next term in the expansion in μ^2 of (III.B–9) with the choice (III.B–11), Watling (1986) chose Y of the form

$$Y_s = -S_{t-s} + \mu^2 \log U_{t-s}$$

with

$$U_s(a) = T_0(\Phi_t^{-1}(a))\,\big(\det T_a\Phi_s^{-1}(a)\big)^{1/2}\ .$$

So far, we have assumed the classical flow $\Phi_s : M \to M$ to be a diffeomorphism, at least for $s < \tau$. What happens if the classical flow does not have an inverse?

2. "When strict WKB breaks down". Classical paths are critical points of the action. When the critical points are degenerate, strict WKB breaks down. This can happen for several reasons, depending on the nature of the degeneracy of the critical point. For instance, the classical flow on the configuration space can be caustic forming.

In the example quoted in Section I.A, glory scattering, the critical points of the action (i.e., the classical paths) are degenerate for two reasons:

i) The classical paths are characterized by an initial momentum and a final momentum constrained by the energy conservation law—a consequence of the invariance of the system under time translation.

ii) The geometry of glory scattering: Two paths with different impact parameters, in the neighborhood of the glory impact parameter can have the same initial momentum and final momenta equal "to first order," i.e., a caustic situation in phase space.

The degeneracies have been classified (DeWitt-Morette, 1984) and much is known about WKB on and beyond the caustics. Such WKB approximations are not "strict" WKB approximations. They give quantitative answers to several phenomena: glories, rainbows, conservation laws. See references in DeWitt-Morette (1984) and DeWitt-Morette (1987).

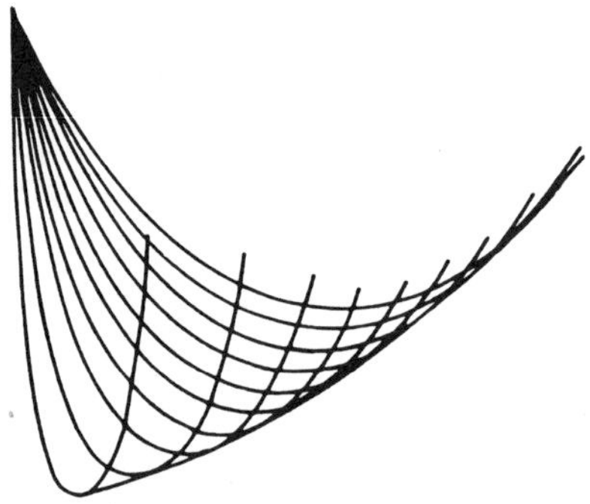

Figure 6. For a point in the "dark" side of the caustic there is no classical path; for a point on the "bright" side there are 2 classical paths that coalesce into a *single* one as the intersection of the 2 paths approaches the caustic. Note that the paths do not arrive at an intersection at the same time, the paths do not intersect in a space-time diagram.

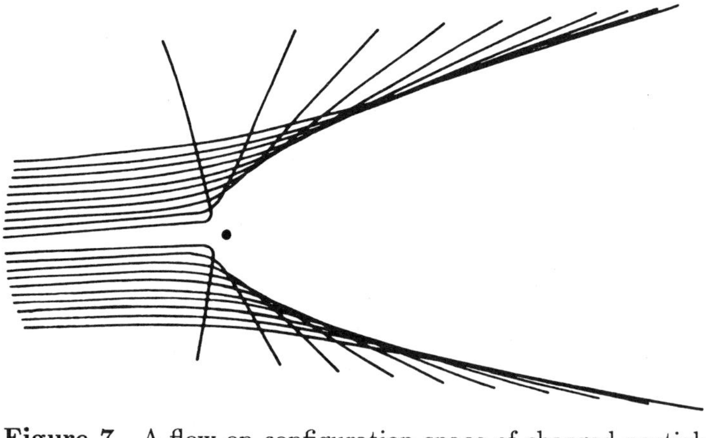

Figure 7. A flow on configuration space of charged particles in a repulsive Coulomb potential.

3. Nonanalyticity of WKB approximations. In this course we have discussed diffusion and Schrödinger equations. Their solutions can be analytically continued one in the other—which, as we have emphasized in Section I.A, is not to say that their integrators can be analytically continued. But the WKB approximations of their solutions cannot be analytically continued because the limit $\hbar = 0$ is a singular limit. A very instructive example has been analyzed by Michael Berry (to appear)—concerns the sex life of moths: male moths use their sense of smell to find their ways to the females. According to a WKB approximation (small diffusion coefficient) of the diffusion equation for the usual environment, the males can be successful sufficiently often. According to an analytic continuation starting from the WKB approximation of the corresponding Schrödinger equation, they cannot.

IV. CONCLUSION

I have presented the tools to solve problems of quantum mechanics in curved spacetimes:

i) Choose the stochastic process on a frame bundle that describes your system.

ii) Choose your initial wave function.

iii) A stochastic process together with an initial wave function defines a path integral on the projective system $(\Omega, \mathcal{O}, \mu)$.

iv) Compute the path integral.

But I have presented only one application, namely, the glory scattering of waves by black holes—a classical wave problem identical to a quantum mechanical problem when the waves are monochromatic. There are many challenging problems waiting to be done. To mention but one: paths which take their values in spacetime have unique properties; e.g., some preliminary calculations indicate that when the flow of spacetime classical paths is caustic forming, the system may undergo pair productions and pair annihilations. It would be interesting to compute the corresponding path integral—i.e., the path integral with a potential and an initial wave function such that the corresponding classical system defines a caustic forming flow.

I am sure that many problems presented in this course can be stated in terms of stochastic processes on fibre bundles and solved accordingly. Thank you for your attention.

ACKNOWLEDGEMENT

Discussions with Askell Hardarson and Ubirajara van Kolck have been enjoyable and profitable.

APPENDIX: SELECTED REFERENCES TO OTHER APPROACHES

I had hoped to include an appendix titled "Other Approaches to Defining Feynman Path Integrals." I had planned to begin with (in no particular order) Itô's definition, nonstandard analysis, Nelson-Trotter formulae, product integrals (Dollard-Friedman), Green functions (loop expansions), oscillatory integrals (Albeverio-Hoegh Krohn), polygonal approximations (lattice approximations, time slicing, discretization), analytic continuations (in time, in mass, in Planck's constant), coherent states, group theoretical approaches, partition of unity, Meiman's rough measures, Brownian motion on the circle, spinor chain—to name only the ones that come readily to my mind without going to the library. It quickly became clear that this project was far too ambitious, given the time available. I shall only mention a few references, which I have found particularly helpful as a starting point; most of them include excellent references for further (or prior) study. In chronological order:

K. Itô (1961), Generalized Uniform Complex Measures in the Hilbertian Metric Space with their Application to the Feynman Integral, *in:* "Proceedings of the Fifth Berkeley Symposium on Mathematical Statistics and Probability," Vol. II, pp. 145–161 (University of California Press, Berkeley) .

J. B. Keller and D. W. McLaughlin (1975), The Feynman Integral, *Am. Math. Monthly* **82**, 451–465.

A. Truman (1978), The polygonal path formulations of the Feynman path integral, *in:* "Feynman Path Integrals," S. Albeverio, et al., eds., Lecture Notes in Physics 106 (Springer-Verlag, Berlin).

Ph. Combes, R. Hoegh-Krohn, R. Rodriguez, M. Sirugue, and M. Sirugue-Collin (1980), Poisson Processes on Groups and Feynman Path Integrals, *Commun. Math. Phys.* **77**, 269–288.

G. W. Johnson (1982), The equivalence of two approaches to the Feynman integral, *J. Math. Phys.* **23**, 2090–2097 (a very useful comparison of Cameron and Storvick, on the one hand, and Albeverio and Hoegh-Krohn, on the other).

Marc Henneaux and Claudio Teitelboim (1982), Relativistic Quantum Mechanics of Supersymmetric Particles, *Annals of Physics* **143**, 127–159.

G. Kallianpur, D. Kannan, and R. L. Karandikar (1985), Analytic and sequential Feynman integrals on abstract Wiener and Hilbert spaces, and a Cameron-Martin formula, *Ann. Ins. Henri Poincaré* **21**, 323–361.

I. Bakas and H. LaRoche (1986), Path Integral Quantization and Coherent States, *J. Phys. A.: Math. Gen.* **19**, 2513–2523.

Ted Jacobson (1989), Bosonic Path Integral for Spin-1/2 Particles, *Phys. Lett.* **B216**, 150–154 (Spinor chain path integrals).

The approaches of the two following papers are not as developed as the previous ones but are intriguing.

N. N. Meiman (1983), Functional integrals as integrals on locally noncompact groups with generalized measures, *J. Math. Phys.* **25**, 1412–1433.

Thomas J. S. Taylor (1987), Applications of harmonic analyses on the infinite dimensional torus to the theory of the Feynman integral, *in:* "Functional integration with emphasis on the Feynman integral," Rendi Circ. Mat. Palermo (2), Suppl. **17**, pp. 349–362.

The two following papers do not belong to this list because they define path integrals for diffusion equations ("quantum mechanics with imaginary time"!) but suggest ideas that could be used for Feynman path integrals.

Alice Rogers (1988), Path Integraton in Superspace, *in:* "Complex differential geometry and supermanifolds in strings and fields," pp. 114–149, eds., P. J. Boongaarts and R. Nartoni, Lecture Notes in Physics 311 (Springer-Verlag, Berlin).

René Carmona, to appear, Path integrals for relativistic Schrödinger operators, *in:* the "Proceedings of the 1988 Nordic Summer School in Mathematics" (to appear).

REFERENCES

Berry, M., to appear, Scaling and nongaussian fluctuation in the catastrophe of waves, *Prometheus,* (UNESCO publication).

Bichteler, K., 1984, Lecture Notes. Stochastic Differential Equations. Preliminary Version. Reprint Department of Mathematics, University of Texas at Austin (quoted in II.B.1).

Bourbaki, N., 1969, *Eléments de Mathématiques* Livre VI "Integration," Chap. IX, Integration sur les espaces topologiques separés, also referred to as Fascicule 35 or No. 1343 of the Actualités Scientifiques et Industrielles, Hermann, Paris (quoted in I.A).

Carverhill, A. P., 1985, Flows of Stochastic Dynamical Systems: Ergodic Theory, *Stochastics* **14**, 273–317.

Carverhill, A. P., and Elworthy, K. D., 1983, Flows of Stochastic Dynamical Systems: The Functional Analytic Approach, *Z. Wahrscheinlichkeitstheorie verw. Gebeite,* 245–267 (quoted in III.B.1).

Choquet-Bruhat, Y., and DeWitt-Morette, C., 1982, "Analysis, Manifolds and Physics (Part I)," Revised Edition (North-Holland, Amsterdam) (quoted in II.A.2).

Darwin, Sir Charles, 1959, The gravity field of a particle, *Proc. Roy. Soc. London* **A249**, 180–194 (quoted in I.A).

DeWitt, B. S., 1957, Dynamical theory in curved spaces. I. A review of the classical and quantum action principles, *Rev. Mod. Phys.* **29**, 337–397 (quoted in III.A).

DeWitt-Morette, C., 1976, The semiclassical expansion, *Ann. of Physics,* 367–399 (quoted in I.B, III.B).

DeWitt-Morette, C., 1984, Feynman Path Integrals—From the Prodistribution Definition to the Calculation of Glory Scattering, *Acta Physica Austriaca Suppl.* **26**, 101–170 (quoted in I.A, I.B, III.B).

DeWitt-Morette, C., 1987, Path Integration in "Functional Integration with Emphasis on the Feynman Integral", Suppl. Rendiconti del Circolo Mat. dé Palermo, Serie II, no. 17, 211–228 (quoted in I.B.2, III.B.2).

DeWitt-Morette, C., and Elworthy, K. D., 1978, Stochastic Differential Equations. Lecture Notes and Problems. Center of Relativity preprint. Reproduced 1981 *in:* "Stochastic Differential Equations," Proceedings of the "5-Tage Kurs" Universität Bielefeld (quoted in I, II, III).

DeWitt-Morette, C., Elworthy, K. D., Nelson, B. L., and Sammelman, G. S., 1980, A stochastic scheme for constructing solutions of the Schrödinger equation, *Ann. Institut H. Poincaré* ***A*32**, 327–341. (quoted in II, III).

DeWitt-Morette, C., Maheshrvari, A., and Nelson, B., 1979, Path Integration in Nonrelativistic Quantum Mechanics, *Physics Reports* **50**, 255–372 (quoted here in I.A, I.B, III.B).

Dyson, F. J., 1972, Missed opportunities, *Bulletin of the American Math. Soc.* **78**, 635–652 (quoted in I.A).

Elworthy, K. D., 1982, "Stochastic Differential Equations on Manifolds," London Math. Soc. Lecture Note Series 70 (Cambridge University Press, Cambridge) (quoted in III.A.1, III.B.1).

Elworthy, K. D., 1988, Geometric Aspects of Diffusions on Manifolds *in:* "Ecole d'été de Probabilités de Saint-Flour", ed., P. L. Hennequin, Lecture Notes in Mathematics, No. 1362 (Springer-Verlag, Berlin).

Elworthy, K. D., and Truman, A., 1981, Classical Mechanics, the diffusion (heat) equation and the Schrödinger equation on a Riemannian manifold, *J. Math. Phys.* **22**, 2144–2166 (quoted in III.A, III.B).

Futterman, J. A. H., Handler, F. A., and Matzner, R. A., 1988, "Scattering from Black Holes" (Cambridge University Press, Cambridge) (quoted in I.A).

Laidlaw, M. G. G., and DeWitt-Morette, C., 1971, Feynman Functional Intgrals for Systems of Indistinguishable Particle, *Phys. Rev.* **D3**, 1375–1378 (quoted in II).

Morette, Cecile, 1951, On the definition and approximation of Feynman's path integral, *Phys. Rev.* **81**, 848–852 (quoted in III.A).

Nelson, Edward, 1967, "Dynamical Theories of Brownian Motion," Princeton University Press, Princeton.

Nelson, Edward, 1987, "Radically Elementary Probability Theory," Princeton University Press, Princeton.

Pauli, W., 1952, Ausgewahte Kapitel aus Der Feldquantisierung in "Pauli Lectures in Physics," ed. C. P. Enz, Massachusetts Institute of Technology Press, CITY (quoted in III.A).

Truman, A., 1976, Feynman path integrals and quantum mechanics as $\hbar \to 0$, *J. Math. Phys.* **17**, 1852–1862 (quoted in III.B).

Truman, A., 1977, The classical action in nonrelativistic quantum mechanics, *J. Math. Phys.* **18**, 1499–1509 (quoted in III.B).

van Kampen, N. G., 1976, Stochastic Differential Equations, *Physics Reports* **24C**, 172–228 (quoted in I.A, footnote 1).

Watling, K. D., 1986, Formulae for solutions to (possibly degenerate) diffusion equations exhibiting semiclassical and small time asymptotics, Ph.D. thesis, University of Warwick (quoted in III.B.1).

Young, A., and DeWitt-Morette, C., 1986, Time Substitutions in Stochastic Processes as a Tool in Path Integration, *Annals of Physics* **169**, 140–166 (quoted in I.B).

PARTICLES AND FIELDS

W. G. Unruh
C.I.A.R. Cosmology Program
Dept of Physics, University of British Columbia
Vancouver, B. C., Canada V6T 2A6

Abstract

How do particles arise in quantum field theories? The concept of particles is examined, and is related to specific forms for the interaction between the field and other localised forms of matter. For certain types of interaction, and more importantly, for studying specific interactions of the matter, it is shown that the quantum field can be regarded as though it were actually a quantum mechanics of particles.

This paper represents a follow-on to the series of lectures I gave at the Erice summer school in May, 1989. In those lectures, the time was principally spent in reviewing the results obtained by me 12 years ago[1] on the behaviour of accelerated particle detectors, including some of the more recent experimental suggestions for observing the effect[2]. That result was: An accelerated particle detector responds as if immersed in a thermal bath with a temperature of

$$T = 2\pi a \frac{\hbar}{ck} \qquad [1]$$

where a is the acceleration of the detector and k is Boltzman's constant. Since that result has been the subject of two excellent recent reviews by Frolov and Ginsburg[3], and by Takagi[4], I will concentrate here on the nature of the relation between quantum field theories, particles, and the notion of a particle detector.

Quantum mechanics began in the 1920's as a new theory of particles. Wave theories (or field theories) had been recognised already in the early 1800's as being crucial to the understanding of the light, and found their supreme expression in the theory of electromagnetism of Maxwell. However, the idea

Quantum Mechanics in Curved Space-Time
Edited by J. Audretsch and V. de Sabbata
Plenum Press, New York, 1990

that particles underlie the description of nature was dominant in physics. The notion of particles had, to a limited extent, even been restored to electromagnetism by Einstein in 1905, with his analysis of the photoelectric effect. He showed that the electromagnetic field, wonderfully described by Maxwell's field equations, behaved in some instances in its interactions with matter, as though made up of particles or discrete lumps of energy.

Schroedinger, with his wave mechanics, developed the idea of DeBroglie that wave equations had some relevance even to the physics of particles. In the development of quantum mechanics of the late 20's and 30's, it was realised however that the wave function, Ψ, which occurs in Schroedinger's equation, is not a field in the sense of, for example, the electromagnetic field. The electromagnetic field is something physical and has real existence; its amplitude is measurable, at each and every point in space. It has a continuous range of possible values at any one point in space, and the values it takes at different points in space are, in principle, independent of each other (the dynamics may of course produce relations between the values at different points). A particle, on the other hand, exists at one and only one point in space at any time. The observable qualities of the particle are not its value at each point, but its discrete existence or non-existence, at any given point in space at any time. It is either here, or there, and if here, it cannot be elsewhere.

The wave function associated with a particle, on the other hand shares the property with the field, that it has a value at each and every point in space at any time. That value is however not measurable. No other system can couple to its value, and it has no direct influence on the developement of other physical systems. Rather, its square is the probability of finding the particle at a particular position at a given time. The underlying reality is still that of particle physics, of an entity which can be at one and only one position at any given time, but our knowledge of that reality comes via the probabilistic wave function.

The introduction of probabilities into the fundamental theory of matter has caused much confusion. Because the equations of classical physics describe the evolution of real quantities, of real attributes of the world, attempts have been made to regard the wave function as having a real existence as well. Attempts to do so have in general led to innumerable arguments and confusion. This confusion has been most pronounced in the discussions of the change of the wave function on the receipt of new knowledge. If one regards the wave function as real, the "collapse of the wave packet" on the receipt of new knowledge leads one to ask what caused the collapse- what is the dynamics which governs the collapse. The obtaining of new knowledge is not considered a sufficient cause for the change, since how can my mental state (the new knowledge I have) influence the real world (the wave function)? To me this is another example the "Mind Projection fallacy" castigated by Jaynes[5] in his discussions of the nature of probability in probabilistic classical theories.

In the 1930's and 40's, in the attempts to unify quantum theory with both electromagnetism and with special relativity, field theories commpletely replaced particle mechanics as the underlying fundamental theory of nature. All phenomena of nature were thereafter to be described not by particles, but by fields. This change in the view of the world was forced on physics by the apparent impossibility of finding a consistent relativistic particle quantum mechanics. The difficulty of finding a consistant relativistic particle theory occurs not least because time and space play very different roles in particle quantum mechanics. In particle theories, Time is a parameter which labels stages of development of a system, and is not itself a measurable attribute of the system, while space exists only as the set of potential values that the measurements of the position of the particle can give. In field theories, on the other hand, time and space have the same status as parameters of the theory. This unity of role of time and space in field theories makes the development of a relativistic theory (in which space and time are inherently intermingled) much easier. The special role that time plays in the interpretation of the quantum field theory can still lead to problems, but I will ignore these here.[10]

This victory of field theory however opens the question of why particle quantum mechanics was as successful as it was in describing many physical phenomena. If the field represents the true reality, and if it can take on an arbitrary value at each and every point in space, how are we led to the idea that when we measure the electron in a hydrogen atom, we measure it to have one and only one position at any particular time? Since the field can be non-zero at a large number of places at once, why cannot a "single" electron be seen to be in many places at once, or why does the concept of particles make any sense at all? In addition, if particles do make sense in quantum field theory, and those particles are reasonably described by a Schroedinger equation, what is any relation between the value of the field of which the particle is a part, and the wave function in the Schroedinger quantum mechanics of that particle? How does the real field bear any relation to the probabilistic wave function? The term "second quantization", which has been inacurately used to characterise the development of quantum field theories from quantum particle theories, leads one to think that there is some relationship between the field and the wave function, but what, if anything, is that relation?

I will argue that in certain circumstances, the field can behave as though it has particle-like properties, and that the quantum field can be regarded as playing the role of the wave function for that particle. The probability of finding a particle at some position is related to certain properties of the field in terms of making certain particular measurements on the field.

Let us begin by examining a non-relativistic field theory, with a complex scalar field $\Psi(x)$ as the field of interest. The action will be taken to be the first order (in time) action

$$S = \int \{ i\Psi^* \dot{\Psi} - H(\Psi,\Psi^*) \, d^3x dt \qquad [2]$$

where H is some real (Hermitian) local function of Ψ and Ψ^* (the Hermitian conjugate of Ψ). I will furthermore assume that H is left invariant if one transforms Ψ by a constant phase.

The equation of motion of Ψ is given by

$$i\dot{\Psi} = \partial H/\partial \Psi^* \qquad [3]$$

The form of this action is clarified by defining new real variables

$$q(x) = \text{Real}(\Psi)\sqrt{2} \; ; \qquad p(x) = \text{Imag}(\Psi)\sqrt{2} \qquad [4]$$

so that the action takes the form (after an integration by parts in time of the first term)

$$S = \int \{ p(x)\dot{q}(x) - H((q+ip)/\sqrt{2},(q-ip)/\sqrt{2})\} \qquad [5]$$

From this we see that p is just the momentum conjugate to q, as the symbolism implies, and $\hat{H}(p,q) = H(\sqrt{2}(p+iq),\sqrt{2}(p-iq))$ is just the Hamiltonian. (Note, that q(x) and p(x) are the field configuration and momenta, not the configuration and momentum of "particles"). The phase invariance of the system is a global gauge transformation, namely $\Psi \rightarrow e^{i\theta}\Psi$ where θ is some constant, and corresponds in the q,p representation as a canonical transformation on the canonical variables.

The Noether charge associated with the phase transformation is

$$N=\int \Psi(x)^*\Psi(x)\ d^3x \qquad [6]$$

and, because of the invariance of the action under the transformation, H commutes with N. N is a positive definite operator. Writing it as

$$N=\int (p^2+q^2-1))/2\ d^3x \qquad [7]$$

we see that N is just a sum of "simple harmonic oscillator Hamltonians" and thus has a discrete spectrum consisting of the natural integers and bounded below by 0.

Now, choose the subspace of states of this field theory which obeys $N|s\rangle=1|s\rangle$. For a theory of the above form, we can regard this subspace of states as a "one particle" subspace of the system in a sense that will become clearer below. Furthermore the expectation value

$$P(x)= \langle s|\Psi^*(x)\Psi(x)|s\rangle \qquad [8]$$

will be found to represent the probability of finding that particle at position x. Since the field exists everywhere, how can we associate a location for the "particle" that the $|s\rangle$ state is supposed to represent? To find where the particle is,I will use a "particle detector". This particle detector is some physical system, with special properties, which is coupled to the field Ψ at a specific point in space at a specific time. Since the field is the only dynamical variable we have, we must couple to it. We cannot couple to attributes of the particle like its position operator. The ability to regard the field as having particle aspects will depend on the behaviour of certain types of detectors.

One can define a class of detectors so that the probability that it will be excited will just be proportional to P(x) (eqn. 8). Furthermore, it will turn out that the probability that two such detectors, located at different points x and x', will be excited at the same time, when the field is in its one particle state, is zero. Thus measurements made on the field by these detectors can be viewed as if the field were actually a single particle, which has one and only one position any one time, and possesing a probability of being found somewhere proportional to certain attributes of the field. Thus in this state, and as viewed by this particular type of detector, the field acts like a quantum particle.

For a model of a detector, I take a two level system, with levels labeled 0 and 1. Define an operator which takes the detector from level 0 to to level 1, and call it $A^\dagger$, i.e., the operator is defined by

$$A^\dagger|0\rangle = |1\rangle \; ; \qquad A^\dagger|1\rangle = 0. \qquad [9]$$

while the operator from level 1 to level 0 will be denoted by A. These can be most naturally chosen to obey $[A,A^\dagger]_+=1$, where the + subscrpt indicates the anticommutator. The coupling between the field and the detector will be taken to be of the form

$$\varepsilon[\Psi(x_0)A^\dagger + \Psi^*(x_0)A]\ \delta(t-t_0) \qquad [10]$$

where ε is a coupling constant. This detector will thus respond only to the value of the field at the point x_0 at time t_0. To lowest order in ε, the amplitude for finding the detector in its excited state 1, and the field in state $|p\rangle$ after the interaction is given, in the interaction picture

$$\varepsilon\langle p|\langle 1|\Psi(x_0,t_0)A(t_0)^\dagger + \Psi(x_0,t_0)A(t_0)|0\rangle|s\rangle$$

$$= \varepsilon\langle p|\langle 1|\Psi(x_0,t_0)\ |1\rangle|s\rangle \qquad [11]$$

since $A|0\rangle = 0$, and $A^\dagger|0\rangle = |1\rangle$. Thus the probability of finding the detector excited irrespective of the final state of the field is given by

$$P_{01} = \varepsilon^2 \Sigma_p \{ \langle s|\Psi^*(x_0,t_0)|p\rangle\langle p|\Psi(x_0,t_0)|s\rangle \} \qquad [12]$$

where the sum is over a complete set of intermediate states for the field $|p\rangle$. This finally gives for the transition probability from state 0 to 1

$$P_{01} = \varepsilon^2 P(x_0). \qquad [13]$$

where P(x) is defined in eqn. 8. We can interpret this by saying tht the detector will be excited only if the particle is located at the point x_0. Thus we can interpret the probability of excitation as a constant (the probability that if the particle is at x_0, it will excite the detector) times the probability of finding the particle at x_0. The probability of finding the particle at x_0 is thus just proportional to $P(x_0)$ as claimed. Furthermore, if we place a second detector at $x=x_1$, with lowering operator B and interaction strength ε' say, then the amplitude for finding both detectors in the state 1 if they started in 0 is given by

$$\varepsilon\varepsilon'\langle p|\langle 1|\langle 1'|(A^\dagger\Psi_0 + A\Psi^\dagger{}_0)(B^\dagger\Psi_1 + B\Psi^*{}_1)|0'\rangle|0\rangle|s\rangle$$

$$= \varepsilon\varepsilon'\langle p|\Psi_0\Psi_1|s\rangle \qquad [14]$$

where Ψ_0 refers to $\Psi(x_0,t_0)$ and Ψ_1 to $\Psi(x_1,t_0)$. The probability that both detectors make a transition becomes

$$P_{00'11'} = \langle s|\Psi^*{}_0\Psi^*{}_1\Psi_0\Psi_1|s\rangle \qquad [15]$$

But $|s\rangle$ obeys $N|s\rangle = |s\rangle$, and Ψ obeys $[N,\Psi] = -\Psi$. Ψ thus acts as a lowering operator on N. Since N has 0 as a minimum eigenvalue, we have

$$\Psi_0\Psi_1|s\rangle = 0 \qquad [16]$$

Thus the probability that both detectors are found to be excited is zero. (this is also true of more than 2 detectors located at different places and triggered at the same times). It is this feature which strengthens our interpretation of the behaviour of the field as that of a particle. It is typical of a particle that it have one and only one position at any given time. The detectors thus behave as though they were interacting with a particle, rather than with a field. Thus we can regard the N=1 state of the field as a one particle state, with P(x) giving the probability of finding that particle at the position x.

One important feature to note is that the interpretation of this state of the field as a single particle state depended specifically on the choice we made for detector. The detector has two discrete states, and our measurement of the detector is designed so that we will measure it as being in one or the other of the two states. It is this discreteness of the possible outcomes of the measurement on the detector which result in the discreteness of the particles either being at that position or not. Secondly, the interaction has the specific form. The "annihilation" field Ψ, rather than Ψ^*, is coupled to the "creation" operator, $A^\dagger$, for the detector.

The specific form of the interaction chosen is not the only one possible for the interpretation of the field as a particle however. In particular, the above interaction is somewhat unsatisfactory in that it results in the destruction of the particle. The state the field is left in after the interaction is the N=0 state, rather than another N=1 state. The detector measures the particle by destroying it. Although this is typical of some processes, there are also processes in which the "particle" survives its measurement (measurements in quantum mechanics are typically assumed to be of this survival form). We can also design detectors which do not destroy the "particle", but still allow us to regard the field as though made of particles. One specific model of this form has an interaction between the field and the detector of the form

$$\varepsilon(A^\dagger+A) \int j(x-x_0)\Psi(x,t)^*\Psi(x',t)d^3x\ \delta(t-t_0) = \varepsilon\ (A^\dagger+A)J(x_0,t) \quad [17]$$

where j is a function sharply peaked around zero, and J is defined by this equation. The probability of finding the detector in the state $|1\rangle$ is found to be

$$P_{01} = \varepsilon^2 \ \langle s|J(x_0)J(x_0)|s\rangle \quad [18]$$

Because $\Psi\Psi|s\rangle = 0$, we can commute the central pair of $\Psi(x')\Psi^*(x)$ in JJ, to give

$$P_{01} = \varepsilon^2\langle s|\int\Psi^*(x)j^2(x-x_0)\Psi(x)|s\rangle \ d^3x$$

$$= \varepsilon^2 \int j^2(x-x_0)\ P(x)\ d^3x \quad [19]$$

Thus P(x) again be interpreted as representing the probability that the detector will be excited, and thus the probability that the particle was at x at time t_0. Furthermore, because Ψ^* and Ψ commute if evaluated at separated points, the probability that two detectors at two different points will both be excited is again found to be zero- ie, the field can be interpreted as a single particle having a unique position at that time.

However, not all couplings between the field and some simple other system can be interpreted as though the field were a particle. Had we coupled the field to some detector with a continuous readout variable, we would not have had the same results. For example, letting the detector have a "free particle" Hamiltonian in some internal configuration variable and momentum Q and P

$$H_d = P^2/2 \qquad [20]$$

setting up the initial state of the detector to be a sharply peaked state around P=0 say, with width ΔP, and coupling to the field by

$$H_I = \varepsilon\, Q\, (\Psi^*(x_0) + \Psi(x_0)) \qquad [21]$$

and measuring the momentum of the detector afterwards, one would have found a result centered around P=0, but with a spread of possible outcomes with a width greater than the initial width by a term of the form

$$\delta P^2 = \varepsilon^2 \langle s|(\Psi^*(x_0) + \Psi(x_0))^2|s\rangle. \qquad [22]$$

The outcome of the experiment can have a continuous range of values, which is inconsistent with the idea that the particle is either at the point x_0 or not. Furthermore, there will be a finite probability that two separate detectors at different points would register the changes (i.e., would give a value of P after the interaction larger than the initial spread in the values of P). For this type of detector, the characterisation of the field in terms of "particles" would be inappropriate, and would not lead to reasonable predictions for the behaviour of the detectors.

We thus see that the notion of particles in a field theory depends both on the state of the field, and also on the detectors one uses to examine the field. One can encode this insight in an aphorism:

Particles are what particle detectors detect.

This apparent tautology expresses the idea that particles and particle detectors are closely intertwined in a quantum field theory. Not only must one have a notion of particles in order to define particle detectors (i.e., those measuring devices which detect the properties of the entities we call particles), but particles are really only defined in terms of the devices and situations used to detect them. Particles are not primitive notions in a field theory, but rather are derived notions which are sometimes useful in describing the interactions of the fields with the rest of matter (particle detectors).

The expectation value P(x) of eqn. 8 represents the probability of finding the particle at $\dot{x}$. Can Ψ in some sense be regarded as the amplitude, the wave function, of that particle? Is, as the notion of "second quantisation" implies, the field Ψ really just a wave-function for the particle. Can we regard the field equation (eqn 3) as being the Scroedinger equation for the particle?

The equation for the field Ψ (which is just the equation of motion of the quantum field in the Heisenberg representation) is

$$i\dot{\Psi} = \partial H(\Psi^*,\Psi)/\partial\Psi^*. \qquad [23]$$

If H is bilinear in Ψ and Ψ^*, this equation can be rewritten as

$$i\dot{\Psi} = \int h(x,x')\Psi(x')d^3x' \qquad [24]$$

where h is some real kernel function. The Heisenberg representation solution of these equations of motion can be written as

$$\Psi(x,t) = \Sigma_i \; a_i \; \psi_i(x,t) \qquad [25]$$

where a_i are "annihilation" operators obeying $[a_i,a_j^\dagger]=\delta_{ij}$, and the ψ_i are a complete set of functions under the norm $\int\psi^*\psi d^3x$. The functions ψ_i can be taken to obey the classical Scroedinger equation , namely eqn 23 with Ψ replaced with ψ_i. Furthermore, the functions ψ_i must be normalised such that

$$\int\psi_i^*(x)\psi_j(x)dx = \delta_{ij} \qquad [26]$$

in order that Ψ and Ψ^* obey the correct commutation relations in terms of q and p the canonical variables. Finally, any single particle state $|s\rangle$ can be defined (with an appropriate choice of the functions ψ_i) so that

$$|s\rangle = \Sigma_i\alpha_i a_i^\dagger|0\rangle \qquad [27]$$

with $\Sigma_i|\alpha_i|^2=1$, where $|0\rangle$ is the lowest eigenstate of N. The probabilty function P(x) will then just be given by

$$P(x)= \langle s|\Psi(x)^*\Psi(x)|s\rangle =[\Sigma_i\alpha_i\psi_i(x)]^*[\Sigma_j\alpha_j\psi_j(x)] \qquad [28]$$

Since $\psi = \Sigma_i\alpha_i\psi_i$ obeys the Schroedinger like wave equation 23, we can regard it as corresponding to the Schroedinger wave function for the particle. Measurements of the field with particle detectors will therefor lead one to find that the field behaves as though it were a particle obeying Schroedinger mechanics with the wave function ψ.

On the other hand, if H is not bilinear, but contains higher order terms, this correspondence does not hold. The field does not correspond to a classical system obeying the non-linear generalisation of the Schroedinger equation. There is in general no simple relationship between the non-linear classical theory derived from eqn. 3 and a single particle Schroedinger quantum mechanics.[6]

We have thus found that in this non-relativistic quantum field theory, there exists a well defined relation between certain measurements made in certain states of the quantum field theory, and the notion of a particle obeying an ordinary non-relativistic Schroedinger quantum mechanics.

Let us leave this simple model field theory, and proceed to an example of the more usual relativistic field theories. Is the notion of particle a useful one in these field theories as well? If so, what are the states and the particle detectors, or the types of interactions for which they are useful? As a paradigm I will use a massless real scalar field theory, with Lagrangian action (in a general spacetime) of

$$S= \int \tfrac{1}{2} \sqrt{g}\, \{g^{\mu\nu}\partial_\mu\Phi\, \partial_\nu\Phi \}\, dt d^3x \qquad [29]$$

We can cast this theory into the form we have already considered. The procedure goes under the name of Hamiltonian diagonalization. I will use the usual Arnowitt, Deser, and Misner (ADM) form for the metric[7], in which we define functions N, N^i, and a three dimensional metric γ_{ij} by

$$\begin{aligned} g_{00} &= N^2 - N^iN^j\gamma_{ij} \\ g_{0i} &= N^j\gamma_{ij} \\ g_{ij} &= \gamma_{ij}. \end{aligned} \qquad [30]$$

We rewrite the action in terms of these metric variables, and then define the momentum conjugate to the field Φ by

$$\pi(x) = \delta S/\delta\dot{\Phi} \qquad [31]$$

where the dot denotes time derivative as usual. The Hamiltonian action now becomes

$$S = \int \{\pi\dot{\Phi} - N(\ \pi^2/\sqrt{\gamma} + \sqrt{\gamma}\ (\gamma^{ij}\partial_i\Phi\partial_j\Phi))/2$$

$$- N^i\pi\partial_i\Phi\ \}d^3x dt \qquad [32]$$

(One can always choose spatial coordinates such that the functions N_i are zero, and I assume we have done this.) Now define the operator

$$\nabla_0{}^2\phi \ =\{\ [\ \tfrac{N}{\sqrt{\gamma}}]^{1/2}\ \partial_i\ (N\sqrt{\gamma}\ \gamma^{ij}\ \partial_j)\ \ [\ \tfrac{N}{\sqrt{\gamma}}]^{1/2}\ +m^2\ \}\phi \qquad [33]$$

where the subscript 0 indicates that the functions N and γ_{ij} are to be taken as having their value on the hypersurface $t=t_0$. Now define a canonical transformation on the field by taking as the new canonical variables

$$\hat{\Phi} = \nabla_0{}^{-1/2}\ [\tfrac{N}{\sqrt{\gamma}}]_0^{-1/2}\ \Phi \qquad [34]$$

$$\hat{\pi} = \nabla_0{}^{1/2}\ [\tfrac{N}{\sqrt{\gamma}}]_0^{1/2}\ \pi$$

Note that since N_0 and γ_{ij0} are time-independent by definition, so is ∇_0 , and so is the canonical transformation. The Hamiltonian action on the $t=t_0$ hypersurface now becomes

$$S_0 = \int \{\ \hat{\pi}\ \partial\hat{\Phi}/\partial t\ -\ \{\ \hat{\pi}\nabla_0\hat{\pi}\ +\ \hat{\Phi}\nabla_0\hat{\Phi}\}/2\ \}\ d^3x \qquad [35]$$

Defining new complex scalar fields Ψ by

$$\Psi = (\hat{\Phi} + i\ \hat{\pi})/\sqrt{2} \quad ; \quad \Psi^* = (\ \hat{\Phi} - i\ \hat{\pi})/\sqrt{2} \qquad [36]$$

the action on that hypersurface becomes

$$S_0 = \int \Psi^*\partial\Psi/\partial t\ -\ \{\ \nabla_0{}^{1/2}\ \Psi^*\ \nabla_0{}^{1/2}\ \Psi\ \} \qquad [37]$$

which is just of the form $H(\Psi^*,\Psi)$ of eqn. 2. The operator $\nabla_0{}^{1/2}$ is highly non-local, even though $\nabla_0{}^2$ is a local operator. Unless the metric is time-independent, this form will only be valid at the time t_0. At other times in addition to the bilinear terms, one will also have terms quadratic in Ψ and in Ψ^*.

At t_0, the number operator

$$N=\int \Psi^*(x)\Psi(x)\ d^3x \qquad [38]$$

commutes with the Hamiltonian, and can be used to define the single particle states of the system just as in our non-relativistic case. Thus the probability that the particle would be found at position x will, if we follow our former procedure, be given by $P(x) = \langle\Psi^*(x)\Psi(x)\rangle$.

Going through this procedure in flat spacetime, ($N=1$, $\gamma_{ij}=\delta_{ij}$) for the massless scalar field, we have

$$\Psi(x) = \int K(x-x')\ \Phi(x') + i\ L(x-x')\ \pi(x')\ d^3x' \qquad [39]$$

where

$$K(x-x') = \int e^{-ik\cdot(x-x')}/[k^2]^{1/4}\ d^3k/(2\pi)^3 \\ \propto |x-x'|^{-5/2}, \qquad [40]$$

$$L(x-x') = \int e^{-ik\cdot(x-x')}\ [k^2]^{1/4}\ d^3k/(2\pi)^3 \\ \propto |x-x'|^{-7/2}.$$

Ψ is thus a highly non-local function of Φ and π. Furthermore, if we choose the state $|s\rangle$ so that the probability of finding the particle at any place other than x_0 is zero, those mode functions Ψ are just the eigenfunctions of the Pryce-Newton-Wigner(PNW) position operator[8] for a relativistic theory.

However, is this definition of particle actually useful? As I argued previously, particles are useful only to the extent that their use makes the behaviour of "particle detectors" (physical systems which interact with the field) more transparent. The particle detectors defined for the field Ψ are however not very useful in that they do not correspond to any realistic interaction with the true physical field Φ. The Ψ detectors are defined to have local interactions with the field Ψ, which correspond to very non-local interactions with the physical field Φ.

Can we define a type of particle detector as a physical system which interacts locally with the field Φ, and has at least some of the properties we would expect a particle detector to have? It was this question I was implicitly trying to answer in my paper of 1976[1]. First, let us look at some ideas which do not work. The detector will again be taken to be a two level system, with lowering operator A. One of the lessons we have learned is that we need the discreteness of this type of system to hope to get a particle interpretation. Let us try to couple it directly to the field Φ via an interaction of the form $\varepsilon(A^{\dagger}+A)\Phi(x_0)\delta(t-t_0)$. We then get the probability of excitation to be proportional to

$$\rho(x)= \langle\Phi(x)\Phi(x)\rangle \qquad [41]$$

The "particle" operator

$$n=\int\Phi^2(x)d^3x \qquad [42]$$

however is not conserved by the Hamiltonian, nor does it have a discrete spectrum. There seems to be no way in which one could base a notion of particle on this type of detector.

We could try to couple to the local function

$$\hat{\Psi}(x) = \Phi(x) + i\hat{\pi}(x) \qquad [43]$$

at an instant of time. The associated number operator $\int\hat{\Psi}^*(x)\hat{\Psi}(x)d^3x$ does have a discrete positive integer spectrum. However, it does not commute with H. Furthermore, its zero eigenstate (and one-particle eigenstates) have expectation values of energy which diverge as Ω^4 where Ω is the cutoff. Surely one of the attributes of a particle is that it be linked with the carrying of a finite amount of energy.

A better definition of particle detector can be obtained by thinking about how real detectors work. An example of a photon detector would be, say, a hydrogen atom. We start the atom in its ground state. In the absence of radiation, the atom remains in its ground state. If, however, there are "photons" present, the atom may find itself in one of its excited states through the absorption of energy from a photon. Thus if we find it in such an excited state, we can say that it has interacted with a particle of light--a photon. It has detected the presence of a photon. We are thus led to tie the absorption of energy by the atom with its detection of a particle.

How can we translate this vague idea into the model of a particle detector? We keep the two level system as the model detector. However, we will now demand that the two levels are at different energies- zero for the state $|0\rangle$ and E for the state $|1\rangle$. We will assume a coupling of the detector directly with the field Φ, but will not assume that the interaction is instantaneous. (This assumption is certainly valid for realistic detectors). In particular, if the detector is located at position x_0, the interaction with the field is assumed to be via $\Phi(t,x_0)$ (where we are working in the Heisenberg representation for the field) over a time much longer than 1/E.

The interaction is assumed to be of the form

$$H_I = \varepsilon\,\Phi(x_0,t)\,(A+A^{\dagger}) \qquad [44]$$

where A is the lowering operator taking you from the higher energy state to the lower state. The Hamiltonian for the detector is given by

$$H_d = E\,A^{\dagger}A \qquad [45]$$

and, as a model, the field is again taken to be a massless scalar field with Lagrangian

$$L = \frac{1}{2}\int[\,\dot{\Phi}^2 - (\nabla\Phi)^2]\,d^3x \qquad [46]$$

and thus with Hamiltonian

$$H_f = \frac{1}{2}\int[\pi^2 + (\nabla\Phi)^2]\,d^3x\ . \qquad [47]$$

The Heisenberg equations of motion for A to lowest order are

$$i\dot{A} = E\,A + \varepsilon\Phi(x_0,t). \qquad [48]$$

The detector is initially taken to be in the lowest energy state $|0\rangle$, which is annihilated by the operator A_0, the initial value of the operator A. The probability of finding the detector in the upper state $|1\rangle$ is just given by the expectation value of the operator $A^{\dagger}A$. This operator has eigenvalue of zero in the lowest state, and of one in the higher state. The expectation value, which is the probability of being in a state times the eigenvalue in that state, is thus just the probability of being in the upper state. In the Heisenberg representation, the state remains constant, while the operators evolve, so the expectation of $A^{\dagger}A$ is , to lowest order in ε

$$\langle s|\langle 0|A^{\dagger}A|0\rangle|s\rangle = \qquad [49]$$

$$\varepsilon^2\int_0^t e^{iE(t'-t'')}\langle s|\Phi_0(x_0,t'')\Phi_0(x_0,t')|s\rangle dt'dt''$$

where $|s\rangle$ is the initial state on the field Φ. I have used the fact that $A_0|0\rangle = \langle 0|A_0^\dagger = 0$ where the subscript $_0$ is to indicate the operators obtained to zero order in ε. If the integral over the times is sufficiently large, only the parts of Φ that go as e^{-iEt} in the second factor, and e^{iEt} in the first will survive. Now, define "single particle" state $|s\rangle$ by

$$|s\rangle = \int \alpha_k\, a_k^\dagger d^3k\ |vac\rangle \qquad [50]$$

where the vacuum state is defined by

$$a_k|vac\rangle = 0 \quad \text{for all } k \qquad [51]$$

and a_k are defined by

$$\Phi_0(x,t) = \int\{ e^{-i(\omega t - k.x)}\, a_k/[(2\pi)^3\omega]^{1/2} \qquad [52]$$

$$+ e^{i(\omega t - k.x)}\, a_k^\dagger/[(2\pi)^3\omega]^{1/2}\}\ d^3k$$

with $\omega^2 = k\cdot k$. (Note that these a_k are just the Fourier transforms of the fields Ψ defined previously for this system). Over times much greater than 1/E, we find that the transition probability of eqn 31 just becomes

$$\langle 0|A^\dagger A|0\rangle \sim \varepsilon^2 \int_0^t \phi^*(x_0,t'')\phi(x_0,t')e^{iE(t'-t'')}dt'dt'' \qquad [53]$$

where ϕ is a complex field amplitude

$$\phi(x,t) = \int \alpha_k e^{-i(\omega t - k\cdot x)}/[(2\pi)^3\omega]^{1/2}\ d^3k \qquad [54]$$

If $\phi(x_0,t')$ is non-zero only within the time interval 0 to t, the excitation probability becomes

$$\langle 0|A^\dagger A|0\rangle = \varepsilon^2 \int \alpha_k^* \alpha_{k'} e^{-i(k-k')\cdot x_0}\ \delta(\omega - E)\delta(\omega' - E) d^3k d^3k'/2\pi E \qquad [55]$$

The δ functions restrict the k integrals to be only over the $\omega(k)=E$ subspaces. If on the other hand, α_k is almost a δ-function in ω (in particular if the width of α as a function of ω is much less than 1/t, the time over which the detector is on), the probability integral becomes of order t^2. This is troubling since with a long wave-train in the field one would have expected the detector probability to increase as order t, not t^2. The answer to this apparent paradox is that to see the response increase as order t, one must continually observe the detector to see if it has been excited.

During an observation of the detector to see if it has been excited, the phase correlation between the amplitude to be in the lower $|0\rangle$ state and the upper $|1\rangle$ state is lost. It is this correlation which leads to the increase as order t^2 rather than order t. To see this, let us assume that the detector has an energy given by $E + e(t)$ rather than just E for the state $|1\rangle$. $e(t)$ is a stochastic function of t, such that the phase, $\varphi(t)=\int_0^t(E+e)dt'$ becomes uncertain by of order unity a time period of order $\tau<<t$ (i.e., $\Delta\varphi^2(\tau)=\langle(\int_0^\tau e(t')dt')^2\rangle\sim 1$). I will for simplicity assume that e(t) is a uniform Gaussian stochastic process with zero norm. This is not supposed to be an accurate model of the measurement process, but is rather a very simple model to investigate the effect that the phase loss during measurement will have on the detection process for the field. The time τ is supposed to represent the time between successive measurements to determine whether the detector is in the state $|1\rangle$, the time over which the phase correlation of the two states is lost.

Let us take the single particle state of the field to be such that α_k is a δ-function in ω at $\omega=\omega_0$. We then find that the probability of finding the detector in its upper state is given by

$$\langle 0|A^\dagger A|0\rangle\sim\varepsilon^2\int\alpha_k{}^*\alpha_{k'}e^{-i(k-k')\cdot x_0}d^3kd^3k'/[(2\pi)^3\omega_0] \qquad [56]$$

$$\times\int_0^t e^{-i(E-\omega_0)(t'-t'')}\exp(-\langle(\int_{t'}^{t''}e(t''')dt''')^2\rangle_s/2)\ dt'dt''.$$

where $\langle\rangle_s$ is the statistical expectation value of the term in the brackets. I have used the fact that for a Gaussian process the statistical expectation value of the exponential of a linear term in the random function is the exponential of one half of the expectation value of the function squared. For this Gaussian process we have

$$\langle(\int_{t'}^{t''}e(t''')dt''')^2\rangle_s = |t'-t''|/\tau \qquad [57]$$

and for times $t>>\tau$, the second term(the time integral) becomes $t\tau/2[1+(\omega_0-E)^2]$. The probability of finding the detector in its excited state when the field is in a single particle state of definite frequency is proportional to t, the total time of observation of the field in the plane wave state, and to τ, the time between measurement on the detector. Note that as τ goes to 0, corresponding to a very frequent reading of the state of the detector, the probability that the detector is excited goes to zero. This is just an example of the "Watch-Dog" (Watch dogs prevent changes in the state of ones home) or "Watched-Pot" (A watched pot never boils) effect[9].

What kind of particles does this detector detect? In this case, they are in fact the same as the particles defined by Ψ. In terms of the annihilation operators, a_k, the field Ψ becomes

$$\Psi(x,t) = \int 2a_k\, e^{-i(\omega t-k\cdot x)} /(2\pi)^3\, d^3k \qquad [58]$$

since for this problem $\nabla^{1/2} e^{ik\cdot x}$ is just $\omega^{1/2}e^{ik\cdot x}$. Thus the state $|s\rangle=\int\alpha_k a_k{}^\dagger dk|vac\rangle$ can be written as the single particle state defined in terms of Ψ by

$$|s\rangle = \int\psi(x,t)\ \Psi^*(x,t)d^3x\ |vac\rangle \qquad [59]$$

where $|vac\rangle$ is the zero particle state for $N=\int\Psi^*\Psi d^3x$, and ψ is just $\int\alpha_k\, e^{i(\omega t-k\cdot x)}d^3k$. The key to the behaviour of the detector coupled to Ψ was that the field Ψ was expressible purely in terms of annihilation operators, leading to the fact that $\Psi\Psi|s\rangle=0$. In the case of the detector coupled to the Φ field, it is the conservation of energy, expressed in the integral $\int e^{iEt}\Phi(x_0,t)dt$, which leads to this expression also being expressible in terms purely of annihilation operators, which allows the detector to work as a particle detector.

We have however traded off the highly non-local but instantaneous coupling to the field Φ for the Ψ detectors, for a detector coupled locally to the physical field Φ but over a non-local period in time. I feel however that the latter is more physical than the former in that more natural interactions between the field and other matter take that form. Note that the two types of detectors operate very differently. For the detector coupled locally to Ψ, the relation of the particles to energy is incidental. For the detector coupled locally to the Φ field, the energy content of the particles is crucial.

As a passing comment I would also point out that, as in the case of the Ψ field, one can also design particle detectors which do not, in the process of detecting the particle also destroy it. In particular, coupling the detector to the square of the Φ field at the point x_0 also produces a reasonable particle detector. (This would correspond physically to particle detectors like cloud chambers say in which the particle carries on its existence after the interaction.) The interaction is taken to be of the form

$$H_I = \varepsilon(A^\dagger + A)\,\Phi^2(x_0,t) \qquad [60]$$

The probability of detection now takes the form

$$\langle s|A^\dagger A|s\rangle = \langle s|\int_0^t e^{-iE(t'-t'')}\Phi^2(x_0,t')\Phi^2(x_0,t'')dt'dt''|s\rangle \qquad [61]$$

Writing Φ in terms of creation and annihilation operators, and after some calculation one gets

$$\langle s|A^\dagger A|s\rangle = \int_0^t \phi^*(x_0,t')\phi(x_0,t'')e^{i(E+\omega)(t'-t'')}/[(2\pi)^3\omega]d^3k\,dt'dt'' \qquad [62]$$

where ϕ was defined previously, which is of the same form as eqn 53. However, in this case, the probability that two detectors will both be excited will not in general be zero. This is easily understood physically since over the time period t, the particle could travel from one detector to the other, and excite both. This does not happen when the detector is linearly coupled to Φ since then the detector absorbs the particle at the same time as it measures it, leaving nothing for the other detector to detect. (Note that this explanation is cast in particle language, since that description is the simplest way of understanding the interaction of the field with this type of detector.)

We finally come to the crux of the matter. We have two different ways of defining particle detector in terms of their interaction with the field, either instantaneously and locally with Ψ and thus non-locally with Φ, or locally but over a long time period with Φ. Since Φ is the physical field, I believe that the latter is the physically more reasonable way. The non-local coupling through Ψ is in some ways more elegant in terms of ordinary quantum measurement theory, since it allows measurements to be made at a single instant of time. The usual formulations of the measurement problem in quantum mechanics assume that measurements can always be made instantaneously. However, the spatial non-locality of the process seems to me to be much more artificial than the temporal non-locality of the second definition of detector.

Let us therefore accept the second definition. We are then led to ask some interesting questions about the behaviour of these detectors in various situations. Since the detection process takes place over time, it is not surprising that the behaviour of the detector could depend on the state of motion of the detector. This leads to the result obtained in my 1976 paper. Instead of having the detector stationary at the point x_0, let us allow the detector to move along some trajectory

$$t=T(\tau) \;;\quad x = X(\tau) \qquad [63]$$

where I take τ to be the proper time along the path the detector follows. Let us also assume that the field is not in a single particle state, but in the vacuum state $|vac\rangle$. For a detector at rest, this would lead to a zero probability of the detector being excited. Since it is the proper time τ which governs the development of the internal structure of the detector, the equations of motion for the operator A now becomes

$$\frac{dA}{d\tau} = -iEA + \varepsilon\Phi(X(\tau),T(\tau)) \qquad [64]$$

so to lowest order in ε we again find that the probability of excitationis just given by

$$\langle vac,0|A^{\dagger}A|0,vac\rangle = \qquad [65]$$

$$\int_0^T e^{-iE(\tau'-\tau'')}\langle vac|\Phi_0(X(\tau'),T(\tau'))\Phi_0(X(\tau''),T(\tau''))|vac\rangle d\tau' d\tau''$$

In general, the integral over τ'' does not project out the "annihilation" operator (in terms of the usual expansion of Φ_0) part, and the detector has a finite probability of excitation even in the $|vac\rangle$ state. In the case in which the detector is in uniform acceleration, that probability of excitation just corresponds to what one would expect if the detector were immersed in a thermal bath of particles with temperature proportional to the acceleration.

Is this thermal bath real? I have emphasised that the description of the field in terms of particles depends on how the particles are detected. If one's particle detector is modelled by the kind of detector system I have described (ie, a two or multi-level system) in which the energy levels are relatively stable "pointer" variables, an analysis like the above will lead to the probability of excitation just being thermal. The system will tend toward the state of thermal equilibrium with a temperature proportional to the acceleration ($a/2\pi$ in Planck units). You could cook your steak by accelerating it (if the minor problem, that a temperature of 300°C requires an acceleration of about 10^{24}cm/sec^2, did not make the technique somewhat impractical).

In conclusion, the theme has been that our fundamental theories of matter are field, not particle theories. Particles can however be a useful concept in that the interaction of the field with localised systems can mimic the behaviour of particles. At all times one must however remember that the fundamental theory is not a particle theory, and that describing the system in terms of particles may be misleading. It is however amazing that there are any situations in which a field theory, defined in terms of a quantity, the field, which has no location, which exists everywhere at once, can mimic the behaviour of particles, which do have a definite location, and are defined in terms of their existence at specific points in space.

Acknowledgements

I would like to thank the organisers of the Erice summer School for inviting me to talk on the subject of particle detectors, which forced me to rethink an old subject. The Canadian Institute for Advanced Research has supported me with the LAC Minerals Fellowship, and the Natural Science and Engineering Research Council of Canada has been generous in its grant (580441) for support of my research.

References

1: W.G. Unruh, Phys. Rev. D14, 870 (1976)

2: There have been a number of suggestions of ways in which the effect could be measured. Some are J.S. Bell, J.M. Leinaas, Nucl. Phys. B212, 131 (1983); J. Rogers, Phys. Rev. Lett. 61, 2113 (1988); E. Yablonovitch, Phys. Rev. Lett.62, 1742 (1989)

3: V.L. Ginsburg, V.P. Frolov, Sov.Phys. Usp. 30, 1073 (1987) (original Russian version in Usp. Fiz Nauk 153, 633 (1987)

4: S. Takagi, Prog. Theor. Phys. Supp. 88, 1 (1986)

5: E. T. Jaynes "Probability Theory as Logic" in Proceedings of the Workshop on Maximum Entropy and Bayesian Methods ed P.Fouger (Kluwer, Holland 1990)

6: S. Weinberg, Phys. Rev. Lett. 62, 485 (1989) Ann.Phys.(N.Y.)194, 336 (1989) has recently suggested a non-linear variant of Schroedinger quantum mechanics based on a non-linear classical field theory for the wave function. A paper investigating the relation between such a non-linear quantum mechanics and the particle descrition of a non-linear field theory is at present in preperation

7: See for example chapter 10 and appendix E in R.M. Wald, General Relativity (U. Chicago Press, Chicago, 1984) or chapter 21.7 in C.W. Misner, K. Thorne, J.A. Wheeler Gravitation (W.H. Freeman, New York, 1973).

8 M.H.L. Pryce, Proc. Roy. Soc. 195A, 62 (1948) ; T.D. Newton, E.P. Wigner, Rev. Mod. Phys. 211, 400 (1949)

9: K. Kraus, Found. Phys. 11, 547 (1981) B. Misra, E.C.G. Sudarshan, J. Math. Phys 18, 756 (1977)

10: As an example of the problems caused by the interpretive importance of time in quantum gravity see W.G. Unruh, R.M. Wald, Phys. Rev. D 40, 2598 (1989)

QUANTUM MECHANICS OF BLACK HOLES IN CURVED SPACE-TIME

Zhang Zhenjiu[†], Huang Huanran[†], Bao Gong[†] and He Changbai[‡]

[†]Center of Relativity Research,
Huazhong Normal University, Wuhan, 430070, China
[‡]Beijing Observatory, Academia Sinica, Beijing, 100080, China

INTRODUCTION

In quantum theory, all physical fields are described on a fundamental level by the principle of quantum theory. In the Heisenberg picture, the quantum states of systems are represented by vector in a Hilbert space. A convenient basis in this space is the Fock representation. The normalized basis ket vector $|>$, can be constructed from the vector $|0>$, called vacuum. Observable will not always have a definite value and one, in general, can only predict probabilities for the outcome of measurements.

The observable quantities in general relativity always have definite values. Up to now, all of the known exact quantum field theories associated with general relativity run into difficulties. The full quantum theory of gravity, or quantum gravity in 4-dimensions, remains a goal for future, but might be on the horizon with superstring theory (Gross et al. 1985, a.b), or with a canonical transformation to new variables for the phase space of general relativity recently discovered by Ashtekar (1987), Jocobson and Smolin (1988).

We do have a theory of quantum fields, "quantized matter fields", which propagate in a classical (unquantized), external (non-dynamical), curved spacetime (gravitational field) and describes certain states where both general relativity and quantum theory are important, as an approximation to a full quantum theory of gravity (see, Sciama et al. 1981; Birrell et al. 1982; Wald, 1984; Audretsch, 1989; Frolov, 1989). This theory predicts the creation of particle pairs out of vacuum by a gravitational field in the vicinity of a black hole and the emission of a thermal spectrum by a black hole formed by collapse (Hawking, 1975). Actually, in the early 1930's, Schrodinger realized (see, Halpern, 1988) that a time-dependent gravitational field would give rise to pair creation. However, the field only really took off with the discovery of Hawking effect in 1974.

In these lectures, we will restrict attention to the case of a real scalar field. The analysis of other linear fields of spin s=1 or s<1 is very similar, although important differences occur in the fermion case. Fields of s>1 do not have a natural generalization to curved spacetime. We assume spacetime $(M,g_{\mu\nu})$ to be a c^{∞} 4-dimensional, globally hyperbolic, pseudo-Riemannian manifold M with metric $g_{\mu\nu}$.

The lectures are organized as follows. At first, we give a brief description of spacetime structure we consider (Section I). We discuss

Quantum Mechanics in Curved Space-Time
Edited by J. Audretsch and V. de Sabbata
Plenum Press, New York, 1990

the quantum field in flat spacetime, in a static homogeneous gravitational field, and in curved spacetime; and describe the semiclassical back-reaction program, conformal trace anomalies and quantum vacuum in the gravitational field of a black hole in Section II. The algebraic approach to quantum field theory which, at least in some cases, appears to admit a consistent conceptual and mathematical framework is briefly discussed in Section III. After we discuss particle detector, particle creation by a collapsing spherical body and Hawking radiation in Section IV, V and VI respectively, we discuss, in Section VII, in more details, the black hole thermodynamics, including the four laws of black hole thermodynamics, general covariant thermodynamics and the approach to black hole entropy, Euclidiean Einstein action approach, membrane paradigm and entropy generation in the interaction.

We use units in which $\hbar$(Planck constant/2π)=c(speed of light in vacuum)=G(gravitational constant)=K_B(Boltzman constant)=1.

Some of the recent development on the subject can be found in the references listed alphabetically.

I. SPACETIME STRUCTURE

We assume spacetime, which is considered as the background and describes the gravitational field, to be c^∞ 4-dimensional, globally hyperbolic pseudo-Riemannian manifold(M, $g_{\mu\nu}$). The differentiability ensures the existence of diffBerential equations; while the global hyperbolicity ensures the existence of Cauchy hypersurfaces, which is required by causality. The pseudo-Riemannian metric $g_{\mu\nu}$ associated with the line element

$$ds^2 = g_{\mu\nu}(x)\,dx^\mu dx^\nu \tag{1.1}$$

has signature (-1, 1, 1, 1)=2. It should be reduced to the case of Lorentz metric when the spacetime is flat, and be asymptotic flat when we get far away from matter.

We shall make use of Penrose conformal diagrams for depicting the causal structure of spacetime. A conformal transformation of the metric may be described by

$$g_{\mu\nu}(x) \rightarrow \bar{g}_{\mu\nu}(x) = \Omega^2(x)\,g_{\mu\nu}(x) \tag{1.2}$$

where $\Omega(x)$ is a continuous, non-vanishing, finite, real function. Then

$$\Gamma^\rho_{\mu\nu} \rightarrow \bar{\Gamma}^\rho_{\mu\nu} = \Gamma^\rho_{\mu\nu} + \Omega^{-1}(\delta_\mu^\rho \Omega_{;\nu} + \delta_\nu^\rho \Omega_{;\mu} - g_{\mu\nu} g^{\rho\alpha} \Omega_{;\alpha}) \tag{1.3}$$

$$R_{\mu\nu} \rightarrow \bar{R}_{\mu\nu} = \Omega^{-2} R_{\mu\nu} - 2\Omega^{-1}(\Omega^{-1})_{;\mu}{}^{\rho} g_{\nu\rho} + \tfrac{1}{2}\Omega^{-4}(\Omega^2)_{;\rho\sigma} g^{\rho\sigma} \delta_{\mu\nu} \tag{1.4}$$

$$R \rightarrow \bar{R} = \Omega^{-2} R + 6\Omega^{-3}\Omega_{;\mu\nu} g^{\mu\nu} \tag{1.5}$$

$$(\Box + \tfrac{1}{6}R)\Phi = (\bar{\Box} + \tfrac{1}{6}\bar{R})\bar{\Phi} = \Omega^{-3}(\Box + \tfrac{1}{6}R)\Phi \tag{1.6}$$

$$\Box\Phi = g^{\mu\nu}\Phi_{;\mu\nu} = (-g)^{-1/2}[(-g)^{1/2} g^{\mu\nu}\Phi_{,\nu}]_{,\mu} \tag{1.7}$$

$$\bar{\Phi} = \Omega^{-1}(x)\Phi. \tag{1.8}$$

The properties of Penrose diagram are:

1. All null rays remain at 45° : conformal transformation leave the null cones invariant.
2. All null rays terminate on the diagonal boundary lines: J^+ — future null infinity, J^- — past null infinity.
3. Asymptotically timelike lines converge on points: i^+ — future timelike infinity, i^- — past timelike infinity.
4. Asymptotically spacelike lines converge on i^0 — spacelike infinity.
5. Event horizon H : The null asymptote is an event horizon for the accelerated particle.

The spacetime with special geometric symmetries can be described by using Killing vectors ξ^{μ}, which are solutions of Killing's equation

$$\mathcal{L}_{\xi} g_{\mu\nu}(x)=0 \tag{1.9}$$

where $\mathcal{L}_{\xi}$ is Li derivatives with respect to ξ^{μ}.

The spacetime with some geometric symmetries is associated with conformal flatness, that is, the spacetime is conformal to Minkowski spacetime.

$$\mathcal{L}_{\xi} g_{\mu\nu}(x)=\lambda(x)\, g_{\mu\nu}(x) \tag{1.10}$$

where ξ^{μ} is called conformal Killing vector and $\lambda(x)$ is a non-singular, non-vanishing scalar function.

In the semiclassical level, $g_{\mu\nu}$ should be the solutions of Einstein field equations

$$R_{\mu\nu}-\tfrac{1}{2}g_{\mu\nu}R = 8\pi T_{\mu\nu}. \tag{1.11}$$

II. QUANTUM FIELD IN CURVED SPACETIME, BACK-REACTION AND ANOMALIES

A great deal of the formalism of the quantum field theory in Minkowski spacetime can be extended to curved spacetime with little modification. But a number of striking and fundamental phenomena have been predicted, such as that natural vacuum states possess thermal properties, and that each of the inequivalent field quantizations treated represents a different physical situation and there is no question of one of the vacuum states being the correct or true one (Sciama et al, 1981).

2.1 Quantum Field in Flat Spacetime

In flat spacetime, the metric is

$$ds^2=\eta_{\mu\nu}dx^{\mu}dx^{\nu}, \quad \mu,\nu=0,\ 1,\ 2,\ 3; \quad \text{dig}\ \eta_{\mu\nu}=(\ -1,\ 1,\ 1,\ 1\ ) \tag{2.1}$$

The equation satisfied by massless scalar field ϕ is

$$\Box\phi=0 \tag{2.2}$$

The solutions can be expressed as

$$\phi(x)=\int [\,\phi_K(x)\,a_K+\bar{\phi}_K(x)\,a_K^{\dagger}\,]\,d^3k \tag{2.3}$$

where the plane wave $\phi_K= A_K e^{i K_\mu x^\mu}$, $A_K = (2\pi)^{-2/3}(2\omega_K)^{-1/2}$, $\omega_K= (\,m^2 + k_i k^i)^{1/2}$. For a hermitian quantum field, ϕ_K and $\bar{\phi}_K$ are hermitian conjugated field operators, while a_K and $a_K^{\dagger}$ are annihilation and creation operators which satisfy the commutation relations

$$[\ a_{K'}^{\dagger}, a_K\ ] = \delta(\ K,K'\ ) \tag{2.4}$$

$$[\ a_K, a_{K'}\ ] = [\ a_K^{\dagger}, a_{K'}^{\dagger}\ ] = 0. \tag{2.5}$$

For convenience, we use Hadamard function

$$G(x,x')=\tfrac{1}{2}\langle\ \phi(x)\ \phi(x') + \phi(x')\ \phi(x)\ \rangle \tag{2.6}$$

and define the vacuum state $|o\rangle$ in Minkowski spacetime as

$$a_K|0|M\rangle=0\quad . \tag{2.7}$$

For a homogeneous space,

$$< a^{\dagger}_{k}, a_{k'} > = n_k \delta(k-k') \tag{2.8}$$

where n_k is the number density of mode $\vec{k}$. And we have

$$G(x,x') = \int \phi^2(x,x'|k^i)(n_i + \tfrac{1}{2})d^3k \tag{2.9}$$

After renormalization, we have

$$G^{ren} = G - G_0 = <\phi^2>^{ren} = \int \phi^2 (x|k) n_k d^3k \tag{2.10}$$

which is finite for physically reasonable states, can be measured and is of interest to physicists.

The stress-energy tensor for a classical field ϕ can be written as

$$T_{\mu\nu} = \lim_{x \to x'} D_{\mu\nu}(x,x') \, \phi(x) \, \phi(x') \tag{2.11}$$

where

$$D_{\mu\nu}(x,x') = (1-2\xi)\tfrac{1}{2}(\nabla_\mu \nabla_{\nu'} + \nabla_{\mu'} \nabla_\nu) + (2\xi - \tfrac{1}{2}) g_{\mu\nu} g^{\alpha\beta'} \nabla_\alpha \nabla_{\beta'}$$
$$- \xi(\nabla_\mu \nabla_\nu + \nabla_{\mu'} \nabla_{\nu'}) + \tfrac{3}{4} g_{\mu\nu}(\nabla^\alpha \nabla_\alpha + \nabla^{\alpha'} \nabla_{\alpha'}) \tag{2.12}$$

The stress-energy tensor for a quantum field can be written as

$$< T_{\mu\nu}(x) > = \lim_{x \to x'} D_{\mu\nu}(x,x') G(x,x'), \tag{2.13}$$

$$< T_{\mu\nu}(x) >^{ren} = \int d^3k \, t_{\mu\nu}(x,|k) n_k \tag{2.14}$$

$$t_{\mu\nu}(x|k) = \lim_{x' \to x} D_{\mu\nu}(x,x') \phi^2(x,x'|k). \tag{2.15}$$

For a plane-wave, we have

$$t_{\mu\nu} = \frac{k_\mu k_\nu}{(2\pi)^3 \omega_k} \tag{2.16}$$

Summary:

1. We use inertial frame and use $t = x_0$ to distinguish positive and negative frequencies in solutions of the field equations.
2. a, a^{+} are defined by means of decomposition of the quantum field operator over the positive and negative frequency modes.
3. We define vacuum state and then make renormalization.

2.2 Quantum Field in a Static Homogeneous Gravitational Field

According to the equivalence principle, the laws of physics (including the quantum ones) in any uniform gravitational field should take on the same forms as they do in a uniformly accelerated reference frame in flat space-time.

I-Frame. In inertial frame (I-frame), we use Cartesian coordinates x^α as

$$x^\mu = (x^0, x^i), \quad i = 1, 2, 3 \tag{2.17}$$

where x^i are the spatial orthogonal coordinates. The distance between any two points with fixed spatial coordinates (x^i_1) and (x^i_2) measured in this reference frame is time independent. Because the metric of the flat space-time is time independent.

A-Frame. In a uniformly accelerated frame (A-frame), we use Rindler coordinates (τ, ρ, x^2, x^3), where (ρ, x^2, x^3) are the spatial orthogonal coordinates. The relation between Cartesian coordinates and Rindler coordinates is

$$x^0 = \rho \sinh\eta\ , \quad x^1 = \rho\cosh\eta \tag{2.18}$$

The world line of a uniformly accelerated particle is described by $\rho = \rho_0 =$ constant and $x^3 = x^4 =$ constant. Its acceleration is $w = (w_\mu w^\mu)^{1/2} = \rho_0^{-1}$. Its proper time is $\tau = \eta\rho_0$. The hypersurface, $\tau =$ constant, describes a set of simultaneous events for a co-moving observer at the origin of this frame. The Rindler coordinates cover the part of R_+ of the Minkowski space where $x^1 > |x^0|$. The distance between two fixed points in this frame is time-independent as that in the flat spacetime.

G-Frame. In a homogeneous gravitational field, such as the regions of size l<<L, Where L is a characteristic scale of inhomogenity of a static gravitational field created by a massive black hole, an observer at rest, in G-frame, in this field, should have the same metric.

According to the equivalence principle, one can obtain the renormalized expressions for $<\phi^2>$ and $<T_{\mu\nu}>^{ren}$. In the simplified case when the average number of Rindler quanta depends only on the frequency ν, by using symmetry properties of $<T_{\mu\nu}>^{ren}$, invariant under transformations along the Rindler time η, under transformations and rotations of the coordinates x^3, x^4, and under reflection $\eta \to -\eta$, $x^3 \to -x^3$, $x^4 \to -x^4$. We have

$$<T_{\mu\nu}(x)>^{ren} = \mathrm{diag}(-\varepsilon, \tfrac{\varepsilon}{3}, \tfrac{\varepsilon}{3}, \tfrac{\varepsilon}{3}) \tag{2.19}$$

where $\varepsilon \sim \rho^{-4}$.

Here, the quantization scheme in homogeneous gravitational field is developed by Boulware(1975). We use concepts of Boulware particle and Boulware vacuum. This quantum state possesses the lowest possible energy in the static gravitational field.

For the equilibrium thermal radiation in a homogeneous gravitational field, the average number of the Boulware quanta is

$$n = [\exp(2\pi\beta\omega) - 1]^{-1} \tag{2.20}$$

where $\beta = (2\pi\theta)^{-1}$. The temperature θ is registrated at the point $\rho = \rho_0$. The local temperature

$$\theta_l = \frac{\theta\rho_0}{\rho} \tag{2.21}$$

and ω is the proper frequency of Boulware quantum measured by the clock at the origin of the G-frame. This equation really describes the thermal distribution. It should be stressed that the temperature and frequency are dependent to the red-shift factor in the same way, so $\beta\omega$ is invariant. when $\theta = (2\pi\rho_0)^{-1}$, so called Unruh temperature, $<T_{\mu\nu}>^{ren}$ and $<\phi^2>^{ren}$ vanish. The corresponding quantum state is known as the Hartle-Hawking vacuum state $|0,H>$ which corresponding to $|0,M>$.

In flat spacetime, renormalization can be done by subtracting an unobservable contribution of the vacuum null fluctuations due to high symmetry of the Minkowski spacetime. In the curved spacetime, there arises an ambiquity, which may be restricted by Wald's axioms(1977, 1978). In calculation of $<T_{\mu\nu}>^{ren}$ in the gravitational field of an isolated uncharged black hole, all possible renormalization procedures obeying the Wald's axioms give the same answer. The point-spliting method is the most useful for some purposes (see, Birrell and Davies, 1982). The renormalization of the graviton contribution to $<T_{\mu\nu}>$ is discussed by Allen et al(1988). Unruh(1976) studied an interaction of a uniformly accelerated detector with a quantum field(see lectures by Unruh in this proceedings).

2.3 Quantum Field in Curved Spasetime

Lagrangian density. It can be written as

$$\mathcal{L}(x) = \tfrac{1}{2}[-g(x)]^{1/2}\{g^{\mu\nu}(x)\phi_{,\mu}\phi(x)_{,\nu} - [m^2 + \xi R(x)]\phi^2(x)\} \tag{2.22}$$

where $\phi(x)$ is the scalar field, and m is the mass of the field quanta. The coupling between the scalar field and the gravitational field is represented by the term $\xi R(x)\ \phi(x)$, where ξ is a numerical factor and R(x) is the Ricci scalar curvature. The action is

$$S=\int \mathcal{L} d^4x. \tag{2.23}$$

The variation of S equal to zero , then we have

$$[\Box + m^2 + \xi R(x)]\phi(x)=0 \tag{2.24}$$

Scalar Product. We can define the scalar product as

$$(\phi_1, \phi_2)=-i\int \phi_1(x) \overleftrightarrow{\partial_\mu} \phi_2^*(x)\ [-g_\Sigma(x)]^{1/2} d\Sigma^\mu \tag{2.25}$$

where $d\Sigma^\mu=n^\mu d\Sigma$, n^μ is a future-directed unit vector orthogonal to Σ, and Σ is a spacelike hypersureface, which is a Cauchy surface in our case. The value of (ϕ_1, ϕ_2) is independent of Σ.

There exists a complete set of mode solutions $U_i(x)$, which are orthonormal:

$$(U_i, U_j)=\delta_{ij}, \quad (U_i^*, U_j^*)=-\delta_{ij}, \quad (U_i, U_j^*)=0. \tag{2.26}$$

The field ϕ may be expanded as

$$\phi(x)=\sum_i[(a_i U_i(x) + a_i^\dagger U_i^*(x)]. \tag{2.27}$$

The commutation relation are

$$[a_i, a_j^\dagger]=\delta_{ij}, \quad \text{etc.} \tag{2.28}$$

Vacuum States and Fock Space. This decomposition of ϕ defines a Fock space and a vacuum state $|0\rangle$ as

$$a_j|0\rangle=0 \quad \forall j. \tag{2.29}$$

In Minkowski spacetime, the vacuum is invariant under the action of the Poincare group. In curved spacetime, the Poincare group is no longer a symmetric group of the spacetime, in general, thus there is no Killing vector at all with which to define positive frequency modes. The coordinate systems are physically irrelevant. There also exists a second complete set of mode solutions $\bar{U}_j(x)$, in which the field ϕ can be expanded as

$$\bar{\phi}(x)=\sum_j [\bar{a}_j \bar{U}_j(x) + \bar{a}_j^\dagger \bar{U}_j^*(x)]. \tag{2.30}$$

It defines a new Fock space and also a new vacuum state $|\bar{0}\rangle$ as

$$\bar{a}_j|\bar{0}\rangle=0, \quad \forall j. \tag{2.31}$$

Bogolubov Transformation. $\bar{U}_j$ can be expanded in terms of U_i,

$$\bar{U}_j=\sum_i(\alpha_{ij} U_i + \beta_{ji} U_i^*), \tag{2.32}$$

Conversely, U_i can be expanded in terms of $\bar{U}_j$,

$$U_i=\sum_j(\alpha_{ji}^* \bar{U}_j - \beta_{ji} \bar{U}_j^*). \tag{2.33}$$

These relations are known as Bogolubov transformation. The matrices α_{ij}, and β_{ij} are called Bogolubov coefficients

$$\alpha_{ij}=(\bar{U}_j, U_i), \quad \beta_{ij}=-(\bar{U}_j, U_i^*), \tag{2.34}$$

which possess the following properties:

$$\sum_K(\alpha_{iK}\alpha_{jK}^* - \beta_{iK}\beta_{jK}^*)=\delta_{ij} \tag{2.35}$$

$$\sum_K (\alpha_{iK}\beta_{jK} - \beta_{iK}\alpha_{jK}) = 0. \tag{2.36}$$

We also have

$$a_i = \sum_j (\alpha_{ji}\bar{a}_j + \beta_{ji}^* \bar{a}_j^\dagger) \tag{2.37}$$

$$\bar{a}_j = \sum_i (\alpha_{ji}^* a_i - \beta_{ji}^* a_i^\dagger) \tag{2.38}$$

Vacuum. Therefore, we have the expectation value of the operator $N_i = a_i^\dagger a_i$ for the number of U_i-mode particles in the state $|\bar{0}\rangle$, such as

$$\langle \bar{0}|N_i|\bar{0}\rangle = \sum_j |\beta_{ji}|^2, \tag{2.39}$$

which is to say that the vacuum of the $\bar{U}_i$ modes contains $\sum|\beta_{ji}|^2$ particles in the U_i mode. Thus, the two sets of modes U_i and $\bar{U}_j$ do not share a common vacuum state.

In order to answer the question of which set of modes furnishes the best description of a physical vacuum, it is necessary to specify the details of the quantum measurement process. In particular, the state of motion of the detector can affect whether or not particles are observed to be present. (see Section 2.6 in details).

2.4 Semi-classical Back-reaction Program

Physical Description. The interaction of gravity with other fields can be described at three different levels:

1. Classically gravitational field(g) plus other classical fields(f) obey classical equations;

2. A full quantum description of both g and f by means of a wavefunction $\psi(g,f)$, which obeys the Wheeler-DeWitt equation. Including the energy of a scalar matter field, the Wheeler-DeWitt equation reads

$$[-\tfrac{1}{2}l\ \nabla_x^2 + \tfrac{1}{2}l^{-2}h^{1/2}(2\Lambda - {}^3R) + h^{1/2}T_{nn}(\Phi, -i\tfrac{\delta}{\delta\Phi})]\psi(h_{ij},\Phi) = 0. \tag{2.40}$$

Here $T_{nn}(\Phi,\pi)$ is the stress-energy of the matter field expressed in terms of the field's value and momentum and projected onto the normals of the space-like hypersurface, l is Planck length and h_{ij} are 3-metric, $h=\det|h_{ij}|$. But we have no means of solving this equation.

3. $\bar{g}$ is still a classical field, while f are quantized in field($\bar{g}$) and are described by some wavefunction $X(\bar{g},f)$. In this description, we need determination of $X(\bar{g},f)$ and $\bar{g}$. $X(\bar{g},f)$ is determined by the functional Schrodinger equation in quantum field theory in curved spacetime,

$$[i\partial/\partial t - H(\bar{g},f)]X(\bar{g},f) = 0 \tag{2.41}$$

where $H(\bar{g},f)$ is the Hamiltonian for the field f in the background metric $\bar{g}$.

We need a c-number equation for $\bar{g}$,

$$R_{\mu\nu} - \tfrac{1}{2}g_{\mu\nu}R = 8\pi G\langle T_{\mu\nu}\rangle \tag{2.42}$$

where $\langle T_{\mu\nu}\rangle$ stands for a "suitable quantum average". It may take as the expectation value of $T_{\mu\nu}$ in, for example, some quantum vacuum state $|\psi\rangle$.

For the backreaction in finite system, approximate solutions can be found by the simplifying approach. According to York(1985, 1986, 1987), in the first approximation, the back-reaction program is as follows.

1. Let a curved background spacetime with metric $\bar{g}_{\mu\nu}$ be Ricci flat

$$\bar{R}_{\mu\nu} = \bar{G}_{\mu\nu} = 0. \tag{2.43}$$

2. There is an external(non-gravitational) free field ϕ on the background, and ϕ is in a vacuum state

$$\langle\phi\rangle = 0 \tag{2.44}$$

but

$$<\phi^2>\neq 0 \quad (2.45)$$

for quantum fluctuation of ϕ. The expectation value of a renormalized symmetric stress-energy tensor $<T_{\mu\nu}>^{ren}$ of ϕ is simply denoted by $T_{\mu\nu}$ and satisfies

$$\bar{\nabla}_\mu T^{\mu\nu}=0, \quad (2.46)$$

where $\bar{\nabla}_\mu$ denotes the covariant derivative with respect to $\bar{g}_{\mu\nu}$.

3. The metric

$$g_{\mu\nu}=\bar{g}_{\mu\nu}+\psi_{\mu\nu} \quad (2.47)$$

where $\psi_{\mu\nu}$ represents the effect of quantum fluctuation of the metric. $\psi_{\mu\nu}$ corresponds to a field ψ on the background, and one has

$$<\psi>=0 \quad (2.48)$$

$$<\psi^2>\neq 0. \quad (2.49)$$

The effective stress-energy tensor $\tau_{\mu\nu}$ of ψ satisfies

$$\bar{\nabla}^\mu \tau_{\mu\nu}(\psi)=0. \quad (2.50)$$

4. The back-reaction problem is to solve the Einstein equation

$$G_{\mu\nu}(g)=8\pi[\tau_{\mu\nu}(\psi)+T_{\mu\nu}(\phi)] \quad (2.51)$$

for a classical metric

$$g_{\mu\nu}=\bar{g}_{\mu\nu}+\Delta g_{\mu\nu}. \quad (2.52)$$

If we write

$$G_{\mu\nu}(g)=\bar{G}_{\mu\nu}(\bar{g})+\Delta G_{\mu\nu}(\bar{g},\Delta g)=\Delta G_{\mu\nu}(\bar{g},\Delta g) \quad (2.53)$$

and denote the linear part of $\Delta G_{\mu\nu}$ by $\delta G_{\mu\nu}$, the linearization is carried out with respect to $\epsilon=\hbar M^{-2}$, where M is the Schwarzschild mass of the background field, because of the Bianchi identity in the background metric, we have

$$\bar{\nabla}^\mu \delta G_{\mu\nu}(\bar{g},\Delta g)=0. \quad (2.54)$$

5. To ignore $\tau^{\mu\nu}$, use Page's (1982) closed-form expression for $T^{\mu\nu}$, and consider as boundary an ideal massless perfectly reflecting spherical wall of area $4\pi r_0^2$, the microcanonical boundary condition can be specified as the total effective energy at r_0

$$m(r_0)=M+E_{rad}(r_0) \quad (2.55)$$

where E_{rad} is the energy of the radiation.

6. We get the result of the equilibrium temperature distribution

$$T_{loc}(r)|g_{tt}(r)|^{1/2}=T_\infty=(K_{EH}\hbar)(2\pi)^{-1}=\text{constant} \quad (2.56)$$

where the surface gravity

$$K_{EH}=\frac{1}{4M}[1+\epsilon(\frac{K_0+12}{3840\pi})] \quad (2.57)$$

and $K_0=-3840\pi\rho_0(r_0)$.

In further study of this direction, we would know more about the role of metric fluctuations near r=2M, where they appear to play a decisive role in the dynamical origin of the temperature of black holes.

2.5 Conformal Trace Anomalies

For massless fields of spin s=0, $\frac{1}{2}$, and 1 in a 4-dimension curved spacetime,the equations describing these fields are invariant under conformal transformation

$$g_{\mu\nu} \rightarrow \bar{g}_{\mu\nu} = \Omega^2(x)\, g_{\mu\nu} \tag{2.58}$$

and $T_\mu{}^\mu = 0$, i.e. trace-free.

In quantum mechanics in curved spacetime, the conformal invariance is broken after renormalization. The trace of the renormalized stress-energy tensor which remain invariant under conformal transformation does not vanish, known as conformal trace anomalies, and can be written as

$$\langle T_\mu{}^\mu \rangle^{ren} = \alpha(H + \tfrac{2}{3}\Box R) + \beta J + r\Box R \tag{2.59}$$

where

$$H = C_{\alpha\beta\gamma\delta}C^{\alpha\beta\gamma\delta} = R_{\alpha\beta\delta\gamma}R^{\alpha\beta\delta\gamma} - 2R_{\alpha\beta}R^{\alpha\beta} + \tfrac{1}{3}R^2 \tag{2.60}$$

$$J = R_{\alpha\beta\gamma\delta}R^{\alpha\beta\gamma\delta} - 4R_{\alpha\beta}R^{\alpha\beta} + R^2 \tag{2.61}$$

$$\alpha = \frac{1}{2^9\, 45\,\pi^2}[12h(0) + 18h(\tfrac{1}{2}) + 72h(1)] \tag{2.62}$$

$$\beta = \frac{1}{2^9\, 45\,\pi^2}[-4h(0) - 11h(\tfrac{1}{2}) - 124h(1)], \tag{2.63}$$

for point-splitting and Zeta function renormalization

$$\gamma = \frac{1}{2^9\, 45\,\pi^2}[-120h(1)], \tag{2.64}$$

and $\gamma = 0$ for dimensional renormalization. [h(s) is the number of helicities of the field of spin s.]

From the study of the behavior of the effective action under the conformal transformation gives the quantity which is conformally invariant (Douker,1986),

$$|g|^{1/2}\{\langle T_\nu{}^\mu\rangle^{ren} + \alpha[(C^{\alpha\mu}{}_{\beta\nu}\ln g)^{;\beta}{}_{;\alpha} + \tfrac{1}{2}R_\alpha{}^\beta C^{\alpha\mu}{}_{\beta\nu}\ln g] + \beta[2H_\nu{}^\mu - 4R_\alpha{}^\beta C^{\alpha\mu}{}_{\beta\nu}] + \tfrac{1}{6}I_\nu{}^\mu\gamma\} \tag{2.65}$$

where

$$H_{\mu\nu} = -R_\mu{}^\alpha R_{\alpha\nu} + \tfrac{2}{3}RR_{\mu\nu} + (\tfrac{1}{2}R_\beta{}^\alpha R_\alpha{}^\beta - \tfrac{1}{4}R^2)\, g_{\mu\nu}, \tag{2.66}$$

$$I_{\mu\nu} = 2R_{;\mu\nu} - 2RR_{\mu\nu} + (\tfrac{1}{2}R^2 - 2R^{;\alpha}{}_{;\alpha})\, g_{\mu\nu}. \tag{2.67}$$

2.6 Quantum Vacuum in the Gravitational Field of a Black Hole (Frolov, 1989)

For an observer which is at rest in a static gravitational field, its velocity is $U^\mu = \xi^\mu/|\xi_\mu\xi^\mu|^{1/2}$, and acceleration is $w_\mu = -\frac{1}{2}\nabla_\mu \ln f$.

For a quantum field ϕ in the Schwarzschild field, the equation for the field is

$$\Box\phi = 0. \tag{2.68}$$

Its solution is

$$\phi_{\omega n\beta} = e^{-i\omega t}\, \psi_{\omega n}(r)\, Y_{n\beta}(\vec{n}) \tag{2.69}$$

where $Y_{n\beta}(\vec{n})$ are the spherical harmonics on S^2, and $\psi_{\omega n}(r)$ are radial functions, obeying the equation

$$[\frac{d^2}{dr^{*2}} + \omega^2 - V]\,\psi_{\omega n} = 0 \tag{2.70}$$

where $dr^* = dr/f$, and the effective potential is

$$V = [6/r^2 + r_g/r^3]\, f. \tag{2.71}$$

We use $\phi_{\omega n\beta\epsilon}$ ($\epsilon=\pm1$) to describe the in-modes and out-modes respectively. The quantum field $\bar{\Phi}$ outside the black hole can be expanded in these modes

$$\bar{\Phi}(x)=\sum_{\omega\beta n\epsilon}(\phi\bar{a}+\phi\bar{a}^{*}). \tag{2.72}$$

The modes $\phi_{\omega\beta n\epsilon}$ are chosen to be of positive frequency with respect to the Killing time t.

Boulware Vacuum. It is the state of the minimal Killing energy. This state is important in a static spacetime without horizon. $<T_{\mu\nu}>_B^{ren}$ and $<\phi^2>_B^{ren}$ in B-vacuum are divergent at the horizon. In the presence of a black hole, the B-vacuum state is physically badly defined and it can not met in real situation.

Hartle-Hawking-Vacuum. The most preferable choice of the quantum state in the homogeneous gravitational field is the H-vacuum. For the state, the average number of B-quanta out-going from the horizon is described by the thermal distribution

$$n=[\exp(\omega/\theta)-1]^{-1} \tag{2.73}$$

where, $\theta=\frac{\kappa}{2\pi}$, is the black hole temperature. The frequency ω is defined with respect to time t connected with the Rindler time η by the relation $t=\kappa^{-1}\eta$. This state describes the black hole in equilibrium with a thermal radiation inside the thermal bath. The equilibrium is possible if the radiation temperature measured at far distance from the black hole coincides with the black hole temperature ($\kappa/2\pi$).

The physical situation is a black hole surrounded by a mirror shell which reflects the out-going quanta and sends them back to the black hole. The number of B-particles in the in-modes will be the same as the number of B-particles in the out-modes. $<\phi^2>_H^{ren}$ and $<T_{\mu\nu}>_H^{ren}$ for H-vacuum state are finite everywhere outside black hole including both future and past event horizons, while at infinite they coincide with that of the equilibrium thermal radiation in flat spacetime.

Unruh Vacuum. The quantum state, which describes the situation when the black hole is placed in an empty space and there is no in-coming particles in in-modes, is known as the Unruh vacuum states. Such a black hole is a source of stationary thermal radiation

$$<\phi^2>_U^{ren} \approx A/r^2 \tag{2.74}$$

$$<T_{tr}>_U^{ren} \approx B\theta^2/r^2 \tag{2.75}$$

where A and B can be found by numerical calculation (see, Elster,1983).

One can assume that the black hole is surrounded by the thermal atmosphere of B-particles. A part of these particles penetrates the gravitational barrier and forms the Hawking radiation at infinity (see, Frolov and Thorne, 1988). The description for the Hawking radiation of a rotating black hole can be found in Bolashenko and Frolov (1989).

III. THE ALGEBRAIC APPROACH

During the last few years, a particularly suitable mathematical framework along the way of the algebraic approach to quantum field theory has reached a stage where is able to shed new light on the Hawking effect. (see Kay 1988; Kay, Wald, 1987, 1988, and references therein; Dimock, Kay, 1978; also see the talk of Wald at this course). Haag.et.al(1984) discusses an axiomatic framework for interacting feilds in curved spacetime for not-necessarily-linear, while the work done by K.Fredenhagen, et.al(1987) is an interesting attempt to incorporate some results of QFT in CST in a possible axiomatic framework for quantum gravity.

We restrict ourselves here to a linear model field theory mainly according to Wald(1988).

Given a spacetime structure (M.g) and the covariant Klein-Gordon equation

$$(\square + m^2)\Phi = 0 \tag{3.1}$$

on (M.g), we construct a suitable *algebra A generated by objects-smeared quantum feilds

$$\Phi(F) = \int_M \phi(x) F(x) \; |g|^{1/2} d^4x \tag{3.2}$$

Satisfying

$$\phi(F)^* = \phi(F) \tag{3.3}$$

$$[(\square + m^2)\phi](F) = \phi[(\square + m^2)F] = 0 \tag{3.4}$$

$$[\phi(F_1), \Phi(F_2)] = i\Delta(F_1, F_2) I, \tag{3.5}$$

where F is real valued, I is the identity, and $\Delta(F_1, F_2)$ is the smeared (advanced minus retarded) fundamental solution to the K-G equation, which can be expressed by

$$\Delta(F_1, F_2) = \iint_{M\times M} \Delta(x,y) F_1(x) F_2(y) \, |g(x)|^{1/2} |g(y)|^{1/2} d^4x \, d^4y \tag{3.6}$$

thus the algebra A contains products of smeared fields at different points such as $\Phi(x)\Phi(y)$. The covariant Dirac equation or Maxwell equation could be treated similarly, but some aspect of the theory of massless fields in curved spacetime requires special treatment because of infra-red problem.

Then an admissible state ω on A is defined to be positive, normalized and linear functional on A and by specifying all smeared n-point functions $\omega[\phi(F_1), \ldots \Phi(F_n)]$.

Given a algebra A together with a state ω, one can find a representation ρ of A as operators on a Hilbert space H and a cyclic vector Ω in H such that

$$\omega(A) = \langle \Omega | \rho(A) \Omega \rangle, \qquad \text{for all A in A} \tag{3.7}$$

and can define the folium of a state ω to consist of the set of all states

$$\omega_\sigma(A) = \mathrm{tr}\,[\sigma \rho(A)] \tag{3.8}$$

on H.

For a linear field theory, the attention is restrited to the union of the folia of some suitable set of quasi-free states, which can be completely characterized by their smeared anticommutator functions

$$G(F_1, F_2) = \omega[\phi(F_1)\phi(F_2) + \phi(F_2)\phi(F_1)] \tag{3.9}$$

They imply the conditions:

$$G[(\square + m^2)F_1, F_2] = G[F_1, (\square + m^2)F_2] = 0, \tag{3.10}$$

$$G(F_1, F_2) \geq 0, \tag{3.11}$$

$$\Delta(F_1, F_2) \leq G(F_1, F_2)^{1/2} G(F_2, F_1)^{1/2}. \tag{3.12}$$

Further more, by the modification of the equivalence principle and renormolization, one demands " the short distance behaviour of the states should be what it ought to be". Therefore, the set of admissible states one defined to be the union of the folia of the quasi-free globally Hadamard states.

The definition of admissible states is good because there do exists "many" quasi-free globally Hadamard states, for which the expectation value of a suitable renormolized energy momentum tensor can be defined, and moreover, on the Kruskal spacetime there is an "essentially unique" Schwarzschild isometry-invariant quasi-free globally Hadamard state which when restricted to an exterior Schwarzschild wedge and viewed with respect to Schwarzschild time evolution, is a thermal equilibrium state at the Hawking temperature $T=(8\pi M)$. In certain cases, the unique quasi-free global Hadamard state presumably not only exists and also is respectively the "Hartle-Hawking state", the Schwarzschild vacuum and the "Euclidiean de Sitter vacuum".

Suppose kay's two conjecture(on any bounded open region, the globally Hadamard states define a unique folium; Among the class of quasi-free states on A the local and global Hadamard notions are equivalent) were true, the ultra-violet behavior is the same for all the states in question which on any infra-red problems, and we could say that the laws of quantum field theory in curved spacetime were local.

Moreover, Kay and Wald (1988) have showed that existence of states satisfying their theorems does not hold in general, especial, any stationary Hadamard states on the Schwarzschild-de Sitter spacetime and on the Kerr spacetime.

IV. PARTICLE DETECTOR

According to Sciama(1981), the model detector is essentially a single atom, initially in its ground state, and weakly coupled to the quantum field under consideration. Whenever it is excited to a high energy state we may usually say that the atom has detected a particle of appropriate energy. If the detector is unaccelerated, it moves on as timelike geodesic. In the rest inertial frame of the detector, t is the ordinary inertial time coordinate and the frequency is associated with it. More generally, as a freqency variable for a moving inertial detector, we have quantity $u_\mu k^\mu$, where u^μ is the four velocity of the detector and k^μ is the propagation vector of the field being measured. The Doppler shift is included. But this interpretation becomes problematic for accelerated motion. As we will see later that the response of detector undergoing uniform acceleration is the same as it were immersed in a heat bath at a temperature T. It can be proved formally that the Minkowski vacuum has all the properties of the thermal state when considered from the point of view of accelerated observers. Now, we consider the universal kinematic effect of a uniformly accelerated detector. We assume that the time measured by an arbitrarily moving detector along its world line, this is a physical hypothesis. We wish to compare the responses of two detectors, one moves inertially while the other is uniformly accelerated, which means a motion which acceleration measured in the instantanous rest frame of the body is a constant. For the detector which moves inertially, in I-frame, the energy absorption rate of the detector is determined by

$$\Pi_i = \frac{-\omega}{2\pi}\theta(-\omega) \tag{4.1}$$

where θ denote the step function. The detector can not be excited to a higher energy state. It means no particle being detected.

For the uniformly accelerated detector, in A-frame, we use Rindler coordinate(see section 2.2).

We refer to the remote part and future as the in-and out-regions respectively. We suppose that in-and out-regions spacetime admits natural particle states and a such kind of privileged quantum vacuum can be called Minkowski vacuum. Since we work in the Heisenberg picture, if we choose the state of the quantum field in the in-region to be the vacuum state, then it will remain that state. However, at later times in the out-region, free falling particle detectors may still register particles. The in-vacuum may not coincide with the out-vacuum. We can therefore say that particles have been "created" by the time dependent external gravitational field. When gravitational fields are present, inertial observers become free falling observers, and in general no two free-falling detectors will agree on choice of vacuum.

Rindler Detector (Rindler, 1969). Consider that detector moves along a hyperbolic trajectory in the (z,t) plane:

$$x=y=0, \quad z=(t^2+\alpha^{-2})^{1/2} \qquad \alpha\text{—constant.} \tag{4.2}$$

this represents a detector that accelerates uniformly with acceleration α in the frame of the detector. We have the transition propability per unit proper time

$$\frac{c^2}{2\pi}\sum_E \frac{(E-E_0)|\langle E_0|m(0)|E\rangle|}{e^{2\pi(E-E_0)/\alpha}-1} \tag{4.3}$$

The appearance of the Plank factor $[e^{2\pi(E-E_0)/\alpha}-1]^{-1}$ indicates that it can be thought that the detector remains unaccelerated, but immersed in a bath of thermal radiation at the temperature

$$T=\alpha/2\pi K_B=\text{acceleration}/2\pi k_B \tag{4.4}$$

where k_B -Boltzmann's constant.

Physical Meaning. A uniformly accelerated observer will "see" thermal radiation, even so the field ϕ is in the vacuum state $|o\rangle$ and no particles are detected by inertial observers. Both the detector and the field gain energy. The detector transites to the high energy level, while the field causes the emission of quanta. The energy comes from the external field which overcomes the resistance to accelerate the particle.

When the quantum field is not in a vacuum state but a many particles state, i.e

$$|{}^1n_{k_1},{}^2n_{k_2},\ldots\ldots{}^jn_{k_j}\rangle=({}^1n!\ {}^2n!\ \ldots\ldots{}^jn!)^{-\frac{1}{2}}(a^+_{k_1})^{{}^1n}(a^+_{k_2})^{{}^2n}\ldots\ldots(a^+_{k_j})^{{}^jn}|0\rangle \tag{4.5}$$

the G^+ is replaced by

$$G^+(x,x')+\int d^{n-1}k\ n_k\,u_k(x)\,u_k^*(x')+\int d^{n-1}k\ n_k u_k^*(x)\,u_k(x') \tag{4.6}$$

where n_k is the number density of quanta in k-space. The transition probability to all possible E and ϕ is

$$c^2\sum_E |\langle E|m(0)|E_0\rangle|^2\,\mathcal{F}(E-E_0) \tag{4.7}$$

where

$$\mathcal{F}(E)=\int_{-\infty}^{\infty}d\tau\int_{-\infty}^{\infty}d\tau'\ e^{-iE(\tau-\tau')}\,G^+(x(\tau),\ x(\tau')) \tag{4.8}$$

is detector response function, which is independent of the details of the detector and is determined by the positive frequency Wightmam Green function G^+. For an inertial detector

$$x^i=x_0^i+v^it=x_0^i+v^i\tau(1-v^2)^{-1/2} \tag{4.9}$$

if v=0 and if the quanta are distributed isotropically, the absorption of a single quantum of mass m by the detector will not occur unless the energy level spacing $E-E_0$ in the detector is at least equal to the particle rest mass m, and the transition response rate of the detector to the bath of quanta is proportional to the number of quanta in the mode of interest. If v=0,

$$\frac{\mathcal{F}(E)}{\tau}=\frac{1}{4\pi^{1/2}\Gamma(3/2)}(E^2-m^2)^{1/2}\, n_{(E^2-m^2)^{1/2}}\, \theta(E-m), \tag{4.10}$$

where T is the total duration for which the detector is switched on. If $v \neq 0$, in the massless case (also if the quanta are distributed isotropically)

$$\frac{\mathcal{F}(E)}{T}=\frac{1}{4\pi}\left(\frac{1-v^2}{v^2}\right)^{1/2}\int_{E_1^-}^{E^+} n_K dk \tag{4.11}$$

where $E^{\pm}=E\,[(1\pm v)/(1\mp v)]^{\frac{1}{2}}$, the transition with $E-E_0$ will select quanta from a whole range. The factor $[(1+v)/1-v]$ and $[(1-v)/(1+v)]$ are recognized as the usual Dopler blue- and redshift factors respectively.

In general, there is no simple relation between $\langle N_i \rangle$ and the particle number as measured by a detector, even if it is free falling. But, as a special case, in a spacetime that is asymptotically static in the remote past and future, the vacuum state in the in-region will be detected by an inertial detector no quanta. However, in the out-region, the detector will generally register the presence of some quanta even if the hield is in the in-vacuum state. In the simple case of a homogeneous universe (as the asymplotically static, spacially flat Robertson-Walker model), G^+ will be invariant under spatial translation and rotation. The Bogolubov transformation will be diagonal and isotropic

$$\alpha_{KK'}=\alpha_K \delta_{KK'} \tag{4.12}$$

$$\beta_{KK'}=\beta_K \delta_{-KK'} \tag{4.13}$$

$$u_K^{in}(x)=\alpha_K u_K^{out}(x)+\beta_K u_{-K}^{out}(x) \tag{4.14}$$

The detector response function per unit time is

$$\frac{\mathcal{F}(E)}{\tau}=\frac{1}{4\pi^{1/2}\Gamma(3/2)}(E^2-m^2)^{1/2}\,\left|\beta_{(E^2-m^2)^{1/2}}\right|^2\theta(E-m) \qquad (v=0) \tag{4.15}$$

where we have chosen the Robertson-Walker scalar factor to be unity in the out-region. The response function is identical to that associated with an isotropic bath of quanta with $|\beta_K|^2$ particles in modes k in a permanently static spacetime.

V. PARTICLE CREATION BY A COLLAPSING SPHERICAL BODY

For a spherically symmetric ball of matter surrounded by empty space, the unique spherically symmetric vacuum solution in the exterior region is described by the Schwarzschild metric

$$ds^2=-(1-2M/r)dt^2+(1-2M/r)^{-1}dr^2+r^2(d\theta^2+\sin^2\theta\, d\varphi^2). \tag{5.1}$$

Schwarzschild coordinates are singular at r=2M, but the curvature of the manifold is not. It is usual to introduce Kruskal coordinates in which the metric takes the form

$$ds=\frac{32M^3}{r}e^{-r/2M}(-dv^2+du^2) \tag{5.2}$$

in which r is to be understood as a function of u and v given implicitly by

$$\left(\frac{r}{2M}-1\right)e^{r/2M}=u^2-v^2. \tag{5.3}$$

This metric is singular only at the curvature singularities where r= 0, and with the coordinate ranges

$$-\infty < v < \infty,\ -\infty < u < \infty,\quad u^2-v^2>-1, \tag{5.4}$$

represents the maximal analytic extension of the Schwarzschild manifold.

In the remote past, one can construct the standard Minkowski space quantum vacuum state. After collapse, the spacetime will have the Schwarzschild form in the out region, but the modes of any quantum field propagating through the interior of the ball will be seriesly disrupted, so this vacuum will no longer correspond to the Minkowski space vacuum, constructed in the in-region. One must calculate the Bogolubov transformation between the in- and out-vacuum states.

When we only consider the massless scalar field ϕ, the direct interaction between quantum field and the collapse matter is being ignored, and the presence of the matter in the model is used simply to produce an appropriate gravitational field. In this case, we have the wave equation

$$\Box\phi=0. \tag{5.5}$$

The solution of it is

$$\phi\sim r^{-1}R_{\omega\ell}(r)\ Y_{\ell m}(\theta,\varphi)e^{-i\omega t} \tag{5.6}$$

where $Y_{\ell m}$ is a spherical harmonic, and $R_{\omega\ell}$ satisfies

$$\frac{d^2}{dr^{*2}}R_{\omega\ell}+\{\ \omega^2-\ [l(l+1)r^{-2}+2Mr^{-3}][1-\ 2Mr^{-1}]\}R_{\omega\ell}=0 \tag{5.7}$$

where $r^*=r+2M\ln|(r/2M)-1|$.

In the asymptotic regions, $(r\to\infty)$ ϕ reduces to

$$\phi\sim r^{-1}Y_{\ell m}e^{-i\omega u}, \qquad u=t-r^*, \text{ outcoming waves} \tag{5.8}$$

$$\phi\sim r^{-1}Y_{\ell m}e^{-i\omega v}, \qquad v=t+r^*, \text{ incoming waves} \tag{5.9}$$

where u and v are the null coordinates. We decompose ϕ into a complete set of positive freqency modes $f_{\omega l m}$

$$\phi=\sum_{l,m}\int d\omega\ (a_{\omega l m}f_{\omega l m}+a^{\dagger}_{\omega l m}f^{*}_{\omega l m}) \tag{5.10}$$

$f_{\omega l m}$ are normalized by the condition

$$(f_{\omega_1 l_1 m_1},f_{\omega_2 l_2 m_2})=\delta(\omega_1-\omega_2)\delta_{l_1 l_2}\delta_{m_1 m_2} \tag{5.11}$$

The in-vacuum state is defined by

$$a_{\omega l m}|0\rangle=0 \qquad \forall\ \omega,l,m. \tag{5.12}$$

which corresponds to the absence of incoming(advanced) radiation from $\mathcal{J}^-$.

In the asymptotic region, $(r\to\infty)$, in the remote part, there is no incoming radiation from $\mathcal{J}^-$. The incoming waves $r^{-1}Y_{\ell m}e^{-i\omega v}$, approach the surface of the ball and suffer a blueshift, convrege on the center of the ball, and pass on through to become outgoing waves $r^{-1}Y_{\ell m}e^{-i\omega u}$ with a red shift. If the ball is static, these two effects exactly compensate. If the ball is collapsing, the net redshift becomes appreciable.

Compute the form of the redshift modes reaching $\mathcal{J}^+$. For a two dimensional model of collapsing ball,

$$ds^2=c(r)du\,dv \qquad \text{outside the ball} \tag{5.13}$$

$$u=t-r^*+R_0^* \tag{5.14}$$

$$v=t+r^*-R_0^* \tag{5.15}$$

$$r^*=\int c^{-1}dr \qquad c=1-2M/r \qquad r=r+2M\times\ln|r/2M-1|. \tag{5.16}$$

For the case of R_0^* being constant, we have

$$ds^2 = A\ (U,v)\,dU\,dv \qquad \text{inside the ball} \tag{5.17}$$

$$U = \tau - r + R_0 \tag{5.18}$$

$$V = \tau + V - R_0 \tag{5.19}$$

$$R_0^* = R_0 + 2M \times |\ln| r/2M - 1| \tag{5.20}$$

The center of radial coordinates is the line

$$v = U - 2R_0 \qquad (\text{ at } r=0) \tag{5.21}$$

We desire the solution of the two dimensional wave equation

$$\Box \phi = 0 \tag{5.22}$$

that $\phi = 0$ at $r=0$, and reduce to standard exponential on $\mathcal{J}^-$. We obtain the asymptotic modes

$$i(4\pi\omega)^{-1/2}(e^{-i\omega v} - e^{i\omega(ce^{-\kappa v}+d)}) \tag{5.23}$$

where c and d are constants. The calculation shows that an inertial particle detector, which will register no particles at $\mathcal{J}^-$, will register particles at $\mathcal{J}$ in the out region for the vacuum state in the in-region.

We exam the vacuum expectation values of the stress-energy operator. In the Unruh vacuum, we find that the renormalized expectation value of the stress-energy $\langle T_{\mu\nu}\rangle^{ren}$ is regular, in a free falling frame, on the future horizon but not on the past horizon. At infinity this vacuum corresponds to an out going flux of black-body radiation at the black-hole temperature. It is the Unruh vacuum that best approximates the vacuum relevant to the gravitational collapse of a massive body(see Section 2.6).

VI. HAWKING RADIATION

In January 1974, Hawking announced:

Black Hole formed by collapse are not completely black, but emit radiation with a thermal spectrum due to quantum effects.

The gravitational disturbance produced by collapsing star induces the creation of an outgoing flux of radiation. This implies that the " in vacuum " state contains a thermal flux of outgoing particles.

6.1 Hawking Radiation and Spacetime

Soon after Hawking's discovery, it became clear that the thermal effects in their "pure and idealized form" have to do with spacetimes which admit a one-parameter group of isometries possessing a bifurcate Killing horizon which consists of a pair of interacting null hypersurfaces which are orthogonal to the Killing field, such as schwarzschild spacetime of an "eternal black hole". In the analysis of quantum mechanics on the black hole spacetime, there have emerged certain "preferred vacuum states" of the quantum field, such as the Hartle-Hawking states, which are invariant under the full isometry group of the spacetime, and are nonsingular on the full Killing horizon, but some known vacuum states, such as Boulware(1975) vacuum states and Unruh vacuum states (Unruh 1976, Hawking 1976) are singular on past or all of the Killing horizon. The striking feature of Hartle-Hawking vacuum of Schwarzschild spacetime is their thermal properties with respect to the isometries which generate the Killing horizon. Kay and Wald (1988) (also see the talk of Wald in this Proceedings) showed that this thermal properties continue to hold in general spacetime which possess an appropriate reflection isometry and in which the surface gravity κ is constant over the Killing horizon. However the constancy of the κ automatically holds if the spacetime satisfied Einstein's equations with matter obeying the dominant energy condition.

6.2 Schwarzschild Black Hole

Now we consider two simplified models:

1. Two dimensional analogue of the gravitational collapse, the metric is chosen to correspond to a spherically symmetric ball of matter imploding across its event horizon in an arbitrary way;
2. Four-dimensional case but ignore the effects of backscattering.

The expected spectrum is Planckian, corresponding to a thermal spectrum from a black body of temperature

$$T=\frac{K}{2\pi K_B} \tag{6.1}$$

The four dimensional calculation(Hawking, 1975) is essentially the same as for the two dimensional model described here. A flux of the particles from the vicinity of the hole with a thermal spectrum corresponding to the temperature given by $T=K/2\pi K_B$. The number of particles per unit time in the frequency range ω to $\omega+d\omega$ passing out through the surface of the sphere is

$$(d\omega/2\pi)\,(e^{8\pi M\omega}-1)^{-1}=(d\omega/2\pi)\,(e^{\omega/TK_B}-1) \tag{6.2}$$

where

$$T=\frac{K}{2\pi K_B},\quad K=\frac{1}{4M},\ \text{(surface gravity)} \tag{6.3}$$

This appears Planck spectrum.

If we consider backscattering, the spectrum is not precisely Planckian, but can be regarded as "thermal" in the following sense: the B.H is in thermal equilibrium with the surrounding heat bath. This is essentially true for photons neutrinors and linearized graviton fields.

The total luminosity of the S.B.H has been estimated by Page

$$L=(3.4\times10^{46})\,(M/1\text{gm})^{-2}\ \text{erg.s}^{-1},\ (m>>10^{17}\text{gm}) \tag{6.4}$$

and

$$T=(1.2\times10^{26}\text{k})\,(1\text{gm}/M), \tag{6.5}$$

$$T_{\odot}\sim6\times10^{-8}\,\text{k} \tag{6.6}$$

(In this case, only massless quantum emission is relevant.

For $M\leq10^{17}$ gm, $T\geq10^{9}$ k, the creation of thermal electron-positron pairs become possible. For $M\sim10^{15}$ gm, $R\sim10^{-13}$ cm(fermi) within the range of the strong interaction.

6.3. Kerr Black Hole

(i) At large r, ω is replaced by $(\omega-m\Omega)$ in the corresponding formula of Schwarzschild hole, where m is the azimuthal quantum number of the spheroidal harmonics, Ω is the angular speed of the event horizon.

(ii) The Planck factor at J^+ becomes

$$\{\exp[2\pi k^{-1}(\omega-m\Omega)]\pm1\}^{-1}\qquad +\text{Fermion},\ -\text{Boson} \tag{6.7}$$

(iii) The rotation of the hole greatly enhances the emission of higher-spin particles.

(iv) The emission causes its rotation rate to slow. The emission is stronger for positive m than negative.

(v) Super-radiance.—The hole induces stimulated emission: In the Boson case , when $\omega<m\Omega$,

$$\{\exp[2\pi k^{-1}(\omega-m\Omega)]-1\}^{-1}<0,\ \text{(negative)}. \tag{6.8}$$

And even $M\to\infty(T\to0)$, this factor remains finite. This means that the effect of radiation amplify the incoming, classical wave(with positive m).

6.4. Reissner-Nordstrom Hole

When an electrically neutral field propagating in the background of a R-N hole with charge e,

$$ds^2=[1-(2M/r)+(e^2/r^2)]dt^2-[1-(2M/r)+(e^2/r^2)]^{-1}dr^2-r^2(d\theta^2+\sin^2\theta d\varphi^2). \quad (6.9)$$

(i) The even horizon is at

$$r=r_+=M+(M^2-e^2)^{1/2}. \quad (6.10)$$

(ii) The surface gravity

$$K=(1-16\pi^2e^4/A^2)/4M \quad (6.11)$$

where $A=4\pi r_+^2$ is the area of event horizon.

(iii) And the temperature of the charged black hole is

$$T=(1-16\pi^2e^4/A^2)/8\pi K_B M. \quad (6.12)$$

It means that the presence of the charge depresses the temperature of the hole.

(iv) The third law of thermaldynamics applied to black holes ---cosmic censorship hypothesis(naked singularities can not form from gravitational collapse.) That is, T=0 could not in principl be achieved.

(v) The spontaneous creation of charged particle pairs is possible in the background electric field even in the absence of a gravitational background field.

(vi) There will be a "charge super-radiance phenomenon".

(vii) Electron -positron can dominate for $M\lesssim 10^5 M_\odot$ and $M \gtrsim 10^{15}$ gm.

6.5. Physical Aspects of Black Hole Emission

From the inspection of

$$(d\omega/2\pi)\,[\exp(8\pi M\omega)-1]^{-1}, \quad (6.13)$$

it can be shown that the average wavelength of emitted quanta is $\sim M$, comparable with the size of the hole. The particle concept is only useful near J^+. In the vicinity of black hole, the concept of locally defined particle breaks down.

Hawking radiation from the continuous spontaneous creation of virtual particles and antiparticles around the black hole is independent of the details of the collapse.

For the static metric

$$ds^2=c(r)\,du\,dv \quad (6.14)$$

no particle are created, but there will be nonzero vacuum 'polarization' stress due to spacetime curvature :

$$\langle 0|T_{uu}|0\rangle=\langle 0|T_{vv}|0\rangle=(2cc''-c'^2)/192\pi \quad (6.15)$$

$$\langle 0|T_{uv}|0\rangle=cc''/96\pi \quad (6.16)$$

where a prime denotes differentiation with respect to r.

An observer who crosses the event horizon along a constant Kruskal position line measures a finite energy density.

The event horizon is a global construct and has no local significance. The notion of energy does have a local significance.

Hawking flux and the static vacuum polarization diverge as the horizon is approached.

In the Hawking radiation, there is no energy flux crosses the horizon, because the hole absorbs negative energy, and its area decreases, so does the mass. But the temperature and the luminosity rise, s.b.h has a negative specific heat.

The effect of initial quanta fades out exponentially, so the Hawking effect is independent of any physically reasonable initial quantum state.

Black Hole Evaporation. When

$$\frac{1}{M}\frac{dM}{dt} \sim M^{-1} \sim K_B T \qquad (6.17)$$

where $M \sim 10^{-5}$ gm is Planck mass and $R_0 \sim 10^{-33}$ cm is Planck length, higher order quantum gravity effects will be important. The end result of Hawking evaporation will be: explosion, naked-singularity, or a Planck mass object. A study of b.h. evaporation could provide a good opportunity for us to probe the physics of ultra-high energy particles. The lifetime of a hole is $10^{-26}(M/1gm)^3$ sec. The law of baryon number conservation is transcended.

6.6. Look Deeper

The black hole entropy has a statistical mechanical origin as the logarithm of the number of "internal states" of the hole which could correspond to its externally observed states.

But, what is relation between black hole temperature and any real, physical, thermal effects? How to understand spontaneous thermal emission from black hole?

Unruh(1976) showed that an accelerated particle detector in flat empty spacetime should behave as it were bathed in a perfect bath of thermal radiation with temperature $T=\hbar a/2\pi K_B$, where a is the detector's acceleration.

A static observer just above a Schwarzschild horizon can be viewed, in the Rindler approximation, as completely analogous to an accelerated observer in flat spacetime with acceleration $a=K_H/\alpha$, where K_H is the surface gravity on the horizon, and α is the red-shift factor.

The locally measured temperature

$$T=\hbar a/2\pi K_B=\hbar(K_H/\alpha)/2\pi K_B=T_H/\alpha. \qquad (6.18)$$

where T_H is the temperature on the horizon.

An accelerated observer in flat spacetime sees a thermal bath, but freely falling observers see pure vacuum. A static observer just above a Schwarzschild horizon sees a thermal atmosphere, but freely falling observers see no such atmosphere at all.

Then, Unruh and Wald(1982) showed that when an accelerated observer absorbs a quantum from the surrounding thermal bath, a freely falling observer sees him emit a quantum. Both observers agree that absorption/emission has increased the energy in the radiation field.

If we look at things from the back-reaction(Sciama, Candeles and Deutsch, 1981), the situation would be as follows. An evaporation of a Schwarzschild black hole would correspond an inward flux of negative energy. The hole's classical spacetime curvature $G_{\mu\nu}$ is produced by the renormalized expectation value $\langle T_{\mu\nu}\rangle$ of the stress-energy tensor of the quantized fields that are evaporating.

$$G_{\mu\nu}=\langle T_{\mu\nu}\rangle^{ren}. \qquad (6.19)$$

The static observer above the horizon of a s.b.h. sees a negative renormalized energy density

$$\langle \varepsilon\rangle=\langle T^{00}\rangle=-\langle T^{0r}\rangle=-L/(4\pi r_H^2\alpha^2)<0 \qquad (6.20)$$

where L is the evaporative luminosity of the hole.

How to understand that a b.h. atmosphere looks thermal but has a negative renormalized energy density?

Very near the horizon, vacuum polarization(Zurek and Thorne 1985) gives a contribution to $\langle T_{\mu\nu}\rangle$ which is precise the negative of that of thermal bath with local temperature $T=T_H/\alpha$. If the atmosphere measured by a static observer above the horizon were thermal, its contribution to $\langle T_{\mu\nu}\rangle$ would be precisely canceled by that of vacuum polarization, giving zero. But, evaporation slightly depletes it from thermality after renormalization, the near-horizon energy density is slightly negative.

VII. BLACK HOLE THERMODYNAMICS

7.1 The Four Laws of Black Hole Thermodynamics

The Zeroth Law. The surface gravity K_H is a constant over horizon of a stationary black hole.

The First Law.

$$dM=\frac{1}{8\pi}KdA+\Omega_H dJ+\Phi dq$$

where $(\Omega_H dJ+\Phi dq)$ is the work term for charged rotating hole.

The Second Law. We recall its history.
1. Hawking's area theorem(1971). The event horizon area never decreases

$$dA\geq 0 \tag{7.1}$$

for which the weak energy condition is satisfied. This strongly suggests the identification of A with entropy S.

2. The connection between entropy and information. Assuming one bit of information per subatomic particle, then

$$S\sim MK_B/m \tag{7.2}$$

is the total information loss. The low bound of entropy, as $m\rightarrow 0$, can be found by considering that the Compton wavelength of the constituent particles should be $\lesssim M$, the black hole radius, (Hawking,1972). So the maximum entropy (Bekenstein,1972)

$$S_{max}\sim M^2K_B/\hbar =M^2K_B\sim K_BA\,. \tag{7.3}$$

$$S=\frac{1}{4}K_BA.\ \text{(by Hawking, DeWitt, 1975)}. \tag{7.4}$$

3. Black hole evaporation is in violation of the weak energy condition, thus it violates the area theorem. But if we take account of the entropy change in the environment of the hole, The second law of thermodynamics for a system containing slowly evolving black holes is just a special case of the standard second law of thermodynamics. In such a system the total entropy, including that of the holes and that of matter and fields outside the holes' stretched horizons, can never decrease.(see, Thorne, 1986).

The Third Law. It is impossible to achieve K=0 by a physical process.

7.2 General Covariant Thermodynamics and the Approach to Black Hole Entropy

The collapsing processes of self-gravitating system are the non-equilibrium, thermodynamical, irreversible processes, where any kind of dissipation might accompany, particularly, when the systems collapse near to their own Schwarzschild radius for spherically symmetric systems.

Recently, the covariant, causal and transient relativistic thermodynamics, applicable to strong gravitational and rotating fields, have been proposed by Israel (1976, 1984), Hiscock and Lindblom (1985), and Carter (1988).

The first law of the general covariant thermodynamics can be written as

$$\nabla_\mu S^\mu = A^\nu \nabla_\mu B^\mu{}_\nu, \tag{7.5}$$

where

$$A = \left(\frac{1}{T}, \frac{P}{T}, -\frac{\mu}{T}, 1\right) \tag{7.6}$$

$$B^\mu{}_\nu = (\rho^\mu, u^\mu, n^\mu, Q^\mu), \qquad Q^\mu = q^\mu/T. \tag{7.7}$$

$$q^\mu = -KP^{\mu\nu}(\nabla_\nu T + Ta_\nu) \tag{7.8}$$

$$S^\mu = su^\mu + Q^\mu \tag{7.9}$$

where T-temperature, s-entropy density, μ-chemical potential, ρ-energy density and P-pressure are measured in local co-moving frame.

The second law of the general covariant thermodynamics can be written as

$$\nabla_\mu S^\mu > 0, \quad \text{(for irreversible processes)}, \tag{7.10}$$

$$\nabla_\mu S^\mu = 0, \quad \text{(for local thermodynamic equilibrium)}. \tag{7.11}$$

The entropy production can be expressed as

$$T\nabla_\mu S^\mu = \zeta\theta^2 + 2\eta\sigma^{\mu\nu}\sigma_{\mu\nu} - \mu\nabla_\mu u^\mu + \frac{1}{KT}q^\mu q_\mu. \tag{7.12}$$

The stress-energy tensor can be expressed by

$$T^{\mu\nu} = \rho u^\mu u^\nu + (P - \zeta\theta)P^{\mu\nu} - 2\eta\sigma^{\mu\nu} + q^\mu u^\nu + q^\nu u^\mu, \tag{7.13}$$

which satisfies the local conservation law

$$\nabla_\mu T^{\mu\nu} = 0. \tag{7.14}$$

Based on these, the studies to support that $\frac{1}{4}K_B A$ is the thermodynamic entropy of a black hole have been carried out by Sorkin, Wald and Zhang (1981), Zhang (1983) and Zhang et al. (1988).

Time-Symmetry. The globally hyperbolic spacetime can be foliated by Cauchy surfaces Σ_t. On Σ_t, we have

$$P_{\mu\nu} = g_{\mu\nu} + u_\mu u_\nu. \tag{7.15}$$

The external curvature of Σ_t can be expressed by

$$K_{\mu\nu} = \frac{1}{2}\pounds_v P_{\mu\nu} = \sigma_{\mu\nu} + \frac{1}{3}\theta h_{\mu\nu}. \tag{7.16}$$

For the cases of expansion-free $\theta=0$ and shear-free $\sigma_{\mu\nu}=0$, we have

$$K_{\mu\nu} = 0, \tag{7.17}$$

which is known as time-symmetry. The dynamical metric of a spherically symmetric system can be found from Einstein's constrain equations for the case of the time-symmetry.

Total Entropy Production. In the region U of the spacetime between two hypersurfaces Σ_1 and Σ_2, we can defined the total entropy production by

$$\Delta S = \int_U (\nabla_\mu s^\mu)\tilde{\omega} = \oint_{\partial U} s^\mu n_\mu \tilde{\alpha} \tag{7.18}$$

where $\tilde{\omega}$ -differential 4-form, $\tilde{\alpha}$ -differential 3-form, ∂U-boundary of U, n_μ-

1-form orthogonal to ∂U, which satisfies

$$n_\mu \xi^\mu = 0 \tag{7.19}$$

where ξ^μ-any vector tangent to ∂U. For the asymptotically flat spacetime, equation (7.18) becomes

$$\Delta S = \int_{\Sigma_2} s^\mu n_\mu \tilde{\alpha} - \int_{\Sigma_1} s^\mu n_\mu \tilde{\alpha} . \tag{7.20}$$

We can define a quasi-stationary state if in the neighbor ΔU of Σ we have

$$\nabla_\mu s^\mu = 0 . \tag{7.21}$$

Then, the total entropy of the quasi-stationary state of the system is

$$S = \int_\Sigma s^\mu n_\mu d\Sigma \tag{7.22}$$

which is independent of Σ in ΔU. The total entropy production between the interval of time t_1 and t_2 is

$$\Delta S = S_1 - S_2 . \qquad (t_1 > t_2) \tag{7.23}$$

It has also been shown (by Zhang et al, 1988) that a consistency condition relating Planck length should enter the consideration.

7.3 Membrane Paradigm. (Thorne, Price and Macdonald, 1986)

1. Basic Idea. We can think of each layer of black hole atmosphere as acquiring, when it sinks through the stretched horizon, the contemporary values of the stretched horizon's mass, angular momentum, angular velocity, surface gravity, surface temperature, and entropy. Each layer then remains those values forever thereafter as it sinks, at the local speed of light, toward the true horizon. Correspondingly, at a universal time we can regard the above quantities as functions of height in the hole's atmosphere. At fixed universal time, the mass and angular momentum of the hole can be regared as sums over contributions from each very thin layer of atmosphere.

2. Statistical Origin of Entropy. We can regard the entropy S_H of a black hole as Boltzmann's constant K_B times the logarithm of the total number N_H of quantum mechanically distinct ways that the black hole could have been made.

$$S_H = K_B \ln N_H . \tag{7.24}$$

York has proposed Euclidiean Einstein action approach to study the black hole thermodynamics (1985, 1986, 1987), while Hu and Kandrup (1987) has studied entropy generation in interaction by using a statistical subdynamics analysis.

That whether S_H is a measure of the number of internal states of a black hole and what is the underlying basis of the laws of black hole thermodynamics are not fully understood. Nevertheless, the existence of the laws indicates the likelihood of a deep connection between gravitation, quantum theory and thermodynamics.

ACKNOWLEDGEMENTS

One of us (Z.Z) is grateful for the hospitality of "Ettore Majorana" Centre For Scientific Culture, Bologna University, International Centre For Theoretical Physics and SISSA during the course of this work and to Professors V. P. Frolov, B. L.Hu, D. W. Sci W. Unruh, and R. Wald for helpful discussions.

REFERENCES

Allen B., Folacci A., and Ottewill A. C. 1988. Phys. Rev. D38,1069.
Ashtekar A. 1987. Phys. Rev. D36,1587.
Audretsch J. 1989. in this Proceedings.
Bekenstein J. D. 1975. Phys.Rev. D12,3077.
Birrell N. D., and Davies P. C. W. 1982. Quantum Fields in Curved Space. Cambridge University Press.
Bolashenko P. A., and Frolov V. P. 1989.in Quantum Theory and Gravitation. Proc. of the Lebedev Institute,v.197,(ed. M. A. Markov,Nova Science Publ. Commack)
Boulware D. G. 1975. Phys. Rev. D11,1404.
Carter B. 1988. preprint.
Dimock J., and Kay B. S. 1987. Ann. Phys. (NY),175,366.
Dowker J. S. 1986. Phys. Rev. D33,3150.
Elster T. 1983. Phys. Lett. A94,205.
Frolov V. P. 1989. in this Proceedings.
Frolov V. P., and Thorne K. S. 1988. Phys. Rev. D (to appear).
Hredenfagen K., and Haag R. 1987. Commun. Math. Phys. 108,91.
Gross D.J., Harvey J. A., Martinec M. A., and Rohm R. 1985. a. Nucl. Phys. B256,253; b. Phys. Rev. Lett. 54, 502.
Haag R.,Narnnhofer H., and Stein U. 1984. Commun. Math. Phys. 94,219.
Halpern L. 1988. in Differential Geometrical Methods in Theoretical Physics, eds. K. Bleuler and M. Werner, Kluwer Academic Publishers.
Hartle J. B. 1987. Prediction in Quantum Cosmology Gravitation in Astrophysics, Cargese. 1986. ed: J. B. Hartle and B. Carter (New York, Plenum).
Hawking S. W. 1975. Commun. Math. Phys. 43,199.
1976. Phys. Rev. D14,2460.
Hiscock W. A.,and Lindblom L. 1985. Phys. Rev. D31,725.
Israel W. 1976. Ann. Phys. 100,310
1984. Ann. Phys. 152,30.
Jacobson T., and Lee Smolin. 1988. Class. Quantum Grav. 5, 583.
Kay B. S., and Wald R. M. 1988. Theorem on the uniqueness and thermal properties of stationary, nonsingular, quasi-free states on space-time with a bifurcate Killing horizon, (preprint).
Padmanabhan T. 1989. Class. Quantum Grav. 6,533.
Page D. N. 1982. Phys. Rev. D25,1499.
Rindler W. 1969. Essential Relativity (New York: Van Nostrond).
Ross D. K. 1987. Class. Quantum Grav. 4,995.
Sciama D.W., Candelas P., and Deutsch D. 1981. Adv. Phys. 30, 327.
Sorkin R., Wald R. M.,and Zhang Z. 1981. Gen. Rel. Grav. 12, 1127.
Thorne K. S. ,Zurek W. H., and Price R. H. 1986. in "Black Holes: The Membrane Paradigm".eds: Thorne K. S. Price R. H. and Macdonald P. A. Yale University Press. (New Haven and London)
Unruh W. G.1976. Phys. Rev. D14,870.
Unruh W. G., and Wald R. 1982.Phys. Rev. D25, 942.
Wald M. R. 1977. Commun. Math. Phys.54,1.
1978. Phys. Rev. D17,1477.
1984. General Relativity, University of Chicago Press, Chicago.
1989. in this proceedings.
Zhang Z. 1983. in Proceedings of MG3,North-Holland,1289.
Zhang Z., Wang X., and Shi J. 1988. in Proceedings of ICGC87, Cambridge University Press,(in press).
Zurek W. H., and Thorne K. S. 1985. Phys. Rev. Lett. 54,2171.

ABSORPTION CROSS SECTION OF A MINI BLACK HOLE

Li Zhi FANG

Beijing Astronomical Observatory
Chinese Academy of Sciences
Beijing, People's Republic of China

Absorption of particles by a stellar formed black hole is mainly a classical problem, because the de Broglie wavelength λ of the particle is generally much smaller than the gravitational radius r_s of such a black hole. However, quantum effects should be considered for the absorption process of decoupled massive particles by the primordial black hole in the early universe, because the wavelength of such a non-relativistic particle can sometimes be larger than r_s of the mini black hole.

Theoretically, the main purpose in calculating the absorption of a black hole is to find a non-Hermitian Hamiltonian or a complex effective potential from the real space-time metric. This has been discussed in many articles(Matzner 1968, Sanches 1976, 1977, Fang and Wang 1984). In the case of the absorption of massless particles, all those theories are rather sophisticated. We will show that, in the case of non-relativistic particle absorption, it can be described more simply. This is mainly because the non-relativistic particle satisfies the Schrodinger equation, so that the effective potential can easily be derived. Spallucci(1977) also calculated the quantum mechanical absorption of a small black hole. However, he assumed that the gravitational field in the inner part of a black hole can be replaced by a potential well with a given radius and depth. This does not seem an appropriate treatment for the problem of a black hole.

We consider a Schwarzschild black hole with mass M and radius $r_s=2GM/c^2$. In Finkelstein coordinates, the metric is

$$ds^2=-(1-\frac{r_s}{r})c^2dt^2+\frac{2r_sc}{r}\,drdt+(1+\frac{r_s}{r})dr^2+r^2(d\theta^2+\sin^2\theta\, d\varphi^2) \qquad (1)$$

namely, the metric tensor is given by

$$g^{\mu\nu}=\begin{pmatrix} -1/c^2(1+r_s/r) & r_s/cr & 0 & 0 \\ r_s/cr & (1-r_s/r) & 0 & 0 \\ 0 & 0 & r^{-2} & 0 \\ 0 & 0 & 0 & r^{-2}\sin^{-2}\theta \end{pmatrix} \qquad (2)$$

and

$$g^{-1}=\det g^{\mu\nu}=-c^2r^4\sin^2\theta \qquad (3)$$

In curved space, the Klein-Gordon equation can be written as

Quantum Mechanics in Curved Space-Time
Edited by J. Audretsch and V. de Sabbata
Plenum Press, New York, 1990

$$\hbar^2 g^{\mu\nu}\Psi_{;\mu\nu} = m^2c^2\Psi \tag{4}$$

where m denotes the mass of the particle. Considering eq.(1), we have

$$\begin{aligned} g^{\mu\nu}\Psi_{;\mu\nu} &= \frac{1}{\sqrt{-g}}\frac{\partial}{\partial x^\mu}\sqrt{-g}\, g^{\mu\nu}\frac{\partial}{\partial x^\nu}\Psi \\ &= \frac{1}{r^2\sin\theta}\,\frac{\partial}{\partial x^\mu}\, r^2\sin\theta\, g^{\mu\nu}\frac{\partial}{\partial x^\nu}\Psi \end{aligned} \tag{5}$$

Therefore, eq.(4) now becomes

$$\hbar^2\Big[-\big(1+\frac{r_s}{r}\big)\frac{1}{c^2}\frac{\partial^2}{\partial t^2} + \frac{2r_s}{cr}\frac{\partial}{\partial t}\frac{\partial}{\partial r} + \frac{r_s}{cr^2}\frac{\partial}{\partial t} + \frac{1}{r^2}\frac{\partial}{\partial r}r^2\big(1-\frac{r_s}{r}\big)\frac{\partial}{\partial r} + \frac{1}{r^2\sin\theta}\frac{\partial}{\partial\theta}\sin\theta\frac{\partial}{\partial\theta} + \frac{1}{r^2\sin^2\theta}\frac{\partial^2}{\partial\varphi^2}\Big]\Psi = m^2c^2\Psi \tag{6}$$

If we consider a stationary solution, Ψ can be expressed as

$$\Psi = \exp(-i\epsilon t/\hbar)\psi \tag{7}$$

In the non-relativistic approximation, i.e.

$$\epsilon \simeq mc^2 + E, \quad E \ll mc^2 \tag{8}$$

one finds the equation for stationary wavefunction ψ as follows

$$-\frac{\hbar^2}{2m}\Big[\frac{1}{r^2}\frac{\partial}{\partial r}r^2\frac{\partial}{\partial r} + \frac{1}{r^2\sin\theta}\frac{\partial}{\partial\theta}\sin\theta\frac{\partial}{\partial\theta} + \frac{1}{r^2\sin^2\theta}\frac{\partial}{\partial\varphi^2}\Big]\psi + V\psi = E\psi \tag{9}$$

where the effective 'potential' is given by

$$V = V_N + V_1 + iV_2 \tag{10}$$

and

$$V_N = -\frac{GMm}{r} \tag{11}$$

$$V_1 = r_s\frac{\hbar^2}{2m}\Big(\frac{1}{r^2}\frac{\partial}{\partial r}r\frac{\partial}{\partial r}\Big) \tag{12}$$

$$V_2 = r_s ch\Big(\frac{1}{r}\frac{\partial}{\partial r} + \frac{1}{2r^2}\Big) \tag{13}$$

where V_N is the Newtonian potential, V_1 and V_2 are due to the correction of the black hole. In this case a complex effective 'potential' is obtained naturally.

One can solve the problem of absorption in the 'potential' (13) by using standard methods of quantum mechanics. Firstly, we should find the expression of the absorption cross section. From eqs.(9)-(13), we have the following relation

$$\frac{d}{dt}\int \rho\, dr + \oint \vec{J}\cdot d\vec{s} = -\int W dr \tag{14}$$

where the integral of the second term on the left-hand side runs over the whole surface of the black hole, and

$$\rho = \psi\psi^* \tag{15}$$

$$\vec{J} = (i\hbar/2m)(\psi\nabla\psi^* - \psi^*\nabla\psi) \tag{16}$$

$$W = -(1/\hbar)[\,\psi(V_1 + V_2)\psi^* + \psi^*(V_2 - V_1)\psi\,] \tag{17}$$

From eq.(12), it is easily to show that

$$\psi V_1\psi^* - \psi^* V_1\psi = 0 \tag{18}$$

Therefore, W depends only on the imaginary potential V_2

$$W = -(1/\hbar)(\,\psi V_2\psi^* + \psi^* V_2\psi\,) \tag{19}$$

We consider the exact solution under the Newtonian potential as the zero-order wavefunction and V_1+iV_2 as a perturbation term of the Hamiltonian. Namely, the unperturbated wavefunctions ψ_o are the solutions of the following equation

$$\left[-\frac{h^2}{2m}\frac{1}{r^2}\left(\frac{\partial}{\partial r} r^2 \frac{\partial}{\partial r} + \frac{1}{\sin\theta}\frac{\partial}{\partial\theta}\sin\theta\frac{\partial}{\partial\theta} + \frac{1}{\sin^2\theta}\frac{\partial}{\partial\varphi^2} \right) - \frac{GMm}{r} \right]\psi_o = E\psi_o \tag{20}$$

When ψ is taken to be any stationary solutions of eq.(20), the term $\vec{J}$ in eq.(14) equals to zero. Therefore, in the first-order approximation of perturbation theory, the absorption is given only by the term of W in eq.(14). The absorption cross section can then be expressed as follows

$$\sigma = \frac{1}{v}\frac{1}{\hbar}\left(\int \psi_o V_2 \psi_o^* d\tau + \int \psi_o^* V_2 \psi_o d\tau \right) \tag{21}$$

where $v = kh/m$ is the incoming velocity of the particle and the integral in eq.(21) runs over the whole space out of the black hole.

The solution of unperturbated equation (20) in the form of a partial wave expression is the same as in the Coulomb field, which is

$$\psi_o = \sum_{l=0}^{\infty} C_l r^l e^{ikr} F(l+1-i, 2(l+1); -2ikr) P_l(\cos\theta) \tag{22}$$

where

$$C_l = (2ikr)^l e^{\alpha\pi/2}\,\Gamma(l+1-i\alpha)/2l! \tag{23}$$

$$k^2 = 2mE/\hbar^2, \qquad \alpha = Gm^2M/\hbar^2 k \tag{24}$$

The absorption cross section can be found by substituting eq.(22) into eq.(21). It is

$$\sigma = -\frac{4\pi}{\hbar v}\sum_{l=0}^{\infty}\frac{1}{2l+1}\int_{r_s}^{\infty}(R_l V_2 R_l^* + R_l^* V_2 R_l) r^2 dr \tag{25}$$

where

$$R_l = C_l r^l e^{ikr} F(l+1-i\alpha, 2(l+1); -2ikr) \tag{26}$$

After the integration in eq.(25), we find finally the expression of absorption cross section of a black hole as

$$\sigma = \sum_{l=0}^{\infty} \sigma_l \tag{27}$$

and

$$\sigma_l = -C_l C_l^* \frac{4\pi}{v}\frac{r_s c}{(2l+1)} r^{2l+1} F_l F_l^* \Big|_{r=r_s}^{\infty} \tag{28}$$

The confluent hypergeometric function F(a, b; z) has the following asymptotic expression when $|z| \to \infty$

$$F(a,b;z) \sim \frac{\Gamma(b)}{\Gamma(b-a)} (-z)^{-a} \sum_{n=0}^{\infty} \frac{\Gamma(n+a)}{\Gamma(a)} \frac{\Gamma(n+a-b+1)}{\Gamma(a-b+1)} \frac{(-z)^{-n}}{n!}$$

$$+ \frac{\Gamma(b)}{\Gamma(a)} e^{z} z^{a-b} \sum_{n=0} \frac{\Gamma(n+1-a)}{\Gamma(1-a)} \frac{\Gamma(n+b-a)}{\Gamma(b-a)} \frac{z^{-n}}{n!} \tag{29}$$

Therefore, when $r \to \infty$, we have

$$|F(l+1-i\alpha, 2(l+1); -2ikr)| < Ar^{-(l+1)} \tag{30}$$

or

$$|r^{2l+1} F_l F_l^*| < r^{-1} A^2 \tag{31}$$

One finds then

$$r^{2l+1} F_l F_l^* |_{r \to \infty} = 0 \tag{32}$$

From eqs.(28) and (32), it can be easily be found that in the case of mini black hole, or $kr_s \ll 1$, the cross section σ_l can be rewritten as

$$\sigma_l = c\pi r_s^2 \, \frac{2^{2(2l+1)} e^{\alpha\pi} \Gamma(l+1-i\alpha)\Gamma(l+1+i\alpha)}{2l!(2l+1)!} (kr_s)^{2l} \tag{33}$$

It is obvious from eq.(33) that the dominat term in eq.(27) is given by the $l=0$ wave. Thus we have

$$\sigma \simeq \sigma_0 = \frac{c}{v} \frac{8\pi^2 \alpha}{1-\exp(-2\pi\alpha)} r_s^2 \tag{34}$$

When $\alpha \ll 1$, eq.(34) becomes

$$\sigma \sim \frac{c}{v} 4\pi r_s^2 \tag{35}$$

The (1/v) dependence of the cross section indicates that the final result of a particle bound by a black hole is absorbed by the black hole.

As an application of the absorption cross section, we can discuss the evolution of a mini black hole in a early universe. In principle, a black hole will disappear due to Hawking's radiation. It shows that in the case of that the time scale of the disappearance is less than the age of the universe, the radiation mechanism will play an important role in the evolution of mini black holes. This is, such a mini black hole formed in the early universe will not be able to survive until present time. A mini black hole can, however, survive if its absorption rate is larger than the Hawking's radiation. Hawking's radiation is given by

$$-\left(\frac{dM}{dt}\right)_R \sim \frac{\hbar c^4}{G^2} \frac{1}{M^2} \tag{36}$$

If in the early universe there is a type of non-relativistic particle with mass m and number density n, then the absorption rate is

$$\left(\frac{dM}{dt}\right)_A \sim \sigma mnv = cnmr_s^2 \tag{37}$$

Therefore, from the condition of

$$-\left(\frac{dM}{dt}\right)_R = \left(\frac{dM}{dt}\right)_A , \tag{38}$$

we can find a lower limit of the mass of black hole, which can survive in the universe. It is

$$M > \left(\frac{m_{pl}}{n m l_{pl}^3}\right)^{1/4} m_{pl} \tag{39}$$

where $m_{pl} = (\hbar c/G)^{1/2}$ and $l_{pl} = (G\hbar/c^3)^{1/2}$ are, respectively, the Planck mass and the Planck length.

References

Fang L.Z. and Wang R.C., 1984, Class Quantum Grav. 1, 403

Matzner R.A., 1968, J. Math. Phys., 9, 163

Sanches N., 1976, J. Math. Phys., 17, 688

Sanches N., 1977, Phys. Rev., D16, 937

Spallucci E., 1977, Lett. Nuovo Cim., 20, 391

PARTICLE CREATION AND VACUUM POLARIZATION NEAR BLACK HOLES

V.P. Frolov

P.N. Lebedev Physical Institute

Moscow, USSR

INTRODUCTION

The effect of the particles creation by black holes discovered by Hawking (1974, 1975) is a result of the vacuum instability in a strong gravitational field. The action of such a field on vacuum virtual particles provides a part of them with enough energy to become real. Some of the created real particles are swallowed by the black hole while the other escape to infinity and form the Hawking radiation. This effect has been studied in detail.

The particle creation does not exhaust all possible quantum effects. The state of those vacuum virtual particles which do not become real is also affected by the action of the external field. As a result their contribution to local observables (e.g. to the mean value of the stress-energy tensor $\langle T_{\mu\nu} \rangle$) and hence the value of these observables changes and becomes depending on parameters of the external field. This effect is known as the vacuum polarization by an external field. The investigation of local observables and in particular of $\langle T_{\mu\nu} \rangle$ in a strong gravitational field of a black hole is an important problem. These quantities are needed for studying the back reaction of quantum fields on the black-hole's metric and for construction of a self-consistent model of a quantum evaporating black hole. One may hope that this model can help us to answer the intriguing questions concerning the final state of an evaporating

Quantum Mechanics in Curved Space-Time
Edited by J. Audretsch and V. de Sabbata
Plenum Press, New York, 1990

black hole and the spacetime structure inside a black hole expecially in the regions where according to the classical theory there must exist singularities.

The aim of these lectures is twofold: (i) to give a tutorial introduction into the theory of quantum effects in the gravitational field of a black hole, and (ii) to provide a brief summary of the main results concerning the vacuum polarization by black holes which have been obtained till now.

The lectures are organized as follows. We start by considering the simplest possible case: the quantum theory of a scalar massless field in a flat spacetime (Section I). We briefly discuss the field quantization in an inertial frame of reference and the calculations of the local observables $\langle \hat{\varphi}^2 \rangle$ and $\langle T_{\mu\nu} \rangle$ which describe the quantum field fluctuations and its stress-energy tensor correspondingly. In this Section a simple geometrical model of a heated black body is also described and used for studying its thermal radiation.

The quantum effects in a uniformly accelerated reference frame in a flat spacetime and in a homogeneous gravitational field are discussed in Section II. The equivalence principle is used to relate these phenomena. The problem of renormalization of local observables $\langle \hat{\varphi}^2 \rangle$ and $\langle T_{\mu\nu} \rangle$ in a curved spacetime and conformal trace anomalies are briefly discussed in Section III. Section IY contains the summary of results concerning the massless fields contribution to the vacuum polarization near black holes. Possible approximations for $\langle T_{\mu\nu} \rangle^{ren}$ in the spacetime of a black hole are described and compared in Section Y.

We use the sign conventions of Misner, Thorne, and Wheeler (I973) and Planck's units $\hbar = c = G = 1$.

I. QUANTUM FIELD THEORY IN FLAT SPACETIME

I.I. Quantum field theory in inertial frame. Local observables

Before considering the theory of quantum effects in the gravitational field of black holes (which is the main topic of the lectures) we begin by discussing the theory of a scalar massless field in a flat spacetime. This theory is expounded in many textbooks and there is no need to reproduce its details. We just use it as a simple example in order to illustrate the problems which arise in a curved spacetime in all their completeness. It is not necessary now to specify the number of the spacetime dimensions. Many formulas can be obtained directly in n -dimensional case without any additional complications. That is why we prefer to work up to some moment in n -dimensional spacetime which has $D = n - 1$ spatial dimensions. In order to specify the results to a real four-dimensional spacetime it is enough to put $D = 3$ into the final formulas.

We shall use an inertial reference frame and standard Cartesian coordinates X^μ in it so that the spacetime line element is

$$ds^2 = dX_\mu dX^\mu = -(dX^0)^2 + \sum_{i=1}^{D} (dX^i)^2 . \qquad (I.I)$$

The massless scalar field $\varphi(X)$ in these coordinates satisfies the equation

$$\Box\varphi \equiv -\partial_0^2\varphi + \sum_{i=1}^{D} \partial_i^2\varphi = 0 , \qquad (I.2)$$

where $\partial_\mu \equiv \partial/\partial X^\mu$. It is convenient to write the general solution of this equation using the standard plane-wave expansion

$$\varphi(X) = \int d^D\kappa [\varphi_{\vec{\kappa}}(X) a_{\vec{\kappa}} + \bar{\varphi}_{\vec{\kappa}}(X) a^*_{\vec{\kappa}}] \qquad (I.3)$$

where

$$\varphi_{\vec{\kappa}}(X) = \frac{e^{-i\omega_\kappa X^0}}{[(2\pi)^D 2\omega_{\vec{\kappa}}]^{1/2}} e^{i\vec{\kappa}\cdot\vec{X}} \qquad (I.4)$$

Here $\vec{K}=(K^1,\dots,K^D)$ and $\vec{X}=(X^1,\dots,X^D)$ are D -vectors, $\vec{K}\cdot\vec{X}=\sum_{i=1}^{D} K^i X^i$ and $\omega_{\vec{K}}=|\vec{K}|=\sqrt{\vec{K}^2}$. (This expansion is also valid for a massive (with mass m) field if only the frequency $\omega_{\vec{K}}$ is changed by $\omega_{\vec{K}}=\sqrt{\vec{K}^2+m^2}$).

For a real classical field φ the coefficients $a_{\vec{K}}$ and $a^*_{\vec{K}}$ are complex conjugated numbers while for a hermitian quantum field $\hat{\varphi}$ they are hermitian conjugated operators. These operators $\hat{a}_{\vec{K}}$ and $\hat{a}^*_{\vec{K}}$ are known as operators of annihilation and creation of a scalar quantum in the mode $\vec{K}$. They satisfy the following commutation relations

$$[\hat{a}^*_{\vec{K}}, \hat{a}_{\vec{K}'}] = \delta(\vec{K}-\vec{K}'), \tag{I.5}$$

other possible commutators vanish. These relations are a consequence of the canonical commutation relations

$$[\hat{\varphi}(X^0,\vec{X}), \hat{\varphi}(X^0,\vec{X}')] = [\dot{\hat{\varphi}}(X^0,\vec{X}), \dot{\hat{\varphi}}(X^0,\vec{X}')] = 0 \tag{I.6}$$

$$[\dot{\hat{\varphi}}(X^0,\vec{X}), \hat{\varphi}(X^0,\vec{X}')] = i\,\delta(\vec{X}-\vec{X}')$$

and of the normalization choice of the solutions $\varphi_{\vec{K}}$.

For calculation of local observables such as the expectation values $\langle\hat{\varphi}^2\rangle$ and $\langle T_{\mu\nu}\rangle$ in some quantum state it is convenient to introduce the following function

$$G(X,X') = \frac{1}{2}\langle\hat{\varphi}(X)\hat{\varphi}(X') + \hat{\varphi}(X')\hat{\varphi}(X)\rangle . \tag{I.7}$$

which is known as the Hadamard function. Its value depends on the particular choice of a state over which averaging is taken. For the vacuum state $|0;M\rangle$, defined by the equation $\hat{a}_{\vec{K}}|0;M\rangle=0$ the Hadamard function reads

$$G_0(X,X') = \frac{\Gamma(\frac{D-1}{2})}{4\pi^{(D+1)/2}[(X-X')^2]^{(D-1)/2}} . \tag{I.8}$$

For an arbitrary choice of the state this function cannot be calculated exactly. We restrict ourselves by considering the states for which $<\hat{a}^*_{\vec{k}}\ \hat{a}^*_{\vec{k}'}> = <\hat{a}_{\vec{k}}\ \hat{a}_{\vec{k}'}> = 0$. Then using the expansion (I.3) and the commutation relations (I.5) one may write the following representation for this quantity

$$G(X,X') = \int d^D k\, d^D k'\, \varphi^2(X,X'|\vec{k},\vec{k}')[<\hat{a}^*_{\vec{k}}\ \hat{a}_{\vec{k}'}> + \frac{1}{2}\delta(\vec{k}-\vec{k}')]$$

(I.9)

where

$$\varphi^2(X,X'|\vec{k},\vec{k}') = \varphi_{\vec{k}}(X)\bar{\varphi}_{\vec{k}'}(X') + \bar{\varphi}_{\vec{k}}(X)\varphi_{\vec{k}'}(X').$$

Now we make an assumption that the system is homogeneous in space. In this case

$$<\hat{a}^*_{\vec{k}}\ \hat{a}_{\vec{k}'}> = n_{\vec{k}}\,\delta(\vec{k}-\vec{k}') \qquad \text{(I.10)}$$

and the Hadamard function reads

$$G(X,X') = \int d^D k\, \varphi^2(X,X'|\vec{k})(n_{\vec{k}} + \frac{1}{2}) \qquad \text{(I.11)}$$

where $\varphi^2(X,X'|\vec{k}) \equiv \varphi^2(X,X'|\vec{k},\vec{k})$. The quantity $n_{\vec{k}}$ describes the average density of particles with the momentum $\vec{k}$ in a chosen quantum state. For $n_{\vec{k}} = 0$ the function $G(X,X')$ coincides with $G_0(X,X')$.

The Hadamard function (I.11) is finite untill the points X and X' either coincide or are separated by null interval. But any attempt to take a coincidence limit for its arguments results in divergence. That is why the direct calculation of $<\hat{\varphi}^2(X)>$ gives a meaningless infinite result. The reason of this infinity is the integration over infinitely large $\vec{k}$ in (I.11) due to the presence of 1/2-term in the square bracket. The substraction of $G_0(X,X')$ from $G(X,X')$

$$G^{ren}(X,X') = G(X,X') - G_0(X,X') = \int d^D\kappa\, \varphi^2(X,X'|\vec{\kappa})\ \hat{n}_{\vec{\kappa}} \tag{I.12}$$

eliminates this divergence. The quantity

$$\langle\hat{\varphi}^2(X)\rangle^{ren} = G^{ren}(X,X) = \int d^D\kappa\, \varphi^2(X|\vec{\kappa})\ \hat{n}_{\vec{\kappa}} \tag{I.13}$$

is finite if only the average number of particles $\hat{n}_{\vec{\kappa}}$ decreases rapidly enough with growing of $|\vec{\kappa}|$. It happens for physically reasonable states and this provides the necessary cut-off of the integral. (In Eq.(I.11) and later on we use the following notation $\varphi^2(X|\vec{\kappa}) \equiv \varphi^2(X,X|\vec{\kappa}) = 2|\varphi_{\vec{\kappa}}(X)|^2$.)

The finite quantity $\langle\hat{\varphi}^2(X)\rangle^{ren}$ is the difference between the mean value of $\hat{\varphi}^2(X)$ for a given state and the vacuum mean value of $\hat{\varphi}^2(X)$. Namely this difference can be measured and hence is of interest to physisists. The procedure of the substraction of the unobservable null fluctuations contribution is known as a renormalization. Using the explicit expression (I.4) for plane waves one can rewrite (I.13) in a more explicit form

$$\langle\hat{\varphi}^2(X)\rangle^{ren} = \int \frac{d^D\kappa}{(2\pi)^D\,\omega_{\vec{\kappa}}}\ \hat{n}_{\vec{\kappa}} \ . \tag{I.14}$$

The quantity $\langle\hat{\varphi}^2(X)\rangle^{ren}$ may be considered as a measure of fluctuations of the quantum field $\hat{\varphi}$. In the same manner one may calculate the contribution of the field $\hat{\varphi}$ into the average of the stress-energy tensor. The stress-energy tensor $T_{\mu\nu}(X)$ for a classical field φ can be written as

$$T_{\mu\nu}(X) = \lim_{X'\to X} D_{\mu\nu}(X,X')\,\varphi(X)\,\varphi(X') , \tag{I.15}$$

where

$$D_{\mu\nu}(X,X') = \frac{1}{2}(1-2\xi)(\nabla_\mu \nabla_{\nu'} + \nabla_{\mu'}\nabla_\nu) + (2\xi - \frac{1}{2}) g_{\mu\nu} \cdot$$
$$\cdot g^{\alpha\beta'} \nabla_\alpha \nabla_{\beta'} - \xi(\nabla_\mu \nabla_\nu + \nabla_{\mu'}\nabla_{\nu'}) + \tag{I.16}$$
$$+ \frac{\xi}{D+1} g_{\mu\nu}(\nabla^\alpha \nabla_\alpha + \nabla^{\alpha'}\nabla_{\alpha'}) .$$

The formulas (I.15)-(I.16) require some explanations. The operators ∇_μ and $\nabla_{\nu'}$ in the Cartesian coordinates read $\partial/\partial X^\mu$ and $\partial/\partial X^{\nu'}$. An arbitrary parameter ξ reflects a possible ambiguity in the choice of the stress-energy tensor. The stress-energy tensor $T_{\mu\nu}$ for $\xi = 0$ is known as a canonical one. However a more preferable choice is

$$\xi = \frac{(n-2)(n-1)}{4} \equiv \frac{D(D-1)}{4} . \tag{I.17}$$

For this choice $T_{\mu\nu}(X)$ is traceless

$$T^\mu_\mu = 0 , \tag{I.18}$$

and the theory possesses additional (conformal) invariance. The expression (I.15) means that the operations of differentiation are performed at first and only after the coincidence limit is taken. One can verify that this procedure gives the standard expression for $T_{\mu\nu}(X)$.

One may try to define the average value $< T_{\mu\nu}(X) >$ of the stress-energy tensor for a quantum field φ by the relation

$$<T_{\mu\nu}(X)> = \lim_{X' \to X} D_{\mu\nu}(X,X') G(X,X') . \tag{I.19}$$

But this coincidence limit does not exist for the same reason as it happened in the case of $<\hat{\varphi}^2(X)>$. In order to obtain the finite physically observable quantity one needs to substract the contribution of unobservable null fluctuations, i.e. to substitute $G^{ren}(X,X')$ into Eq. (I.17) instead of $G(X,X')$. This procedure gives the renormalized stress-energy tensor $<T_{\mu\nu}(X)>^{ren}$ which can be written as follows

$$<T_{\mu\nu}(X)>^{ren} = \int d^D\kappa\, t_{\mu\nu}(X|\vec{\kappa})\ \hat{n}_{\vec{\kappa}} \quad . \tag{I.20}$$

The quantity

$$t_{\mu\nu}(X|\vec{\kappa}) \equiv \lim_{X'\to X} D_{\mu\nu'}(X,X')\varphi^2(X,X'|\vec{\kappa}) \tag{I.21}$$

may be considered as a contribution of the single quantum in a mode $\vec{\kappa}$ into the stress-energy tensor. For plane-waves (I.4)

$$t_{\mu\nu}(X|\vec{\kappa}) = \frac{\kappa_\mu \kappa_\nu}{(2\pi)^D \omega_{\vec{\kappa}}} \tag{I.22}$$

It is easy to see that $t_\mu^{\ \mu}(X|\vec{\kappa}) = 0$ and hence in the flat spacetime one has

$$<T_\mu^{\ \mu}(X)>^{ren} = 0 \tag{I.23}$$

I.2 Equilibrium thermal radiation

Now we discuss some properties of $<T_{\mu\nu}(X)>^{ren}$ and in particular its dependence on the state of a quantum field. The calculations of this quantity are greatly simplified for a special choice of a state for which the average number of particles $n_{\vec{\kappa}}$ depends only on their energy $\omega_{\vec{\kappa}}$

$$n_{\vec{\kappa}} = n(\omega_{\vec{\kappa}}) \tag{I.24}$$

For this case the stress-energy tensor $< T_{\mu\nu}(X) >^{ren}$ is of the form

$$< T_{\mu\nu}(X) >^{ren} = \mathrm{diag}\left(\varepsilon, \frac{\varepsilon}{D}, \frac{\varepsilon}{D}, \ldots, \frac{\varepsilon}{D} \right), \tag{I.25}$$

where $\mathrm{diag}(a_0, a_1, \ldots, a_D)$ denotes a diagonal tensor with the diagonal elements $a_0, a_1, \ldots, a_D$. The energy density ε is

$$\varepsilon = \frac{\Omega_{D-1}}{(2\pi)^D} \int_0^\infty dk\, k^D\, n(k) . \tag{I.26}$$

Here

$$\Omega_{D-1} = \frac{2\pi^{D/2}}{\Gamma(D/2)} \tag{I.27}$$

is the surface of a unit $(D-1)$-dimensional sphere S^{D-1} and $k \equiv |\vec{k}|$. Eqs. (I.25)-(I.26) immediately follow from (I.20), (I.22) and (I.24) if only one takes into account that

$$\int d^D k\, k_i k_j\, f(k) = D^{-1} \delta_{ij} \int d^D k\, k^2 f(k) = \\ = D^{-1} \Omega_{D-1} \int_0^\infty dk\, k^{D+1} f(k) . \tag{I.28}$$

The calculation of $< \hat{\varphi}^2(X) >^{ren}$ is analogous and gives

$$< \hat{\varphi}^2 >^{ren} = \frac{\Omega_{D-1}}{(2\pi)^D} \int_0^\infty dk\, k^{D-2}\, n(k) . \tag{I.29}$$

The relation (I.24) is valid for the most important for applications case of equilibrium thermal radiation. The Bose distribution $n_\theta(\omega)$ for the temperature θ reads

$$n_\theta(\omega) = [\exp(\omega/\theta) - 1]^{-1} . \qquad (I.30)$$

and we have

$$\langle \hat{\varphi}^2 \rangle_\theta^{ren} = \frac{1}{2^{D-1}\pi^{D/2}} \frac{\Gamma(D-1)\zeta(D-1)}{\Gamma(D/2)} \theta^{D-1} , \qquad (I.31)$$

$$\varepsilon \equiv \langle T_{00} \rangle_\theta^{ren} = \frac{1}{2^{D-1}\pi^{D/2}} \frac{\Gamma(D+1)\zeta(D+1)}{\Gamma(D/2)} \theta^{D+1} . \qquad (I.32)$$

Here $\zeta(z)$ is the Riemann zeta-function. We do not reproduce here these (as well as many other) calculations which can be found in the textbooks or in the cited references. Some useful formulas which are required for such calculations are collected in Appendix. In particular Eqs.(I.31) and (I.32) are obtained by using the formula (A.I) of the Appendix.

Now we make rather simple but useful remark concerning the structure of $\langle T_{\mu\nu} \rangle^{ren}$ for thermal equilibrium radiation in the flat spacetime. A lot of information about this quantity can be obtained before calculations. The problem under consideration possesses a rather large group of symmetries. In particular the state is invariant under the transitions and reflections of the space and time coordinates as well as space rotations. These properties imply that the traceless tensor $\langle T_{\mu\nu} \rangle^{ren}$ must be of the form (I.25), the energy density ε being independent of space coordinates and time. In the units we use ($\hbar = c = G = k_B = 1$) the stress-energy tensor possesses a dimension of $(\text{length})^{-(D+1)}$. The temperature θ is the only dimensional quantity in our problem and its dimension is: $(\text{length})^{-1}$. Hence one has $\varepsilon \sim \theta^{D+1}$. The above consideration allows one to show that the massless fields contribution to the thermal average of the stress-energy tensor can be written as (I.25) where

$$\varepsilon = h(s)\,\sigma_s\,\theta^{D+1} . \qquad (I.33)$$

Here $h(s)$ is the number of helicities of the field of spin s and σ_s is the only unknown dimensionless constant which depends on the statistics of the fields and must be calculated. The calculations give

$$\sigma_{Bose} = \frac{1}{2^{D-1}\pi^{D/2}} \frac{\Gamma(D+1)\zeta(D+1)}{\Gamma(D/2)},$$

$$\sigma_{Fermi} = \frac{1}{2^{D-1}\pi^{D/2}} \frac{\Gamma(D+1)\zeta(D+1)}{\Gamma(D/2)}(1-2^{-D}). \tag{I.34}$$

I.3 Thermal radiation of a heated body

If there is a heated body it becomes a source of a thermal radiation. The distribution of emitted quanta over the energies is described by Eq.(I.30). But the system as a whole is far from equilibrium and the energy density distribution of the radiation in space is highly inhomogeneous. In order to be able to compare later a black-hole radiation with a radiation of a heated black body we consider now the latter in more details.

At first we describe a simple geometrical model of an ideal radiating black body. For this purpose we consider the following possible realization of a source of the black body radiation proposed by Kirchhoff (1860). Let us imagine a spherical isolated shell with a small hole in it (see Fig.1). Almost any light ray which enters through this hole into the cavity inside the shell cannot go out. After a number of reflections it is finally absorbed by the inner surface of the shell. It means that such a hole plays the role of a completely absorbing ("black") body. On the other hand if the inner surface of the shell is heated and there is a thermal radiation inside the cavity it can go out through the hole into the external space. It means that if only the shell is large enough the small hole in it may be considered as an ideal source of the black body radiation.

Instead of this physical model of the black body we consider now its pure geometrical version. Let us consider the geometry of space which is schematically shown in Fig.2.

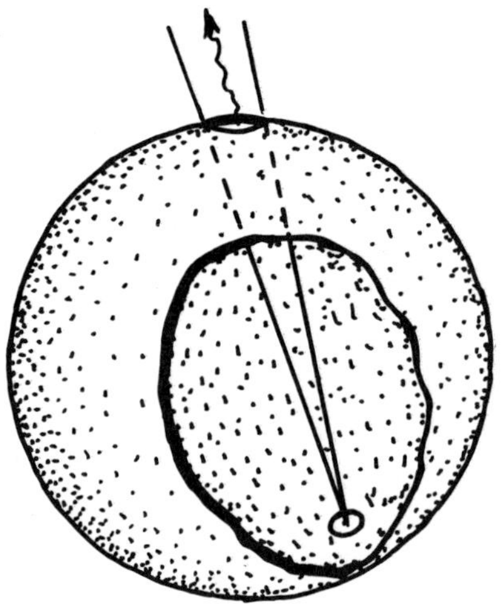

Fig.I. Kirchhoff model of a black body

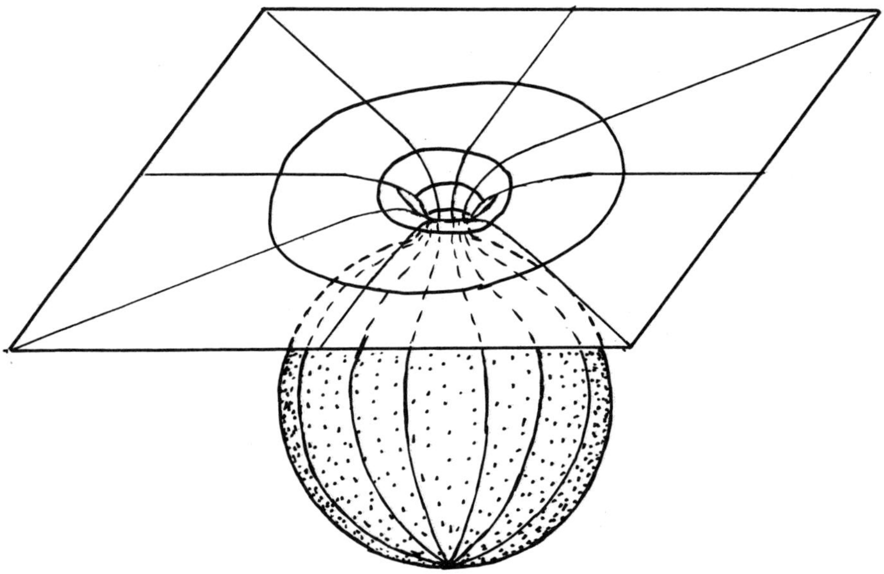

Fig.2. Space geometry for line element (I.35)

The metric for such a spacetime can be written as follows

$$ds^2 = -dT^2 + dq^2 + r^2(q)\, d\omega^2_{D-1} \quad . \qquad (I.35)$$

As earlier the number of spacetime dimensions is taken to be equal to $D+1$ and $d\omega^2_{D-1}$ is the line element on a unit sphere S^{D-1}. We suppose that the smooth function $r(q)$ possesses the properties

$$r(q) = \begin{cases} a \sin q/a \;, & 0 \leq q \leq q_0 = a(\pi - \delta) \;; \\ q - c \;, & q \geq q_1 > q_0 \;. \end{cases} \qquad (I.36)$$

If the parameter δ is small enough then the "internal space" is an almost closed D -dimensional sphere. This "internal space"is connected with the external flat space through a narrow spherical "mouth". The radius of the "mouth" sphere coincides with the minimum of the function r. We denote this minimum radius b.

If the size of the "internal space" is large enough and it contains some matter then a light ray which passes through the "mouth" into the "internal space" is captured and finally is absorbed by matter. It means that from the point of view of an external observer such a system is a "black body" of a radius b. On the other hand if there is thermal radiation in the "internal space" then part of it will be going out through the "mouth" and our "black body" becomes thermally radiating.

It should be stressed that the metric (I.35) does not describe any kind of black holes. In particular there is no red-shift at all in this space. The metric (I.35) is not a solution of the Einstein equations and in this sense it does not describe any physical spacetime.

The above described model may be greatly simplified if we assume that the size of the "internal space" tends to infinity. In this limit the spacetime manifold R_b can be described as follows. Let us take a flat spacetime R and introduce spherical coordinates in which the line element (I.I) reads

$$ds^2 = -dT^2 + dr^2 + r^2 d\omega^2_{D-1} . \tag{I.37}$$

Denote by R_+ the spacetime which is obtained from R by cutting out a ball of a radius $r = b$. Let us take another copy of the flat spacetime R with the line element

$$ds^2 = -dT^2 + dr'^2 + r'^2 d\omega^2_{D-1} \tag{I.38}$$

and obtain R_- by cutting out a ball of a radius $r' = b$ from it. We denote by R_b the result of gluing of R_+ and R_- together along their internal sphere boundaries (see Fig.3).

Now we consider the quantum field $\hat{\varphi}$ in the space R_b. It satisfies the wave equation

$$\Box \hat{\varphi} = 0 \tag{I.39}$$

and the boundary conditions

$$\hat{\varphi}|_{r=b} = \hat{\varphi}|_{r'=b}, \quad \partial_r \hat{\varphi}|_{r=b} = -\partial_{r'} \hat{\varphi}|_{r'=b} \tag{I.40}$$

at the "mouth" surface S_b. Due to these boundary condi-

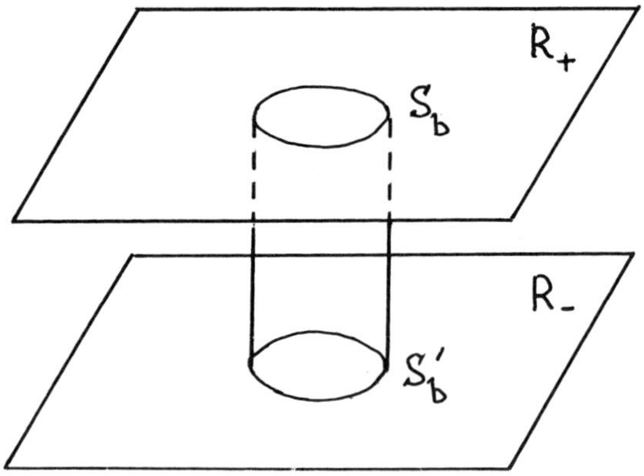

Fig.3. Geometrical model of a "black body"

tions it is natural to use the spherical modes expansion for the quantization instead of the plane waves expansion (I.3)-(I.4). The equation (I.39) in the spherical coordinates (I.37) reads

$$[-\partial_T^2 + r^{1-D}\partial_r(r^{D-1}\partial_r) + r^{-2}\Delta_{D-1}]\hat{\varphi} = 0 \qquad (I.41)$$

where Δ_{D-1} is the Laplace operator on the unit sphere S^{D-1}. We shall not use the explicit form of Δ_{D-1}. We need only to know that its eigenfunctions $Y_{N\beta}(\vec{n})$ ($N = 1,2,\ldots$)

$$\Delta_{D-1} Y_{N\beta}(\vec{n}) = -(N-1)(N+D-3)Y_{N\beta}(\vec{n}) \qquad (I.42)$$

form a complete set of functions on S^{D-1}. A unit vector $\vec{n}$ is used to indicate a point on the unit sphere S^{D-1}. The index β distinguishes the finitely many, linear independent, solutions of Eq.(I.42). The spherical harmonics $Y_{N\beta}(\vec{n})$ satisfy the equations

$$\int_{S^{D-1}} d\omega_{D-1}\, Y_{N\beta}(\vec{n})\, Y_{N'\beta'}(\vec{n}) = \delta_{NN'}\,\delta_{\beta\beta'} \qquad (I.43)$$

$$\sum_{N,\beta} Y_{N\beta}(\vec{n})\,\overline{Y}_{N\beta}(\vec{n}') = \delta(\vec{n},\vec{n}'), \qquad (I.44)$$

where $\delta(\vec{n},\vec{n}')$ is an invariant δ-function on S^{D-1}.

Now we describe a set of basic solutions which is used for the field quantization in R_b. Denote by J a collective index $\omega N\beta$. One can verify that the functions φ_{J+}

$$\varphi_{J+}(p) = \frac{1}{2\sqrt{2}}\, e^{-i\omega T}\, Y_{N\beta}(\vec{n}) \begin{cases} r^{1-D/2}[H_\nu^{(2)}(\omega r) + R_J H_\nu^{(1)}(\omega r)], & p \in R_+; \\ (r')^{1-D/2}\, T_J H_\nu^{(1)}(\omega r'), & p \in R_-. \end{cases} \qquad (I.45)$$

are the solutions of Eq.(I.39) obeying the boundary conditions (I.40) provided $\nu = N - 2 + D/2$ and

$$R_J = -\frac{1}{2}\left[\frac{H_\nu^{(2)}(z)}{H_\nu^{(1)}(z)} + \frac{H_\nu^{(2)\prime}(z)}{H_\nu^{(1)\prime}(z)}\right], \qquad z = b\omega; \qquad (I.46)$$

$$T_J = \frac{1}{2}\left[\frac{H_\nu^{(2)}(z)}{H_\nu^{(1)}(z)} - \frac{H_\nu^{(2)'}(z)}{H_\nu^{(1)'}(z)}\right] \equiv \frac{4i}{\pi z[(H_\nu^{(1)}(z))^2]'}$$

Here $H_\nu^{(i)}(z)$ is a Hankel function. We denote by φ_{J-} the functions which are obtained from φ_{J+} by a simple change $r \leftrightarrow r'$. The interpretation of the solutions $\varphi_{J\varepsilon}$ $(\varepsilon = \pm)$ is rather simple. The mode φ_{J+} describes an incoming spherical wave in R_+ with quantum numbers $\omega N \beta$ which after partial absorption by the "black body" is outgoing to infinity. The coefficients T_J and R_J are the transmission and reflection coefficients correspondingly and they obey the relation

$$|R_J|^2 + |T_J|^2 = 1 . \tag{I.47}$$

One can write the following decomposition of the field operator

$$\hat{\varphi}(p) = \sum_{J,\varepsilon} [\varphi_{J\varepsilon}(p)\,\hat{a}_{J\varepsilon} + \bar{\varphi}_{J\varepsilon}(p)\,\hat{a}^*_{J\varepsilon}] . \tag{I.48}$$

For the chosen normalization of the mode functions $\varphi_{J\varepsilon}$ the canonical commutation relations (I.6) imply the following equations

$$[\hat{a}_{J\varepsilon}\, \hat{a}^*_{J'\varepsilon'}] = \delta_{JJ'}\, \delta_{\varepsilon\varepsilon'} , \tag{I.49}$$

other possible commutators vanish. Here and later we use the following evident abbreviations

$$\sum_{J,\varepsilon} \equiv \sum_{\varepsilon} \int_0^\infty d\omega \sum_{N,\beta} , \qquad \delta_{JJ'} = \delta(\omega-\omega')\,\delta_{NN'}\,\delta_{\beta\beta'} . \tag{I.50}$$

Before considering the thermal radiation of the "black body" in the framework of our model let us discuss briefly what happens when the size of the body is made negligibly small. In this limit the probability $|T_J|^2$ of the propagation of the wave through the "mouth" becomes also negligible. In the limit each of the spaces $R_\pm$ coincides with the complete Minkowsky space and one can use the following flat-space modes

$$\varphi^{(0)}_{J}(p) = \frac{1}{\sqrt{2}} e^{-i\omega T} Y_{N\beta}(\vec{n}) r^{1-D/2} J_{\nu}(\omega r) , \qquad (I.51)$$

where $J_{\nu}(z)$ is a Bessel function. The expansion of the field operator $\hat{\varphi}$ in the flat spacetime $R = R_{+}$

$$\hat{\varphi}(p) = \sum_{J} (\varphi^{(0)}_{J}(p) \hat{a}^{(0)}_{J} + \varphi^{(0)}_{J}(p) \hat{a}^{*(0)}_{J}) \qquad (I.52)$$

can be used for the calculations of the quantum effects in R in the same manner as it was done by using the plane-waves decomposition (I.3) .

Now let us consider what happens when the "black body" is present. If we are interested in the characteristics of the local observables at points lying beyond the "black body" (in the region R_{+}) we must renormalize them by substracting the usual flat space contribution of the null fluctuations. We shall deal with the states for which

$$\langle \hat{a}^{*}_{J\varepsilon} \hat{a}_{J'\varepsilon'} \rangle = n_{J\varepsilon} \delta_{JJ'} \delta_{\varepsilon\varepsilon'} , \qquad (I.53)$$
$$\langle \hat{a}_{J\varepsilon} \hat{a}_{J'\varepsilon'} \rangle = \langle \hat{a}^{*}_{J\varepsilon} \hat{a}^{*}_{J'\varepsilon'} \rangle = 0 .$$

For such states the renormalized expectation values $\langle \hat{\varphi}^{2}(X) \rangle^{ren}$ and $\langle T_{\mu\nu}(X) \rangle^{ren}$ in R_{+} are

$$\langle \hat{\varphi}^{2}(X) \rangle^{ren} = \sum_{J,\varepsilon} \varphi^{2}(X|J\varepsilon) n_{J\varepsilon} + \Delta\varphi^{2}(X) , \qquad (I.54)$$

$$\langle T_{\mu\nu}(X) \rangle^{ren} = \sum_{J,\varepsilon} t_{\mu\nu}(X|J\varepsilon) n_{J\varepsilon} + \Delta T_{\mu\nu}(X) , \qquad (I.55)$$

where $\varphi^{2}(X|J\varepsilon) = 2|\varphi_{J\varepsilon}(X)|^{2}$, $t_{\mu\nu}(X|J\varepsilon)$ is a contribution of a quantum in the mode $J\varepsilon$ into the stress-energy density and

$$\Delta\varphi^{2}(X) = \frac{1}{2} \sum_{J} [\varphi^{2}(X|J+) + \varphi^{2}(X|J-) - \varphi^{2}(X|J0)] , \qquad (I.56)$$

$$\Delta T_{\mu\nu}(X) = \frac{1}{2} \sum_{J} [t_{\mu\nu}(X|J+) + t_{\mu\nu}(X|J-) - t_{\mu\nu}(X|J0)] . \qquad (I.57)$$

The last terms in the square brackets in Eqs.(I.56) and (I.57) are the quantities calculated for the flat spacetime modes (I.51).

If there is no incoming particles falling down into the "black body" ($n_{\mathcal{J}+} = 0$) and the "black body" is cold ($n_{\mathcal{J}-} = 0$) the quantities $<\hat{\varphi}^2(X)>^{ren}$ and $<T_{\mu\nu}(X)>^{ren}$ coincide with $\Delta\varphi^2(X)$ and $\Delta T_{\mu\nu}(X)$ given by Eqs.(I.56) and (I.57), correspondingly. These terms are known as the vacuum polarization contribution. In a more general case the average values of the local observables in a given quantum state are changed under the action of an external field or under the change of the boundary conditions. This effect is known as the vacuum polarization.

If the "black body" is heated up to some temperature θ then $n_{\mathcal{J}-}$ describes the thermal distribtuion and reads

$$n_{\mathcal{J}-} = n_\theta(\omega_{\mathcal{J}}) \equiv [\exp(\omega_{\mathcal{J}}/\theta) - 1]^{-1} . \tag{I.58}$$

The calculation of the local observables is troublesome especially for points lying close to the "black body" where the behaviour of the mode functions is basically defined by the boundary conditions of the field at its surface. The vacuum polarization contribution decreases rapidly enough at far distances ($r \to \infty$). In this limit one can also use the asymptotic behaviour of the Hankel function (A.3)-(A.4) and get

$$<\hat{\varphi}^2(r)>^{ren} \simeq \frac{1}{r^{D-1}} \int_0^\infty \frac{d\omega}{2\pi\omega} \sum_{N,\beta} |Y_{N\beta}(\vec{n})|^2 |T_{\omega N}|^2 \frac{1}{e^{\omega/\theta} - 1}, \tag{I.59}$$

$$Q(r) \equiv <T_{Tr}(r)>^{ren} \simeq$$

$$\simeq \frac{1}{r^{D-1}} \int_0^\infty \frac{\omega d\omega}{2\pi} \sum_{N,\beta} |Y_{N\beta}(\vec{n})|^2 |T_{\omega N}|^2 \frac{1}{e^{\omega/\theta} - 1} . \tag{I.60}$$

The quantity $Q(r)$ describes the density of the energy flux from the heated body. For small values of the parameter ωb ($\omega b \ll 1$) one has $|T_{\omega N}|^2 \ll 1$. On the other hand for $\omega b > N \gg 1$ $|T_{\omega N}|^2 \simeq 1$. These properties and Eq.(A.5) allow one to show that

$$<\hat{\varphi}^2(r)>^{ren} \sim \frac{b^{D-1}}{r^{D-1}}\,\theta^{D-1}, \qquad (I.61)$$

$$Q(r) \sim \frac{b^{D-1}}{r^{D-1}}\,\theta^{D+1} \qquad (I.62)$$

This approximation is good enough when the wave-length of the thermal radiation ($\sim \theta^{-1}$) is much smaller than the size of a heated body b. In the physical four-dimensional spacetime Eq.(I.62) reproduces the expression for the energy radiation by a spherical heated (with the temperature θ) body.

I.4 Summary and additional remarks

In conclusion of this section we shall make some remarks concerning the general structure of the quantum field theory in a flat spacetime and its possible generalization to curved metrics. Our first step was the choice of some inertial frame (or I -frame as we call it for abbrevation). We use time coordinate T in this I -frame to distinguish between positive and negative frequences in solutions of the field equations. The operators of creation and annihilation of the particles in the Minkowski space (M -particles) are defined by means of the decomposition of the quantum field operator over positive and negative frequency modes. The action of a creation operator on a given quantum state increases its energy, the quantity which is conjugated to T and is conserved because of the spacetime symmetry under time T translations. The quantization scheme and in particular the vacuum state $|0;M>$ definition remains the same if one chooses another inertial frame or another (not necessary plane wave) mode decomposition provided the frequencies are separated in positive and negative parts with respect to time T in a chosen I -frame. In order to guarantee the standard form of the commutation relations for the creation and annihilation operators one needs to choose the properly normalized modes. One can do it in the general case (curved spacetime and many component boson hermitian field φ_i, $i = 1,\dots, r$) as follows. If $D^i[\varphi] = 0$ are linear field equations for the field φ_i in a curved spacetime obtained by a variation of the corresponding field action then one has

$$\varphi_{1\,i}\, D^i[\varphi_2] - \varphi_{2\,i}\, D^i[\varphi_1] = \nabla_\mu B^\mu(\varphi_1, \varphi_2) \tag{I.63}$$

where φ_1 and φ_2 are two arbitrary functions. The quantity

$$B(\varphi_1, \varphi_2) \equiv \int_\Sigma B^\mu(\varphi_1, \varphi_2)\, d\sigma_\mu \tag{I.64}$$

defined for any two solutions φ_1 and φ_2 of the field equations does not depend on the particular choice of the complete Cauchy surface Σ provided the solutions decrease rapidly enough at infinity. (The surface element on Σ is denoted by $d\sigma_\mu$).

It can be shown that the canonical commutation relations for a quantum field $\hat{\varphi}_i$ are equivalent to the relations

$$[B(\hat{\varphi}, \varphi_1), B(\hat{\varphi}, \varphi_2)] = iB(\varphi_1, \varphi_2) \tag{I.65}$$

which are to be satisfied for any two solutions φ_1 and φ_2. If $\{\varphi_i^J, \bar{\varphi}_i^J\}$ is a complete set of the solutions satisfying given boundary conditions and normalized as follows

$$\begin{aligned} B(\varphi^J, \varphi^{J'}) &= B(\bar{\varphi}^J, \bar{\varphi}^{J'}) = 0 \\ B(\varphi^J, \bar{\varphi}^{J'}) &= i\delta_{JJ'} \end{aligned} \tag{I.66}$$

then one can decompose the quantum field $\hat{\varphi}_i$ over these modes

$$\hat{\varphi}_i = \Sigma(\varphi_i^J \hat{a}_J + \bar{\varphi}_i^J \hat{a}_J^*) . \tag{I.67}$$

Eqs. (I.65) and (I.66) guarantee that the operators $\hat{a}_J$ and $\hat{a}_J^*$ obey the standard commutation relations

$$[\hat{a}_J, \hat{a}_{J'}^*] = \delta_{JJ'} , \quad [\hat{a}_J, \hat{a}_{J'}] = [\hat{a}_J^*, \hat{a}_{J'}^*] = 0. \tag{I.68}$$

The formulas for the scalar field quantization in the flat spacetime we used in this Section may be considered as the simpliest possible application of the general approach. For more details concerning the general quantization scheme in curved spacetime see e.g. (DeWitt, 1975; Birrell and Davies, 1982; Frolov, 1987).

II. QUANTUM FIELD THEORY IN HOMOGENEOUS GRAVITATIONAL FIELD

2.I Uniformly accelerated reference frame. Reference frame at rest in homogeneous gravitational field

Now we make our first step in studying the quantum effects in a static gravitational field. We consider the simplest possible case of a static homogeneous field. In some sense this particular example plays the role of the "Rosetta stone" and it helps us in understanding the general situation.

According to the equivalence principle the laws of physics (including the quantum ones) in any uniform gravitational field should take on the same form as they do in a uniformly accelerated reference frame in flat spacetime. The application of the equivalence principle gives us an important tool for studying the influence of the gravitational field on quantum effects.

We begin by describing two types of reference frames. The first one is a uniformly accelerated reference frame in flat spacetime. For short we call it A -frame. The other one is a reference frame which is at rest in a homogeneous gravitational field which we call G -frame.

In order to describe the properties of the A -frame it is convenient to introduce new (Rindler) coordinates $x^\mu = (\eta, \rho, \xi^2, \ldots, \xi^D)$ in the flat spacetime which are related to the Cartesian coordinates X^μ as follows

$$X^0 = \rho \sinh\eta, \quad X^1 = \rho\cosh\eta, \quad X^A = \xi^A \quad (A = 2, \ldots, D) \,. \tag{2.I}$$

The Rindler coordinates cover the part R_+ of the Minkowski space where $X^1 > |X^0|$ (see Fig.4). The world line $\rho = \rho_0 = \text{const}$, $\xi^A = \xi^A_0 = \text{const}$ describes the motion of a uniformly accelerated particle, $w = |w_\mu w^\mu|^{1/2} = \rho_0^{-1}$ being its acceleration and $\tau = \rho_0 \eta$ being the proper time parameter. One can connect a rigid reference frame with such a particle. The origin of this frame coincides with the particle. The parameter $\tau = \rho_0 \eta$ plays the role of the time. The plane $\tau = \text{const}$ is the set of events which are simultaneous from the point of view of the observer placed at the

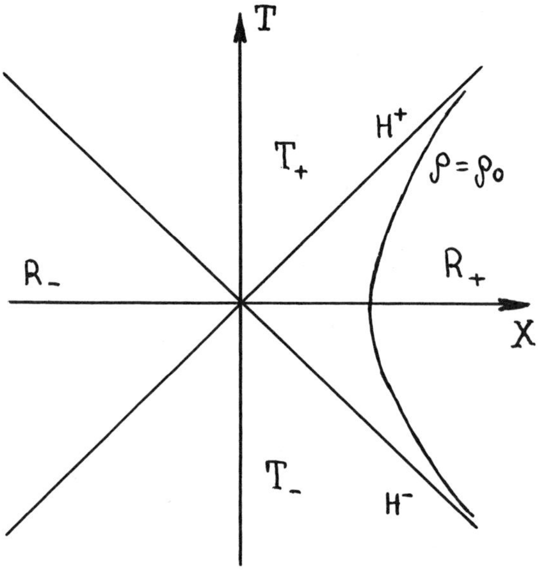

Fig.4. Minkowski spacetime

particle. The coordinates (ρ, ξ^A) are the spatial orthogonal coordinates in the frame under consideration. The constructed reference frame is rigid in the following sense. The distance between any two points with fixed space coordinates (ρ_1, ξ_1^A) and (ρ_2, ξ_2^A) measured in this reference frame is time independent. The Rindler coordinates (η, ρ, ξ^A) are connected with the described A-frame in the same manner as the Cartesian coordinates X^μ are connected with the inertial I-frame. In the three other parts R_-, T_+ and T_- of the Minkowski space (see Fig.4) the Rindler-like coordinates can be introduced as follows

$$\left.\begin{array}{l} X^0 = \varepsilon\rho \sinh\eta, \quad X^1 = \varepsilon\rho\cosh\eta, \\ X^A = \xi^A \end{array}\right\} \text{ in } R_\varepsilon \text{ where } \varepsilon X^1 > |X^0| ,$$
$$\left.\begin{array}{l} X^0 = \varepsilon\rho\cosh\eta, \quad X^1 = \varepsilon\rho \sinh\eta, \\ X^A = \xi^A \end{array}\right\} \text{ in } T_\varepsilon \text{ where } \varepsilon X^0 > |X^1| . \tag{2.2}$$

The flat spacetime metric (I.I) in these coordinates reads

$$ds^2 = \varepsilon(-\rho^2 d\eta^2 + d\rho^2) + \sum_{A=2}^{D} (d\xi^A)^2 . \tag{2.3}$$

For $\varepsilon = +1$ this metric coincides with the metric for a homogeneous gravitational field. Let us consider a test body which is at rest at a point $\rho = \rho_0$, $\xi^A = \xi^A_0$ in this gravitational field. The four-acceleration of such a body is $w = \rho_0^{-1}$ and hence it must be affected by some external (non-gravitational) force. Freely moving test particles will have an acceleration $\vec{g} = -\vec{w}$ with respect to the body. One may relate a rigid reference frame with this body in the same manner as it was done for the uniformly accelerated particle. In the so constructed G-frame $\tau = \rho_0 \eta$ is the time coordinate and (ρ, ξ^A) are orthogonal space coordinate.

G-frame of reference has the most obvious meaning in the case when the static gravitational field is created by a massive body. The static G-frames are singled out by the property that they are not moving with respect to the massive body. If L is a characteristic scale of inhomogeneity of the gravitational field then the gravitational field in the regions of size $\ell << L$ may be considered as homogeneous. For such regions the static G-frame coincides with the above described G-frame in a homogeneous gravitational field.

2.2. Equivalence principle for quantum phenomena. $\langle\hat{\varphi}^2\rangle^{ren}$ and $\langle T_{\mu\nu}\rangle^{ren}$ in the homogeneous gravitational field

For quantization of a scalar massless field $\hat{\varphi}$ in the uniformly accelerated frame one uses its decomposition in terms of basic solutions which are of positive and negative frequency with respect to time τ in A-frame. This choice of the positive frequency definition provides the following specification of the general quantization scheme described in I.4. The properly normalized solutions of the wave equation (I.2) are

$$\varphi_{\nu\vec{q}}(x) = \frac{\sinh^{1/2}(\pi\nu)}{\pi^{\frac{D+1}{2}}\, 2^{\frac{D-1}{2}}}\, e^{-i\nu\eta + i\vec{q}\vec{\xi}}\, K_{i\nu}(q\rho)\,. \qquad (2.4)$$

Here $\vec{\xi} = (\xi^2, \dots, \xi^D)$, $\vec{q} = (q^2, \dots, q^D)$ and $K_\mu(z)$ is a Macdonald function. The field operator $\hat{\varphi}$ in the region R_+ can be written as follows

$$\hat{\varphi}(x) = \sum_{y} [\varphi_{y}(x)\hat{b}_{y} + \bar{\varphi}_{y}(x)\hat{b}^{*}_{y}] , \tag{2.5}$$

where y is a collective index $y = \nu\vec{q}$ and

$$\sum_{y} \equiv \int_{0}^{\infty} d\nu \int d^{D-1} q . \tag{2.6}$$

The operators of creation ($\hat{b}^{*}_{y}$) and annihilation ($\hat{b}_{y}$) of a quantum in a mode y (of the so called Rindler quantum or briefly R -quantum) obey the commutation relations

$$[\hat{b}_{y}, \hat{b}^{*}_{y'}] = \delta_{yy'} \equiv \delta(\nu-\nu')\, \delta^{D-1}(\vec{q}-\vec{q}') \tag{2.7}$$

the other commutators vanish.

We restrict ourselves by considering the quantum states for which

$$\begin{aligned} &< \hat{b}^{*}_{y}\, \hat{b}^{*}_{y'} > = < \hat{b}_{y}\, \hat{b}_{y'} > = 0 , \\ &< \hat{b}^{*}_{y}\, \hat{b}_{y'} > = n^{R}_{y}\, \delta_{yy'} . \end{aligned} \tag{2.8}$$

The states with finite number of Rindler quanta and thermal equilibrium states possess these properties. For such a state one has

$$< \hat{\varphi}^{2}(x)> = \sum_{y} \varphi^{2}(x|y)(n^{R}_{y} + \tfrac{1}{2}) . \tag{2.9}$$

Here and later we suppose that a point x lies in R_{+}

$$\varphi^{2}(x|y) = 2|\varphi_{y}(x)|^{2} , \tag{2.10}$$

and n^{R}_{y} is a mean value of Rindler quanta in the mode y for a given quantum states.

The quantity $<\hat{\varphi}^{2}(x)>$is divergent and it must be renormalized. Fortunately it is not difficult to do because till now we are considering the quantum theory in a flat space time and we know that

$$< 0;M\,|\,\hat{\varphi}^{2}(x)\,|\,0;M >^{ren} = 0$$

for the Minkowski vacuum state $|0;M\rangle$. The usual substraction of the null fluctuation contribution in the Minkowski space gives

$$\langle\hat{\varphi}^2(x)\rangle^{ren} = \sum_{\mathcal{J}} \varphi^2(x|\mathcal{J})(n^R_{\mathcal{J}} - n_{0,\mathcal{J}}), \tag{2.11}$$

where

$$n_{0,\mathcal{J}} = n_0(\nu_{\mathcal{J}}) \equiv [\exp(2\pi\nu_{\mathcal{J}}) - 1]^{-1} \tag{2.12}$$

is the average number of Rindler quanta in the state $|0;M\rangle$. In order to prove this formula it is sufficient to use Eqs.(A.6)-(A.7) and verify that

$$\sum_{\mathcal{J}} \varphi^2(x,x'|\mathcal{J})(n_{0,\mathcal{J}} + \tfrac{1}{2}) = \frac{\Gamma(\frac{D-1}{2})}{4\pi^{\frac{D+1}{2}}[s(x,x')]^{D-1}}, \tag{2.13}$$

where

$$\varphi^2(x,x'|\mathcal{J}) = \tfrac{1}{2}[\varphi_{\mathcal{J}}(x)\bar{\varphi}_{\mathcal{J}}(x') + \bar{\varphi}_{\mathcal{J}}(x)\varphi_{\mathcal{J}}(x')],$$

$$s^2(x,x') = \rho^2 + \rho'^2 - 2\rho\rho'\cosh(\eta-\eta') + \sum_{A=2}^{D}(\xi^A - \xi'^A)^2. \tag{2.14}$$

The quantity in the right-side of Eq.(2.13) coincides with the vacuum Hadamard function (1.8) for points x and x' lying in the R_+-region written in the Rindler coordinates.

The renormalized value of the stress-energy tensor can be written in the same manner

$$\langle T_{\mu\nu}(x)\rangle^{ren} = \sum_{\mathcal{J}} t_{\mu\nu}(x|\mathcal{J})(n^R_{\mathcal{J}} - n_{0,\mathcal{J}}), \tag{2.15}$$

$t_{\mu\nu}(x|\mathcal{J})$ being a contribution of a Rindler quantum in the mode $\mathcal{J}$ into the stress-energy tensor. This contribution is obtained by applying the operator $D_{\mu\nu}(x,x')$ given by Eq.(1.16) to $\varphi^2(x,x'|\mathcal{J})$ and taking the coincidence limit. It should be stressed that it is more convenient to calculate $t_{\mu\nu}(x|\mathcal{J})$ not in the Cartesian but directly in the Rindler coordinates. In this case the operators ∇_μ and $\nabla_{\mu'}$ in Eq.(1.16) are to be understood as covariant derivatives and $g_{\mu\nu'}(x,x')$ is an operator of a parallel transport from a point x' to a point x. Eqs.(2.11) and (2.15) allow one to calculate $\langle\hat{\varphi}^2(x)\rangle^{ren}$ and $\langle T_{\mu\nu}(x)\rangle^{ren}$ in the

A -frame if only the number of the Rindler quanta n^R_J for a given state is given.

Now it is time to use the equivalence principle. According to this principle Eqs.(2.11) and (2.15) give also the expressions for $\langle \hat{\varphi}^2(x)\rangle^{ren}$ and $\langle T_{\mu\nu}(x)\rangle^{ren}$ in a static homogeneous gravitational field. This statement is correct but it requires some explanations. The equivalence principle guarantees the identity of the physical laws in A - and G -frames provided the initial and boundary conditions in the both frames are the same. For quantum phenomena it means that the quantum states in the A - and G -frames must be chosen in a similar manner. The simplest way to do this is to use for the quantization in G -frame the same mode expansion (2.5), $\hat{b}^*_J$ and $\hat{b}_J$ being interpreted as the operators of the creation and annihilation of a quantum in a mode J in a static homogeneous gravitational field.

The described quantization scheme in the homogeneous gravitational field is a particular case of the general approach developed by Boulware (1975 a,b; 1976) for static gravitational field. In order to define positive and negative frequencies he used the standard Killing time. The so defined quanta in a static gravitational field are known as Boulware particles (or B -quanta) and the state $|0;B\rangle$ without any B -particles is called the Boulware vacuum (or B-vacuum). This quantum state possesses the lowest possible energy in the static gravitational field. We preserve the same notations (B-quanta, B-vacuum and so on)for the static homogeneous field we are considering now.

Thus according to the equivalence principle one has the following expressions for $\langle \hat{\varphi}^2(x)\rangle^{ren}$ and $\langle T_{\mu\nu}(x)\rangle^{ren}$ for the homogeneous gravitational field

$$\langle \hat{\varphi}^2(x)\rangle^{ren} = \sum_J \varphi^2(x|J)(n^B_J - n_{0,J}), \tag{2.16}$$

$$\langle T_{\mu\nu}(x)\rangle^{ren} = \sum_J t_{\mu\nu}(x|J)(n^B_J - n_{0,J}). \tag{2.17}$$

These formulas are obtained by changing the superscript R in Eqs.(2.11) and (2.15) by the superscript B , the functions $\varphi^2(x|J)$ and $t_{\mu\nu}(x|J)$ being the same as earlier.

This guarantees the "same states" choice in A- and G-frames which is required by the equivalence principle.

2.3. Thermal radiation in homogeneous gravitational field. Vacuum polarization in the gravitational field of a cosmic string

The calculation of $\langle \hat{\varphi}^2 \rangle^{ren}$ and $\langle T_{\mu\nu} \rangle^{ren}$ in a homogeneous gravitational field is not a simple problem. It is greatly simplified in the special case when n_y^B depends only on the frequency ν $n_y^B = n^B(\nu_y)$. In this case the integrations over $\vec{q}$ can be fulfiled in closed form and one has

$$\langle \hat{\varphi}^2(x) \rangle^{ren} = \frac{1}{\rho^{D-1}} \int_0^\infty \frac{d\nu}{\nu} \Pi_D(\nu)[n^B(\nu) - n_0(\nu)] \tag{2.18}$$

where

$$\Pi_D(\nu) = \frac{1}{2\pi^{\frac{D+1}{2}}} \frac{\Gamma(\frac{D-1}{2})}{\Gamma(D-1)} P_D(\nu) . \tag{2.19}$$

For the even number of the spacetime dimensions ($D = 2k+1$, $k = 1, 2, \ldots$) the function $P_D(\nu)$ reads

$$P_{2k+1}(\nu) = \nu^2(\nu^2+1) \ldots [\nu^2 + (k-1)^2] , \tag{2.20}$$

while for the odd number of the spacetime dimensions ($D = 2(k+1)$, $k = 0, 1, \ldots$) it has the form

$$P_{2(k+1)}(\nu) = \nu \tanh(\pi\nu) \cdot (\nu^2 + \tfrac{1}{4}) \ldots [\nu^2 + (k-\tfrac{1}{2})^2] . \tag{2.21}$$

(Eqs. (A.8)-(A.11) are needed to obtain these results).

The expression for $\langle T_{\mu\nu} \rangle^{ren}$ can be obtained in the same way but the calculations are more complicated. As earlier a lot of information about $\langle T_{\mu\nu} \rangle^{ren}$ can be obtained before calculations by using its symmetry properties. The problem under consideration possesses the following symmetries. It is invariant under translations along the Rindler time η, under translations and rotations of the coordinates ξ^A $(A = 2, \ldots, D)$ and under reflections $\eta \to -\eta$ and $\xi^A \to -\xi^A$. The traceless conserved stress-energy tensor with these symmetry properties and which possesses the dimension of $(\text{length})^{-4}$ must be of the form

$$\langle T_\mu^\nu(x)\rangle^{ren} = \operatorname{diag}(-\varepsilon, D^{-1}\varepsilon, \ldots, D^{-1}\varepsilon) \tag{2.22}$$

where $\varepsilon \sim \rho^{-4}$. The explicit calculation gives

$$\varepsilon(\rho) = \frac{1}{\rho^{D+1}} \int_0^\infty \nu d\nu \, \Pi_D(\nu)[n^B(\nu) - n_0(\nu)] . \tag{2.23}$$

The function $\Pi_D(\nu)$ is given by Eq.(2.19). For more details concerning the calculations of $\langle\hat{\varphi}^2\rangle^{ren}$ and $\varepsilon(\rho)$ see (Takagi, 1986).

The case of equilibrium thermal radiation in a homogeneous gravitational field is of particular interest. It should be noted that the local temperature Θ_{loc} of such radiation in a static gravitational field depends on the point so that the value $\Theta_{loc}(x)\sqrt{-g_{00}(x)}$ is invariant. To characterize the temperature of the radiation one needs to indicate the registration point. We shall choose the origin $\rho = \rho_0 = g^{-1}$ of the G-frame as such a registration point. If the temperature of the radiation in the G-frame is Θ then the local temperature at a point ρ is

$$\Theta_{loc}(\rho) = \frac{\Theta \rho_0}{\rho} . \tag{2.24}$$

The average number $n_\Theta^B(\nu)$ of the Boulware quanta for thermal radiation in the homogeneous gravitational field reads

$$n_\Theta^B(\nu) = [\exp(2\pi\beta\nu) - 1]^{-1} , \tag{2.25}$$

where

$$\beta = g/2\pi\Theta . \tag{2.26}$$

It is easy to see that $\omega = g\nu$ is the proper frequency of the B-quantum in the G-frame (i.e. the frequency measured by the clock at the origin of the G-frame) and Eq.(2.25) really describes the thermal distribution

$$n^{B}_{\theta}(\nu) \equiv [\exp(\omega/\theta) - 1]^{-1} . \tag{2.27}$$

It should be stressed that the point dependence of the frequency due to the red-shift is the same as the local temperature dependence so that the ratio $\omega_{loc}/\theta_{loc}$ is invariant.

For the thermal radiation in a homogeneous gravitational field one has

$$\langle \hat{\varphi}^2(x)\rangle^{ren} = \frac{1}{\rho^{D-1}} [\Phi_D(\beta) - \Phi_D(1)] , \tag{2.28}$$

$$\varepsilon(x) = \frac{1}{\rho^{D+1}} [E_D(\beta) - E_D(1)] , \tag{2.29}$$

where

$$\Phi_D(\beta) = \int_0^\infty \frac{d\nu}{\nu} \Pi_D(\nu) [\exp(2\pi\beta\nu) - 1]^{-1} , \tag{2.30}$$

$$E_D(\beta) = \int_0^\infty \nu d\nu \, \Pi_D(\nu) [\exp(2\pi\beta\nu) - 1]^{-1} . \tag{2.31}$$

It is evident that the quantities $\langle \hat{\varphi}^2(x)\rangle^{ren}$ and $\langle T_{\mu\nu}\rangle^{ren}$ vanish for the thermal radiation if only it has the so called Unruh temperature

$$\theta_U = g/2\pi . \tag{2.32}$$

The corresponding quantum state is known as the Hartle-Hawking vacuum state $|0;H\rangle$. It is this state which according to the equivalence principle corresponds to the Minkowski vacuum $|0;M\rangle$.

For completeness we present here the results concerning the contribution of the massless fields with spin s to $\langle T_{\mu\nu}\rangle^{ren}$ in four-dimensional spacetime

$$\langle T_\mu^\nu(x)\rangle^{ren} = \mathrm{diag}(-\varepsilon, \tfrac{1}{3}\varepsilon, \tfrac{1}{3}\varepsilon, \tfrac{1}{3}\varepsilon) , \tag{2.33}$$

$$\varepsilon(\rho) = \frac{h(s)}{2\pi^2\rho^4} [I_s(\beta) - I_s(1)] ,$$

$$I_s(\beta) = \int_0^\infty \nu d\nu (\nu^2 + s^2) [\exp(2\pi\beta\nu) - (-1)^{2s}]^{-1} . \tag{2.34}$$

Here $h(s)$ is a number of helicities of the field of spin s. The integrals (2.34) can be calculated in closed form and one has

$$\varepsilon = \frac{3}{g^4 \rho^4} [\theta^2 - (\frac{g}{2\pi})^2][a\theta^2 + b(\frac{g}{2\pi})^2], \tag{2.35}$$

where g is the acceleration of a G-frame, θ is the temperature of thermal radiation in this frame and

$$a = \frac{\pi^2}{90}\tilde{a}, \quad \tilde{a} \equiv h(0) + \frac{7}{8}h(1/2) + h(1), \tag{2.36}$$

$$b = \frac{\pi^2}{90}\tilde{b}; \quad \tilde{b} \equiv h(0) + \frac{17}{8}h(1/2) + 11h(1). \tag{2.37}$$

Using Eq.(2.28) it is also easy to write down an explicit expression for $\langle\hat{\varphi}^2(x)\rangle^{ren}$ in the four-dimensional homogeneous gravitational field

$$\langle\hat{\varphi}^2(x)\rangle^{ren} = \frac{1}{12 g^2 \rho^2}[\theta^2 - (g/2\pi)^2]. \tag{2.38}$$

The expressions(2.33), (2.35) and (2.38) for $\langle T_{\mu\nu}\rangle^{ren}$ and $\langle\hat{\varphi}^2\rangle^{ren}$ are written in special coordinates connected with the chosen G-frame. It is instructive to rewrite them in a general covariant way. For this purpose we note that

$$\xi^\mu \partial_\mu \equiv \frac{1}{\rho_0}\frac{\partial}{\partial\eta} \equiv g\frac{\partial}{\partial\eta} \tag{2.39}$$

is a Killing vector normalized to unit at the origin of the G-frame. Denote by

$$w_\mu = \frac{1}{2}\nabla_\mu \ln|\xi^2| \tag{2.40}$$

the four-acceleration of a Killing observer. In the homogeneous gravitational field one had $w(\rho) \equiv |w_\mu w^\mu|^{1/2} = \rho^{-1}$. The sought general covariant expressions for $\langle\hat{\varphi}^2\rangle^{ren}$ and $\langle T_{\mu\nu}\rangle^{ren}$ are

$$\langle\hat{\varphi}^2(x)\rangle^{ren} = \frac{1}{12}[\theta^2_{loc} - \frac{w_\mu w^\mu}{4\pi^2}], \tag{2.41}$$

$$\langle T_{\mu\nu}(x)\rangle^{ren} = \left(g_{\mu\nu} - \frac{4\xi_\mu \xi_\nu}{\xi^2}\right)\left[\theta^2_{loc} - \frac{w_\mu w^\mu}{4\pi^2}\right]\left[a\theta^2_{loc} + \right. \qquad (2.42)$$

$$\left. + b\,\frac{w_\mu w^\mu}{4\pi^2}\right],$$

where

$$\theta^2_{loc} = -\theta^2/\xi^2 \quad . \qquad (2.43)$$

In the high-temperature limit the leading part of these expressions depends only on the local temperature and this dependence reproduces the results obtained for the thermal radiation in I -frame in the flat spacetime. The additional terms which are defined by the acceleration and do not depend on the temperature are related with the vacuum polarization by the homogeneous gravitational field. (w may be considered as a measure of the strength of the field).

It is interesting to note that there is another physical problem (namely the vacuum polarization in the gravitational field of a cosmic string) which is formally closely related with the problem under consideration. We have no place to discuss this relation in very details. So we restrict ourselves by making a few remarks on the origin of this connection.

Let us change η in the Rindler line element (2.3) ($\varepsilon = +1$) by $i\varphi$. Considering φ as a real coordinate we get an Euclidean metric of the $(D+1)$ -dimensional space. The above described quantum field theory in the homogeneous gravitational field can be obtained by analytical continuation of the Euclidean field theory in this space. If one chooses the imagine time (in our case φ) to be periodic with the period $2\pi\beta$ then after returning back to the physical spacetime one gets the quantum theory describing the system with the temperature β^{-1} . In particular the stress-energy tensor (2.42) for such a state can be obtained by the analytical continuation from the analogous quantity calculated in the framework of the Euclidean formulation.

Let us consider now another possible analytical continuation $\xi^D \to -it$ which gives the line element

$$ds^2 = -dt^2 + d\rho^2 + \rho^2 d\varphi^2 + \sum_{A=2}^{D-1} (d\xi^A)^2 \quad . \qquad (2.44)$$

The spacetime described by this metric is locally flat but it possesses (for $\beta \neq 1$) a cone-like singularity at $\rho = 0$. In the four dimensional case this metric is the solution of the vacuum Einstein equation which describes the gravitational field of a stright-line cosmic string. The angle deficit $\Delta\varphi = 2\pi(1-\beta)$ of the cone singularity is related with the mass per a unit length of a cosmic string μ as follows $\Delta\varphi = 8\pi\mu$. By analytical continuation of Eqs.(2.33) and (2.35) one can get the following expression for the contribution of the massless fields of spin s into the vacuum stress-energy in the gravitational field of a straight-line cosmic string

$$< T_{\mu}^{\nu} >^{ren}_{\substack{cosmic\\string}} = diag(\varepsilon/3 , \varepsilon/3 , \varepsilon/3 , -\varepsilon) , \tag{2.45}$$

$$\varepsilon = \frac{1}{480\pi^2\rho^4}(\beta^{-2} - 1)(\tilde{\alpha}\beta^{-2} + \tilde{b}) . \tag{2.46}$$

In this Section the main attention was paid to the calculations of the local observables $<\hat{\varphi}^2>^{ren}$ and $<T_{\mu\nu}>^{ren}$. This is one of a number of problems which arise when studying the physics in the uniformly accelerated frame and in the homogeneous gravitational field. Some other problems are discussed in other lectures in this Proceedings. Here we make only brief remarks. We also give here the references to the papers where different aspects of the vacuum polarization in the homogeneous gravitational field and in the gravitational field of cosmic strings have been discussed.

Unruh (1976) studied an interaction of a uniformly-accelerated detector (i.e. a system with internal degrees of freedom) with a quantum field. He showed that such a detector registers Rindler quanta and hence it becomes thermally excited when moving in the M-vacuum state. In I -frame this process is accompanied by radiation of M -quanta. (Unruh and Wald, 1984). These effects were studied in very details in a number of papers. For general review and references see (DeWitt, 1979; Sciama et al, 1981 ; Birrell and Davies, 1982; Takagi, 1986; Ginzburg and Frolov, 1987). See also lecture by Unruh in this Proceedings.

The discussion of the equivalence principle in quantum

domain can be found in (Candelas and Sciama, 1984; Ginzburg and Frolov, 1987; Frolov and Thorne, 1988).

The study of the vacuum polarization in the homogeneous gravitational field has rather long history. The formula (2.34) for the energy density of conformal massless fields in the zero-temperature limit was obtained by Candelas and Deutsch (1977, 1978). The finite temperature corrections can be found by conformal transformations from that on the open Einstein universe (Candelas and Dowker, 1979; Dowker, 1983; Brown et al, 1986). The generalization to N-dimensional case was made by Takagi (1986).

The effect of the vacuum polarization in the gravitational field of a straight cosmic string was investigated by Helliwell and Konkowski (1986), Linet (1987), Frolov and Serebriany (1987) and Dowker (1987a). The Casimir effect around the cone in N-dimensional spacetime was studied by Dowker (1987b).

III. RENORMALIZATION AND CONFORMAL ANOMALIES

3.1. Stress-energy tensor renormalization. Wald's axioms.

The local observables $<\varphi^2>$ and $<T_{\mu\nu}>$ in a curved spacetime are divergent for the same reason as in the flat spacetime. Their renormalization can be done by substracting some standard expressions which correspond to an unobservable contribution of the vacuum null fluctuations. In the flat spacetime this contribution can be unambiguiously singled out due to high symmetry of the Minkowski space. In the curved spacetime there arises an ambiguity. Fortunately this ambiguity can be drastically restricted if one requires that the physically observable renormalized values $<\hat{\varphi}^2>^{ren}$ and $<T_{\mu\nu}>^{ren}$ satisfy some natural requirements. These requirements were formulated by Wald (1977, 1978) as follows:

(i) The covariant conservation law $<T_\mu{}^\nu>^{ren}{}_{;\nu} = 0$.

(ii) In Minkowski spacetime $<T_{\mu\nu}>^{ren}$ reduces to the standard expression.

(iii) The matrix element of $T_{\mu\nu}$ between any two orthogonal states agrees with the formal expression (which yields finite, unambiguous values).

(iv) Causality holds: for fixed "in" ("out") state, $<T_{\mu\nu}>^{ren}$ at point p depends only on the spacetime geometry to the past (future) of p.

According to the theorem proved by Wald (1977) any two prescriptions satisfying these axioms can differ at most by a conserved local curvature term.

An important corollary of this theorem is that the renormalized value $<T_{\mu\nu}>^{ren}$ for massless fields is unambiguously defined in vacuum spacetimes where $R_{\mu\nu} = 0$. In vacuum spaces the only nonvanishing local curvature tensor of dimension $(\text{length})^{-4}$ must be quadratic in the Weyl tensor. But it is not difficult to show that such a tensor cannot satisfy the conservation law. In other words if we are interested in calculation of $<T_{\mu\nu}>^{ren}$ in a given vacuum gravitational field (e.g. in the gravitational field of an isolated uncharged black hole) then all possible renormalization procedures obeying the Wald's axioms give the same answer.

The situation with $<\hat{\varphi}^2>^{ren}$ is analogous. The only difference is that the axiom (i) has no sence and must be omitted and the ambiguity in a definition of $<\hat{\varphi}^2>^{ren}$ is a local curvature scalar. In the four-dimensional spacetime the only curvature scalar of the proper dimension $(\text{length})^{-2}$ is R. Thus in the vacuum gravitational field in the absence of matter or in the gravitational field created by the matter with the traceless stress-energy tensor $<\hat{\varphi}^2>^{ren}$ is defined unambiguiously.

A number of renormalization methods are known. The most useful for our purposes is the point-splitting method. This method is a direct generalization of the procedure used in the previous sections. In order to obtain the finite result in the flat spacetime we have substracted from the Hadamard function the uniqely defined null fluctuations contribution (I.8) $(n \equiv D+1)$

$$G_0(x,x') = \frac{\Gamma(\frac{n}{2}-1)}{2(2\pi)^{n/2}\,\sigma^{n/2-1}}, \qquad (3.1)$$

where $\sigma = \frac{1}{2}s^2(x,x')$ and $s(x,x')$ is a geodesic interval in a flat metric. The function (3.1) is universal. It does not depend on the boundary conditions or on the choice of a quantum state.

The idea of the point-splitting method is to renormalize the Hadamard function $G(x,x')$ for a quantum field in a curved spacetime by substracting some universal function which possesses the same divergences as $G(x,x')$ and which depends only on the local geometry near a given point. For a scalar field in four-dimensional spacetime this method can be illustrated as follows. (For general description of this method see Birrell and Davies (1982)). The Hadamard function for this field allows the following representation (Hadamard, 1923; DeWitt and Brehme, 1960)

$$G(x,x') = \frac{\Delta^{1/2}(x,x')}{8\pi^2}\left[\frac{1}{\sigma} + v(x,x')\ln\sigma + w(x,x')\right], \quad (3.2)$$

where 2σ is the square of the geodesic distance between x and x' and Δ is the Van Vleck-Morette determinant

$$\Delta(x,x') = -\frac{1}{\sqrt{g(x)\,g(x')}}\det[\nabla_\mu \nabla_{\nu'} \sigma(x,x')] . \quad (3.3)$$

The function $v(x,x')$ is unambiguiously defined by the local geometry and $w(x,x')$ is the only function in this representation which depends on the boundary conditions. If x' is close to x one can write the series expansion of $\Delta^{1/2}$, v and w in terms of σ and to subsract a few first terms of this expansion from $G(x,x')$. This procedure gives the renormalized value $G^{ren}(x,x')$. In order to find out the coefficients of the required series expansions one may either use the system of recursion relations which arises after substitution (3.2) into the field equations or take advantage of the Schwinger-DeWitt's proper-time technique. The details concerning the former method which is often called the Hadamard regularization can be found in (Adler et al, 1977; Wald, 1978; Brown and Ottewill, 1983; Brown, 1984; Bernard and Folacci, 1986; Tadaki, 1987). The latter is described in (Christensen, 1976, 1978). The explicit form of the terms which are to be substracted to renormalize $\langle T_{\mu\nu}\rangle$ for massless fields of spin $s = 0$, $1/2$ and 1 in an arbitrary four-dimensional curved spacetime is given in (Christensen, 1978). The renormalization of the graviton contribution to $\langle T_{\mu\nu} \rangle$ is discussed in (Allen et al, 1988).

The expressions which are needed for the renormalization of $\langle \hat{\varphi}^2 \rangle$ are much simpler. In particular in order to renormalize $\langle \hat{\varphi}^2 \rangle$ in four- and five-dimensional Ricci-flat spacetimes it is sufficient to substract from $G(x,x')$ the expressions

$$G^{(4)}_{div}(x,x') = \frac{1}{8\pi^2 \sigma} ,$$

$$G^{(5)}_{div}(x,x') = \frac{1}{8\pi^2 (2\sigma)^{3/2}} . \tag{3.4}$$

Later (in Section IY) we shall use these formulas.

3.2. Conformal trace anomalies

Let us consider now massless fields of spin $s = 0, 1/2$ and 1 in a four-dimensional curved spacetime. The equations describing these fields are conformal invariant, i.e. invariant under conformal transformations of a metric

$$g_{\mu\nu} \to \tilde{g}_{\mu\nu} = \Omega^2(x)\, g_{\mu\nu} . \tag{3.5}$$

In the case of a scalar field it happens if one writes the field equation in the form

$$(\Box - \frac{1}{6} R)\varphi = 0 . \tag{3.6}$$

This choice corresponds to the choice $\xi = 1/6$ of the parameter ξ in Eq.(I.16). In the classical theory the direct corollary of the conformal invariance is vanishing of the trace of the stress-energy tensor for such a field.

In quantum theory it appears to be impossible to preserve the conformal invariance after a renormalization. The conformal invariance is broken by the terms which are to be substracted in the renormalization procedure. In particular the quantity $|g|^{1/4}\varphi^2$ for a conformally invariant classical scalar field does not transform under conformal transformations (3.5). This property is violated if we substitute $\langle \hat{\varphi}^2 \rangle^{ren}$ instead of φ^2. One can show that in the quantum case the quantity which remains invariant under conformal transformations reads (Page, 1982)

$$|g|^{1/4}\,(\langle\hat{\varphi}^2\rangle^{ren} + R/288\pi^2)\,. \qquad (3.7)$$

For the same reason the trace of the renormalized stress-energy tensor for a conformal massless field does not vanish. For the corresponding values of $\langle T_\mu^\mu\rangle^{ren}$ known as the conformal trace anomalies one has

$$\langle T_\mu^\mu\rangle^{ren} = \alpha(\mathcal{H} + \tfrac{2}{3}\Box R) + \beta\mathcal{J} + \gamma\Box R\,, \qquad (3.8)$$

where

$$\mathcal{H} \equiv C_{\alpha\beta\gamma\delta}C^{\alpha\beta\gamma\delta} = R_{\alpha\beta\gamma\delta}R^{\alpha\beta\gamma\delta} - 2R_{\alpha\beta}R^{\alpha\beta} + \tfrac{1}{3}R^2\,, \qquad (3.9)$$

$$\mathcal{J} = {}^*R_{\alpha\beta\gamma\delta}\,{}^*R^{\alpha\beta\gamma\delta} = R_{\alpha\beta\gamma\delta}R^{\alpha\beta\gamma\delta} - 4R_{\alpha\beta}R^{\alpha\beta} + R^2\,. \qquad (3.10)$$

The different (dimensional, point-splitting and zeta-function) renormalizations agree that

$$\alpha = \frac{1}{2^9\,45\pi^2}\,[12h(0) + 18h(1/2) + 72h(1)]\,, \qquad (3.11)$$

$$\beta = \frac{1}{2^9\,45\pi^2}\,[-4h(0) - 11h(1/2) - 124h(1)]\,.$$

There is disagreement on the value of γ :dimensional renormalization gives $\gamma = 0$ while point-splitting and zeta-function renormalization give

$$\gamma = \frac{1}{2^9\,45\pi^2}\,[-120h(1)]\,. \qquad (3.12)$$

This difference is not important in the Ricci-flat space-times. (For references see Birrell and Davies (1982). See also recent papers by Pascual et al (1988) and Allen et al (1988) where the contributions of $s = 3/2$ and $s = 2$ to $\langle T_\mu^\mu\rangle^{ren}$ were considered).

The other corollary of the conformal invariance breaking is that the quantity which is invariant under conformal transformations is not $|g|^{1/2}\langle T_\nu^\mu\rangle^{ren}$ but the following rather complicated expression (Page, 1982)

$$|g|^{1/2}\{\langle T^{\mu}_{\nu}\rangle^{ren} + \alpha[(C^{\alpha\mu}{}_{\beta\nu}\ln g)^{;\beta}{}_{;\alpha} + \frac{1}{2}R^{\beta}_{\alpha}C^{\alpha\mu}{}_{\beta\nu}\ln g]^{+} \tag{3.13}$$

$$+\beta[2H^{\mu}_{\nu} - 4R^{\beta}_{\alpha}C^{\alpha\mu}{}_{\beta\nu}] + \frac{1}{6}\gamma I^{\mu}_{\nu}\},$$

where

$$H_{\mu\nu} = -R^{\alpha}_{\mu}R_{\alpha\nu} + \frac{2}{3}RR_{\mu\nu} + (\frac{1}{2}R^{\alpha}_{\beta}R^{\beta}_{\alpha} - \frac{1}{4}R^{2})g_{\mu\nu}, \tag{3.14}$$

$$I_{\mu\nu} = 2R_{;\mu\nu} - 2RR_{\mu\nu} + (\frac{1}{2}R^{2} - 2R^{;\alpha}{}_{;\alpha})g_{\mu\nu}.$$

This can be verified by direct calculations (Tadaki, 1988). But much simpler way to obtain this result is the study of the behaviour of the effective action under the conformal transformations (Brown, 1984; Riegert, 1984; Brown and Ottewill, 1985; Dowker, 1986).

After these brief remarks about the general scheme of the renormalization in a curved spacetime we return to the main topic: the theory of quantum effects in the gravitational field of a black hole.

IY QUANTUM EFFECTS IN THE GRAVITATIONAL FIELD OF BLACK HOLE

4.1. Black hole in a thermal bath and thermal radiation of a black hole

The spherically symmetric solution of the vacuum Einstein equations in $(D+1)$ -dimensions describing a static black hole in $(D+1)$ -dimensional spacetime can be written as follows (Tangherlini, 1963; Myers and Perry, 1986; Gibbons and Wiltshire, 1986)

$$ds^{2} = -f\,dt^{2} + f^{-1}dr^{2} + r^{2}d\omega^{2}, \tag{4.1}$$

$\xi^{\mu}\partial_{\mu} = \partial_{t}$ being the Killing vector and

$$f \equiv -\xi^{2} = 1 - (\frac{r_g}{r})^{D-2}. \tag{4.2}$$

The mass of such a black hole, which can be defined by comparing of the asymptotic behaviour (at $r \to \infty$) of the solution (4.I) with the solutions of the linearized gravitational equations in $(D+1)$-dimensional spacetime is given by

$$M = \frac{\pi^{\frac{D}{2}-1} (D-1)}{8\,\Gamma(D/2)} r_g^{D-2} . \qquad (4.3)$$

The surface H defined by the equation $r = r_g$ where $f = 0$ is a horizon. It is easy to show that the surface gravity of a $(D+1)$-dimensional black hole defined as

$$(\xi^{\mu} \xi^{\nu}{}_{;\mu})\Big|_H = \varkappa \xi^{\nu}\Big|_H \qquad (4.4)$$

can be written as follows

$$\varkappa = (w^2 f)^{1/2}\Big|_H = \frac{1}{2}\frac{df}{dr}\Big|_{r=r_g} = \frac{D-2}{2r_g} . \qquad (4.5)$$

Here $w^2 = w_\mu w^\mu$ and $w_\mu = -\frac{1}{2}\nabla_\mu \ln f$ is the acceleration of the Killing observer (i.e. an observer which is at rest in a static gravitational field and possesses the $(D+1)$-velocity $u^\mu = \xi^\mu / |\xi \cdot \xi|^{1/2}$).

It is evident that in the physical four-dimensional spacetime (for $D = 3$) Eqs.(4.I) and (4.2) describe a Schwarzschild black hole of mass $M = r_g/2$ and with the surface gravity $\varkappa = 1/4M$.

We consider now a quantum scalar massless field $\hat{\varphi}$ in the gravitational field of a static spherically symmetric $(D+1)$-dimensional black hole. The wave equation

$$\Box \varphi = 0 \qquad (4.6)$$

in the metric (4.I) allows the separation of variables and its solutions can be written in the form

$$\varphi_y = e^{-i\omega t} y_{\omega N}(r) Y_{N\beta}(\vec{n}) . \qquad (4.7)$$

Here $y = \omega N \beta$ is a collective index, $Y_{N\beta}(\vec{n})$ are the spherical harmonics on S^{D-1} and $y_{\omega N}(r)$ are radial functions obeying the equation

$$[r^{-(D-1)}\frac{d}{dr}(r^{D-1} f \frac{d}{dr}) + \frac{\omega^2}{f} - \frac{(N-1)(N+D-3)}{r^2}] y_{\omega N} = 0. \quad (4.8)$$

We introduce two sets of solutions $\varphi_{y\varepsilon}$ ($\varepsilon = \pm$) which are given by Eq.(4.7) with the radial functions $y_{\omega N}(r)$ defined by the following asymptotic conditions

$$y_{y+}(r) \sim \frac{1}{2\sqrt{\pi\omega}\, r^{\frac{D-1}{2}}} \begin{cases} e^{-i\omega_y r} + R_y e^{i\omega_y r}, & r \to \infty; \\ T_y e^{-i\omega_y r^*}, & r \to r_g; \end{cases} \quad (4.9)$$

$$y_{y-}(r) \sim \frac{1}{2\sqrt{\pi\omega}\, r^{\frac{D-1}{2}}} \begin{cases} T_y e^{i\omega_y r}, & r \to \infty; \\ e^{i\omega_y r^*} + R_y e^{-i\omega_y r^*}, & r \to r_g; \end{cases} \quad (4.10)$$

where $dr^* = f^{-1} dr$.

The in-modes φ_{y+} describe the incoming from infinity spherical waves which are partly absorbed by the black hole and partly scattered back to infinity. The coefficients T_y and R_y are transmission and reflection coefficients, correspondingly. The up-modes φ_{y-} describe the spherical waves propagating from the black hole to infinity and being partly scattered back to the black hole by the gravitational and centrofugal barrier. The quantum field $\hat{\varphi}$ outside the black hole can be expanded in these modes as follows

$$\hat{\varphi}(x) = \sum_{y\varepsilon} (\varphi_{y\varepsilon} \hat{a}_{y\varepsilon} + \bar{\varphi}_{y\varepsilon} \hat{a}^*_{y\varepsilon}). \quad (4.11)$$

For our choice of the basic solutions $\varphi_{y\varepsilon}$ the operators $\hat{a}_{y\varepsilon}$ and $\hat{a}^*_{y\varepsilon}$ satisfy the standard commutation relations

$$[\hat{a}_{y\varepsilon}, \hat{a}^*_{y'\varepsilon'}] = \delta_{yy'} \delta_{\varepsilon\varepsilon'}, \quad (4.12)$$

other commutators vanish.

The modes φ_y are chosen to be of positive frequency with respect to the Killing time t. The particles created and annihilated by the operators $\hat{a}^*_{y\varepsilon}$ and $\hat{a}_{y\varepsilon}$ are known as Boulware particles (or B -quanta). The Boulware vacuum state $|0;B>$ (or B -vacuum) is defined by relations $\hat{a}_{J\varepsilon}|0;B> = 0$. The action of $\hat{a}_{J\varepsilon}$ on a quantum state decreases its Killing energy. B -vacuum is the state of the minimal Killing energy. This state plays an important role in the quantum theory in a static spacetime if only there is no event horizon. The presence of the event horizon changes the situation drastically. In order to show this let us study the metric (4.I) near $r = r_g$ in more details.

According to Eq.(4.5)

$$f = 2\varkappa(r - r_g) + O((r - r_g)^2) . \tag{4.I3}$$

Let ρ be a new radial coordinate

$$\rho = \int_{r_g}^{r} \frac{dr}{\sqrt{f}} \tag{4.I4}$$

then near the horizon one has $f \approx \varkappa^2 \rho^2$ and the metric (4.I) has a Rindler-like form

$$ds^2 \simeq -\rho^2 d\eta^2 + d\rho^2 + r_g^2 d\omega^2_{D-1} , \tag{4.I5}$$

where $\eta = \varkappa t$. In a spacetime region of a space size $l << r_g$ and lying in the strip $|r_g^{-1}(r - r_g)| << 1$ the sphere of the radius r_g can be approximated by a plane. The line element (4.I5) in this approximation coincides with the Rindler line element (2.3) ($\varepsilon = +1$).

We have seen that the mean value of the renormalized stress-energy tensor $<T_{\mu\nu}>^{ren}$ in the B -vacuum in a homogeneous gravitational field is divergent at the horizon $\rho = 0$. The presence of the finite curvature near the horizon in the case of a black hole does not change this result. The leading divergent term of $<T_{\mu\nu}>^{ren}$ in the B -vacuum near the event horizon of a black hole is of the form

$$<T_{\mu\nu}>^{ren}_B \approx \varepsilon(\rho) D^{-1} [g_{\mu\nu} - \frac{(D+1)\xi_\mu \xi_\nu}{\xi^2}] , \tag{4.I6}$$

$$\varepsilon = -\rho^{-(D+1)} \int_0^\infty \nu d\nu \Pi_D(\nu) [\exp(2\pi\nu) - 1]^{-1} , \tag{4.I7}$$

where $\Pi_D(\nu)$ is given by Eq.(2.I9). The expectation value $<\hat{\varphi}^2>^{ren}_B$ is also divergent at the horizon

$$\langle \hat{\varphi}^2 \rangle_B^{ren} \simeq - \frac{1}{\rho^{D-1}} \int_0^\infty \frac{d\nu}{\nu} \Pi_D(\nu) [\exp(2\pi\nu) - 1]^{-1} . \qquad (4.18)$$

In the physical four-dimensional spacetime the cotribution of the conformal invariant massless fields of spin s to the divergent near the horizon part of $\langle T_{\mu\nu} \rangle_B^{ren}$ is

$$\langle T_{\mu\nu} \rangle_B^{ren} \simeq - \frac{b}{48\pi^4} \left(g_{\mu\nu} - \frac{4 \xi_\mu \xi_\nu}{\xi^2} \right) \frac{1}{\rho^4} , \qquad (4.19)$$

where the coefficients b are given by Eq.(2.37). These results indicate that in the presence of a black hole the B -vacuum state is physically badly defined and it can not be met in real situations.

The most preferable choice of the quantum state in the homogeneous gravitational field is the H -vacuum for which $\langle T_{\mu\nu} \rangle_H^{ren} = 0$ and $\langle \hat{\varphi}^2 \rangle_H^{ren} = 0$. For this state the average number of B -quanta out-going from the horizon is described by the thermal distribution. The corresponding boundary condition for the black hole reads

$$n_{J-} = n_{\theta_{BH}}(\omega_J) \equiv [\exp(\omega_J/\theta_{BH}) - 1]^{-1} , \qquad (4.20)$$

where $n_{J\varepsilon}$ is defined by the relation

$$\langle \hat{a}^*_{J\varepsilon} \hat{a}_{J'\varepsilon'} \rangle = n_{J\varepsilon} \delta_{JJ'} \delta_{\varepsilon\varepsilon'} , \qquad (4.21)$$

and

$$\theta_{BH} = \varkappa / 2\pi \qquad (4.22)$$

is the black-hole temperature. (The factor $\varkappa$ arises because we define the frequency ω with respect to time t connected with the Rindler time η by the relation $t = \varkappa^{-1}\eta$.)

In order to understand better the meaning of these boundary conditions for a black hole it is instructive to compare the quantum field propagation in the vicinity of the black hole and in the homogeneous gravitational field. The spacetime picture of the particles propagation in the gravitational field can be obtained by constructing the wave-packet like solutions from the field modes φ_y or simply by using the geometric optic approximation. Under this approximation the massless particles are moving along null geodesics.

The null geodesics in the homogeneous gravitational field can be represented by null straight lines in coordinates $(T, \vec{X})$ and all of them (exluding only those moving exactly in the X^1 -direction) cross the past (H^-) and future (H^+) event horizons (see Fig.4). A trajectory of a massless particle in the Rindler coordinates is shown in Fig.5a. A particle goes out of the horizon ($\rho = 0$), reaches a point of the maximum value of ρ and after falls down to the horizon. The main features of this behaviour follow directly from the wave equation $\Box\varphi = 0$. After the separation of variables $\varphi \sim e^{-i\nu\eta} e^{i\vec{q}\vec{\xi}} \Phi_y(\rho)$ in the Rindler coordinates the corresponding radial equation is

$$\left[\frac{d^2}{dr^{*2}} + \nu^2 - V_y \right] \Phi_y = 0 , \qquad (4.23)$$

where $r^* = \ln \rho$ and

$$V_y(r^*) = \rho^2 q^2 = q^2 e^{2r^*} . \qquad (4.24)$$

For $\vec{q} \neq 0$ the effective potential V infinitely grows at infinity. (see Fig.5b). That is why the particles in the

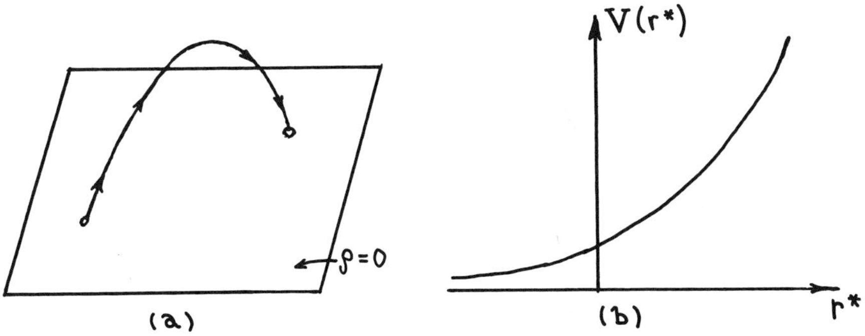

Fig.5. Massless particle propagation in the homogeneous gravitational field
(a) Particle trajectory in Rindler coordinates
(b) Effective potential

homogeneous gravitational field are trapped and cannot reach infinity. In other words in such a field one has the effect of the gravitational confinement.

In the case of a black hole the situation is quite different. The homogeneous field approximation is good enough only in the vicinity of the horizon while at far distances the gravitational field is practically absent. The radial wave equation (4.8) can be written in the form analogous to Eq.(4.23)

$$\left[\frac{d^2}{dr^{*2}} + \omega^2 - V_y \right] \Phi_y = 0 , \tag{4.25}$$

where $\Phi_y = r^{\frac{D-1}{2}} y_y$, $dr^* = dr/f$ and

$$V_y(r) = \left[\frac{(N-\frac{D}{2})^2 - \frac{1}{4}}{r^2} + \left(\frac{D-1}{2}\right)^2 r_g^{D-2} \frac{1}{r^D} \right] f . \tag{4.26}$$

This effective potential is schematically shown in Fig.6. Due to the scattering on this potential a part of the B -particles returns back and a part escapes to infinity.

If the black hole is surrounded by a mirror shell which reflects the out-going quanta and sends them back to the black hole the number of B -particles in the in-mode y will be the same as the number of B -particles in the up-mode y and hence

$$n_{y+} \equiv n_{\theta_{BH}}(\omega_y) . \tag{4.27}$$

The state for which in-modes and up-modes are thermally excited is known as the Hartle-Hawking vacuum state (or H -vacuum). This state describes the black hole in equilibrium with a thermal radiation inside the thermal bath. We suppose that the boundaries of the thermal bath are taken to infinity. The equilibrium is possible if the radiation temperature measured at far distances from the black hole coincides with the black hole temperature (4.22).

The Hartle-Hawking vacuum state possesses the following properties. The renormalized expectation values $\langle \hat{\varphi}^2 \rangle^{ren}_H$ and $\langle T_{\mu\nu} \rangle^{ren}_H$ for this state are finite everywhere outside the black hole including both (future and past) event horizons, while at infinity they coincide with the analogous values for a thermal (with the temperature θ_{BH}) equilibrium

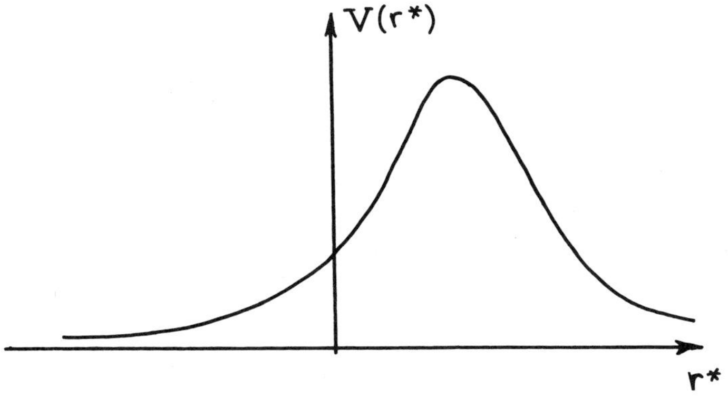

Fig.6. Effective potential for massless particle propagation in the gravitational field of a black hole

radiation in the flat spacetime given by Eqs. (I.3I) and (I.32).

The quantum state which describes the situation when the black hole is placed in an empty space and there is no incoming particles in in-modes ($n_{\mathcal{J}+} = 0$) is known as the Unruh vacuum state. Such a black hole is a source of stationary thermal radiation. One can write for the Unruh vacuum

$$\langle \hat{\varphi}^2(x) \rangle_U^{ren} = \sum_{\mathcal{J}} \varphi^2(x|\mathcal{J}-)\, n_{\Theta_{BH}}(\omega_{\mathcal{J}}) + \langle \hat{\varphi}^2(x) \rangle_B^{ren} , \tag{4.28}$$

$$\langle T_{\mu\nu}(x) \rangle_U^{ren} = \sum_{\mathcal{J}} t_{\mu\nu}(x|\mathcal{J}-)\, n_{\Theta_{BH}}(\omega_{\mathcal{J}}) + \langle T_{\mu\nu}(x) \rangle_B^{ren} . \tag{4.29}$$

The sums over modes in the right-hand-side of Eqs.(4.28) and (4.29) describe the contribution of the thermal flux. They define the asymptotic behaviour of $\langle \hat{\varphi}^2 \rangle_U^{ren}$ and $\langle T_{\mu\nu} \rangle_U^{ren}$ at far distances ($r \to \infty$) where the quantities $\langle \hat{\varphi}^2(x) \rangle_B^{ren}$ and $\langle T_{\mu\nu}(x) \rangle_B^{ren}$ describing the vacuum polarization contribution are negligibly small. Using the asymptotics (4.I0) one gets

$$\langle \hat{\varphi}^2 \rangle_U^{ren} \simeq \frac{1}{r^{D-1}} \int_0^\infty \frac{d\omega}{2\pi\omega} \sum_{N,\beta} |Y_{N\beta}(\vec{n})|^2 |T_{\omega N}|^2 \frac{1}{e^{\omega/\Theta_{BH}} - 1} , \tag{4.30}$$

$$\langle T_{tr} \rangle_U^{ren} \simeq \frac{1}{r^{D-1}} \int_0^\infty \frac{\omega\, d\omega}{2\pi} \sum_{N,\beta} |Y_{N\beta}(\vec{n})|^2 |T_{\omega N}|^2 \frac{1}{e^{\omega/\Theta_{BH}} - 1} . \tag{4.3I}$$

It is instructive to compare these formulas with Eqs.(I.59) and (I.60) for a thermal radiation of a black body in the flat spacetime. For a particular choice of the black body temperature $\theta = \theta_{BH}$ there is a close similarity between them. The difference is that the transmission coefficients $T_{\omega N}$ in Eqs. (4.30) and (4.3I) are to be calculated in the gravitational field of a black hole and they differ from $T_{\omega N}$ in the flat spacetime. The other and more important difference is that the temperature θ_{BH} of the black hole is uniquely defined by its size r_g

$$\theta_{BH} = \frac{D-2}{4\pi r_g}, \tag{4.32}$$

while the temperature and the size of a black body are two completely independent parameters. It means that only one dimensional parameter enters the integrals (4.30) and (4.3I) and one has

$$\langle \hat{\varphi}^2 \rangle_U^{ren} \simeq \frac{A}{r^{D-1}}, \tag{4.33}$$

$$\langle T_{tr} \rangle_U^{ren} \simeq \frac{B}{r^{D-1}} \theta_{BH}^2 \tag{4.34}$$

The exact values of the dimensionalless constants A and B can be found by numerical calculations. The values in the physical four-dimensional spacetime are (Fawcett and Whiting, I982); Page, I982; Elster, I983)

$$B \approx 0{,}0037 h(0) + 0{,}002\, h(1/2) + \tag{4.35}$$

$$+ 0{,}0082 h(1) + 0{,}0001 h(2) .$$

According to the picture described in this section one may assume that the black hole is surrounded by the thermal atmosphere of B -particles. A part of these particles penetrates the gravitational barier and forms the Hawking radiation at infinity. The properties of the thermal atmosphere are described in (Thorne et al, 1986) (see also Frolov, and Thorne, 1988). In the four-dimensional case the expression for the energy and angular momentum flux of a quantum massless fields of spin s from a rotating black hole can be written as a simple generalization of (4.31)

$$\langle T_{tr} \rangle = \frac{1}{4\pi r^2} \sum_{l,m,h} \int_0^\infty d\omega \left| {}_h S_{lm}(\theta; a\omega) \right|^2 n_J (1 - |R_J|^2) \frac{\omega}{2\pi} , \quad (4.36)$$

$$\langle T_{t\varphi} \rangle = - \frac{1}{4\pi r^2} \sum_{l,m,h} \int_0^\infty d\omega \left| {}_h S_{lm}(\theta; a\omega) \right|^2 n_J (1 - |R_J|^2) \frac{m}{2\pi} ,$$

where l and m are spheroidal harmonic indices, $h = \pm s$ is a helisity index,

$$n_J = [\exp(\omega - m\Omega)/\theta_{BH} - (-1)^{2s}]^{-1} , \quad (4.37)$$

and ${}_h S_{lm}(\theta; a\omega)$ is the "spin-weighted spheroidal harmonic" that carries the angular dependence of the field modes. For spinless particles in the gravitational field of a non-rotating black hole these spheroidal harmonics come to the usual spherical harmonics Y_{lm}. In (Eq.4.37) Ω is the angular velocity of a black hole of mass M and angular momentum $J = aM$. For more information about properties and parameters of the Hawking radiation see Thorne et al (1986); Novikov, and Frolov (1986) and references therein. The description of a general partial functional approach for quick calculations of probabilities, correlations functions and polarization properties for the Hawking radiation of a rotating black hole can be found in Frolov (1986); Bolashenko, Frolov (1989).

4.2 Vacuum polarization. $\langle \hat{\varphi}^2 \rangle^{ren}$ and $\langle T_{\mu\nu} \rangle^{ren}$ near a black hole

The calculation of $\langle \hat{\varphi}^2 \rangle^{ren}$ and $\langle T_{\mu\nu} \rangle^{ren}$ in the gravitational field of a black hole for different choices (H, U and B) of vacuum states is a rather complicated problem. In general case one must use numerical calculations. Before describing the results of these calculations we consider some particular cases when the exact answer can be found analytically.

Let us restrict ourselves by considering the value $\langle\hat{\varphi}^2(x_0)\rangle_H^{ren}$ for the points x_0 lying on the event horizons of a spherically symmetric $(D+1)$ -dimensional black hole. The conformal spacetime diagram for this case is shown in Fig.7. The Killing vector $\xi^\mu\partial_\mu=\partial_t$ is tangent to the horizons $H^\pm$ so that under the time translations $t\to t+t_0$ the points of the horizons $H^\pm$ are moving along their generators. The points lying at the sphere S_0 of bifurcations of the horizons are stable under the action of these transformations. The time independence of the Hartle-Hawking vacuum state and its invariance under the time reflection $t\to -t$ together with the spherical symmetry allow one to show that $\langle\hat{\varphi}^2(x_0)\rangle_H^{ren}$ does not depend on a location of the point x_0 on the horizon.

Let us choose the point x_0 on the bifurcation sphere and consider the Hadamard function $G_H(x,x_0)$ for the scalar field $\hat{\varphi}$ and the Hartle-Hawking vacuum state:

$$G_H(x,x_0)=\frac{1}{2}\langle 0;H|\hat{\varphi}(x)\hat{\varphi}(x_0)+\hat{\varphi}(x_0)\hat{\varphi}(x)|0;H\rangle\,. \qquad (4.38)$$

This function does not depend on the time coordinate t of a point x . It can be expressed in terms of the solutions of the radial equation (4.8) with $\omega=0$. These solutions are known exactly and are of the form

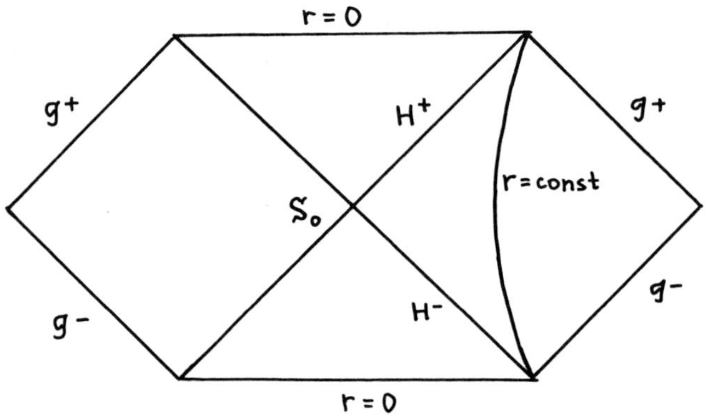

Fig.7. Conformal diagram for a static spherically symmetric $(D+1)$-dimensional eternal black hole

$$y_{0N} = a_1 P_{\frac{N-1}{D-2}}(\cosh 2\eta) + a_2 Q_{\frac{N-1}{D-1}}(\cosh 2\eta) \ , \tag{4.39}$$

where $P_\nu(z)$ and $Q_\nu(z)$ are Legendre functions, a_1 and a_2 are arbitrary constants and

$$\cosh\eta = (r/r_g)^{\frac{D}{2}-1} \ . \tag{4.40}$$

For points separated along the radial direction the calculations give (Frolov et al, 1989)

$$G_H(x,x_0) = \frac{\Gamma(D/2)}{2^{D+\frac{1}{2}}\pi^{\frac{D}{2}+1} r_g^{D-1}} \int_{2\eta}^{\infty} \frac{d\alpha \ \cosh[\alpha/2(D-2)]}{(\cosh\alpha - \cosh 2\eta)^{1/2} \sinh^{D-1}[\frac{\alpha}{2(D-2)}]} \tag{4.41}$$

where η is a parameter related by Eq.(4.40) with the radius r of the point x .

We illustrate the calculations of the explicit value of $\langle\hat{\varphi}^2(x_0)\rangle_H^{ren}$ in four- and five-dimensional cases where the integrals (4.4I) can be expressed in terms of the elementary functions. The calculation of these integrals gives

$$G_H^{(4)}(\eta) = \frac{1}{16\pi^2 r_g^2 \sinh^2\eta} \ , \tag{4.42}$$

$$G_H^{(5)}(\eta) = \frac{1}{8\pi^3 r_g^3 \sinh^3\eta}\left\{\sinh\eta + \frac{\pi}{2} + \arcsin[\cosh^{-1}\eta]\right\} . \tag{4.43}$$

The functions $G_{div}^{(D+1)}$ which are to be substracted according to the renormalization prescription in these spaces are

$$G_{div}^{(4)}(x,x_0) = \frac{1}{8\pi^2\sigma(x,x_0)} = \frac{1}{4\pi^2 r_g^2(\eta+\frac{1}{2}\sinh 2\eta)^2} \tag{4.44}$$

$$G_{div}^{(5)}(x,x_0) = \frac{1}{8\pi^2(2\sigma(x,x_0))^{3/2}} = \frac{1}{8\pi^2 r_g^3 \sinh^3\eta} \ . \tag{4.45}$$

By substracting $G_{div}(x,x_0)$ from $G_H(x,x_0)$ and taking the limit $x \to x_0$ one finally obtains the following results (Frolov et al, 1989):

$$\langle\hat{\varphi}^2(x_0)\rangle_H^{ren\,(4)} = \frac{1}{48\pi^2 r_g^2} \ , \tag{4.46}$$

$$\langle \hat{\varphi}^2(x_0) \rangle_H^{\text{ren}(5)} = \frac{1}{24\pi^3 r_g^3} . \qquad (4.47)$$

Eq.(4.46) reproduces the result obtained by Candelas (1980) for the Schwarzschild black hole in four-dimensions. Its generalization to the Kerr-Newmann black hole can be found in (Frolov, 1982).

A similar method can be used to obtain the contribution of the electromagnetic field into $\langle T_{\mu\nu}(x_0) \rangle_H^{\text{ren}}$ at the horizon of a four-dimensional black hole. Elster (1984) calculated the value of $\langle T_{\mu\nu} \rangle_H^{\text{ren}}$ on the surface of the Schwarzschild black hole. Frolov, and Zel'nikov (1985) found this quantity for the pole of the event horizon of the Kerr black hole. These results were corrected by Jensen et al (1988) who have shown that an additional term must be included in the renormalization scheme used in the mentioned papers.

Now we briefly discuss the numerical calculations of $\langle \hat{\varphi}^2 \rangle^{\text{ren}}$ and $\langle T_{\mu\nu} \rangle^{\text{ren}}$. All such calculations fulfiled till now are restricted to the case of the four-dimensional Schwarzschild black hole. The calculations begin by writing the modes expansion for these quantities with the points separated along the time coordinate. The quantities which are to be substracted according to the renormalization prescription are also presented as a similar series. The renormalization is made by term by term substracting of the obtained series expansions and by taking the coincidence limit. After singling out some finite analytical expression which speeds up the convergence of the obtained series the numerical calculations are used to determine the radial functions and to get the sum of the series.

The numerical calculations of $\langle \hat{\varphi}^2 \rangle_H^{\text{ren}}$ were done by Fawcett, and Whiting (1982); Elster (1982); Candelas and Howard (1984) and Candelas and Jensen (1986). The conformal massless scalar field contribution to $\langle T_{\mu\nu} \rangle_H^{\text{ren}}$ was numerically calculated by Howard (1984). The electromagnetic field contribution into $\langle T_{\mu\nu} \rangle_H^{\text{ren}}$ was found by Jensen and Ottewill (1988).

V. VACUUM POLARIZATION NEAR BLACK HOLES: ANSATZES AND APPROXIMATIONS

5.1 Killing ansatz and Page approximation for $\langle\hat{\varphi}^2\rangle^{ren}$

The numerical calculations of $\langle\hat{\varphi}^2\rangle^{ren}$ and $\langle T_{\mu\nu}\rangle^{ren}$ in the Schwarzschild metric give important information about the properties of these quantities. But if one wants to investigate the backreaction of quantized fields on the geometry of a black hole it is highly desirable to get some analytical expressions for these quantities. It looks like impossible to solve the problem exactly. The best thing which can be done is to use some approximations. Such approaches are discussed and compared in this last Section. All the results obtained in the framework of these approaches are applied to the four-dimensional spacetime. That is why from now on we are considering only the four-dimensional case.

We begin by describing the so called Killing ansatz proposed in (Frolov and Zel'nikov 1987 a,b). The main idea of this approach is as follows. It was shown in Section II that the quantities $\langle\hat{\varphi}^2\rangle^{ren}$ and $\langle T_{\mu\nu}\rangle^{ren}$ for the conformally invariant massless fields in the equilibrium thermal state in the homogeneous gravitational field can be found exactly. The corresponding expressions written in the general covariant form (2.41) and (2.42) are polynomials of the temperature Θ . Only even powers of Θ are present. The coefficients of these polynomials are functions of the metric $g_{\mu\nu}$ the Killing vector ξ^μ and its covariant derivatives. It is evident that if the static gravitational field is slightly inhomogeneous the expressions for $\langle\hat{\varphi}^2\rangle^{ren}$ and $\langle T_{\mu\nu}\rangle^{ren}$ must contain curvature dependent corrections. The main idea of the Killing ansatz is to approximate $\langle\hat{\varphi}^2\rangle^{ren}$ and $\langle T_{\mu\nu}\rangle^{ren}$ in general static spacetimes by expressions which are the local functions of the metric, the curvature, the Killing vector, and their covariant derivatives up to some given order. These expressions can be uniquely defined up to a small number of arbitrary constant parameters if only we assume that they are polynomials in the temperature and possess the same properties (such as conservation law and conformal trace anomaly) as the exact quantities. The arbitrary constant parameters can be chosen to satisfy the required boundary conditions.

We illustrate now this approach by applying it to the calculation of $\langle\hat{\varphi}^2\rangle^{ren}$ for the conformal massless scalar field in a static spacetime. Let ξ^μ be the Killing vector in this spacetime. The exact value of $\langle\hat{\varphi}^2\rangle^{ren}$ for a thermal equilibrium state in a static spacetime possesses the following properties:

(i) The dimension of $\langle\hat{\varphi}^2\rangle^{ren}$ is $(\text{length})^{-2}$.

(ii) $\langle\hat{\varphi}^2\rangle^{ren}$ does not depend on time ($\xi^\mu\partial_\mu\langle\hat{\varphi}^2\rangle^{ren}=0$) and is invariant under the reflection of time.

(iii) In the high-temperature limit $\langle\hat{\varphi}^2\rangle^{ren}$ has the following asymptotic behaviour

$$\langle\hat{\varphi}^2\rangle^{ren} \approx -\frac{1}{12}\frac{\theta^2}{\xi^2} \,. \qquad (5.I)$$

(iy) Under the conformal transformations it transforms in such a way that the quantity

$$|g|^{1/4}\left(\langle\hat{\varphi}^2\rangle^{ren} + \frac{1}{288\pi^2}R\right) \qquad (5.2)$$

remains invariant.

It can be shown (Frolov and Zel'nikov, I987b) that the most general expression $\langle\hat{\varphi}^2\rangle_K$ which is a polynomial in θ is constructed from ξ^μ, curvature and their covariant derivatives and satisfies the conditions (i)-(iy) reads

$$\langle\hat{\varphi}^2\rangle_K = -\frac{1}{12}\frac{\theta^2}{\xi^2} - \frac{R}{288\pi^2} + c\left(w^2 + R_{\alpha\beta}\zeta^{\alpha\beta} - \frac{1}{6}R\right), \qquad (5.3)$$

where

$$w^2 = w_\mu w^\mu \,, \qquad w_\mu = \frac{1}{2}\nabla_\mu \ln|\xi^2| \,,$$
$$\zeta^{\alpha\beta} = \frac{\xi^\alpha\xi^\beta}{\xi^2} \,, \qquad (5.4)$$

and c in an arbitrary dimensionless constant. If one requires that Eq.(5.3) in the homogeneous gravitational field coincides with Eq.(2.4I) then the value of this constant is

$$c = -\frac{1}{48\pi^2} . \tag{5.5}$$

Hence in the framework of the Killing ansatz we obtain uniquely defined expression for

$$\langle \hat{\varphi}^2 \rangle_K = -\frac{1}{12}\left[\frac{\theta^2}{\xi^2} + \frac{w^2}{4\pi^2}\right] - \frac{1}{48\pi^2} R_{\alpha\beta} \zeta^{\alpha\beta} , \tag{5.6}$$

which does not contain any arbitrary parameters at all.

In Einstein ($R_{\mu\nu} = \Lambda g_{\mu\nu}$, $\Lambda = \text{const}$) static spaces this expression coincides with the result obtained by Page (1982). The general idea of the ansatz proposed by Page is as follows. He noticed that the conformal trace anomaly for a scalar massless field vanishes in a space with the metric $d\tilde{s}^2 = |\xi^2|^{-1} ds^2$ if only ds^2 is the metric of a static Einstein space and ξ^μ is a Killing vector. Page suggested to use the Gaussian approximation for the Green function in the space $d\tilde{s}^2$ to get $\langle \hat{\varphi}^2 \rangle^{ren}$ and $\langle T_{\mu\nu} \rangle^{ren}$ in it and after to determine the value of these quantities in the physical spacetime by using the conformal transformation. This approximation gives $\langle \hat{\varphi}^2 \rangle^{ren} = \theta^2/12$ in the space $d\tilde{s}^2$ and yields the expression (5.6) in the physical Einstein spaces ($R_{\alpha\beta} = \Lambda g_{\alpha\beta}$). Later Candelas and Howard (1984) and Candelas and Jensen (1986) have shown that this approximation for $\theta = \theta_{BH}$ agrees up to high accuracy with the results of the numerical calculations for $\langle \hat{\varphi}^2 \rangle_H^{ren}$ not only everywhere outside the black hole but also inside it up to the radius $r \sim r_g/2$.

It should be emphasized that the expression (5.6) is also applicable to more general cases. In particular for the Reissner-Nordström metric

$$\begin{aligned} ds^2 &= -F dt^2 + F^{-1} dr^2 + r^2 d\omega_2^2 , \\ F &= r^{-2}(r - r_+)(r - r_-) , \\ r_\pm &= m \pm \sqrt{m^2 - Q^2} , \end{aligned} \tag{5.7}$$

describing a charged (with charge Q) black hole it gives

$$\langle \hat{\varphi}^2 \rangle_K = \frac{1}{12}\left(\frac{\varkappa}{2\pi}\right)^2 \frac{(r + r_+)(r^2 + r_+^2)}{r^2(r - r_-)} , \tag{5.8}$$

where $\varkappa = (r_+ - r_-)/2r_+^2$. At the horizon $r = r_+$ this quantity coincides with exact value $\langle \hat{\varphi}^2 \rangle_H^{ren}$ (Frolov, 1982).

For $\theta = 0$ the expression (5.6) gives the approximation for the expectation value $\langle \hat{\varphi}^2 \rangle_B^{ren}$ in B -vacuum. In particular it reproduces the correct (divergent) behaviour at the horizon of the Schwarzschild black hole.

5.2 $\langle T_{\mu\nu} \rangle^{ren}$ for conformal massless fields in static four-dimensional spacetimes: Killing ansatz and and BOP-approach

Let us consider now the renormalized vacuum or equilibrium thermal average of the stress-energy tensor $\langle T_{\mu\nu} \rangle^{ren}$ for conformal massless fields in a static spacetime. This quantity possesses the following properties:

(i) The dimension of $\langle T_{\mu\nu} \rangle^{ren}$ is $(\text{length})^{-4}$.

(ii) $\langle T_{\mu\nu} \rangle^{ren}$ does not depend on time

$$\mathcal{L}_\xi \langle T_{\mu\nu} \rangle^{ren} = 0 , \tag{5.9}$$

and is invariant under the reflection of time. ($\mathcal{L}_\xi$ is the Lie derivative and ξ is a Killing vector).

(iii) $\langle T_{\mu\nu} \rangle^{ren}$ obeys the covariant conservation law

$$\langle T^\mu{}_\nu \rangle^{ren}{}_{;\mu} = 0 . \tag{5.10}$$

(iy) In the high-temperature limit $\langle T_{\mu\nu} \rangle^{ren}$ has the following asymptotic behaviour

$$\langle T_{\mu\nu} \rangle^{ren} = a \frac{\theta^4}{\xi^4} (g_{\mu\nu} - 4 \zeta_{\mu\nu}) , \tag{5.11}$$

where

$$a = \frac{\pi^2}{90} [h(0) + \frac{7}{8} h(1/2) + h(1)] . \tag{5.12}$$

(y) $\langle T_{\mu\nu} \rangle^{ren}$ possesses the conformal trace anomalies given by Eq.(3.8).

The expression for $\langle T_{\mu\nu} \rangle_K$ in the framework of the Killing ansatz, i.e. the most general expression which is a polynomial in θ, is constructed from ξ^μ, curvature and their covariant derivatives and satisfies the conditions (i)-(y) is given in (Frolov and Zel'nikov, 1987b). In static Einstein spaces $\langle T_{\mu\nu}\rangle_K$ can be obtained by varying the following effective action

$$\langle T^{\mu\nu}\rangle_K = \left\{ \frac{2}{\sqrt{g}} \frac{\delta}{\delta g_{\mu\nu}} \int d^4x \sqrt{g}\, L_K \right\}_{R_{\mu\nu}=\Lambda g_{\mu\nu}} , \tag{5.13}$$

provided $\delta\xi^\mu / \delta g_{\alpha\beta} = 0$. Here

$$\begin{aligned} L_K = {} & a \frac{\theta^4}{\xi^4} + c \frac{\theta^2}{\xi^2}(R + 6w^2) + \frac{2}{3}\alpha E^2 + \\ & + (\alpha+\beta)[-4w^2E + 6E^2 + 2ER - 4w_\alpha w_\beta R^{\alpha\beta}] + \\ & + d[w^2 - E - \tfrac{1}{6}R]^2 + e[-w^4 + 2w^2E + E^2 + w^2R - 2w_\alpha w_\beta R^{\alpha\beta}] + \\ & + f[\tfrac{8}{3}E^2 + (R_{\alpha\beta}R^{\alpha\beta} + \tfrac{1}{3}R^2)\ln|\xi^2|] , \end{aligned} \tag{5.14}$$

and $E \equiv w^\varepsilon{}_{;\varepsilon}$. The coefficient a is given by Eq.(5.12), α and β are the coefficients of conformal anomalies given by Eq.(3.8), and c, d, e and f are arbitrary constants. In order to use the Killing ansatz expression (5.13) as an approximation one needs to specify the value of these constants. It can be done if one exploits the boundary conditions. It should be also noted that in the Ricci-flat spacetimes the contribution to $\langle T_{\mu\nu}\rangle_K$ by the term with the coefficient f in Eq.(5.14) vanishes. In a particular case of an isolated static black hole the unknown constants can be defined if one requires that at the horizon $\langle T_{\mu\nu}\rangle_K$ is finite and coincides with the exact value $\langle T_{\mu\nu}\rangle^{ren}_H$. This uniquely defined tensor $\langle T_{\mu\nu}\rangle_K$ gives rather good approximation for $\langle T_{\mu\nu}\rangle^{ren}_H$ for the scalar and electromagnetic field everywhere outside the black hole and automatically gives the exact answers at the horizon and at infinity.

The Killing ansatz can be slightly generalized in order to obtain an approximation for $\langle T_{\mu\nu}\rangle^{ren}_U$. As we know there is a radiation propagating from a black hole to infinity in the U -vacuum state. Thus $\langle T_{\mu\nu}\rangle^{ren}_U$ is not invariant under the reflection of time though it is still time independent. This violation of the condition (ii) makes it possible the existence of the additional terms in the expression for $\langle T_{\mu\nu}\rangle_K$ Vaz (1988) showed that an addition of the term

$$g \frac{\theta^3}{\xi^2} \xi_{(\mu} w_{\nu)} \tag{5.15}$$

to the approximate value for $\langle T_{\mu\nu}\rangle_H$ allows one to describe the Hawking radiation with rather high accuracy. The property of finiteness of the obtained tensor at the future event horizon is the boundary condition which determines the coefficient g in Eq.(5.I6).

Now we discuss the relation of the Killing ansatz with the Brown-Ottewill-Page (BOP) approach. This approach is a generalization of the original method proposed by Page (I982) and based on using the conformal space $d\tilde{s}^2 = \omega^{-2} ds^2$, $\omega^2 = -\xi^2$. Later Brown and Ottewill (I985) suggested a slightly different approach. They pointed out that there is an ambiguity ($\omega^2 \exp(2\lambda t)$) in the choice of the conformal factor for which the trace anomaly vanishes in the space $d\tilde{s}^2$. Considering the problem of the vacuum polarization near a static black hole they have shown that if one takes λ to be equal to the surface gravity of the black hole and put $\langle \tilde{T}^{\mu\nu}\rangle^{ren} = 0$ in the conformal spacetime one obtaines the approximate value for $\langle T_\mu{}^\nu\rangle^{ren}$ in the Hartle-Hawking vacuum state. The choice $\lambda = 0$ gives the approximation for the Boulware vacuum state. This approach was applied to the conformal massless fields of spin $s = 0, 1/2$ and I. It reproduces the Page's result for $s = 0$. In later papers (Zannias (I984); Brown et al (I986)) it was proposed to use more general ansatz $\langle \tilde{T}^{\mu\nu}\rangle^{ren} = \xi^\mu \xi^\nu \langle \tilde{T}_\alpha{}^\alpha\rangle^{ren}$ for the renormalized stress-energy tensor in the conformal spacetime $d\tilde{s}^2$.

It can be shown that the expression for the stress-energy tensor $\langle T_{\mu\nu}\rangle_{BOP}$ in the physical spacetime in the BOP-approximation coincides with Eq.(5.I3) for the following special choice of the coefficients in Eq.(5.I4)

$$d = -4\alpha - 6\beta + 3\gamma, \quad e = 2\alpha, \quad f = -\alpha, \quad c = \frac{1}{8\pi^2}(\beta - \alpha). \qquad (5.I6)$$

In the homogeneous gravitational field limit $\langle T_{\mu\nu}\rangle_{BOP}$ coincides with Eq.(2.42). But for $s = 1$ this ansatz gives the wrong value for $\langle T_{\mu\nu}\rangle_H^{ren}$ at the horizon of a black hole.

For discussion of other approaches and approximations see (Bernard and Folacci, I986; Castagnino et al I987; Tadaki, I987; Vaz, I988).

The numerical results and the analytical methods based on different approximations described in this Section are only the first step in solving the general problem of constructing a selfconsistent model for a quantum black hole. Only a few results were obtained in this direction till now. The main work is still ahead.

APPENDIX

Some helpful formulas are collected in this appendix. They are refered to in the main text and can be used to verify the calculations omitted in the lectures.

$$\int_0^\infty \frac{x^{z-1}\,dx}{e^x - 1} = \Gamma(z)\zeta(z) , \qquad \text{(A.I)}$$

$$\int_0^\infty \frac{x^{z-1}\,dx}{e^x + 1} = [1 - 2^{(1-z)}]\,\Gamma(z)\zeta(z), \qquad \text{(A.2)}$$

$\zeta(z)$ is the Riemann zeta-function.

The asymptotics of the Hankel functions for a large real z are

$$H^{(1)}_\nu(z) \sim \sqrt{\frac{2}{\pi z}}\; e^{i(z - \nu\pi/2 - \pi/4)} , \qquad \text{(A.3)}$$

$$H^{(2)}_\nu(z) \sim \sqrt{\frac{2}{\pi z}}\; e^{-i(z - \nu\pi/2 - \pi/4)} \qquad \text{(A.4)}$$

The spherical harmonics $Y_{N\beta}(\vec{n})$ on the sphere S^{D-1} satisfy the relation

$$\sum_\beta |Y_{N\beta}(\vec{n})|^2 = \frac{(N-2+D/2)\,\Gamma(N+D-3)\,\Gamma(D/2-1)}{2\pi^{D/2}\,\Gamma(N)\,\Gamma(D-2)} \qquad \text{(A.5)}$$

Some useful integrals containing $K_{i\nu}(z)$ are given below

$$\int d^{D-1}\vec{q}\, f(q)\, e^{i\vec{q}\vec{\xi}} = \sqrt{\pi}\,\Gamma\left(\frac{D}{2}-1\right)\Omega_{D-3}\left(\frac{2}{\xi}\right)^{\frac{D-3}{2}} \cdot$$

$$\cdot\int_0^\infty dq\, q^{\frac{D-1}{2}} J_{\frac{D-3}{2}}(q\xi)\, f(q) \qquad \text{(A.6)}$$

$$\int_0^\infty dq\, q^{\frac{D-1}{2}} J_{\frac{D-3}{2}}(q\xi) K_{i\nu}(q\rho) K_{i\nu}(q\rho') =$$

$$= \frac{\sqrt{\pi}\,\xi^{\frac{D-3}{2}}}{2^{3/2}(\rho\rho')^{\frac{D-1}{2}}}(u^2-1)^{-\frac{D-2}{4}}\,\Gamma\left(\frac{D-1}{2}+i\nu\right)\cdot \qquad \text{(A.7)}$$

$$\cdot\Gamma\left(\frac{D-1}{2}-i\nu\right) P^{1-\frac{D}{2}}_{i\nu-\frac{1}{2}} \quad , \qquad u \equiv \frac{\xi^2+\rho^2+\rho'^2}{2\rho\rho'}$$

$$\int d\vec{q}\, |K_{i\nu}(q\rho)|^2 = \frac{\Omega_{D-2}}{\rho^{D-1}}\, B_D(\nu) \qquad \text{(A.8)}$$

where

$$B_D(\nu) \equiv \int_0^\infty dx\cdot x^{D-2}|K_{i\nu}(x)|^2 = \frac{2^{D-4}}{\Gamma(D-1)}\left[\Gamma\left(\frac{D-1}{2}\right)\right]^2 \left|\Gamma\left(\frac{D-1}{2}+i\nu\right)\right|^2 ,$$

$$B_{D+2}(\nu) = \frac{D-1}{4D}\left[(D-1)^2+4\nu^2\right] B_D \quad , \qquad \text{(A.9)}$$

$$\int dx\cdot x^D |K'_{i\nu}(x)|^2 = \frac{1}{4D}\left[(D-1)^2(D+1)+4\nu^2\right] B_D \, .$$

(A.IO)

$$|\Gamma(k+i\nu)|^2 = \frac{\pi\nu}{\sinh(\pi\nu)}(\nu^2+1)\cdots\left[\nu^2+(k-1)^2\right] ,$$

$$\left|\Gamma\left(k+\frac{1}{2}+i\nu\right)\right|^2 = \frac{\pi}{\cosh(\pi\nu)}\left(\nu^2+\frac{1}{4}\right)\cdots\left[\nu^2+\left(k-\frac{1}{2}\right)^2\right] . \qquad \text{(A.II)}$$

REFERENCES

Adler S.L., Liberman J., and Ng Y.J. 1977. Ann.Phys. (N.Y) <u>106</u>, 279.

Allen B., Folacci A., and Ottewill A.C. 1988, Phys.Rev. D38, 1069.
Bander M., and Itzykson C. 1966. Rev.Mod.Phys. 38, 330, 346.
Bernard D., and Folacci A. 1986. Phys.Rev. D34, 2286.
Bolashenko P.A., and Frolov V.P. 1989. In: Quantum theory and Gravitation. Proc. of the Lebedev Physics Institute, v.197 (Ed. M.A.Markov, Nova Science Publ., Commack).
Boulware D.G. 1975a. Phys.Rev. D11, 1404.
Boulware D.G. 1975b. Phys.Rev. D12. 350.
Boulware D.G. 1976. Phys.Rev. D13, 2169.
Birrell N.D., and Davies P.C.W. 1982. Quantum Fields in curved space. (Cambridge Univ.Press, Cambridge, England).
Brown M.R. 1984. J.Math.Phys. 25, 136.
Brown M.R., and Ottewill A.C. 1983. Proc.Roy.Soc. London A389, 379.
Brown M.R., and Ottewill A.C. 1985. Phys.Rev. D31, 2514.
Brown M.R., Ottewill A.C., and Page D.N. 1986. Phys.Rev. D33, 2840.
Candelas P. 1980. Phys.Rev. D21, 2185.
Candelas P., and Deutsch D. 1977. Proc.R.Soc. London. A254, 79.
Candelas P., and Deutsch D. 1978. Proc.R.Soc. London. A362, 251.
Candelas P., and Dowker J.S. 1979. Phys.Rev. D19, 2902.
Candelas P., and Howard K.W. 1984. Phys.Rev. D29, 1618.
Candelas P., and Jensen B.P. 1986. Phys.Rev. D33, 1596.
Candelas P., and Sciama D.W. 1984. In: Quantum Theory of Gravity: Essays in Honor of the 60 Birthday of Bryce S.DeWitt (Eds. S.M.Christensen, Adam Hilger, Bristol), p.78.
Castagnino M.A., Harari D., and Nunez C.A. 1987. J.Math.Phys. 28, 184.
Christensen S.M. 1976. Phys.Rev. D14, 2490.
Christensen S.M. 1978. Phys.Rev. D17, 946.
DeWitt B.S., and Brehme R.W. 1960. Ann.Phys. (N.Y.) 9, 220.
DeWitt B.S. 1975. Phys.Repts. 19, 297.
DeWitt B.S. 1979. In:General Relativity (Eds. S.W.Hawking and W.Israel, Cambridge Univ.Press), p.680.
Dowker J.S. 1983. Phys.Rev. D28, 3013.
Dowker J.S. 1986. Phys.Rev. D33, 3150.
Dowker J.S. 1987a. Phys.Rev. D36, 3095.
Dowker J.S. 1987b. Phys.Rev. D36, 3742.
Elster T. 1982. Phys.Lett. A93, 58.

Elster T. I983. Phys.Lett. A94, 205.
Elster T. I984. Class.Quant.Grav. I, 43.
Fawcett M.S. , and Whiting B. I982. In: Quantum Theory of Space and Time (Eds. M.J.Duff and C.J.Isham, Cambridge Univ. Press, Cambridge, England).
Frolov V.P. I982. Phys.Rev. D26, 954.
Frolov V.P. I987. In: The Physical Effects in the Gravitational Field of Black Holes. Proc of the Lebedev Physics Institute, v.I69, (Ed.M.A.Markov, Nova Science Publ., Commack), I.
Frolov V.P., Mazzitelli F.D., and Paz J.P. I988. Quantum effects near multidimensional black holes. Preprint, Instituto de Astronomia y Fisica del Espacio, Buenos Aires.
Frolov V.P., and Serebriany E.M. I987. Phys.Rev. D35, 3779.
Frolov V.P., and Sanchez N. I986. Phys.Rev. D33, I604.
Frolov V.P., and Thorne K.S. I988. Preprint GRP-I63, Caltech; Phys.Rev. D (to appear).
Frolov V.P., and Zel'nikov A.I. I985. Phys.Rev. D32, 3I50.
Frolov V.P., and Zel'nikov A.I. I987a. Phys.Lett. BI93, I7I.
Frolov V.P., and Zel'nikov A.I. I987b. Phys.Rev. D35, 303I.
Fulling S.A., and Ruijsenaars S.N.M. I987. Phys.Repts. I52, I35.
Ginzburg V.L., and Frolov V.P. I987. Sov.Phys.Uspekhy, 30, I073.
Hadamard J. I923. Lectures on Cauchy's Problem in Linear Partial Differential Equations (Yale Univ.Press).
Hartle J.P., and Hawking S.W. I976. Phys.Rev. DI3, 2I88.
Hawking S.W. I974. Nature 248, 30.
Hawking S.W. I975. Commun.Math.Phys. 43, I99.
Hawking S.W. I979. In: General Relativity, An Einstein Centenary Survey (Eds. S.W.Hawking and W.Israel, Cambridge Univ. Press, Cambridge, England).
Helliwell T.M., and Konkowski. I986. Phys.Rev. D34, I9I8.
Howard K.W. I984. Phys.Rev. D30, 2532.
Jensen B., and Ottewill A. I988. Renormalized electromagnetic stress tensor in Schwarzschild spacetime. Preprint.
Jensen B., McLaughlin J., and Ottewill A. I988. Renormalized electromagnetic energy density on the horizon of a Kerr black hole. Preprint.

Laflamme R. I988. Geometry and Thermofields. Preprint DAMTP/R--88/2, Cambridge.
Kirchhoff G. I860. Ann.Phys. I9, 275.
Linet B. I987. Phys.Rev. D35, 536.
Loustó C.O., and Sanchez N. I989. Back reaction effects in Black hole space-times. Preprint Meudon, France.
Misner C.W., Thorne K.S., and Wheeler J.A. I973. Gravitation (Freeman, San Francisco).
Novikov I.D., and Frolov V.P. I986. Black Hole Physics (Nauka, Moscow, in Russian). English edition by Reidel Publ.Co., Holland, I989.
Page D.N. 1976. Phys. Rev. D13, 198.
Pascual P., Taron J., and Tarrach R. I988. Phys.Rev. D38, 37I5.
Riegert R.J. I984. Phys.Lett. BI34, 56.
Sciama D.W., Candelas P., and Deutsch D. I98I. Advances in Phys. 30, 327.
Tadaki S. I987. Progr. Theor. Phys. 77, 67I.
Tadaki S. I988. Hadamard Regularization and Conformal Transformation. Preprint KUNS-942, Kyoto Univ., Japan.
Takagi S. I986. Progr.Theor.Phys.Suppl. 88, I.
Tangherlini F.R. I963. Nuov.Cim. 77, 636.
Thorne K.S., Price R.H., and Macdonald D.A. I986. Black Holes: The Membrane Paradigm. (Yale University Press, New Haven,Conn.)
Unruh W.G. I976. Phys.Rev. DI4, 870.
Unruh W.G., and Wald R.M. I984. Phys.Rev. D29, I047.
Vaz C. I988. Approximate stress-energy tensor for evaporating black holes. Preprint JHU-TIPAC-8806, Baltimore, USA.
Wald R.M. I977. Commun.Math.Phys. 54, I.
Wald R.M. I978. Phys.Rev. DI7, I477.
Zannias T. I984. Phys.Rev. D30, II6I.

VACUUM STATES IN SPACETIMES WITH KILLING HORIZONS

Robert M. Wald
Enrico Fermi Institute and Department of Physics
University of Chicago
Chicago, IL 60637

1. INTRODUCTION

Soon after the discovery by Hawking in 1974 that a black hole formed by gravitational collapse will radiate a thermal distribution of field quanta, a number of authors showed that, in particular spacetimes, when a quantum field is in a certain "natural vacuum state," then appropriate observers "see" a thermal distribution of particles. For the ordinary vacuum state of Minkowski spacetime, Unruh (1976) obtained such a result for accelerating observers. In extended Schwarzschild spacetime, Hartle and Hawking (1976) and Israel (1976) defined a natural vacuum state (the "Hartle-Hawking vacuum"), which is a thermal state for a static observer. Similarly, for de Sitter spacetime, Gibbons and Hawking (1977) defined the "Euclidean vacuum state" and showed that it has thermal properties for any inertial (geodesic) observer.

The above spacetimes have in common the presence of a one-parameter group of isometries which have an associated bifurcate Killing horizon (see section 2). The "natural vacuum state" is one which is invariant under these isometries. Thus, it might be expected that the above results could be generalized to other spacetimes.

In these lectures, I shall describe the results of research -- done in collaboration with Bernard S. Kay -- which carries out such a generalization. We consider a linear scalar field in an essentially arbitrary globally hyperbolic spacetime which possesses a group of isometries with a bifurcate Killing horizon. (The consideration of a scalar field is not essential, but our methods rely heavily on the linearity of the field.) In these spacetimes, we prove the uniqueness (but not existence) of isometry invar-

Quantum Mechanics in Curved Space-Time
Edited by J. Audretsch and V. de Sabbata
Plenum Press, New York, 1990

iant vacuum states which are regular in the sense that their two-point distribution is everywhere of the Hadamard form (see section 3 below). Thus, we generalize the notion of a "natural vacuum state" to a wide class of spacetimes. Furthermore, we prove that this state, if it exists, is a thermal (KMS) state -- at least, for a "large" subclass of observables -- at the Hawking temperature for observers following the Killing orbits in one of the "wedges" where these orbits are timelike. Finally, we also prove that for certain classes of spacetimes -- including Kerr and Schwarzschild-de Sitter spacetimes -- there does not exist any isometry invariant, nonsingular state of the quantum field. Thus, the above results of Unruh, Hartle and Hawking, Israel, and Gibbons and Hawking, do not extend to these spacetimes.

Some cautionary remarks are in order with regard to the interpretation of our results. Our results provide precise mathematical statements about possible states of a quantum field in spacetimes with Killing horizons. However, any argument that the "natural vacuum state" of these spacetimes (if it exists) should arise (even approximately) by some natural initial conditions or by some physical process would have to be made separately. In some cases -- such as for the Euclidean vacuum state of de Sitter spacetime -- such an argument can be given. On the other hand, for the case of extended Schwarzschild spacetime, the Hartle-Hawking vacuum possesses a very high degree of initial correlation between the incoming thermal radiation from infinity and the state of the field inside the white hole. It is difficult to imagine how this state could naturally occur as a result of any physical process. Thus, it should be emphasized that the existence of the Hartle-Hawking vacuum in extended Schwarzschild spacetime does not provide a valid argument that a physical Schwarzschild black hole (produced by gravitational collapse) would radiate thermally. Such an argument, however, is provided by the original Hawking derivation. Conversely, our proof of nonexistence of a similar state in (extended) Kerr spacetime does not affect the validity of previously derived results for radiation by Kerr black holes formed by gravitational collapse. Thus, although our results establish a very general relationship between Killing horizons and thermal radiation, considerable caution should be used in interpreting our results as saying, for example, that, physically, Killing horizons must "radiate thermally."

The plan of these lectures is to give a relatively informal account of our methods and results. We begin by defining the class of spacetimes under consideration and discussing some of their properties. Next, we review the theory of quantum fields in curved spacetime (introducing the algebraic viewpoint) and define the class of states we consider. In the final section we describe our theorems on uniqueness and thermal properties of these states.

Finally, I wish to stress that all the results described here were obtained jointly with Bernard S. Kay. Full details of this work have been given in Kay and Wald (1989), and I shall frequently allude to this reference for the proofs and other technical details. The notation and conventions follow that of Wald (1984).

2. KILLING HORIZONS AND GLOBAL HYPERBOLICITY

An isometry $i:M \to M$ on a spacetime (M, g_{ab}) is a diffeomorphism which leaves the metric invariant. A Killing vector field, ξ^a, is the infinitesimal generator of a one-parameter group of isometries. It satisfies

$$0 = \pounds_\xi g_{ab} = 2\nabla_{(a}\xi_{b)} \tag{2.1}$$

where ∇_a is the derivative operator associated with g_{ab} and $\pounds$ denotes the Lie derivative.

An important property of an isometry is that its action on a connected spacetime, (M, g_{ab}), is determined by its action at a single point $p \in M$ together with its induced action, $i^*: V_p \to V_{i(p)}$, on the tangent space V_p at p. This result follows from the fact that a geodesic through p is uniquely determined by its tangent at p, so the above information is sufficient to determine the action of i on all geodesics through p; however, since any point $q \in M$ can be connected to p by a sequence of broken geodesics, this determines i. For a one-parameter group of isometries i_t, this implies that a Killing vector field, ξ^a, is determined by its value and the value of its derivative $F_{ab} = \nabla_a \xi_b = \nabla_{[a}\xi_{b]}$ at any point $p \in M$. This result may also be proven directly from the equations satisfied by a Killing field; see, e.g. Wald (1984).

Consider, first, the case of a two-dimensional manifold and suppose a Killing field, ξ^a, vanishes at a point $p \in M$, $\xi^a(p) = 0$. Then ξ^a is determined by the value of the two-form F_{ab} at p, which is unique up to scaling. The induced action $i_t^*: V_p \to V_p$ of the group of isometries i_t on the tangent space V_p at p is determined by the infinitesimal action of the associated Killing field,

$$\delta v^a = \pounds_\xi v^a = F^a{}_b v^b \tag{2.2}$$

for all $v^a \in V_p$. The nature of this action depends upon the signature of the metric. (The metric enters eq. (2.2) via the raising of an index of F_{ab}.) In the case of Riemannian signature, eq. (2.2) is the same as that of an infinitesimal rotation in flat, Euclidean space. Thus $i_t^*: V_p \to V_p$ is simply an ordinary rotation on the tangent space. In particular, for $t = t_0$ corresponding to a "360° rotation," $i_{t_0}^*$ is the identity transformation on V_p, which implies that $i_{t_0}:M \to M$ also is the identity transformation. Consequently, all the orbits of i_t on M are closed (with period t_0), and in a neighborhood of p, the orbits have the structure shown in Figure 1a.

For a metric of Lorentz signature, eq. (2.2) corresponds to the action of an infinitesimal Lorentz boost in Minkowski spacetime. Thus, $i_t^*:V_p \to V_p$ is simply an ordinary Lorentz boost. It follows that i_t^* takes null vectors at p into multiples of themselves, and hence that i_t takes null geodesics through p into themselves. The orbit structure of i_t in a neighborhood of p is shown in Figure 1b.

In n>2 dimensions with a metric of Riemannian or Lorentz signature, exactly the same analysis applies for any Killing field which vanishes on an (n-2)-dimensional spacelike surface Σ. In the Lorentzian case, the pair of null surfaces (intersecting at Σ) which are generated by the null geodesics orthogonal to Σ is called a bifurcate Killing horizon. As shown in Figure 2, the Killing horizon locally divides the spacetime into the four "wedges" $\mathcal{F}$, $\mathcal{P}$, $\mathcal{L}$, and $\mathcal{R}$. In the following, we shall restrict attention to the physically relevant case n = 4, so that Σ is 2-dimensional.

Since i_t maps the null geodesics generating the Killing horizon into themselves, ξ^a must be everywhere tangent to these geodesics (and hence or-

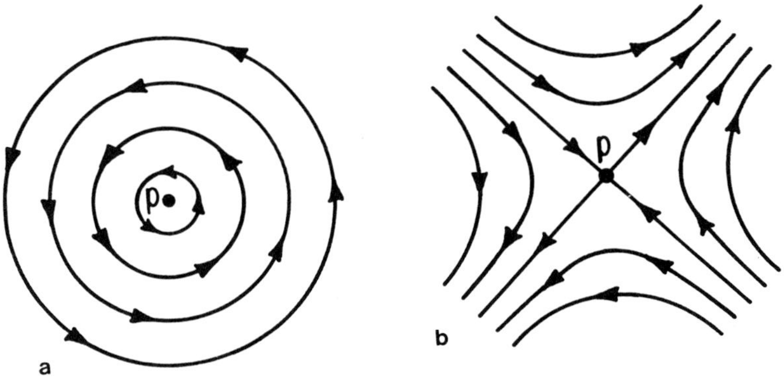

Fig. 1. The orbits of an arbitrary one parameter group of isometries with a fixed point p in two dimensions. In a neighborhood of p, the orbit structure always is as shown in case (a) if the metric has Riemannian signature, whereas it always is as shown in case (b) if the metric signature is Lorentzian.

thogonal to the Killing horizon). Let U be an affine parameter along these geodesics and let l^a be the corresponding affinely parametrized tangent. Then, on the Killing horizon, we have,

$$\xi^a = f\, l^a \tag{2.3}$$

with

$$f = \partial U/\partial u \tag{2.4}$$

where u denotes the Killing parameter. We define the surface gravity, κ, by

$$\kappa = \xi^a \nabla_a \ln f = \partial \ln f/\partial u \tag{2.5}$$

It is not difficult to show that κ must always be constant along each null geodesic generator (see, e.g., Wald, 1984). Equations (2.4) and (2.5) then show that the relationship between affine parameter, U, and Killing parameter, u, on the horizon always is an exponential one,

$$U = e^{\kappa u} \tag{2.6}$$

In a general spacetime, κ can vary from generator to generator of the horizon. However, it is well known (see, e.g., Wald, 1984) that if Einstein's equation holds with matter satisfying the dominant energy condition, then κ must be constant over the horizon.

In order to proceed further we need to make an important assumption about the global causal structure of the spacetime. Recall that a set C

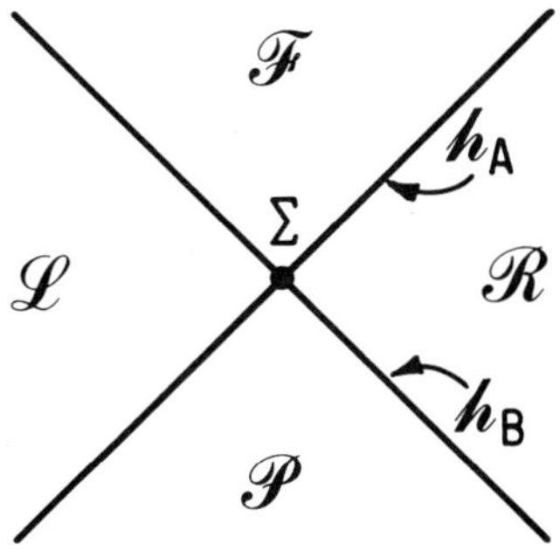

Fig. 2. A bifurcate Killing horizon, consisting of the null surfaces h_A and h_B, which intersect at the two-dimensional surface Σ. The Killing horizon locally divides spacetime into the 4 "wedges" F, P, L, and R, as shown.

in a time orientable spacetime (M, g_{ab}) is said to be a Cauchy surface if every inextendible timelike curve in M intersects $\mathcal{C}$ in precisely one point. (It then follows that $\mathcal{C}$ is a 3-dimensional, C^0, embedded submanifold of M.) A time orientable spacetime is said to be globally hyperbolic if it possesses a Cauchy surface. A number of other equivalent criteria can be given (see, e.g., the remark at the end of Chapter 8 of Wald, 1984). It is well known (Geroch, 1970) that any globally hyperbolic spacetime (M, g_{ab}) can be foliated by Cauchy surfaces of the same topology, and hence that M has topology $\mathbb{R} \times \mathcal{C}$. We require our spacetimes to be globally hyperbolic, and, in addition*, we require that it possess a smooth Cauchy surface $\mathcal{C}$ which contains Σ.

The above requirements allow us to give a global definition of the four "wedges" of spacetime shown in figure 2. We define

$$\mathcal{F} = J^+(\Sigma) \tag{2.7a}$$

$$\mathcal{P} = J^-(\Sigma) \tag{2.7b}$$

$$\mathcal{L} = I^-(h_B{}^{\mathcal{L}}) \cap I^+(h_A{}^{\mathcal{L}}) \tag{2.7c}$$

$$\mathcal{R} = I^-(h_A{}^{\mathcal{R}}) \cap I^+(h_B{}^{\mathcal{R}}) \tag{2.7d}$$

where J^+ denotes causal future, I^+ denotes chronological future, and $h_A{}^{\mathcal{R}}$ denotes the "right wedge portion" of h_A, etc. It follows (Kay and Wald, 1989), that $\mathcal{F}$, $\mathcal{P}$, $\mathcal{L}$, and $\mathcal{R}$ are disjoint (except for the intersection of $\mathcal{F}$ and $\mathcal{P}$ at Σ), and, further, that their union is itself a globally hyperbolic region of spacetime.

The most important reason for imposing global hyperbolicity is that it is necessary and sufficient to ensure a well defined dynamics of a scalar field ϕ satisfying a wave equation of the form,

$$\nabla_a \nabla^a \phi - V\phi = 0 \tag{2.8}$$

or other type of field satisfying a similar linear wave equation. (Here V is any smooth function on spacetime; it could represent a non-minimal coupling term and/or an additional external potential.) Specifically, in any globally hyperbolic spacetime, there exist unique advanced and retarded Green's functions (Choquet-Bruhat, 1968). Closely related to this property is the fact that eq. (2.8) has a well posed initial value formulation for a smooth spacelike Cauchy surface, $\mathcal{C}$, in the following sense (see, e.g., Hawking and Ellis, 1973, or Wald, 1984): Given the pair of smooth functions

---

*If Σ is compact, the existence of a Cauchy surface containing Σ follows from global hyperbolicity alone; see Kay and Wald (1989).

$(\phi_0, \dot{\phi}_0)$ on C, there exists a unique solution to eq. (2.8) such that $\phi|_C = \phi_0$ and $n^a\nabla_a\phi|_C = \dot{\phi}_0$, where n^a denotes the future-directed unit normal to C. Furthermore this solution ϕ depends continuously on $(\phi_0, \dot{\phi}_0)$ in a suitable sense, and has appropriate "domain of dependence" properties.

For a globally hyperbolic spacetime (M, g_{ab}), we define S to be the set of smooth solutions to eq. (2.8) whose initial data, $(\phi_0, \dot{\phi}_0)$, on a smooth spacelike Cauchy surface C has compact support on C. (This property does not depend upon the choice of C by virtue of the "domain of dependence" property, so S is well defined.) Since eq. (2.8) is linear, S has a natural vector space structure. In addition, eq. (2.8) also implies that, for ϕ_1, $\phi_2 \varepsilon S$, the "current vector,"

$$j^a(\phi_1, \phi_2) = \phi_1\nabla^a\phi_2 - \phi_2\nabla^a\phi_1 \tag{2.9}$$

is conserved,

$$\nabla_a j^a = 0 \tag{2.10}$$

By integrating this equation over the spacetime region contained between two Cauchy surfaces and applying Gauss's law, we find that the quantity,

$$\sigma(\phi_1, \phi_2) = \int_C j^a n_a \sqrt{h} d^3x \tag{2.11}$$

is independent of choice of Cauchy surface C. (No "spatial boundary terms" contribute to the surface integral in Gauss's law on account of the compact support property of data for solutions in S.) Thus, we obtain a natural, antisymmetric, bilinear map $\sigma: S\times S \to \mathbb{R}$, which is easily seen to be non-degenerate. This gives S the natural structure of a symplectic vector space, denoted (S, σ).

As we shall discuss further in the next section, in order to make mathematical sense of the notion of a quantum field operator $\hat{\phi}(x)$, it is necessary to view it as a distribution on spacetime. We remind the reader that for a spacetime (M, g_{ab}), we refer to smooth functions of compact support on M (i.e., elements of $C_0^\infty(M)$) as <u>test functions</u>. The set of test functions $C_0^\infty(M)$, has a natural vector space structure, and a topology can be defined on $C_0^\infty(M)$ in the same manner as for $C_0^\infty(\mathbb{R}^n)$ (see, e.g., Reed and Simon, 1972). A (real) distribution D, on spacetime is simply a continuous, linear, real-valued map, $D: C_0^\infty(M) \to \mathbb{R}$. Any locally integrable function $d: M \to \mathbb{R}$ gives rise to a distribution D via,

$$D(F) = \int_M d(x)\, F(x)\, \sqrt{-g}\, d^4x \tag{2.12}$$

for all $F \epsilon C_0^\infty(M)$. Of course, there exist many distributions (such as the "δ-function") which do not arise in this manner.

On a globally hyperbolic spacetime (M, g_{ab}), there is a direct correspondence between the space, $C_0^\infty(M)$, of test functions and the space, S, of solutions to eq. (2.8) defined above. To each $F \epsilon C_0^\infty(M)$ we can associate a solution $\psi \epsilon S$ via the formula,

$$\psi(x) = \int_M [G_{adv}(x, x') - G_{ret}(x, x')] F(x') \sqrt{-g'}\, d^4x' \tag{2.13}$$

where G_{adv} and G_{ret} denote, respectively, the advanced and retarded Green's functions for eqs. (2.8). Conversely, given $\psi \epsilon S$, we can choose two Cauchy surfaces C_1, C_2 with $C_2 \subset I^+(C_1)$ and choose $\chi \epsilon C^\infty(M)$ such that $\chi(x) = 1$ for $x \epsilon I^+(C_2)$ but $\chi(x) = 0$ for $x \epsilon I^-(C_1)$. Then it is not difficult to verify that the function $F \epsilon C_0^\infty(M)$ defined by,

$$F = (\nabla_a \nabla^a - V)(\chi\psi) \tag{2.14}$$

is related to ψ by eq. (2.13). However, many other test functions also satisfy eq. (2.13); indeed, a test function F' will satisfy eq. (2.13) if and only if it differs from F by a term of the form $(\nabla_a \nabla^a - V)H$, where $H \epsilon C_0^\infty(M)$. Thus, we obtain a one-to-one map of S onto the set, $\mathcal{T}$, of equivalence classes, $\{F\}$, of test functions, with the equivalence relation defined by $F_1 \cong F_2$ if and only if there is a test function H such that $F_1 = F_2 + (\nabla_a \nabla^a - V)H$.

A distribution, D, on M is said to satisfy the wave equation (2.8) if

$$D([\nabla^a \nabla_a - V]H) = 0 \tag{2.15}$$

for all $H \epsilon C_0^\infty(M)$. (For a distribution arising from a smooth function d via eq. (2.12), this condition is equivalent to requiring d to satisfy eq. (2.8) in the ordinary sense.) Thus, the condition that D satisfy the wave equation is precisely equivalent to the condition that D be well defined on $\mathcal{T}$, i.e., D(F) is independent of the choice of representative test function, F, in the equivalence class defining an element of $\mathcal{T}$. Since we have a natural, one-to-one, onto correspondence between $\mathcal{T}$ and S, this implies that any distributional solution of eq. (2.8) on M may be viewed as a distribution on S. We will make extensive use of this fact below. Note, in addition, that a distribution on S may be viewed as a distribution defined on C_0^∞ initial data $(\phi_0, \dot{\phi}_0)$ on a Cauchy surface C. Since the quantum field operator $\hat{\phi}$ (to be introduced below) is a distributional solution of eq. (2.8), this means that we can equally well view $\hat{\phi}$ as a ("4-smeared") distribution on M or a ("3-smeared") distribution on initial data on a Cauchy surface for solutions to eq. (2.8).

We conclude this section with the following lemma, based on Green's identity, relating test functions and their corresponding solutions:

Lemma: Let (M, g_{ab}) be a globally hyperbolic spacetime. Let ϕ be any C^∞ solution of eq. (2.8) (not necessarily in S). Let $F \in C_0^\infty(M)$ and let $\psi \in S$ be the corresponding solution given by eq. (2.13). Then, we have,

$$\int_M \phi F \sqrt{-g}\, d^4x = \sigma(\phi, \psi) \tag{2.16}$$

Proof: Choose a Cauchy surface C which lies "below" (i.e., outside the causal future of) the support of F. Let λ be the advanced solution with source F,

$$\lambda(x) = \int_M G_{adv}(x, x')\, F(x') \sqrt{-g'}\, d^4x' \tag{2.17}$$

Then, we have,

$$\begin{aligned} \int_M \phi F &= \int_{J^+(C)} \phi F = \int_{J^+(C)} \phi\, (\nabla^a\nabla_a - V)\lambda \\ &= \int_{J^+(C)} \nabla^a(\phi \overleftrightarrow{\nabla}_a \lambda) + \int_{J^+(C)} \lambda(\nabla^a\nabla_a - V)\phi \end{aligned} \tag{2.18}$$

where the natural spacetime volume element $\sqrt{-g}\, d^4x$ is understood in these integrals. The second term of the last line vanishes since ϕ is a solution to eq. (2.8). The first term can be converted to a surface integral using Gauss's law. Since λ vanishes outside the causal past of F, we get a contribution only from C,

$$\int_M \phi F = \int_C (\phi \overleftrightarrow{\nabla}_a \lambda) n^a \sqrt{h}\, d^3x \tag{2.19}$$

However, since C lies outside the causal future of the support of F, the retarded solution with source F vanishes on C, and hence $\lambda = \psi$ on C. Thus, we have,

$$\begin{aligned} \int_M \phi F &= \int_C (\phi \overleftrightarrow{\nabla}_a \psi) n^a \sqrt{h}\, d^3x \\ &= \sigma(\phi, \psi) \end{aligned} \tag{2.20}$$

as we desired to show.

3. QUANTUM FIELDS IN CURVED SPACETIME

In this section, we shall review the formulation of the quantum theory of a linear scalar field ϕ, satisfying eq. (2.8), in curved spacetime. We begin, however, with a heuristic discussion of free quantum field theory in flat spacetime, followed by a mathematically precise reformulation of this free theory.

The Lagrangian density of a real, free Klein-Gordon field in Minkowski spacetime is,

$$\mathcal{L} = -\frac{1}{2} (\partial^a \phi \, \partial_a \phi + m^2 \phi^2) \tag{3.1}$$

We introduce global inertial coordinates t, x, y, z, and view ϕ evaluated on a t = constant hypersurface as the configuration variable. The conjugate momentum density is then

$$\pi = \frac{\partial \mathcal{L}}{\partial \dot{\phi}} = \dot{\phi} \tag{3.2}$$

We expand ϕ and π in plane wave modes,

$$\phi_{\vec{k}}(t) = \frac{1}{(2\pi)^{3/2}} \int e^{-i\vec{k}\cdot\vec{x}} \, \phi(\vec{x},t) \, d^3x \tag{3.3}$$

$$\pi_{\vec{k}}(t) = \frac{1}{(2\pi)^{3/2}} \int e^{-i\vec{k}\cdot\vec{x}} \, \pi(\vec{x},t) \, d^3x \tag{3.4}$$

Then, the Lagrangian, $L = \int \mathcal{L} \, d^3x$, takes the form

$$L = \frac{1}{2} \int d^3k \, [\,|\pi_{\vec{k}}|^2 - \omega_k^2 \, |\phi_{\vec{k}}|^2] \tag{3.5}$$

where $\omega_k^{\,2} = |\vec{k}|^2 + m^2$. Equation (3.5) suggests that we may view the scalar field as composed of infinitely many decoupled harmonic oscillators* labeled by the continuous index $\vec{k}$.

In the quantum theory, we represent the observables ϕ, π as self-adjoint operators, denoted $\hat{\phi}$, $\hat{\pi}$. We formally introduce annihilation (i.e., "lowering") and creation (i.e., "raising") operators $\hat{a}_{\vec{k}}$, $\hat{a}_{\vec{k}}^{\dagger}$ for each plane wave oscillator mode by,

$$\hat{a}_{\vec{k}} = \sqrt{\frac{\omega_k}{2}} \left(\hat{\phi}_{\vec{k}} + \frac{i}{\omega_k} \hat{\pi}_{\vec{k}}\right) \tag{3.6}$$

$$\begin{aligned} \hat{a}^{\dagger}_{\vec{k}} &= \sqrt{\frac{\omega_k}{2}} \left(\hat{\phi}^{\dagger}_{\vec{k}} - \frac{i}{\omega_k} \hat{\pi}^{\dagger}_{\vec{k}}\right) \\ &= \sqrt{\frac{\omega_k}{2}} \left(\hat{\phi}_{-\vec{k}} - \frac{i}{\omega_k} \hat{\pi}_{-\vec{k}}\right) \end{aligned} \tag{3.7}$$

Then we have,

---

*Note that the oscillators labeled by $\vec{k}$ and $-\vec{k}$ are not independent, since $(\phi_{\vec{k}})^* = \phi_{-\vec{k}}$ by virtue of the reality of ϕ.

$$\hat{\phi}_{\vec{k}} = \frac{1}{\sqrt{2\omega_k}} (\hat{a}_{\vec{k}} + \hat{a}^{\dagger}_{-\vec{k}}) \tag{3.8}$$

and hence,

$$\hat{\phi}(t, \vec{x}) = \frac{1}{(2\pi)^{3/2}} \int \frac{d^3k}{\sqrt{2\omega_k}} (e^{i\vec{k}\cdot\vec{x}} \hat{a}_{\vec{k}} + e^{-i\vec{k}\cdot\vec{x}} \hat{a}^{\dagger}_{\vec{k}}) \tag{3.9}$$

The standard equal time commutation relations

$$[\hat{\pi}(\vec{x}, t), \hat{\phi}(\vec{x}', t)] = -i\delta(\vec{x}, \vec{x}')\hat{I} \tag{3.10}$$

where $\hat{I}$ denotes the identity operator, then imply that the $\hat{a}_{\vec{k}}$ and $\hat{a}^{\dagger}_{\vec{k}}$ satisfy relations appropriate for lowering and raising operators, namely, the $\hat{a}_{\vec{k}}$'s commute among themselves (as do the $\hat{a}^{\dagger}_{\vec{k}}$'s) whereas,

$$[\hat{a}_{\vec{k}}, \hat{a}_{\vec{k}'}^{\dagger}] = \delta(\vec{k}, \vec{k}')\hat{I} \tag{3.11}$$

The Heisenberg equations of motion imply that $\hat{a}_{\vec{k}}$ varies with time as,

$$\hat{a}_{\vec{k}}(t) = e^{-i\omega_k t} \hat{a}_{\vec{k}}(0) \tag{3.12}$$

Thus, from eqs. (3.9) and (3.12), we see that $\hat{a}_{\vec{k}}(0)$ is the coefficient of the positive frequency part of ϕ.

We may formally construct a Hilbert space on which the above operators are defined by paralleling the procedure used for a single harmonic oscillator in ordinary quantum mechanics. We assume existence of state, denoted $|0\rangle$ and called the vacuum state, such that $\hat{a}_{\vec{k}}|0\rangle = 0$ for all $\vec{k}$. The Hilbert space of all states is generated by repeated applications of the operators $\hat{a}^{\dagger}_{\vec{k}}$ to $|0\rangle$. A state of the form $\hat{a}^{\dagger}_{\vec{k}_1} \ldots \hat{a}^{\dagger}_{\vec{k}_n}$ is interpreted as representing an n-particle state.

This Hilbert space construction as well as the above formal expressions for $\hat{a}_{\vec{k}}$ and $\hat{\phi}(t, \vec{x})$, do not make mathematical sense as they stand. However, the main problems stem, in essence, simply from our working with unnormalizable states: the state $\hat{a}^{\dagger}_{\vec{k}}|0\rangle$ formally corresponds to a single particle in the plane-wave, positive frequency mode $\frac{1}{\sqrt{2\omega_k}} e^{-i\omega_k t} e^{i\vec{k}\cdot\vec{x}}$, which does not have a finite norm. Most of the mathematical difficulties can be cured by working with normalized states.

To do so, we proceed as follows. For any solution $\psi \varepsilon S$, we define the norm

$$\|\psi\|^2 = i\sigma(\overline{\psi^+}, \psi^+) \tag{3.13}$$

with σ given by the formula (2.11), where ψ^+ denotes the positive frequency part of ψ (defined by taking the time Fourier transform of ψ and restricting it to positive values of the frequency before Fourier transforming it back) and the bar denotes complex conjugation. We construct the one-particle Hilbert space, $\mathcal{H}$, by taking the completion of S in this norm. We make $\mathcal{H}$ a vector space over $\mathbb{C}$ by defining multiplication by "i" of elements of $\mathcal{H}$ so as to correspond to multiplying the positive frequency part of solutions in S by i. We then define the complex inner product on $\mathcal{H}$ determined by

$$(\psi_1, \psi_2) = i\sigma(\overline{\psi_1{}^+}, \psi_2{}^+) \tag{3.14}$$

for all $\psi_1, \psi_2 \varepsilon S$. Note that we then automatically have,

$$\text{Im } (\psi_1, \psi_2) = \frac{1}{2} \sigma \ (\psi_1, \psi_2) \tag{3.15}$$

It is not difficult to show that $\mathcal{H}$ is isomorphic to the Hilbert space of complex valued, square integrable functions on the positive mass shell, with volume element $d^3k/2\omega_k$.

The Hilbert space of states of the quantum field is then taken to be the symmetric Fock space, $\mathcal{F}(\mathcal{H})$, associated with $\mathcal{H}$,

$$\mathcal{F}(\mathcal{H}) = \mathbb{C} \oplus \mathcal{H} \oplus (\mathcal{H} \otimes_s \mathcal{H}) \oplus \ . \ . \ . \tag{3.16}$$

where $\oplus$ denotes the direct sum and $\otimes_s$ denotes the symmetrized tensor product. Thus, an arbitrary vector $\Lambda \varepsilon \mathcal{F}(\mathcal{H})$ has the form

$$\Lambda = (\lambda_0, \lambda_1, \lambda_2, \ . \ . \ . \ ) \tag{3.17}$$

where $\lambda_0 \varepsilon \mathbb{C}$ is interpreted as the amplitude to be in the vacuum state, $\lambda_1 \varepsilon \mathcal{H}$ is the one-particle amplitude, $\lambda_2 \varepsilon \mathcal{H} \otimes_s \mathcal{H}$ is the two-particle amplitude, etc. We define annihilation and creation operators $\hat{a}(\psi) : \mathcal{F}(\mathcal{H}) \to \mathcal{F}(\mathcal{H})$, $\hat{a}^\dagger(\psi) : \mathcal{F}(\mathcal{H}) \to \mathcal{F}(\mathcal{H})$ for an arbitrary normalized one-particle state $\psi \varepsilon \mathcal{H}$ by,

$$\hat{a}(\psi)\Lambda = (\overline{\psi}\cdot\lambda_1, \ \sqrt{2}\ \overline{\psi}\cdot\lambda_2, \ . \ . \ . \ ) \tag{3.18}$$

$$\hat{a}^\dagger(\psi)\Lambda = (0, \ \lambda_0\psi, \sqrt{2}\ \lambda_1 \otimes_s \psi, \ . \ . \ . \ ) \tag{3.19}$$

It can be verified that $\hat{a}$, $\hat{a}^\dagger$ satisfy commutation relations corresponding to eq. (3.11) for plane wave modes, namely,

$$[\hat{a}(\psi_1), \ \hat{a}^\dagger(\psi_2)] = (\psi_1, \psi_2)\hat{I} \tag{3.20}$$

and that $\mathcal{F}(\mathcal{H})$ is generated by repeated applications of creation operators to the state $|0{>}\varepsilon\mathcal{F}(\mathcal{H})$ (defined as the state with $\lambda_0 = 1$ and $\lambda_i = 0$ for all

$i \geq 1$). Thus, this construction of the Hilbert space of states corresponds to the heuristic discussion given above.

We define the field operator $\hat{\phi}$ on $F(H)$ by,

$$\hat{\phi}(t, \vec{x}) = \sum_i [\psi_i^+(t, \vec{x})\, \hat{a}(\psi_i) + \overline{\psi^+}(t, \vec{x})\, \hat{a}^\dagger(\psi_i)] \tag{3.21}$$

where $\{\psi_i\}$ is an orthonormal basis of H, with each $\psi_i \varepsilon S$, and, again, the "+" denotes positive frequency part. However, although we have improved eq.(3.9) by replacing the continuous integral over unnormalizable plane wave modes with a discrete sum over a basis of normalized states, the series on the right side of eq. (3.21) does not converge pointwise at each t, $\vec{x}$. Nevertheless, we can make mathematical sense out of eq. (3.21) by interpreting it as a distribution. More precisely, for each $F \varepsilon C_0^\infty(\mathbb{R}^4)$, we define the operator $\hat{\phi}(F) = F(H) \to F(H)$ by,

$$\hat{\phi}(F) = \sum_i [(\int F \psi_i^+)\hat{a}(\psi_i) + (\int F\, \overline{\psi_i^+})\hat{a}^\dagger(\psi_i)] \tag{3.22}$$

By using the correspondence (2.13) between test functions and solutions, the definitions (3.18) and (3.19) of $\hat{a}$ and $\hat{a}^+$, and the fact that (3.14) defines an inner product, it is not difficult to show that

$$\hat{\phi}(F) = i(\hat{a}(\psi) - \hat{a}^\dagger(\psi)) \tag{3.23}$$

where ψ is the element of S corresponding to F. (In particular, this shows that eq. (3.22) is independent of the choice of basis, $\{\psi_i\}$, of H.) Since $\hat{\phi}$ is a distributional solution, we also may view it as a distribution on S. In view of the lemma proven at the end of the previous section, we will use the suggestive notation $\sigma(\hat{\phi}, \cdot)$ to denote this distribution on S, i.e., by definition we have, for all $\psi \varepsilon S$,

$$\sigma(\hat{\phi}, \psi) = i(\hat{a}(\psi) - \hat{a}^\dagger(\psi)) \tag{3.24}$$

From eqs. (3.14) and (3.20), it follows that,

$$[\sigma(\hat{\phi}, \psi_1), \sigma(\hat{\phi}, \psi_2)] = i\sigma(\psi_1, \psi_2)\hat{I} \tag{3.25}$$

This completes our review of free quantum field theory in Minkowski spacetime.

It can be seen that the key mathematical structures needed for the above construction of H (and thereby $F(H)$ and $\hat{\phi}$) are the symplectic form, σ, on S and the notion of the "positive frequency part" of a solution in S. As discussed in the previous section, the symplectic form σ is well defined for the wave eq. (2.8) in an arbitrary, globally hyperbolic spacetime. However, a natural generalization of the notion of the "positive frequency part" of a solution is readily available only in the case of a stationary spacetime. In the stationary case, if $V \geq c>0$ and the norm of the stationary Killing field is bounded away from zero (so that a "mass gap" exists), then one can parallel the steps of the flat spacetime construction to define the quantum field theory (Ashtekar and Magnon, 1975; Kay 1978). (If no mass gap is present, infra-red divergences may occur in this procedure; much more severe difficulties may occur if V becomes negative or the Killing field becomes spacelike, since there then may exist solutions in S with unbounded growth in time.) However, in a non-stationary, globally hyperbolic spacetime, there is, in general, no preferred notion of the "positive frequency part" of a solution, and hence there is no preferred construction of $\mathcal{H}$ and $\mathcal{F}(\mathcal{H})$; consequently, there is no preferred notion of a vacuum state $|0\rangle$.

It is not difficult to isolate the mathematical properties of the notion of the "positive frequency part" of a solution in S which are needed in order that the construction of quantum field theory be well defined (see p. 494 of Wald, 1979). Equivalently, these conditions may be formulated in terms of a complex structure, J, on S (see, e.g., Ashtekar and Magnon, 1975). Thus, one may proceed to construct quantum field theory in an arbitrary globally hyperbolic spacetime if one is given a suitable prescription for defining either the "positive frequency part" of solutions or J. Unfortunately, at a technical level there is an awkwardness in proceeding in this manner because normally it will be necessary to enlarge S in order to define these maps; namely, the positive frequency part of a solution in S normally will not lie in the complexification of S, nor will J map S into S. However, typically one will not know what the appropriate enlargement of S must be until the construction of $\mathcal{H}$ is completed, so, in effect, when one proceeds in this manner one must know the final result before one can get started. Fortunately, it is possible to specify the mathematical structure needed to construct $\mathcal{H}$ without the need to enlarge S. Namely, let $\mu: S \times S \to \mathbb{R}$ be any positive bilinear map which for all non-zero $\psi_1, \psi_2 \in S$ satisfies the inequality,

$$\mu(\psi_2, \psi_2) \geq \frac{1}{4} |\sigma(\psi_1, \psi_2)|^2 / \mu(\psi_1, \psi_1) \tag{3.26}$$

and is such that for each $\psi_2 \geq 0$ this inequality is saturated* in the sense

*The construction of $\mathcal{H}$ given in Kay and Wald (1989) does not require saturation of the inequality (3.26). If (3.26) is not saturated, the resulting vacuum vector in $\mathcal{F}(\mathcal{H})$ will not be a pure state. Nevertheless, our theorems on uniqueness and thermal properties (to be described in the next section)

that given $\varepsilon>0$, one can find $\psi_1 \neq 0$ such that equality holds to within ε. Given such a μ, one can complete S in the norm μ to obtain a real Hilbert space H_R. The desired one-particle Hilbert space H is then constructed by using the map $\frac{1}{2}\sigma: H_R \to H_R$ to define a complex structure on H_R and by defining a complex inner product via

$$(\psi_1, \psi_2) = \mu(\psi_1, \psi_2) + \frac{i}{2}\sigma(\psi_1, \psi_2) \tag{3.27}$$

for all $\psi_1, \psi_2 \varepsilon S$. Thus, μ becomes the real part of the inner product on H,

$$\mu(\psi_1, \psi_2) = \mathrm{Re}(\psi_1, \psi_2) \tag{3.28}$$

Details of this construction can be found in appendix A of Kay and Wald, 1989. Conversely, if one constructs a one-particle Hilbert space, H, from S by defining the notion of the "positive frequency part" of a solution or by defining a complex structure, J, then eq. (3.15) will hold automatically, and eq. (3.28) defines a bilinear map $\mu: S\times S \to \mathbb{R}$ which, by the Schwarz inequality, saturates the inequality (3.26). Thus, the specification of such a μ is necessary and sufficient to define a one-particle Hilbert space, H, and thereby $F(H)$ and $|0> \varepsilon F(H)$. Note that in this construction the smeared field operator, $\hat{\phi}(F)$ (or, equivalently, $\sigma(\hat{\phi}, \psi)$) always is given by eq. (3.23) (or, equivalently, eq. (3.24)). Note also that by eq. (3.24) and the definitions of $\hat{a}$ and $\hat{a}^\dagger$, we have

$$(\psi_1, \psi_2)_H = \langle 0 | \sigma(\hat{\phi}, \psi_1)\sigma(\hat{\phi}, \psi_2) | 0 \rangle_{F(H)} \tag{3.29}$$

Hence, we see from eq. (3.28) that μ is simply the real part of the vacuum expectation value of the 2-point distribution,

$$\mu(\psi_1, \psi_2) = \mathrm{Re} \langle 0 | \sigma(\hat{\phi}, \psi_1)\, \sigma(\hat{\phi}, \psi_2) | 0 \rangle \tag{3.30}$$

In an arbitrary globally hyperbolic spacetime, there always exist many bilinear maps μ which saturate (3.26). (One way to show this is to deform the given spacetime so that it is static in the neighborhood of some Cauchy surface. The prescription for defining quantum field theory in static spacetimes then can be used to produce a μ for the original spacetime.) Thus, one always can construct a Fock space, $F(H)$, on which the field operator acts via eq. (3.24). However, in general, the theories obtained by two different choices of μ will not be equivalent, i.e., if H_1 is constructed from μ_1 and H_2 is constructed from μ_2, in general there will not exist a unitary map $U: F(H_1) \to F(H_2)$ such that $\hat{\phi}_2 = U\hat{\phi}_1 U^{-1}$.

---

apply to this more general class of states, which we refer to as "quasi-free states".

Which choice of μ is "the correct one"? As indicated above, apart from some special cases (such as stationary spacetimes), there does not appear to be any preferred answer to this question. This situation frequently has been viewed as a serious difficulty with the formulation of quantum field theory in curved spacetime. However, one can avoid this apparent formulational difficulty simply by widening the notion of states to simultaneously encompass all states arising from any construction of the above type. The question of which is "the correct μ" then becomes on par with the question of which is "the correct state" in $\mathcal{F}(\mathcal{H})$ after μ has been chosen or, in classical theory, which is "the correct solution" to the wave equation; it is to be answered by the physics of the particular problem under consideration.

The algebraic approach to quantum field theory provides an elegant means of accomplishing this widening of the notion of states. We briefly review, now, the basic framework of the algebraic approach to quantum field theory in curved spacetime.

As discussed in the previous section, for a globally hyperbolic spacetime (M, g_{ab}), the smooth solutions to eq. (2.8) with initial data of compact support have the natural structure of a symplectic vector space (S, σ). From this space, we can construct an abstract*-algebra, A, which is isomorphic to the algebra generated by the smeared field operators $\sigma(\hat{\phi}, \psi)$ (see eq. (3.24)) acting on any Hilbert space, $\mathcal{F}(\mathcal{H})$, obtained in the manner described above. In order to notationally distinguish elements of the abstract algebra which we are about to construct from the corresponding ("concrete") operators on a Hilbert space, I will put a tilde under the algebra elements; thus $\underset{\sim}{\sigma}(\hat{\phi}, \psi)$ will denote the algebra element corresponding to the operator $\sigma(\hat{\phi}, \psi)$. One way of constructing a suitable algebra of field observables is to start with the free algebra, A_0, generated by the symbols $\underset{\sim}{\sigma}(\hat{\phi}, \psi)$ (for all $\psi \varepsilon S$) and an identity element $\underset{\sim}{I}$. In other words, A_0 consists of all formal (finite) sums (with complex coefficients) of finite products of the symbols $\underset{\sim}{\sigma}(\hat{\phi}, \psi)$ and $\underset{\sim}{I}$. Thus, an arbitrary element $\underset{\sim}{a} \varepsilon A_0$ has the form,

$$\underset{\sim}{a} = c_0 \underset{\sim}{I} + \sum_{i=1}^{N} c_i \underset{\sim}{\sigma}(\hat{\phi}, \psi_{i1}) \ . \ . \ . \ \underset{\sim}{\sigma}(\hat{\phi}, \psi_{in_i}) \tag{3.31}$$

where each $c_i \varepsilon \mathbb{C}$ and each $\psi_{ij} \varepsilon S$. We define an equivalence relation on A_0 by the condition that $\underset{\sim}{a}_1 \cong \underset{\sim}{a}_2$ if and only if the expression (3.31) for $\underset{\sim}{a}_1$ can be converted to that for $\underset{\sim}{a}_2$ by any formal manipulations which use the linearity of $\underset{\sim}{\sigma}(\hat{\phi}, \psi)$ in ψ together with the commutation relation,

$$[\underset{\sim}{\sigma}(\hat{\phi}, \psi_1), \underset{\sim}{\sigma}(\hat{\phi}, \psi_2)] = i\sigma(\psi_1, \psi_2)\underset{\sim}{I} \tag{3.32}$$

(see eq. (3.25)). The set, A, of equivalence classes of A_0 then has the natural structure of a *-algebra (with the *-map determined by formal complex conjugation), and A comprises an algebra with the desired properties.

Actually, for technical reasons, it is preferable to work with the Weyl algebra, $\mathcal{A}$. This algebra can be constructed by starting with the free algebra generated by the formal symbols

$$\underset{\sim}{W}(\psi) = \exp[-i\underset{\sim}{\sigma}(\hat{\phi}, \psi)] \tag{3.33}$$

and defining an equivalence relation analogous to that which defined A. Specifically, we have

$$\underset{\sim}{W}^*(\psi) = \underset{\sim}{W}(-\psi) \tag{3.34}$$

and the relation (3.32) is now replaced by the corresponding relation,

$$\underset{\sim}{W}(\psi_1)\ \underset{\sim}{W}(\psi_2) = \exp[-i\sigma(\psi_1, \psi_2)/2]\underset{\sim}{W}(\psi_1 + \psi_2) \tag{3.35}$$

Equations (3.34) and (3.35) are known as the Weyl relations. In this case, a natural C*-norm can be put on the resulting set of equivalence classes, and the Weyl algebra, $\mathcal{A}$, then is defined to be the C*-algebra resulting from the completion of the set of equivalence classes under this norm. It is isomorphic to the C*-algebra generated by the operators $\exp(-i\sigma(\hat{\phi}, \psi))$ in any Hilbert space construction of quantum field theory in curved spacetime. (We refer the reader to Simon (1972) for further discussion.) We shall adopt $\mathcal{A}$ as our algebra of "quantum field observables."

Given our algebra of field observables, $\mathcal{A}$, the definition of a state in the algebraic approach is remarkably simple. A <u>state</u>, ω, is nothing more than a positive linear map $\omega:\mathcal{A}\to\mathbb{C}$. Here, by positive we mean that

$$\omega(\underset{\sim}{\alpha}^*\underset{\sim}{\alpha}) \geq 0 \tag{3.36}$$

for all $\underset{\sim}{\alpha}\epsilon\mathcal{A}$. It is convenient to normalize states so that

$$\omega(\underset{\sim}{I}) = 1 \tag{3.37}$$

Given any Hilbert space $\mathcal{F}$ on which there acts a collection of unitary operators $W(\psi)$ satisfying the Weyl relations -- i.e., on which one is given a representation of the Weyl algebra -- any density matrix ρ on $\mathcal{F}$ defines a state, ω, in the algebraic sense by the formula,

$$\omega(\underset{\sim}{\alpha}) = \mathrm{tr}(\rho\alpha) \tag{3.38}$$

for all $\underset{\sim}{\alpha}\epsilon\mathcal{A}$, where $\alpha:\mathcal{F}\to\mathcal{F}$ is the representative of $\underset{\sim}{\alpha}$. Thus, as desired, the collection of all vectors and density matrices arising in all Fock space constructions of quantum field theory in curved spacetime described above are simultaneously encompassed in a unified manner by the algebraic notion of states. In addition, all vectors and density matrices arising in non-Fock representations of the Weyl algebra also are encompassed.

Conversely, given a state ω in the algebraic sense, the GNS construction (see, e.g., Simon, 1972) shows that there always exists a Hilbert space F, a (not necessarily Fock) representation of A on F, and a cyclic vector* $\Omega \varepsilon F$ such that for all $\underset{\sim}{\alpha} \varepsilon A$, we have,

$$\omega(\underset{\sim}{\alpha}) = (\Omega, \alpha\Omega) \tag{3.39}$$

where α is the representative of $\underset{\sim}{\alpha}$. (Here, by cyclicity of Ω we mean that the set of vectors of the form $\alpha\Omega$ is dense in F.) Thus, every algebraic state corresponds to some state in the usual Hilbert space sense.

Given a bilinear map $\mu: S \times S \to \mathbb{R}$ satisfying the inequality (3.26), a state on A is determined by prescribing,**

$$\omega(\underset{\sim}{W}(\psi)) = \exp(-\mu(\psi, \psi)/2) \tag{3.40}$$

The Hilbert space, F, produced by the GNS construction then has the natural structure of a Fock space, $F(H)$, with the Weyl algebra elements represented by operators corresponding to the exponentiated form of eq. (3.24), and with cyclic vector $\Omega = |0\rangle$. Furthermore, this representation will be irreducible -- and, equivalently, $\Omega = |0\rangle$ will be a pure state -- if and only if the

*There is no need to consider density matrices here because every density matrix state on a Hilbert space representation of A is equivalent (in the sense of giving rise to the same algebraic state) to a vector state on an enlarged Hilbert space. In the algebraic approach, the distinction between a pure state and a mixed state is made by defining a pure state to be one which is extremal in the convex set of all states, i.e., a pure state cannot be written as a nontrivial sum of 2 states.

**The inequality (3.26) can be seen to be necessary for the positivity of ω on algebra elements of the form $\underset{\sim}{\alpha}^*\underset{\sim}{\alpha}$ with $\underset{\sim}{\alpha} = (\underset{\sim}{W}(\psi_1) - \underset{\sim}{I}) + i(\underset{\sim}{W}(\psi_2) - \underset{\sim}{I})$ and with ψ_1 and ψ_2 chosen to be "small." That eq. (3.26) also is sufficient for positivity of ω on all algebra elements is less obvious and is most easily demonstrated by verifying that ω corresponds to the vacuum vector of an explicit Fock space construction (see Kay and Wald, 1989).

inequality (3.26) is saturated. (See Kay and Wald, 1989, for proofs of these statements and further details.) Thus, the Hilbert space construction of quantum field theory in curved spacetime described above may be viewed as a special case of the GNS construction for a state of the form (3.40), where μ saturates the inequality (3.26).

The algebraic approach gives a very satisfactory means of formulating quantum field theory in curved spacetime without forcing one to choose a particular Hilbert space representation at the outset. There also are some additional significant formulational advantages of this approach. In particular, it frequently happens that one is able to prove certain properties of a state only with respect to the observables $\sigma(\hat{\phi}, \psi')$, for ψ' in a subspace, S', of S; furthermore, one may not know whether the "modes" in S' are "complete." The statements one can make about the properties of this state are then typically rather awkward to formulate. However, in the algebraic approach, one can conveniently express such properties as follows. Any subspace $S' \subset S$ gives rise to the subalgebra $\mathcal{A}'$ of the Weyl algebra $\mathcal{A}$ generated by elements $\underset{\sim}{W}(\psi')$ with $\psi' \in S'$. The properties one wishes to express then can be formulated in terms of properties of the restriction of the algebraic state ω to the subalgebra $\mathcal{A}'$. Of course, this formulation does not help resolve any substantive issues, such as mode completeness. Nevertheless, it often provides a simple yet precise means of expressing one's results, and we shall make use of this in the next section.

If all physical observables could be expressed in terms of expectation values of products of smeared field operators (or, more precisely, expectation values of elements of $\mathcal{A}$), then the algebraic formulation of quantum field theory in curved spacetime given above would be every bit as satisfactory and complete as the usual formulation of the classical field theory. To elucidate this comment further, consider the classical field theory. The possible states of the system consist of the solutions, ϕ, to eq. (2.8). All physical observables are directly calculable in terms of ϕ. General arguments using the initial value formulation show that in any globally hyperbolic spacetime there always exists an appropriately large class of states. Thus, the only tasks remaining in classical field theory are to identify the states of physical interest and to obtain their physical properties in an explicit form. Similarly, in quantum field theory, for a globally hyperbolic spacetime, the Weyl algebra, $\mathcal{A}$, can be constructed as described above. The possible states of the system consist of the positive linear functionals, ω, on $\mathcal{A}$. All physical observables corresponding to expectation values of products of smeared field operators are directly calculable from ω. Again, there always exists an appropriately large class of states. Thus, if products of smeared field operators encompassed all physical observables, the only tasks remaining in quantum field theory in curved spacetime would be to identify the states of physical interest and obtain their physical properties in an explicit form.

However, in quantum field theory in curved spacetime there do exist observables of physical interest beyond those represented by elements of $\mathcal{A}$. Most prominent among these is the stress-energy tensor, T_{ab}, which is formally expressed as a product of unsmeared field operators. Typically, the expectation value of such additional observables cannot be defined for all algebraic states, i.e., their value is physically infinite on some states. Thus, an additional fundamental problem which arises in the theory is to determine which states possess finite values of the additional physical observables and to formulate a prescription for defining such observables on this restricted class of states.

For the stress-energy tensor, T_{ab}, it appears that the only states which physically possess finite expectation values are those for which the two-point distribution satisfies the Hadamard condition. Roughly speaking, this condition requires that for x_1 "near" x_2, we have

$$<\hat{\phi}(x_1)\hat{\phi}(x_2)> = \frac{1}{(2\pi)^2}\left[\frac{\Delta^{\frac{1}{2}}(x_1,x_2)}{\Gamma} + v(x_1, x_2)\ln\Gamma\right] + w(x_1, x_2) \qquad (3.41)$$

where Γ is the squared geodesic distance between x_1 and x_2, v is given by a formal power series $v = \Sigma v_n(x_1, x_2)\Gamma^n$, and $\Delta^{\frac{1}{2}}$ and each v_n are smooth functions which are constructed from the local spacetime geometry. (Thus, in a given spacetime, all Hadamard states have the same "singular part" of their two-point distribution; the nontrivial information which distinguishes different Hadamard distributions is contained in the smooth function $w(x_1, x_2)$). For states which satisfy the Hadamard condition, the expected value of T_{ab} can be constructed from $w(x_1, x_2)$ (up to ambiguities involving addition of conserved local curvature term) by the "point splitting" prescription; see Wald (1977, 1978) for further discussion. Since the expectation value of T_{ab} should be finite for all physically admissible states, we imposed the Hadamard condition as an additional requirement on states. It may be necessary to impose further conditions on states in order to admit other physical observables, but we will not consider any such additional restrictions here.

Our rough formulation of the Hadamard condition given above is seriously deficient for several independent reasons. We conclude this section by outlining how to give a mathematically precise formulation of the Hadamard condition. First, we must explain what is meant by the left side of eq. (3.41).

Given a state $\omega: \mathcal{A} \to \mathbb{C}$, we define the two-point distribution by λ on SxS by,

$$\lambda(\psi_1, \psi_2) = \frac{\partial^2}{\partial s \partial t} \left\{ \omega[\underset{\sim}{W}(s\psi_1 + t\psi_2)] e^{-ist\sigma(\psi_1, \psi_2)/2} \right\} \Big|_{s,t=0} \quad (3.42)$$

so that for a vacuum state $\lambda(\psi_1, \psi_2)$ corresponds to the right side of eq. (3.29). The first requirement for a state ω to satisfy the Hadamard condition is that the derivatives occurring on the right side of eq. (3.42) exist; for vacuum states, this always is the case. Using the correspondence discussed at the end of the previous section, we may view this bi-distribution λ as a bi-distribution solution Λ acting on test functions, given by,

$$\Lambda(F_1, F_2) = \lambda(\psi_1, \psi_2) \quad (3.43)$$

where F_1 and F_2 correspond, respectively, to ψ_1 and ψ_2 via eq. (2.13). The bi-distribution Λ then corresponds to the smeared form of the left side of eq. (3.41).

There are three independent difficulties with the specification of the right side of eq. (3.41): (1) The formal series defining v need not converge. (2) The right side of eq. (3.41) does not properly define a distribution, i.e., the rules for defining integrals involving the singular terms $1/\Gamma$ and $\ln\Gamma$ have not been specified. (3) Γ is defined and smooth only for sufficiently nearby points, but it is necessary to express the condition that there are no singularities in the distribution for distant, spacelike related points.

Difficulties (1) and (2) are overcome by giving a more precise specification of the "singular part" of the bi-distribution. We define,

$$G^{t,n}_{\varepsilon}(x_1, x_2) = \frac{1}{(2\pi)^2} \left\{ \frac{\Delta^{\frac{1}{2}}}{\Gamma + 2i\varepsilon(t_1 - t_2) + \varepsilon^2} + v^{(n)} \ln[\Gamma + 2i\varepsilon(t_1 - t_2) + \varepsilon^2] \right\} \quad (3.44)$$

where $v^{(n)}$ is the nth partial sum in the series for v (thereby circumventing possible convergence problems) and t is an arbitrary smooth global time function on spacetime. We overcome difficulty (3) by working in a causal normal neighborhood of a Cauchy surface $\mathcal{C}$, i.e., an open neighborhood, N, of $\mathcal{C}$ satisfying the conditions that $\mathcal{C}$ is a Cauchy surface for N and that given any $x_1, x_2 \varepsilon N$, there exists a convex normal neighborhood (in the full spacetime manifold M) which contains $J^-(x_1) \cap J^+(x_2)$. (It is proven in Kay and Wald (1989) that every spacelike Cauchy surface admits a causal normal neighborhood N.) Given a causal normal neighborhood, N, of a Cauchy surface $\mathcal{C}$, we can choose a smooth function $\chi: N \times N \to \mathbb{R}$ such that $\chi(x_1, x_2) = 1$ when $x_1 \varepsilon J^+(x_2)$ or $x_2 \varepsilon J^+(x_1)$ but $\chi(x_1, x_2) = 0$ when x_1, x_2 fail to be contained in a convex normal neighborhood in M (i.e., when $\Gamma(x_1, x_2)$ is ill-defined and/or non-smooth). A mathematically precise statement of the Hadamard condition may now be formulated as follows: A state ω is said to satisfy the Hadamard condition if its two-point distribution λ exists (see eq. (3.42)) and is such that for each positive integer n there exists a C^n-function $H^n: N \times N \to \mathbb{R}$ such that for all test functions F_1, F_2 with support in N, we have

$$\lambda(\psi_1, \psi_2) = \lim_{\varepsilon \to 0} \int_{N \times N} \Lambda_\varepsilon^{t,n}(x_1, x_2) F_1(x_1) F_2(x_2) \tag{3.45}$$

where

$$\Lambda_\varepsilon^{t,n}(x_1, x_2) = \chi(x_1, x_2) G_\varepsilon^{t,n}(x_1, x_2) + H^n(x_1, x_2) \tag{3.46}$$

with $G_\varepsilon^{t,n}$ defined by eq. (3.44) and ψ_1 and ψ_2 are given in terms of F_1 and F_2 by eq. (2.13).

It is proven by Kay and Wald (1989) that this definition is independent of choices of t, C, N, and χ, so the notion of a Hadamard state is well defined. Finally, we note that Fulling, Narcowich, and Wald (1981) have shown that in any globally hyperbolic spacetime, there always exists a large class of Hadamard states.

4. OUR RESULTS ON VACUUM STATES IN SPACETIME WITH HORIZONS

We are now in a position to outline the statements and proofs of the results which we briefly mentioned in section 1. As discussed in section 2, our analysis applies to any globally hyperbolic spacetime which possesses a bifurcate Killing horizon and satisfies the additional requirement that there exists a smooth Cauchy surface which contains the bifurcation surface, Σ. For our theorem on thermal properties, we also will require that the surface gravity, κ, be constant over the horizon and that the spacetime possesses a reflection isometry through Σ. (As shown in Kay and Wald (1989) the local existence of such an isometry automatically holds if the spacetime is analytic.) On such a spacetime we seek all states of the quantum field (in the algebraic sense) which (i) are vacuum states, (ii) satisfy the Hadamard condition, and (iii) are isometry invariant.

As discussed in the previous section, vacuum states are characterized by a bilinear map $\mu: S \times S \to \mathbb{R}$ which saturates* the inequality (3.26). According-

---

*As mentioned in a previous footnote, the proofs of our theorem on uniqueness and thermal properties make no use of the saturation of the inequality (3.26). Thus, our theorems are applicable to the more general class of "quasi-free states."

ing to eq. (3.30), this map μ is directly related to the two-point distribution $\lambda(\psi_1, \psi_2)$ by,

$$\mu(\psi_1, \psi_2) = \mathrm{Re}\lambda(\psi_1, \psi_2) \tag{4.1}$$

where, by requirement (ii), λ has the Hadamard form (3.45). Finally, the isometry invariance of the state requires that

$$\lambda(\psi_{1t}, \psi_{2t}) = \lambda(\psi_1, \psi_2) \tag{4.2}$$

where ψ_t denotes the "time translate" by Killing parameter t of the solution ψ. Thus, we are led to seek all isometry invariant bi-distributions, λ, on S which satisfy the Hadamard condition and whose associated μ, given by eq. (4.1), satisfies (and saturates) the inequality (3.26).

The key idea underlying our results may now be explained. The right side of eq. (3.45) defining the Hadamard condition has a (two-variable) form similar to the (single variable) form of the left side of eq. (2.16). Therefore, by a "double application" of the lemma of section 2, one might expect that,

$$\lambda(\psi_1,\psi_2) = \lim_{\varepsilon\to 0} \int_{C\times C} \psi_1(x_1)\psi_2(x_2)\, \overleftrightarrow{\nabla}_{1a}\, \overleftrightarrow{\nabla}_{2b}\, \Lambda_\varepsilon^{t,n}(x_1,x_2)\, \sqrt{h_1}\, \sqrt{h_2}\; d^3x_1 d^3x_2 \tag{4.3}$$

where C is any Cauchy surface. However (much to our initial surprise!) it turns out that eq. (4.3) fails to hold in general (although a form of eq. (4.3) where the integrals are performed over different Cauchy surfaces does hold -- see eq. (3.60) of Kay and Wald, 1989). The trouble is that, in general, for $\varepsilon>0$, $\Lambda_\varepsilon^{t,n}(x_1,x_2)$)fails to be a bi-solution of eq. (2.8) by a sufficient amount to invalidate the conclusion of the lemma of section 2. The limit on the right side of eq. (4.3) need not exist and, if it does exist, need not equal $\lambda(\psi_1,\psi_2)$. Nevertheless, it is possible to show that eq. (4.3) does hold when the following three conditions are satisfied: (a) A Cauchy surface, C_A, of the form shown in fig. 3 is used. (b) The intersections of C_A with the spacetime support of ψ_1 and ψ_2 are each contained within the "horizon portion" of C_A, i.e., the support of the initial data for ψ_1 and ψ_2 on C_A is contained within h_A. We denote by S_A the subspace of S consisting of solutions whose initial data satisfies this property for some Cauchy surface of the type C_A. (c) The global time function, t, appearing in $\Lambda_\varepsilon^{t,n}$ is chosen so as to agree with affine parameter, U, on h_A. (The lengthy analysis needed to show that eq. (4.3) holds under these conditions can be found in appendix B of Kay and Wald, 1989.) Thus, for all $\psi_1,\psi_2 \in S_A$, we find that any Hadamard bi-distribution on S satisfies,

$$\lambda(\psi_1,\psi_2) = \lim_{\varepsilon\to 0} 4\int_{h_A \times h_A} \psi_1(x_1)\psi_2(x_2) \frac{\partial^2}{\partial U_1 \partial U_2} \Lambda_\varepsilon^{t,n}(x_1,x_2)\sqrt{^{(2)}g_1}\sqrt{^{(2)}g_2}\, d^2y_1\, d^2_2 y\, dU_1 dU_2 \tag{4.4}$$

where y denotes coordinates on the 2-dimensional bifurcation surface Σ and where we integrated by parts to throw the U-derivatives of ψ_1 and ψ_2 occurring in eq. (4.3) onto $\Lambda_\varepsilon^{t,n}$. Remarkably, however, if λ is isometry invariant, then on account of the direct relationship (2.6) between Killing and affine parameters on h_A, the only U-dependence of $\Lambda_\varepsilon^{t,n}$ arises from the regularizing term $2i\varepsilon(t_1-t_2) = 2i\varepsilon(U_1-U_2)$. Thus, the derivatives of $\Lambda_\varepsilon^{t,n}$ can be evaluated explicitly, thereby yielding (Kay and Wald, 1989),

$$\lambda(\psi_1, \psi_2) = -\frac{1}{\pi} \lim_{\varepsilon\to 0} \int \frac{\psi_1(U_1,y)\psi_1(U_2,y)}{(U_1 - U_2 - i\varepsilon)^2} \sqrt{^{(2)}g}\, d^2y\, dU_1\, dU_2 \tag{4.5}$$

for all $\psi_1, \psi_2 \varepsilon S_A$. Equation (4.5) is a universal formula, valid for any isometry invariant Hadamard distribution on the class of spacetimes under consideration. The only dependence upon the Killing horizon in eq. (4.5) arises from the integration over Σ. In particular the integral over affine parameter is of the same form for all Killing horizons.

We may interpret S_A as composed of solutions which "entirely fall through" the horizon h_A. Equations (4.1) and (4.5) define a bilinear map, μ_A, on $S_A \times S_A$. It is not difficult to show that μ_A saturates the inequality (3.26) for $\psi_1,\psi_2 \varepsilon S_A$. Thus, it gives rise to a vacuum state -- denoted ω_A -- on the subalgebra $A_A \subset A$ associated with S_A (i.e., the subalgebra generated by elements of the form $\underset{\sim}{W}(\psi_A)$ with $\psi_A \varepsilon S_A$). This vacuum state ω_A corresponds to defining the "positive frequency part" of solutions in S_A by Fourier

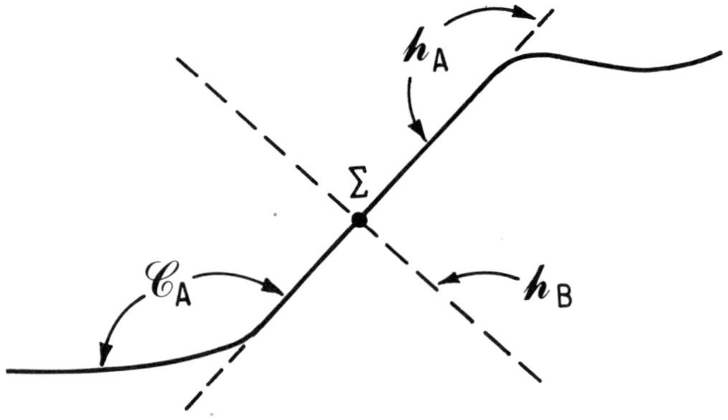

Fig. 3. A Cauchy surface of type C_A, defined as one which contains a portion of the "A" horizon, h_A, as shown.

transformation of their restriction to h_A using affine parameter, U, along the null geodesic generators.

Our uniqueness reqults are obtained as follows. Let ω be an isometry invariant vacuum state on A satisfying the Hadamard condition. Then the two-point distribution of ω must satisfy eq. (4.5) for ψ_1, $\psi_2 \varepsilon S_A$, from which it follows that the restriction of ω to A_A must agree with ω_A. Thus, ω is unique on A_A and also is a pure state on A_A, i.e., it cannot be written as a nontrivial sum of two other states. Similarly, ω is unique and pure on the subalgebra A_B, associated with the subspace, S_B, of solutions which "entirely fall through" the horizon h_B. However, from the uniqueness and purity of ω on A_A and A_B, it follows (see Kay and Wald, 1989) that ω also is unique and pure on the subalgebra A_0 associated with the subspace, S_0, of solutions that can be expressed as sums of elements of S_A and S_B, i.e., $S_0 = S_A + S_B$.

The subspace S_0 is a "large" subspace of S (and hence A_0 is a correspondingly "large" subalgebra of A) in the following sense. Let $\psi \varepsilon S$ and suppose $\sigma(\psi_0, \psi) = 0$ for all $\psi_0 \varepsilon S_0$. Then from the formula for σ on Cauchy surfaces of types C_A and C_B, it follows that ω must be constant along each generator of h_A and of h_B. Consequently, the restriction of ψ to $h_A \cup h_B$ must be isometry invariant. However, the restriction of a solution to $h_A \cup h_B$ uniquely determines the solution in a certain region, $\mathcal{D}$, of spacetime, which we refer to as the domain of determinacy of the horizon. Hence ψ must be isometry invariant in $\mathcal{D}$. Thus, the "symplectic complement" of S_0 in S is composed of the (relatively small -- and, in many cases, empty) set of solutions which are isometry invariant in $\mathcal{D}$.

From the standard null initial value formulation theorems, it follows that $\mathcal{D}$ always contains the "past and future wedges" F and P (see fig. 2). From the Holmgren uniqueness theorem (see, e.g., Treves, 1975), it follows immediately that $\mathcal{D}$ also always contains a portion of the "left and right wedges," L and R. It is shown in Kay and Wald (1989) that in many cases of interest (such as Kerr and de Sitter spacetimes), we have $P \cup F \cup R \cup L \subset \mathcal{D}$. Note that in many such cases (e.g., Schwarzschild and de Sitter spacetimes) $\mathcal{D}$ is the entire spacetime manifold M.

Since our results all are based upon the behavior of the quantum field at the horizon, the largest subalgebra upon which we can expect to prove a uniqueness theorem is $A_{\mathcal{D}}$, the subalgebra associated with $S_{\mathcal{D}}$, where $S_{\mathcal{D}} \subset S$ denotes the subspace of solutions corresponding (via eq. (2.13)) to test functions with support in $\mathcal{D}$. The above result on the symplectic complement of

S_0 suggests that the only possible ambiguity in extending our state ω from A_0 to $A_{\mathcal{D}}$ should be associated with isometry invariant solutions. That this is indeed the case is proven by Kay and Wald (1989): In the absence of "zero modes" (i.e., isometry invariant states on the one-particle Hilbert space) ω is unique on $A_{\mathcal{D}}$. Thus, our uniqueness results may be summarized as follows.

Uniqueness Theorem: Let (M, g_{ab}) be a globally hyperbolic spacetime with a bifurcate Killing horizon and possessing a Cauchy surface containing the bifurcation surface Σ. Let ω be an isometry invariant vacuum state which satisfies the Hadamard condition. Then ω is unique and pure on A_0 (defined above) and, indeed, on A_A and A_B its two-point distribution is given by the explicit formula (4.5). Furthermore, if there are no isometry invariant one-particle states, then ω also is unique on $A_{\mathcal{D}}$. Thus, in cases where $\mathcal{D} = M$, we thereby obtain uniqueness on the full Weyl algebra A.

Note that the term "unique" is used in the above theorem to mean "there is at most one." At the end of this section, we shall mention some examples where it can be proven that no state on A_0 or $A_{\mathcal{D}}$ exists which satisfies the hypothesis of the uniqueness theorem.

We turn now to consideration of the thermal properties of the unique state (if it exists) specified by the above theorem. As indicated in fig. 1b, the orbits of the isometries which generate the horizon are timelike (at least near the horizon) in wedges $\mathcal{R}$ and $\mathcal{L}$. Thus, in region $\mathcal{R}$ we may view the isometries as describing "time translations" and we may view $\mathcal{R}$ as a stationary spacetime in its own right. We define* the algebra $A^{\mathcal{R}}$, of "right wedge observables" to be the Weyl algebra for the spacetime $\mathcal{R}$. (We may, of course, view $A^{\mathcal{R}}$ as a subalgebra of A.) The vacuum state obtained** by the construction for stationary spacetimes described in section 2 will then yield a state on $A^{\mathcal{R}}$. The "time translation isometries" then will be represented by a unitary group acting on the GNS Hilbert space obtained from this state. The Hamiltonian operator, $\hat{H}$, is defined to be the infinitesimal generator of this group. Roughly speaking, a thermal state at temperature β^{-1} with respect to

---

*This definition differs slightly from that given in Kay and Wald (1989), where $A^{\mathcal{R}}$ was defined to be the subalgebra of A associated with the subspace $S^{\mathcal{R}} \subset S$ consisting of solutions which vanish in $\mathcal{L}$.

**For the purpose of this discussion, we assume that this construction can be applied here. This need not be the case, since the Killing field may become spacelike in $\mathcal{R}$ and, in any case, its norm is not bounded away from zero in $\mathcal{R}$.

the given isometries (i.e., "as measured by a family of observers in region $\mathcal{R}$ following orbits of the isometries") is simply the density matrix state $\hat{\rho} = \exp(-\beta\hat{H})$. However, this definition is inadequate because for a field system, $\hat{H}$ typically will have continuous spectrum, so $\exp(-\beta\hat{H})$ will not be trace class and, hence, does not define a normalizable state. Consequently, a considerably more sophisticated procedure is needed to define, in a precise, mathematical manner, the notion of a thermal equilibrium state. In the algebraic approach, this is accomplished by the formulation of the KMS condition. We refer the reader to Kay and Wald (1989) and references cited therein for the definition of the KMS condition and a brief discussion of why a state which satisfies the KMS condition at temperature β^{-1} corresponds formally to the state $\exp(-\beta\hat{H})$.

In Minkowski spacetime, a one-parameter family of Lorentz boosts generates a bifurcate Killing horizon consisting of two intersecting null planes. The ordinary Minkowski vacuum state, $|0_M\rangle$, is then easily seen to satisfy the hypotheses of the above uniqueness theorem. In the "right and left wedges" of Minkowski spacetime, the boost Killing field is timelike, so one can view $\mathcal{R}$ and $\mathcal{L}$ as stationary spacetimes in their own right and define a vacuum state, $|0_R\rangle$, known as the "Rindler vacuum," in those regions as indicated above. (Note that the Rindler vacuum is singular on the horizon, i.e., it does not satisfy the Hadamard condition on all of Minkowski spacetime.) One then may formally express $|0_M\rangle$ as a state in the Fock space of $|0_R\rangle$. (This procedure is formal because the Minkowski and Rindler representations of the field algebra are not unitarity equivalent and the state one obtains is not normalizable.) When one does so, one obtains a state with a very high degree of correlation in the particle content of the right and left wedges. When one traces out over the degrees of freedom of the left wedge to obtain a density matrix describing the state in the right wedge, one obtains the formal expression $\exp(-2\pi\hat{H}/\kappa)$, where $\hat{H}$ is the Hamiltonian conjugate to Lorentz boosts in $\mathcal{R}$ and κ is the surface gravity of the horizon* (see, e.g., Unruh and Wald, 1984, for these formal expressions). Thus, the Minkowski vacuum formally is a thermal state at temperature $\kappa/2\pi$ with respect to a family of observers following the orbits of the boost Killing field. This formal result can be made rigorous (Kay, 1985): In region $\mathcal{R}$, the Minkowski vacuum satisfies the KMS condition at temperature $\kappa/2\pi$ with respect to the notion of "time translation" defined by the boost Killing field.

---

*If we normalize the boost Killing field so that it has unit norm on a particular orbit, then κ is equal to the acceleration of that orbit.

Consider, now, any spacetime (M, g_{ab}) satisfying the hypotheses of the above uniqueness theorem. Then, as discussed above, the unique state ω of that theorem -- if it exists -- corresponds to defining the "positive frequency part" of solutions in S_0 by Fourier transformation using affine parameter on the horizon. However, if κ is constant over the horizon, the relationship between affine and Killing parameters on the horizon of the spacetime (M, g_{ab}) is the same as in Minkowski spacetime (see eq. (2.6)). Thus, for solutions in S_0, the relationship between positive frequency with respect to Killing time in regions L and R and the notion of positive frequency implicit in the state ω is the same as occurs in Minkowski spacetime. Hence, the results on the thermal properties of the ordinary vacuum state of Minkowski spacetime can be carried over to the state ω on the spacetime (M, g_{ab}). More precisely, let $S_0{}^R$ denote the solutions in S_0 whose data has compact support contained in the spacetime R, and let $A_0{}^R$ denote the corresponding subalgebra of A. (Thus, $A_0{}^R$ represents the "right wedge observables" associated with "horizon modes.") Then we have the following theorem, the proof of which is outlined in more detail in Kay and Wald (1989):

Thermal Properties Theorem: Let (M, g_{ab}) be as in the above uniqueness theorem and, in addition, be such that there exists a reflection isometry through Σ. Suppose, further, that the surface gravity, κ, is constant on the horizon and that the state ω specified by the uniqueness theorem exists. Then the restriction of ω to the subalgebra $A_0{}^R$ is a KMS state at temperature $\kappa/2\pi$.

Finally, we remark that, as pointed out previously, the bilinear map μ_A on $S_A \times S_A$ determined by eqs. (4.1) and (4.5) can be verified to satisfy (and, indeed, saturate) the inequality (3.26). The corresponding map $\mu_B : S_B \times S_B \to \mathbb{R}$ also, of course, satisfies eq. (3.26). However, there is no guarantee that the map $\mu_0 : S_0 \times S_0 \to \mathbb{R}$ determined by μ_A and μ_B will satisfy eq. (3.26) for all $\psi_1, \psi_2 \in S_0$. Indeed, one can show (Kay and Wald, 1989) that eq. (3.26) must fail for μ_0 in any spacetime where superradiance occurs in the following sense: There exists a solution $\psi \in S_B{}^R$ such that the "one-particle norm" of ψ (i.e., the Klein-Gordon type norm of the positive frequency part of ψ) on h_A is larger than the one-particle norm of ψ on h_B. Consequently, in Kerr spacetime there cannot exist any isometry invariant, Hadamard, vacuum state ω on A_0. Indeed, by the same argument, for Kerr spacetime there cannot exist any (not necessarily vacuum) state which is isometry invariant and satisfies the Hadamard condition, since positivity of the state, eq. (3.36), requires its two-point distribution to satisfy the same conditions as for a vacuum state. Different arguments can be used to prove non-existence of isometry invariant Hadamard states in certain other spacetimes, such as Schwarzschild-de Sitter spacetime (Kay and Wald, 1989).

ACKNOWLEDGEMENTS

As previously mentioned, all the results described here were obtained in collaboration with Bernard S. Kay. The research was supported, in part, by NSF grant PHY 84-16691 to the University of Chicago and by the Tomalla Foundation.

REFERENCES

Ashtekar, A., and Magnon, A., 1975, Quantum fields in curved spacetimes, Proc. Roy. Soc. Lond., A346:375.

Choquet-Bruhat, Y., 1968, Hyperbolic partial differential equations on a manifold, in: "Battelle Rencontres," C. M. DeWitt and J. A. Wheeler, eds., Benjamin, New York.

Fulling, S. A., Narcowich, F. J. and Wald, R. M., 1981, Singularity structure of the two-point function in quantum field theory in curved spacetime. II, Ann. Phys., 136:243.

Geroch, R. P., 1970, Domain of dependence, J. Math. Phys., 11:437.

Gibbons, G.W., and Hawking, S. W., 1977, Cosmological event horizons, thermodynamics and particle creation, Phys. Rev., D15:2738.

Hartle, J. B., and Hawking, S. W., 1976, Path-integral derivation of black hole radiance, Phys. Rev. D13:2188.

Hawking, S. W., and Ellis, G. F. R., 1973, "The Large Scale Structure of Spacetime," Cambridge University Press, Cambridge.

Israel, W., 1976, Thermo-field dynamics of black holes, Phys. Lett., 57A:107.

Kay, B. S., 1978, Linear spin-zero quantum fields in external gravitational and scalar fields, Commun. Math. Phys., 62:55.

Kay, B. S., 1985, The double-wedge algebra for quantum fields on Schwarzschild and Minkowski spacetimes, Commun. Math. Phys., 100: 57.

Kay, B. S., and Wald, R. M., 1989, Theorems on the uniqueness and thermal properties of stationary, nonsingular, quasi-free states on spacetimes with a bifurcate Killing horizon, to be published.

Reed, M., and Simon, B., 1972, "Methods of Modern Mathematical Physics, Vol. I: Functional Analysis," Academic Press, New York.

Simon, B., 1972, Topics in functional analysis, in "Mathematics of Contemporary Physics," R. F. Streater, ed., Academic Press, New York.

Treves, F., 1975, "Basic Linear Partial Differential Equations," Academic Press, New York.

Unruh, W. G., 1976, Notes on black hole evaporation, Phys. Rev., D14: 870.

Unruh, W. G., and Wald, R. M., 1984, What happens when an accelerating observer detects a Rindler particle, Phys. Rev., D29:1047.

Wald, R. M., 1977, The back reaction effect in particle creation in curved spacetime, Commun. Math. Phys., 54:1.

Wald, R. M., 1978, Trace anomaly of a conformally invariant quantum field in curved spacetime, Phys. Rev., D17:1477.

Wald, R. M., 1979, Existence of the S-matrix in quantum field theory in curved spacetime, Commun. Math. Phys., 70:221.

Wald, R. M., 1974, "General Relativity," University of Chicago Press, Chicago.

MUTUALLY INTERACTING QUANTUM FIELDS IN CURVED SPACE-TIMES: THE OUTCOME OF PHYSICAL PROCESSES

Jürgen Audretsch
Fakultät für Physik
Universität Konstanz
Postfach 55 60
D-7750 Konstanz
West Germany

INTRODUCTION

During the last years many efforts have been made to study quantum field theory in given unquantized space-times. This external field approach for the influence of classical gravitational fields is generally regarded as some sort of semi-classical approximation to a full quantum theory of gravity. It is assumed that characteristic physical traits which show up in this approximation will be at least heuristically important for the construction of the full theory. Furthermore the approximation has its own domain of application during the very early stages of the universe and outside black holes.

In the following we will restrict ourselves to the discussion of the quantum field theory of several quantum fields which are mutually interacting in the presence of an external gravitational field represented by the curvature of space-time. The intention is thereby i.) to contribute to the development of an appropriate conceptual framework (what are the measurable quantities, what is the related calculation scheme?), ii.) to work out the physical characteristics of such a theory (including a discussion of advantages and deficiencies of the respective schemes) and iii.) to answer questions like: How are the minkowskian cross sections and decay rates modified? Exact solvability will be given preference over physical applicability. At present the conceptual framework and the

Quantum Mechanics in Curved Space-Time
Edited by J. Audretsch and V. de Sabbata
Plenum Press, New York, 1990

understanding of the new typical physical effects still need improvement. Examples have therefore to support first of all the intentions i.) and ii.). To keep close to the reliable ground of special relativistic procedures, we restrict to a treatment in the interaction picture using an S-matrix approach whereby the background is always taken into account exactly (Furry picture).

Non-minkowskian situation

Quantum field theory in curved space-time is a special case of a non-minkowskian situation in quantum field theory. This situation is characterized by the existence of two different definitions of particles based on different operational procedures and usually attributed to two different kinds of observers. We will call these observers in- and out-observers because we will specialize below to a scattering situation. For simplicity we discuss real Klein-Gordon fields. A Klein-Gordon inner product (,) is attributed to a class of timelike hypersurfaces.

The in-observer specifies his particle concept in defining a complete orthonormal set $\{u^{in}_{\underline{p}}\}$ of Klein-Gordon solutions as those which describe to particles according to his measuring procedure: $(u^{in}_{\underline{p}}, u^{in}_{\underline{q}}) = \delta_{\underline{p}\underline{q}}$ $\underline{p}$ denotes some set of quantum numbers. The quantized real field may be decomposed with regard to the respective creation and annihilation operators:

$$\Phi = \sum_{\underline{p}} [a^{in}_{\underline{p}} u^{in}_{\underline{p}} + a^{in\dagger}_{\underline{p}} u^{in*}_{\underline{p}}] \tag{1}$$

We quantize the system in the usual way (canonical quantization)

$$[a^{in}_{\underline{p}}, a^{in}_{\underline{q}}] = 0 \quad , \quad [a^{in}_{\underline{p}}, a^{in\dagger}_{\underline{q}}] = \delta_{\underline{p},\underline{q}} \tag{2}$$

set up an in-Fock space and define the vacuum state as usual: $a^{in}_{\underline{p}} |0 \;\; in> = 0$.

We proceed in the same way with regard to the second observer, who bases his particle concept on the solutions $\{u^{out}_{\underline{p}}\}$, which are also orthonormal and complete $(u^{out}_{\underline{p}}, u^{out}_{\underline{q}}) = \delta_{\underline{p},\underline{q}}$. We quantize corresponding to (2) and obtain in addition to (1) a second decomposition of the field operator:

$$\Phi = \sum_{\underline{p}} [a^{out}_{\underline{p}} u^{out}_{\underline{p}} + a^{out\dagger}_{\underline{p}} u^{out*}_{\underline{p}}] \tag{3}$$

The vacuum state of the related Fock space is characterized by
$a^{out}_{\underline{p}} |0 \ out > = 0$.

The physical situation is generically non-minkowskian, if the two particle concepts don't agree: $u^{in}_{\underline{p}} \neq u^{out}_{\underline{p}}$. The immediate consequence is $a^{in}_{\underline{p}} \neq a^{out}_{\underline{p}}$. For the existence of an S-matrix see Wald (1979). Equating (1) and (3), we can write the out-operators as functions of the in-operators (compare (15) and (16) below). Inserting this into the expression for the mean value of the number of out-particles in the in-vacuum

$$N^{(0)}(\underline{p},0) = <0 \ in| \ N^{out}_{\underline{p}} \ |0 \ in> = <0 \ in| \ a^{out\dagger}_{\underline{p}} \ a^{out}_{\underline{p}} \ |0 \ in> \neq 0 \tag{4}$$

we obtain that in general this mean value will not vanish. The important consequence is the observer dependence of the particle concept: The vacuum of the in-particles contains out-particles.

As a particular application of this observer dependence we discuss a scattering situation in which a curved space-time acts as an external field (figure 1). We assume that the curvature vanishes appropriately for the time-asymptotic hyperfaces $\sum^{in}$ and $\sum^{out}$ (cosmological situation). The particle definition is based on the quasi-free Klein-Gordon solutions $u^{in}_{\underline{p}}$ and $u^{out}_{\underline{p}}$ which are defined everywhere (we exclude horizons) and show a particle structure near the hypersurfaces $\sum^{in}$ and $\sum^{out}$ respectively. Usually the in-and out-solutions for the same $\underline{p}$ will differ. Note that

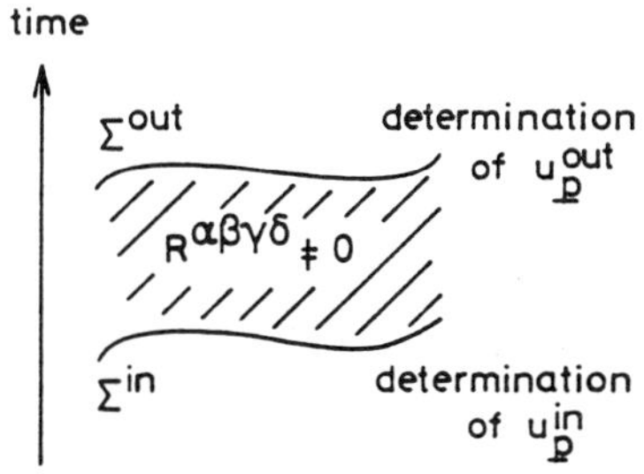

FIGURE 1

particles can only be defined asymptotically when there is no influence of the curved background and the particle concept is stable. This is an important restriction for the application of the scheme.

Equation (4) has a direct consequence in this situation: We assume that the in-region is empty. In the Heisenberg picture the corresponding vacuum state is also the state of the system in the out-region. According to (4) the out-region will then in general contain particles. Physically this means that also in the absence of mutual interaction between particles a non-vanishing space-time curvature radiates particles. As we will show below, this effect which happens in the absence of any possible mutual interaction between particles (zeroth order effect) influences severely all physical effects and the conceptual basis of quantum field theory in expanding universes and curved space-times in general.

Mutually interacting quantum fields in an unquantized cosmological background

We discuss the mutual interaction of two neutral scalar quantum fields:

$$L_I(\lambda, g_{\alpha\beta}, \Phi, \Psi) \tag{5}$$

We assume the field Φ to be massive, the field Ψ is massless. λ is the coupling parameter. The given unquantized curved space-time background with metric $g_{\alpha\beta}$ acts as an external field and is always exactly taken into account.

In the following we present case studies in which all the characteristic structural traits can be studied on the basis of mathematically relatively simple expressions which will be worked out rigorously. Generalizations to complex and fermionic field will be indicated. Physically such an the exact treatment of particular transitions according to the mutual interaction (5) may be regarded as a first step towards a detailed discussion of scalar quantum electrodynamics in curved space-time. Reviews of interacting quantum field theory in this context are given by Birrell (1981), Ford (1984) and Ford (1988).

We consider a 3-flat open Robertson-Walker universe. This topology is chosen because it is very likely to be realized in our universe and because we want to discuss genuine curvature effects leaving the influence of a non-trivial 3-topology apart. The space-time is conformally flat. We introduce the conformal time η

$$ds^2 = a^2(\eta)(d\eta^2 - d\underset{\sim}{x}^2) \tag{6}$$

We treat the scattering processes in the presence of space-time curvature in the interaction picture. Back-reaction is excluded. To construct the physical particle states in the asymptotic regions, we assume according to the S-Matrix scheme that the mutual interaction is switched off adiabatically in the distant past and future:

$$\begin{aligned} S &= \lim_{\varepsilon \to 0} \hat{T} \exp[i \int L_I \exp(-\varepsilon|\eta|) d^4x] \\ &= 1 + iT = 1 + iT^{(1)} + iT^{(2)} + O(\lambda^3) \end{aligned} \tag{7}$$

ε is called the switch-off parameter and T denotes the time-ordering operator with regard to η . The different orders refer to the powers of the coupling parameter λ .This approach is close to the usual scheme in flat space-time. Note that because of (7) the mutual interaction is of infinite duration.

The particle field equations are the massive and massless conformally coupled Klein-Gordon equations $(\nabla_\mu \nabla^\mu + m^2 + R/6)\Phi = 0$. For take $m = 0$. R is the scalar curvature. Solutions of the field equations may be written in the form

$$u_{\underline{p}}(x) = (a(\eta))^{-1} f(\underline{p},\eta)\exp(i\underline{p}x) \tag{8}$$

where $f(\underline{p},\eta)$ satisfies the generalized oscillator equation

$$\partial_\eta^2 f(\underline{p},\eta) + (a^2(\eta)m^2 + \underline{p}^2)f(\underline{p},\eta) = 0 \tag{9}$$

$\underline{p}$ is the conserved momentum parameter. The measured momentum, as registered by the cosmic observer, is $p/a(\eta)$.

We assume that in the in- and out-region $(\eta = -\infty \; , \; \eta = +\infty)$ the curvature diminishes in such a way that the concept of stable particles can be introduced. This means that the solutions $\{u_{\underline{p}}^{in}\}$ and $\{u_{\underline{p}}^{out}\}$ approach essentially plane waves in the in- and out-region repectively. For a more precise discussion of the particle concept see Audretsch (1979) and Birrell and Davies (1982). Because of the conformal invariance of the field equation, the solutions for the massless fields are proportional plane waves

$$f(\underline{k},\eta) \sim \exp(-ik\eta) \quad , \quad v_{\underline{k}}^{in} = v_{\underline{k}}^{out} = v_{\underline{k}} \tag{10}$$

with $k = |\underline{k}|$. In this case there is no influence of the curvature. The in- and out-particle solutions are the same for massless particles. Because of conformal invariance, the physics of the free massless particles agrees completely with the one in flat space-time. The mass of the massive particles on the other hand breaks conformal invariance. This will be important below.

We quantize as above. For Ψ -particles we replace $u \to v$, $\underline{p} \to \underline{k}$ $a \to b$ and have:

$$b_{\underline{k}}^{in} = b_{\underline{k}}^{out} = b_{\underline{k}} \quad , \quad |s^{\Psi} \text{ out} > = |s^{\Psi} \text{ in} > \tag{11}$$

The in-vacuum state is defined according to $a_{\underline{p}}^{in} |0 \text{ in} > = b_{\underline{k}} |0 \text{ in} > = 0$
For a general state vector we write

$$a_{\underline{p}}^{in\dagger} \ldots b_{\underline{k}}^{\dagger} \ldots |0 \text{ in} > = |1_{\underline{p}}^{\Phi} \ldots 1_{\underline{k}}^{\Psi} \ldots \text{ in} > = |e^{\Phi} s^{\Psi} \text{ in} > \tag{12}$$

e and s abbreviate a particular occupation of Φ- and Ψ-modes.

The information regarding the influence of the gravitational background is essentially contained in the Bogoliubov coefficients relating in- and out-particle solutions $\alpha_{\underline{p}} = (u_{\underline{p}}^{in}, u_{\underline{p}}^{out})$, $\beta_{\underline{p}} = (u_{\underline{p}}^{in}, u_{-\underline{p}}^{out*})$ with

$$|\alpha_{\underline{p}}|^2 - |\beta_{\underline{p}}|^2 = 1 \tag{13}$$

There would be a plus sign for fermionic fields. As consequence of the isotropy we have $\alpha_{-\underline{p}} = \alpha_{\underline{p}}$, $\beta_{-\underline{p}} = \beta_{\underline{p}}$. In- and out-particle solutions may be expressed with regard to each other:

$$u_{\underline{p}}^{out} = \alpha_{\underline{p}} u_{\underline{p}}^{in} + \beta_{\underline{p}}^{*} u_{-\underline{p}}^{in*} \quad , \quad u_{\underline{p}}^{in} = \alpha_{\underline{p}}^{*} u_{\underline{p}}^{out} - \beta_{\underline{p}}^{*} u_{-\underline{p}}^{out*} \tag{14}$$

Equating (1) and (3) we may write the in-operators as functions of the out-operators and vice versa. We give this the form which will prove to be useful below:

$$a_{\underline{p}}^{in\dagger} = \alpha_{\underline{p}}^{-1} a_{\underline{p}}^{out\dagger} + \beta_{\underline{p}}^{*} \alpha_{\underline{p}}^{-1} a_{-\underline{p}}^{in} \tag{15}$$

$$a_{\underline{p}}^{out} = \alpha_{\underline{p}}^{-1} a_{\underline{p}}^{in} - \beta_{\underline{p}} \alpha_{\underline{p}}^{-1} a_{-\underline{p}}^{out\dagger} \tag{16}$$

Herewith the mixed commutator relations can be worked out:

$$[a_{\underline{p}}^{out}, a_{\underline{q}}^{in}] = \beta_{\underline{p}} \delta_{\underline{p},-\underline{q}} 1 \tag{17}$$

$$[a_{\underline{p}}^{out}, a_{\underline{q}}^{in\dagger}] = \alpha_{\underline{p}}^{*} \delta_{\underline{p},\underline{q}} \tag{18}$$

Inserting (15) in (4) the Bogoliubov coefficients obtain a direct physical interpretation related to the zeroth order particle creation out of the curved background:

$$N^{(0)}(\underline{p}^{\Phi}|0) = |\beta_{\underline{p}}|^{2} \tag{19}$$

The specifications above (cosmological situation) have the following consequences: i.) Energy is not conserved ii.) There is a conserved 3-momentum parameter. iii.) Massive particles are created out of the vacuum in every momentum mode. iv.) Because of the conformal situation there is no corresponding creation of massless particles. A physical interpretation of quantum field theoretical processes has to be based on this.

MEASURABLE TRANSITION PROBABILITIES

In-out transition amplitudes

One is tempted to base the discussion of cross sections and decay rates as usual on in-out transition amplitudes built with the respective in- and out-states which contain a finite number of particles. If the mutual interaction is switched off completely, such an amplitude can be reduced to the amplitude relating the in-vacuum with the out-vacuum according to

$$< \text{out } d^{\Phi}{}_{s}{}^{\Psi} | c^{\Phi}{}_{r}{}^{\Psi} \text{ in} > \;=\; F(\alpha,\beta) < \text{out } 0|0 \text{ in} > \; \delta\text{-funct.} \tag{20}$$

This result is obtained by transcribing the out-operators according to (16) and using the commutator relations (17) and (18). F is a function of the Bogoliubov coefficients of the modes involved. For higher order we find (Audretsch and Spangehl, 1985):

$$<\text{out } d^{\Phi}{}_{s}{}^{\Psi} | S^{(z)} | \, c^{\Phi}{}_{r}{}^{\Psi} \text{ in}> \;=$$

$$\sum_{g,t} <\text{out } d^{\Phi}{}_{s}{}^{\Psi} | g^{\Phi}{}_{t}{}^{\Psi} \text{ in}><\text{in } g^{\Phi}{}_{t}{}^{\Psi} | S^{(z)} | c^{\Phi}{}_{r}{}^{\Psi} \text{ in}> \tag{21}$$

by inserting

$$1 = \sum |g^{\Phi}{}_{t}{}^{\Psi} \text{ in} > < \text{in } g^{\Phi}{}_{t}{}^{\Psi}| \tag{22}$$

where the sum is taken over all elements of the basis. When discussing (21), on the right hand side the result (20) is to be taken into account.

We introduce a normalization volume V. The function F of (20) proves then to be independent of the volume V. Parker (1969) on the other hand has shown, that in the cosmological case in question

$$|<\text{out } 0|0 \text{ in}>|^2 \xrightarrow{V \to \infty} 0 \tag{23}$$

Accordingly we conclude from (20) and (21): For zero and all higher orders of the mutual interaction, any in-out probability amplitude becomes arbitrarily small for increasing normalization volume V. A single in-out amplitude loses its physical meaning for $V \to \infty$.

With regard to the physical interpretation this can be traced back to the fact that already in zeroth order the cosmological background creates massive particles in all modes. It is therefore impossible to find a completely empty out-state. This explains (23). For the same reason it is also impossible to find outgoing states with only a finite number of massive particle modes occupied and all other massive modes empty, regardless if there has been a mutual interaction between the two quantum fields or not. Consequently the transition probabilities related to (20) and (21) must vanish.

Indicator configurations and added-up probabilities

In addition there is an operational argument why in our case the general quantum field theoretical scheme for transition probabilities cannot be based on in-out amplitudes: A single particle counter for massive particles cannot discriminate between a particle coming out of the background and a particle originating from the mutual interaction. It is therefore impossible to specify the outcome of the mutual interaction on the basis of measurements with single counters for massive particles. The creation of massive particles out of the background necessarily interferes and cannot simply be separated.

To obtain nontrivial statements which refer as close as possible to the mutual interaction only, one has to fix the in-state and to concentrate on out-states which contain a specific indicator configuration, the appearance of which can clearly be determined by particle counters. Apart from the given in-state it is the outgoing indicator configuration by which the particular interaction process is then specified. An indicator configuration is thereby a configuration which cannot be produced by the background. It is therefore a direct consequence of the specific mutual interaction and contains the maximal information about the preceeding scattering process.

Massless particles are in our case good indicators because they are neither influenced nor created by the background. The concept of the general added-up transition probability (Audretsch and Spangehl, 1985) is based on this:

$$w^{add}(s^{\Psi}|c^{\Phi}r^{\Psi}) = \sum_{\text{all } d} |\langle out\; d^{\Phi}s^{\Psi}|S|c^{\Phi}r^{\Psi}\; in\rangle|^2 \tag{24}$$

It answers the question: What is the probability that a particular state of massless particles $|s^{\Psi}\; out\rangle$ will be found in the out-region regardless of what has happened to the massive state and what massive states are going out. With $\sum_{\text{all } d}|d^{\Phi}\; out\rangle\langle out\; d^{\Phi}| = \sum_{\text{all } d}|d^{\Phi}\; in\rangle\langle in\; d^{\Phi}|$ the added-up probability w^{add} can be reduced to in-in amplitudes

$$w^{add}(s^{\Psi}|c^{\Phi}r^{\Psi}) = \sum_{\text{all } d} |\langle in\; d^{\Phi}s^{\Psi}|S|c^{\Phi}r^{\Psi}\; in\rangle|^2 \tag{25}$$

This guarantees that for any given finite order of the mutual interaction the sum over the massive states stops, ending with states containing only a particular number of massive particles. In lower order only very few amplitudes are to be considered similar to flat space-time quantum field theory.

In the expression (24) we sum up over all massive out-states. To obtain more specific answers and to improve the predictive power of the scheme, we may include good massive indicators. In our case we have conservation of the 3-momentum parameter. Accordingly the background creates massive particles only in pairs. Below we will demonstrate in detail, that also processes in higher order are influenced by the appearance of additional massive pairs. All configurations without massive pairs consist therefore of particles which originate from the interaction only (indicator state). The specified added-up probability that a transition into a state containing certain massless particles and unpaired massive particles has occurred - regardless of the creation of massive pairs out of the background or the mutual interaction - is given by:

$$w^{inc}(\hat{d}^{\Phi}s^{\Psi}|c^{\Phi}r^{\Psi}) = \sum_{\text{all } Q} |\langle \text{out } Q^{\Phi}\hat{d}^{\Phi}s^{\Psi}|S|c^{\Phi}r^{\Psi} \text{ in}\rangle|^2 \tag{26}$$

Capital letters denote states which contain only pairs of massive particles. The sum goes over all these states. We call this probability the pair-including probability. It has been shown by Audretsch and Spangehl (1987) that it can be reduced to in-in amplitudes:

$$w^{inc}(\hat{d}^{\Phi}s^{\Psi}|c^{\Phi}r^{\Psi}) = \sum_{\text{all } Q} |\langle \text{in } Q^{\Phi}\hat{d}^{\Phi}s^{\Psi}|S|c^{\Phi}r^{\Psi} \text{ in}\rangle|^2 \tag{27}$$

The probabilities (25) and (27) are the basis of an in-in quantum field theory which can be built up in close analogy to what is known from minkowskian quantum field theory. Essentially the special-relativistic creation and annihilation operators are to be replaced by in-operators and the minkowskian particle solutions (e.g. plane waves) by in-particle solutions. Note that in-in scheme does not mean that it is superfluous to specify the out-particle solutions. The appearance of Bogoliubov coefficients in the results below will demonstrate this. In this scheme too, an in- and a properly defined out-region where particles are stable, is needed.

We can now establish for the in-in scheme a Feynman diagram technique with correspondingly transcribed Feynman rules. The massive Feynman propagator is thereby:

$$i\Delta_F^{in}(x,y) = \langle \text{in } 0|\hat{T}\Phi^{in}(x)\Phi^{in}(y)|0 \text{ in}\rangle = \Phi^{in}(x)\Phi^{in}(y) \tag{28}$$

There is a Wick theorem for in-operators. Disconnected vacuum-vacuum graphs are removed, if necessary, by dividing the in-in amplitude by $\langle in\ 0|S|0\ in\rangle$.

Returning to the discussion of in-out amplitudes, we mention that the relative in-out amplitudes $\langle out\ d^{\Phi}s^{\Psi}|S^{(z)}|c^{\Phi}r^{\Psi}\ in\rangle \langle out\ 0|0\ in\rangle^{-1}$ although possibly finite, have no direct operational meaning. Nevertheless they can be mathematically useful in the in-in scheme because according to (20) and (21). They essentially agree with single or summed in-in amplitudes.

OPTICAL THEOREM

Consequence of the unitarity

We show that the optical theorem for quantum field theory in curved space-time can be transcribed to the in-in scheme thus obtaining a direct physical interpretation by means of added-up probabilities.

The unitarity $S^{\dagger}S = 1$ of the S-operator implies with (7) $T^{\dagger}T = -i(T - T^{\dagger})$ and therefore $\langle in\ a|T^{\dagger}T|a\ in\rangle = 2Im\ \langle in\ a|T|a\ in\rangle$ Inserting (22) this leads to the optical theorem in curved space-time in an in-in formulation (Audretsch, 1988):

$$\sum_{all\ e,s} |\langle in\ e^{\Phi}s^{\Psi}|T|d^{\Phi}r^{\Psi}\ in\rangle|^2 = 2Im\ \langle in\ d^{\Phi}r^{\Psi}|T|d^{\Phi}r^{\Psi}\ in\rangle \tag{29}$$

This can directly be rewritten with added-up probabilities.

$$\sum_{all\ s} w^{add}\ (s^{\Psi}|d^{\Phi}r^{\Psi}) = 2Im\ \langle in\ d^{\Phi}r^{\Psi}|T|d^{\Phi}r^{\Psi}\ in\rangle \tag{30}$$

For a given ingoing state the optical theorem relates the imaginary part of the in-in forward scattering amplitude with the total added-up transition probability into all massless states (sum over s) which can be reached from the in-state according to the interaction. Note that in the cosmological case we don't have energy conservation. In contrast to the minkowskian situation neither the left nor the right hand side will in general reduce to only one term. At the first glance the optical theorem therefore seems to be a rather unspecific relation of restricted use. We demonstrate in the following example that this is not the case.

Example: Emission of a massive particle in the $\Phi\Psi^2$-model

We study in second order the case of one ingoing massless particle and decompose the left hand side of (29) with respect to the contributing Feynman diagrams of figure 2.

$$\sum_{\text{all } e,s} |\langle \text{in } e^{\Phi} s^{\Psi} | T^{(1)} | 1_{\underline{1}}^{\Psi} \text{ in}\rangle|^2 =$$

$$\sum_{\underline{k}} |\langle 2a\rangle|^2 + \sum_{\underline{q},\underline{t}} |\langle 2b\rangle|^2 + \sum_{\underline{q}} |\langle 2c\rangle|^2 \qquad (31)$$

The diagrams of figure 2b and 2c contribute because there is no energy conservation and the ingoing massless particle may pass through and

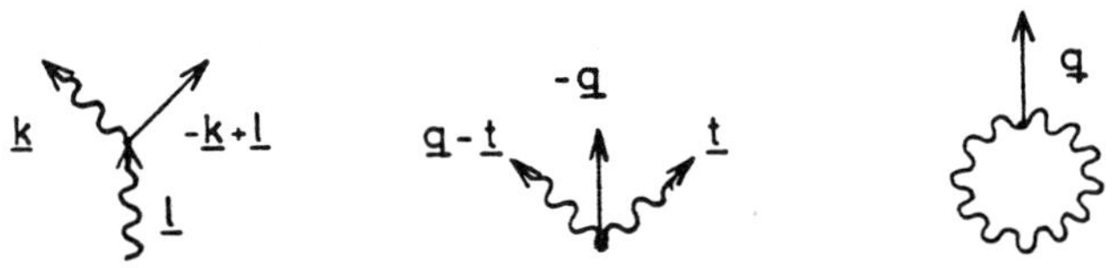

FIGURE 2a FIGURE 2b FIGURE 2c

reappear in the out-region with the same momentum parameter. The squared amplitudes on the r.h.s. of (31) may be given a physical interpretation within our scheme:

$$\sum_{\text{all } \underline{k} \neq \underline{1}} w^{\text{add}} (1_{\underline{k}}^{\Psi} | 1_{\underline{1}}^{\Psi})^{(2)} = \sum_{\underline{k}} |\langle 2a\rangle|^2 \qquad (32)$$

$$\hat{w} (1_{\underline{1}}^{\Psi}\, 1_{\underline{1}}^{\Psi})^{(2)} = \sum_{\text{all } \underline{q},\underline{t}} |\langle 2b\rangle|^2 + \sum_{\text{all } \underline{q}} |\langle 2c\rangle|^2 \qquad (33)$$

The latter is the probability that the ingoing massless particle can be found undisturbed among others in the out-region.

The right hand side of the optical theorem (29) is in our case of the form

$$\text{Im} <\text{in } 1^{\Psi}_{\underline{1}}|T^{(2)}|1^{\Psi}_{\underline{1}} \text{ in}> \;=\; \text{Im } [<\text{in } 0|T^{(2)}|0 \text{ in}> + <3a> + <3b>] \quad (34)$$

where the reference is made to figure 3

FIGURE 3a

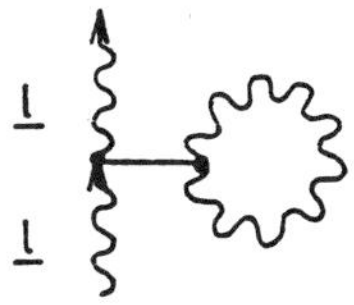

FIGURE 3b

The first term on the r.h.s. of (34) goes back to diagrams, where the ingoing massless particle runs through undisturbed (line without vertex, omitted in the figures). This term can further be elaborated in making use of the optical theorem (29) in second order but now for a vacuum state going in: $|d^{\Phi}s^{\Psi} \text{ in}> \;=\; |0 \text{ in}>$.

$$2\text{Im} <\text{in } 0|T^{(2)}|0 \text{ in}> \;=\; \sum_{\text{all } \underline{q},\underline{t}} |<2b>|^2 + \sum_{\text{all } \underline{q}} |<2c>|^2 \quad (35)$$

Equating (31) and (34) according to (29), we obtain with (35) and (32) the result:

$$\sum_{\text{all } \underline{k} \neq \underline{1}} w^{\text{add}}(1^{\Psi}_{k}|1^{\Psi}_{1})^2 \;=\; 2\text{Im}<3a> + 2\text{Im}<3b> \quad (36)$$

The left hand side may physically be interpreted as the total probability for the emission of a massive particle by a massless particle. Note that only the imaginary parts of the second order in-in selfenergy transitions of massless particles are to be worked out. This relation obtained within the in-in scheme is closest to the relation one is used to in the minkowskian situation.

Additional examples

There are more results which can be obtained in combining relations derived from the optical theorem (Audretsch 1988). For example the added-up probability for the annihilation of a pair of massless particles according to the interaction above can be reduced to the imaginary part of the amplitude for forward scattering of two massless particles where reference is made to figure 4.

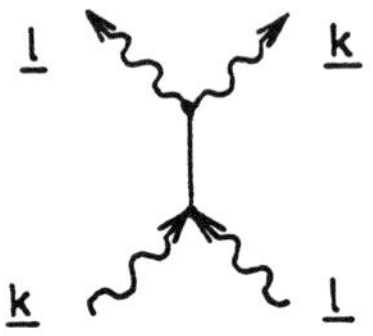

FIGURE 4a

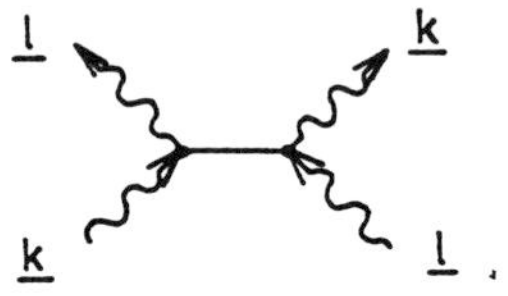

FIGURE 4b

For applications of the optical theorem to scalar quantum electrodynamics in curved space-time see Audretsch (1989).

GRAVITATIONALLY INDUCED AMPLIFICATION AND ATTENUATION

Mean value of the particle number

We continue the discussion of the physics of mutually interacting quantum fields in non minkowskian situations in turning to the discussion of the role of the massive particles, i.e. of particles which show nonvanishing gravitational particle creation out of the vacuum in zeroth order of the mutual interaction. We will show that the curved background also contributes in a very specific way to the outcome of the interaction in higher order. The appropriate concept to study this is the mean number of outgoing particles

$$N(\underline{p}^{\Phi}|a) = \langle \mathrm{in}\ a|S^{\dagger}N_{\underline{p}}^{\Phi\mathrm{out}}\, S|a\ \mathrm{in}\rangle \tag{37}$$

if the ingoing state is $|a\ \mathrm{in}\rangle$. With (22) we rewrite this expression in the form:

$$N(p^{\Phi}|a) = \sum_{\text{all } b,c} \langle\text{in } a|S^{\dagger}|b \text{ in}\rangle \langle\text{in } b|N_{\underline{p}}^{\Phi\text{out}}|c \text{ in}\rangle \langle\text{in } c|S|a \text{ in}\rangle \qquad (38)$$

For later use it is helpful to express the out-particle mean number operator with (15) and (16) by in-operators. Making in addition use of (13) we obtain:

$$N_{\underline{p}}^{\Phi\text{out}} = N^{(0)}(\underline{p}^{\Phi}|0)\,1 + N_{\underline{p}}^{\Phi\text{in}} \genfrac{}{}{0pt}{}{+}{(-)} N^{(0)}(\underline{p}^{\Phi}|0)\,\{N_{\underline{p}}^{\Phi\text{in}} + N_{-\underline{p}}^{\Phi\text{in}}\}$$
$$- \alpha_{\underline{p}}^{*}\beta_{\underline{p}}^{*}\, a_{\underline{p}}^{\text{in}} a_{-\underline{p}}^{\text{in}} - \alpha_{\underline{p}}\beta_{\underline{p}}\, a_{\underline{p}}^{\text{in}\dagger} a_{-\underline{p}}^{\text{in}\dagger} \qquad (39)$$

The minus sign (-) is included to indicate the corresponding result for fermionic fields. For complex fields operators with index $-\underline{p}$ have to be replaced by antiparticle operators.

Induced gravitational amplification and attenuation in the absence of mutual interaction (zeroth order)

Inserting (39) in (37) with $S = 1$ we find

$$N^{(0)}(\underline{p}^{\Phi}|a) = N^{(0)}(\underline{p}^{\Phi}|0) + n(\underline{p}^{\Phi}|a) \genfrac{}{}{0pt}{}{+}{(-)} N^{(0)}(\underline{p}^{\Phi}|0)\,\{n(\underline{p}^{\Phi}|a) + n(-\underline{p}^{\Phi}|a)\} \qquad (40)$$

where the occupation number

$$n(\underline{p}^{\Phi}|\underline{a}) = \langle\text{in } a|N_{\underline{p}}^{\Phi\text{in}}|a \text{ in}\rangle \qquad (41)$$

indicates how many massive particles with parameters $\underline{p}$ are contained in the ingoing state $|a \text{ in}\rangle$. Again the minus sign (-) applies in the fermionic case.

The zeroth order result (40) agrees with the one of Parker (1969, 1971). It shows for vanishing mutual interaction that if the ingoing state is not the vacuum, there are additional more processes in addition to the gravitational particle creation (19) out of the vacuum. The meaning of the three terms in (40) is: particle creation out of the vacuum, particles already contained in the initial state which have passed through the region of external influence and reappear in the out-region and finally, gravitationally induced amplification of the ingoing particle content. Fermions on the other hand show attenuation (minus sign).

Only non-empty modes cause the additional creation of particles out of the background. This makes it an induced process. It happens only if show this clearly: An asymmetry in the particle content of the outgoing $\underline{p}$- and $-\underline{p}$-mode can only be caused by the passed through particles of an

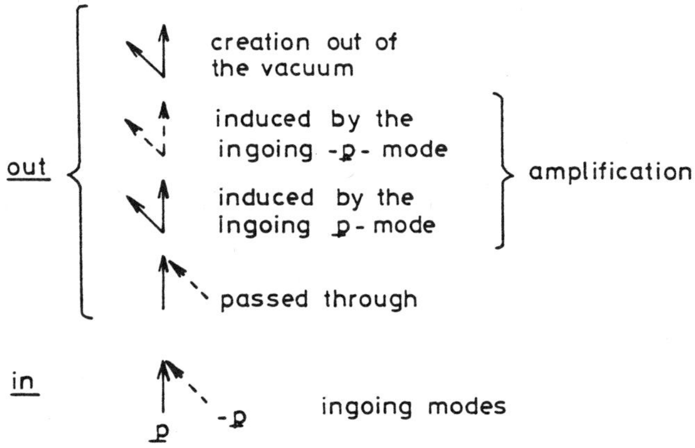

FIGURE 5

there is creation out of the vacuum, thus amplifying this process. Because of momentum conservation the amplification results in additional outgoing pairs. The difference

$$N^{(0)}(\underline{p}^{\Phi}|a) - N^{(0)}(-\underline{p}^{\Phi}|a) = n(\underline{p}^{\Phi}|a) - n(-\underline{p}^{\Phi}|a) \tag{42}$$

asymmetric initial state. Because of this additional creation of pairs ingoing particles in the mode $-\underline{p}$ induce creation in the mode $\underline{p}$. This explains the appearance of $n(-\underline{p}^{\Phi}|a)$. See figure 5 for a schematic representation.

Because of the factor $|\beta_{\underline{p}}|^2$, which describes the spectral behaviour of the vacuum creation, the amplification is mode dependent. This factor demonstrates that the amplification is in our case of gravitational origin or goes back to the non-minkowskian situation whenever this is characterized by a particle creation out of the vacuum. In the example above there is no amplification for the massless particles. We summarize: In the absence of mutual interaction between several quantum fields the initial presence of bosons tends to increase the number of bosons created out of the background. For fermions the situation is reversed. Vacuum creation and amplification/attenuation describe the physics of the zeroth order completely.

Induced gravitational amplification and attenuation in processes of mutual interaction

With (37) and (39) we find in higher order the following result for (S = 0) a mutual or self-interaction:

$$N^{(z)}(\underline{p}^{\Phi}|a) = \sum_{\text{all } b} |\langle \text{in } b|S|a \text{ in}\rangle|^2_{(z)}\, n(\underline{p}^{\Phi}|b) + \tag{43}$$

$$+ N^{(o)}(\underline{p}^{\Phi}|0)\,\Big[\sum_{\text{all } b} |\langle \text{in } b|S|a \text{ in}\rangle|^2_{(z)}\,(n(p^{\Phi}|b) + n(-\underline{p}^{\Phi}|b))\Big] + \mathrm{Re}(\beta^*_{\underline{p}}\sigma_{\underline{p}})$$

With (41) the first term represents a sum over the number of particles with parameter $\underline{p}$ in the state $|b \text{ in}\rangle$ (which can be reached according to the interaction) times the transition probability in the in-in scheme for an initial state $|a \text{ in}\rangle$ to go over into this state $|b \text{ in}\rangle$. This is exactly an expression of the structure one would obtain in the minkowskian situation. The square bracket in the second term contains the first term plus the corresponding one for the pair partners in the mode $-\underline{p}$. Comparison of the total second term in (43) term with the third term in (40) reveals a close structural and physical analogy so that we may interpret this term again as the amplification term. In the fermionic case it has a different sign and represents therefore attenuation. The third term in (43) with

$$\sigma_{\underline{p}} = -2\alpha^{*}_{\underline{p}} \sum_{\text{all } b} \langle \text{in } a|S^{\dagger}|b \text{ in}\rangle \langle \text{in } 1^{\Phi}_{\underline{p}} 1^{\Phi}_{-\underline{p}} \; b|S|a \text{ in}\rangle \tag{44}$$

has no correspondence in the zeroth order formula (40).

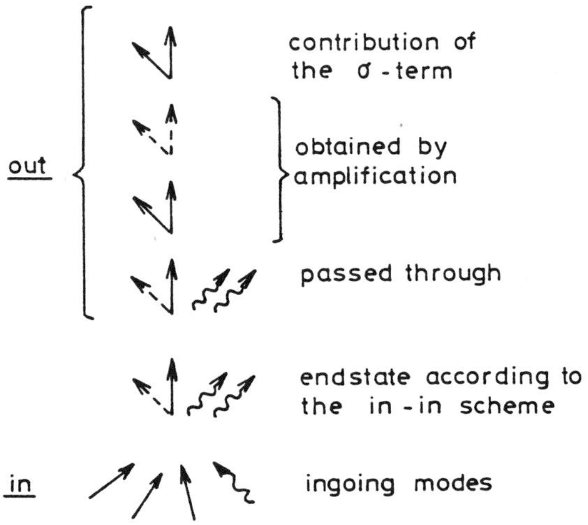

FIGURE 6

Keeping in mind that we are always describing one total coherent in-out process and that there is no particle interpretation as long as the interaction takes place, we may nevertheless visualize the structure of (43) in the following diagrammatic rule: "Let the mutual interaction happen completely within the in-region only and describe it accordingly by in-in transition amplitudes. Process now the corresponding particle outcome as in zeroth order in a twofold way: At first the particles in the $\underline{p}$-mode pass through into the out-region as in (40) to obtain the first term of (43). Secondly, these particles are amplified in the same way as in (40) whereby a possible outcome in the -$\underline{p}$-mode contributes in a symmetric manner. This leads to the second term in (43). Finally the σ-term of (43) is to be added." See figure 6 for a schematic representation of this rule.

Comparison of the $\underline{p}$- with the -$\underline{p}$-mode

$$N(\underline{p}^{\Phi}|a) - N(-\underline{p}^{\Phi}|a) = \sum_{\text{all } b} |<\text{in } b|S|a \text{ in}>|^2[n(\underline{p}^{\Phi}|b) - n(-\underline{p}^{\Phi}|b)] \tag{45}$$

shows that an asymmetry can only be due to the mutual interaction. Amplification and the process leading to the σ-term happen as creation of ($\underline{p}$,-$\underline{p}$) pairs.

Rewriting (43) as

$$N^{(z)}(\underline{p}^{\Phi}|a) =$$

$$\sum_{\text{all } b} |<\text{in } b|S|a \text{ in}>|^2_{(z)} n(\underline{p}^{\Phi}|b)\{1 + N^{(o)}(\underline{p}^{\Phi}|0)[1 + \frac{n(-\underline{p}^{\Phi}|b)}{n(p^{\Phi}|b)}]\} \tag{46}$$

$$+ \mathrm{Re}(\beta^{*}_{\underline{p}}\sigma_{\underline{p}}) \ .$$

we see, that amplification acts as a mode dependent amplification factor. The minkowskian contributions contained in the in-in transition amplitude are altered in a multiplicative way. This fact may be responsible for considerable modifications.

Specified mean value of the particle number

In the expressions (40) and (43) only the in-state has been specified. To make the particle number a concept which describes particular processes, we improve the scheme as we have done it above for transition probabilities. We specify again not only the in- but also the umpaired massive and the massless part of the out-state (indicator state) and sum up over all massive pairs:

$$N(\underline{p}^{\Phi}|\hat{d}^{\Phi}s^{\Psi} \leftarrow c^{\Phi}r^{\Psi}) = \sum_{\text{all } Q} |<\text{out } Q^{\Phi}\hat{d}^{\Phi}s^{\Psi}|S|c^{\Phi}r^{\Psi} \text{ in}>|^2 \, n(\underline{p}^{\Phi}|Q^{\Phi}\hat{d}^{\Phi}s^{\Psi}) \tag{47}$$

We call this the <u>specified mean number</u> $N(\underline{p}^{\Phi}|\leftarrow)$. Operationally it means that we will register the number of massive particles leaving the interaction in the mode $\underline{p}$ only if the specific process $\hat{d}^{\Phi}s^{\Psi} \leftarrow c^{\Phi}r^{\Psi}$ (accompagnied by outgoing massive particles) has happened. The definition

$$n(p^{\Phi}|Q^{\Phi}\hat{d}^{\Phi}s^{\Psi}) = \langle in\ Q^{\Phi}\hat{d}^{\Phi}s^{\Psi}|N_{\underline{p}}^{\Phi in}|Q^{\Phi}\hat{d}^{\Phi}s^{\Psi}\ in\rangle \tag{48}$$

implies that we always count unpaired as well as paired massive particles because this reflects the actual measurement situation. Pairs created by gravitional induction are included and will lead to amplification in the bosonic case.

After some calculations (47) may be rewritten as

$$\begin{aligned} N(\underline{p}^{\Phi}|\hat{d}^{\Phi}s^{\Psi} \leftarrow c^{\Phi}r^{\Psi}) = {} & N^{(o)}(\underline{p}^{\Phi}|0)w^{inc}(\hat{d}^{\Phi}s^{\Psi}|c^{\Phi}r^{\Psi}) + \\ & + \sum_{\text{all } Q} |\langle in\ Q^{\Phi}\hat{d}^{\Phi}s^{\Psi}|S|c^{\Phi}r^{\Psi}\ in\rangle|^2 \{n(p^{\Phi}|Q^{\Phi}\hat{d}^{\Phi}s^{\Psi}) + \\ & + N^{(o)}(p^{\Phi}|0)[n(\underline{p}^{\Phi}|Q^{\Phi}\hat{d}^{\Phi}s^{\Psi}) + n(-\underline{p}^{\Phi}|Q^{\Phi}\hat{d}^{\Phi}s^{\Psi})]\} + \mathrm{Re}(\beta^{*}_{\underline{p}}\tilde{\sigma}_{\underline{p}}) \end{aligned} \tag{49}$$

with

$$\tilde{\sigma}_{\underline{p}} = -2\alpha^{*}_{\underline{p}} \sum_{\text{all } L} \langle in\ a|S^{\dagger}|L^{\Phi}\hat{d}^{\Phi}s^{\Psi}\ in\rangle \langle in\ L^{\Phi}1^{\Phi}_{\underline{p}}1^{\Phi}_{-\underline{p}}\hat{d}^{\Phi}s^{\Psi}|S|a\ in\rangle \tag{50}$$

The latter expression is more simple and vanishes in more cases than (44). We will make use of this fact below.

A detailed analysis shows that the first term in (49) refers to particles which originate solely from zeroth order particle creation out of the vacuum. But in this more specific version these particles are only counted if the indicator configuration is registered simultaneously in the out-region. In non-zero order we have to disregard this contribution. We call the rest of (49) $N^{int}(\leftarrow)$. The various terms in this expression show a clear correspondence to (43), so that the physical interpretation can be transcribed.

Example: Compton scattering in the $\Phi^2\Psi^2$-model and the pure amplification effect

To prove the usefulness of the concepts introduced above for the discussion of the underlying physical structure of processes of mutual interaction, we study second order Compton scattering in the $\Phi^2\Psi^2$-interaction. This will enable us to demonstrate a pure amplification process of higher order, i.e. a process to which the last term in (46) doesn't contribute. To do so we take $1^{\Phi}_{q}1^{\Psi}_{l}$ as initial state, $1^{\Phi}_{p}1^{\Psi}_{k}$ as indicator state for the outgoing particles and compare the number of outgoing massive particles in the mode $\underline{p}$ with the number of outgoing massless particles in the mode $\underline{k}$ (Audretsch and Spangehl, 1987).

The first observation is that $\tilde{\sigma}^{(1)} = 0$ and $\tilde{\sigma}^{(2)} = 0$ vanish, thus making the scattering a pure amplification effect. With (49) and using (19) we end up with two transition amplitudes:

$$N^{int}(\underline{p}^{\Phi}|1^{\Phi}_{\underline{p}}1^{\Psi}_{\underline{k}} \leftarrow 1^{\Phi}_{\underline{q}}1^{\Psi}_{\underline{l}})^{(2)} = \{|\langle in\ 1^{\Phi}_{\underline{p}}1^{\Psi}_{\underline{k}}|T^{(1)}|1^{\Phi}_{\underline{q}}1^{\Psi}_{\underline{l}}\ in\rangle| \qquad (51)$$

$$+ |\langle in\ 1^{\Phi}_{\underline{p}}1^{\Phi}_{-\underline{q}}1^{\Psi}_{\underline{k}}|T^{(1)}|1^{\Psi}_{\underline{l}}\ in\rangle|^2\}(1+|\beta_{\underline{p}}|^2)$$

which are related to the diagrams of figure 7

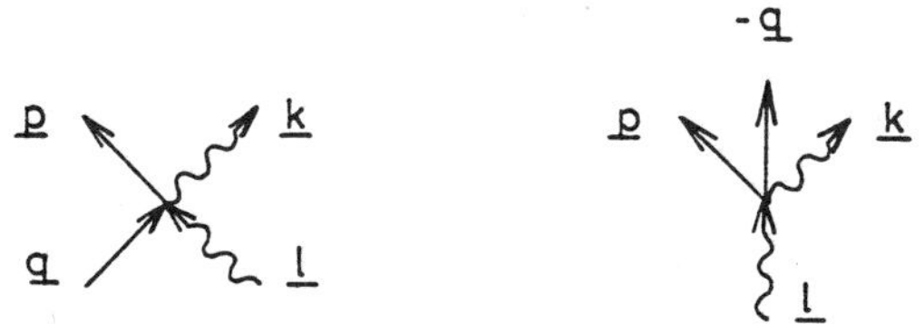

FIGURE 7

Because there is with $|\beta_k|^2 = 0$ no amplification in the massless case, we have $N^{int}(\underline{k}^{\Psi}|\leftarrow)^{(2)} = \{\ldots\}$ with the bracket of (51) and therefore: In Compton scattering, the massive Φ- and the related massless Ψ-particles leaving the mutual interaction are not going out in pairs, as one would expect from the situation in flat space-time. Rather the number of the massive particles is amplified by a momentum dependent factor according to

$$N^{(2)}(\underline{p}^{\Phi}|\leftarrow) = N^{(2)}(\underline{k}^{\Psi}|\leftarrow)[1+|\beta_{\underline{p}}|^2] \tag{52}$$

Relevance of the σ-term

The example of particle creation out of the vacuum for the expansion law

$$a^2(\eta) = 1 + e^{2b\eta}$$

in the different orders of the mutual interaction

$$L_I = -\frac{\lambda\sqrt{-g}}{a^2(\eta)}\Phi\Psi$$

has been discussed rigorously by Audretsch and Spangehl (1986). The results show that the σ-term of equation (46) dominates the second order. This demonstrates that in higher order the process of gravitationally induced amplification/attenuation is only one effect among others. In general we will not have the pure case

Additional examples for induced amplification and attenuation

Close inspection of results obtained by various authors in other contexts reveals contributions of the effect of amplification/attenuation. This is the case for the discussion of particle creation from vacuum according to the interaction $L_I = \lambda\Phi^4$ by Birrell and Ford (1989), according to the interaction $L_I = L_I(h_{\alpha\beta},\Phi)$ (with $h_{\alpha\beta}$ being an unquantized perturbation of the metric) by Friedmann (1989) and according to the QED-interaction $L_I = ie\,\overline{\Psi}\gamma^{\mu}A_{\mu}\Psi$ by Lotze (1985). The latter is an example for attenuation.

Stimulated emission in zeroth order by black holes and accelerated mirrors

Amplification in zeroth order (no mutual interaction between quantum fields) has also been discussed in detail by Wald (1976). He showed for

Schwarzschild black holes that enormous energies are required for the ingoing particles to induce spontaneous emission at late times. If these energies are bounded, there is only stimulated emission (i.e. amplification) at early times and the usual thermal radiation at late times.

For Kerr black holes the same is true for nonsuperradiant modes. In superradiant modes the effect equals classical superradiant scattering if the number of ingoing particles is large (Wald, 1976). Panangaden and Wald (1977) have shown that the results agree with a formula obtained classically by thermodynamical and information theoretic arguments. Finally we have to mention the stimulated emission by accelerated mirrors as discussed by Davies and Fulling (1977).

TRANSITION PROBABILITIES AND THE TWO TIME SCALES

Example: Decay of a massive particle in the $\Phi\Psi^2$-model

As an application of the concept of the added-up probability introduced above, we discuss the decay of a massive particle with the intention to determine its lifetime in the presence of external cosmological gravitational fields. The inherent conceptual difficulties of an S-matrix approach to questions of this type will thereby become evident.

In order to enable rigorous calculations up to a certain order in the coupling parameter and because we are primarily interested in structural discussion, we use toy expansion laws for which the relevant physical effects become very transparent (Audretsch et. al., 1987). The first expansion law is a step

$$\hat{a}(\eta) = \Theta(-\eta)\, a_i + \Theta(\eta)\, a_o \qquad (53)$$

with $a_i = (A-B)^{1/2}$, $a_o = (A+B)^{1/2}$, $A > B > 0$ (see figure 8).

For comparison we then add the corresponding rigorous calculation for the tanh expansion law (see figure 8b)

$$\tilde{a}(\eta) = (A + B \tanh b\eta)^{1/2} \qquad (54)$$

which may be regarded as the prototype of a smoothed out step representing an influence of a curved background of finite duration. Both are statically bounded monotonic expansion laws with flat or approximately flat in- and out-regions allowing simple particle definitions. The Bogoliubov coefficient in the step case is simply

$$\hat{\beta}_{\underline{p}} = -(E_o - E_i)/2(E_o E_i)^{1/2} \tag{55}$$

where we have introduced the energy parameter

$$E_{i/o} = (\underline{p}^2 + m^2 a_{i/o}^2)^{1/2} \tag{56}$$

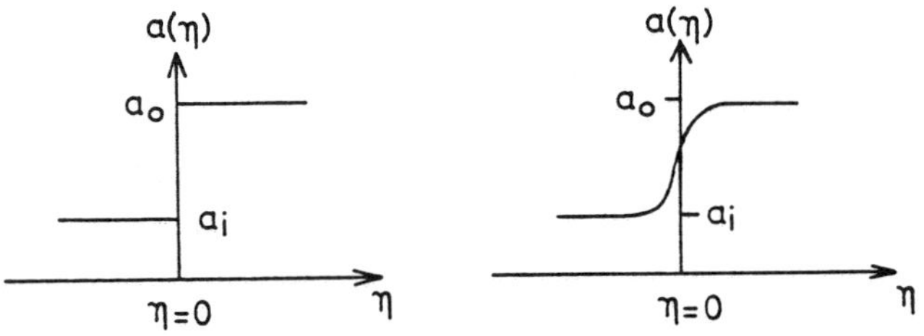

FIGURE 8

We study the decay of one massive into two massless particles according to the mutual interaction

$$L_i = \sqrt{-g}\ (\lambda/a(\eta))\Phi\Psi^2 \tag{57}$$

By including the factor $a(\eta)^{-1}$ the calculations for the rigorous approach are made more simple. The mass of the Φ-particles breaks conformal invariance. There are two in-in diagrams contributing (compare figure 9). The second reflects energy non-conservation, but also the first diagram is to be taken into account outside energy conservation too.

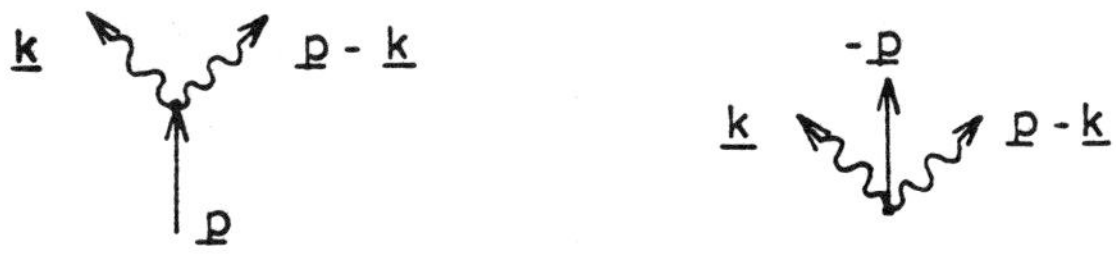

FIGURE 9

After working out the respective transition amplitudes in the step case we obtain for the added-up probability to find two massless Ψ-particles in the out-region if one massive Φ-particle with momentum parameter $\underline{p}$ has been going in:

$$\hat{w}^{add}(1^{\Psi}_{\underline{k}} 1^{\Psi}_{\underline{p}-\underline{k}} | 1^{\Phi}_{\underline{p}}) = \frac{\lambda^2 \pi}{2k|\underline{p}-\underline{k}|V} \left[\frac{1}{2}\frac{1}{E_i} T_\eta \, \delta(\omega_{-i}) + \right.$$

$$\left. + \left(\frac{1}{2} + |\hat{\beta}_{\underline{p}}|^2\right)\frac{1}{E_o} \; T_\eta \, \delta(\omega_{-o})\right] + \hat{\Delta}(k) \tag{58}$$

The momentum parameter is conserved. Because of the asymptotic switching of the mutual interaction according to (7), the infinite conformal time T_η appears. It represents the duration of the interaction between the particles. For the tanh case the result has a similar structure. It is obtained by replacing hat by tilde and by adding a finite correction to the infinite time T_η of the mutual interaction:

$$\tilde{w}^{add} = \hat{w}^{add}(\hat{\ } \to \tilde{\ };\; T_\eta \to T_\eta + \text{finite}) \tag{59}$$

The latter will prove to be important below.

The typical k-dependence of the two terms $\Delta(k)$ and $\Delta(k)$ is shown in figure 10.

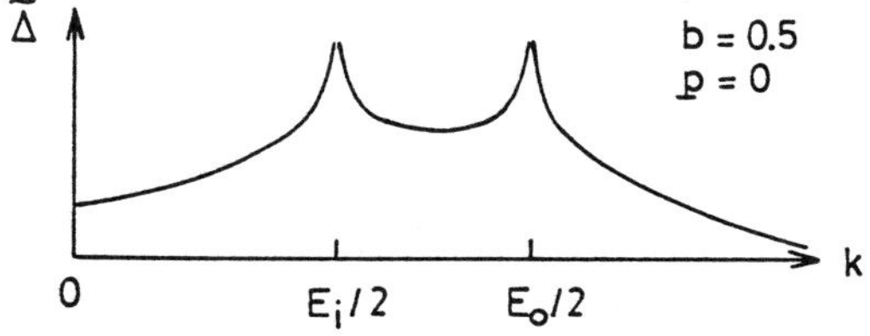

FIGURE 10

These terms as well as the δ-functions become singular at $\omega_{-i} = 0$ and $\omega_{-o} = 0$ defined according to

$$\omega_{-i/o} = E_{i/o} - k - |\underline{k}-\underline{p}| \tag{60}$$

With (56) this means that the decay process becomes resonant when in addition to 3-momentum conservation also the measured energy is conserved. Because the measured energy depends on $a(\eta)$ and we have two asymptotic regions with the different values a_i and a_o, there are two resonances at different values of k. They reflect a sort of minkowskian kernel in the process. Because the cosmological space-time does in fact not demand energy conservation, there is a finite spectrum outside of the resonances.

This is included into the resulting expression when we go over to the <u>total decay probability</u>

$$w^{tot} = \sum_{\underline{k}} w^{add}(1^{\Psi}_{\underline{k}} 1^{\Psi}_{\underline{p}-\underline{k}} | 1^{\Phi}_{\underline{p}}) \tag{61}$$

in summing over all outgoing massless modes. For the step case we find:

$$\hat{w}^{tot} = \frac{\lambda^2}{4\pi m} \left[\frac{1}{2a_i} T_\eta + \frac{1}{2a_o} \left(\frac{1}{2} + |\hat{\beta}|^2 T_\eta\right)\right] \tag{62}$$

This expression refers to a massive particle at rest ($\underline{p} = 0$). Again the result in the tanh case is obtained by replacing hat by tilde and adding a finite term R^{fin}.

$$\tilde{w}^{tot} = \hat{w}^{tot}(\hat{\ } \to \tilde{\ }) + R^{fin}(a_i, a_o, |\tilde{\beta}|^2) \tag{63}$$

with $\beta = \beta_{\underline{p}=0}$.

We turn to the interpretation of the two results (62) and (63) and start with a comparison of the step case with the minkowskian total transition probability

$$w^{tot}_{Mink} = \frac{\lambda^2}{4\pi m} T_t \tag{64}$$

where the infinite duration T_t of the mutual interaction refers to the measured time t. In the step case one half of the particles has the chance to decay in the Minkowski region $\eta \leq 0$ with $a(\eta) = a_i$ and the other half in the region $\eta \geq 0$ with $a(\eta) = a_o$. This explains the factors 1/2. The parameters a_i and a_o appear in the result, because L_I of (57) contains λ/a instead of λ. Introducing in (64), according to $T_t = a_i T_\eta$, the conformal time T_η we obtain the first term of (62).

With regard to the second term in (62) we have to take into account that in addition at $\eta = 0$ massive particles are created out of the vacuum. The creation is according to (19) described by means of $|\beta|^2$. These particles are decaying only in the region $\eta \geq 0$. Therefore, as compared to the first term, the factor 1/2 has to be replaced by $(1/2 + |\beta|^2)$.

For the tanh-case the first term of the result (63) is of exactly the same structure. In this case too, the two nearly static regions are responsible and lead to the factors $1/a_i$ and $1/a_o$ respectively. Again there is a decay of particles created in zeroth order.

But in contrast to the step case, the tanh-case shows a peculiarity: The gravitational influence is not localized to $\eta = 0$ but smeared out over a finite time interval around $\eta = 0$ which we will denote by T^{grav} (gravitational time). This fact is responsible for the appearance of the finite additional term R^{fin} in the result (63).

This latter observation can also be made correspondingly for other expansion laws $a(\eta)$. The following seems to be a general trait of the scheme used above: A gravitational influence of finite duration leads to additive finite corrections of terms which are themselves proportional to the time of the mutual interaction between the particles. This time T_η is infinite because the mutual interaction is switched off according to (7) only asymptotically. The result is that an S-matrix approach to quantum field theory in curved space-time contains inherently two time scales if the influence of the background curvature is of finite duration.

We have to take this into account when going on to the reciprocal <u>lifetime</u> of massive particles at rest. In the minkowskian case this is easily obtained in dividing (64) by T_t:

$$\frac{1}{\tau_{\text{Mink}}} = \frac{\lambda^2}{4\pi m} \tag{65}$$

The corresponding procedure gives in the step case

$$\frac{1}{\hat{\tau}} = \frac{\lambda^2}{4\ m}\left(\frac{1}{2a_i} + \frac{1}{2a_o}\right) + \frac{\lambda}{4\pi E_o}\,|\beta|^2 \tag{66}$$

In the tanh case on the other hand we have to refer to the two time scales. A natural consequence would be to divide R^{fin} by the finite gravitational time T^{grav}.

$$\frac{1}{\tilde{\tau}} = \frac{\tilde{w}^{tot}}{T_\eta} + \frac{R^{fin}}{T^{grav}} \tag{67}$$

This procedure will at least lead to some physically reasonable mean value for the decay rate. Nevertheless it must be stressed that in general T^{grav} cannot be defined exactly. The development of a satisfying method to work out transition probabilities, which are appropriately localized in time, still remains an open problem.

Other expansion laws

The relations above reflect a general structure for all statically bounded expansion laws. The results for expansion laws with one or two infinite regions are contained as limiting cases, e.g. a_i finite, $a_o \to \infty$ with ($E_o \to \infty$), compare Audretsch and Spangehl (1985) for an example.

An example showing the modification of a minkowskian result

A treatment of the particle decay along the lines sketched above for the expansion law

$$a^2(\eta) = a_i^2 + e^{2b\eta}, \quad b > 0 \tag{68}$$

has been given by Audretsch and Spangehl (1985). Again the mean value of the local minkowskian result is gravitationally corrected. Figure 11 shows

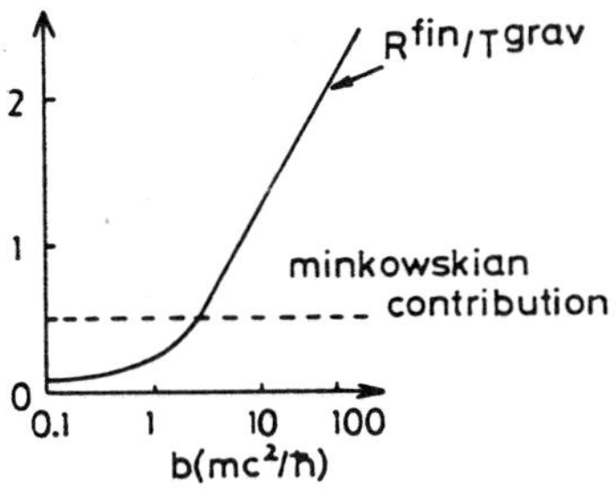

FIGURE 11

that in this case the gravitational influence dominates for large values of the parameter b. Although this rigorous calculation is based on a toy interaction and an unrealistic expansion law, it demonstrates once more severe restrictions of the local applicability of minkowskian results. This is one of the general lessons to be learned.

Additional examples of worked out transition probabilities

Several calculations of transitions probabilities can be found in the literature. The following effects have been discussed:

- For massless scalar and a Maxwell field

$$L_I = \lambda F_{\mu\nu} F^{\mu\nu} \Psi \tag{69}$$

particle decay from vacuum (Birrell, Davies and Ford, 1980) and particle decay and CPT non-invariance (Lotze, 1987).

- For two massive scalar fields

$$L_I = \lambda R(\Phi^* \Lambda(t) \Psi + \Psi^* \Lambda^*(t) \Phi) \tag{70}$$

particle creation from vacuum (Papastamatiou and Parker, 1979).

- For massive and massless scalar fields

$$L_I = \lambda \Phi \Psi^n \tag{71}$$

particle decay and CPT non-invariance (Ford, 1982).

- For Dirac and Maxwell fields (QED)

$$L_I = ie\overline{\Psi}\gamma^u A_\mu \Psi \tag{72}$$

particle creation from vacuum (Lotze, 1985a,b), emission of a photon by an electron (Lotze, 1988), particle decay and CPT non-invariance (Lotze, 1989).

For self-interacting fields see Birrell and Ford (1979) and Leahy and Unruh (1983).

Accelerated observers, black holes

Finally we have to mention discussions of the Green function. Unruh and Weiss (1984) have shown for several interactions that an accelerated observer in thermal equilibrium measures the same Green function as the Minkowski observer in his vacuum state. A similar result has been obtained for black holes by Gibbons and Perry (1978), see also the exact treatment of the Thirring model by Birrell and Davies (1978).

REFERENCES

Audretsch, J., 1979, Cosmological particle creation as above-barrier reflection: approximation method and applications, J. Phys. A.: Math. Gen., 12: 1189.

Audretsch, J., and Spangehl, P., 1985, Mutually interacting quantum fields in an expanding universe: decay of a massive particle, Class. Quantum Grav., 2:733.

Audretsch, J. and Spangehl, P., 1986, Gravitational amplification and attenuation as part of the mutual interaction of quantum fields in curved space-times, Phys. Rev. D, 33:997.

Audretsch, J. and Spangehl, P., 1987, Improved concepts for the discussion of mutually interacting quantum fields in Robertson-Walker universes, Phys. Rev. D, 35:2365.

Audretsch, J., Rüger, A., and Spangehl, P., 1987, Decay of massive particles in Robertson-Walker universes with statically bounded expansion laws, Class. Quantum Grav., 4:975.

Audretsch, J., 1988, On the optical theorem in curved space-time quantum field theory, lectures given at the colloquium in Les Treilles, July 1988 (will appear in Int. J. Theor. Phys.).

Audretsch, J., 1989, Applications of the curved space-time optical theorem, manuscript University of Konstanz.

Birrell, N.D., and Davies, P.C.W., 1978, Massless Thirring model in curved space: Thermal states and conformal anomaly, Phys. Rev. D, 18:4408.

Birrell, N.D., and Ford, L.H., 1979, Self-interacting quantized fields and particle creation in Robertson-Walker universes, Ann. Phys., 122:1.

Birrell, N.D., Davies, P.C.W., and Ford, L.H., 1980, Effects of field interactions upon particle creation in Robertson-Walker universes, J. Phys. A.: Math. Gen., 13:961.

Birrell, N.D., 1981, Interacting quantum field theory in curved space-time, in: "Quantum Gravity 2", C.J. Isham, R. Penrose and D.W. Sciama, eds., Clarendon, Oxford.

Birrell, N.D., and Davies, P.C.W., 1982, "Quantum Fields in Curved Space", CUP, Cambridge.

Davies, P.C.W. and Fulling, S.A., 1977, Radiation from moving mirrors and from black holes, Proc. R. Soc. London A, 350:237.

Ford, L.H., 1982, Particle decay and CPT non-invariance in cosmology, Nucl. Phys. B, 204:35.

Ford, L.H., 1984, Aspects of interacting quantum field theory in curved space-time: renormalization and symmetry breaking, in: "Quantum theory of gravity", S.M. Christensen, ed., Hilger, Bristol.

Ford, L.H., 1988, Recent advances in quantum field theory in curved spacetime, in: Highlights in gravitation and cosmology, B.R. Iyer, A. Kembhavi, J.V. Narlikar, C.V. Vishveshwara, eds. CUP, Cambridge.

Friedman, J.A., 1989, Particle creation in inhomogenuous spacetimes, Phys. Rev. D, 39:389.

Gibbons, G.W. and Perry, M.J., 1978, Black holes and thermal Green Functions, Prod. R. Soc. London A, 358:467.

Leahy, D.A. and Unruh, W.G., 1983, Effects of a ^{4} interaction on black-hole evaporation in two dimensions, Phys. Rev. D, 28:694.

Lotze, K.H., 1985a, Effects of the electromagnetic interaction upon particle creation in Robertson-Walker universes: I. A general framework for the calculation of particle creation, Class. Quantum Grav., 2:351.

Lotze, K.H., 1985b, Effects of the electromagnetic interaction upon particle creation in Robertson-Walker universes: II. A soluble example, Class. Quantum Grav., 2:363.

Lotze, K.H., 1987, Particle decay and violation of CPT invariance in expanding universes: the $^{\circ}$-2 model, Class. Quantum Grav., 4:1437.

Lotze, K.H., 1988, Emission of a photon by an electron in Robertson-Walker universes, Class. Quantum Grav., 5:595.

Lotze, K.H., 1989, Pair creation by a photon and the time-reversed process in a Robertson-Walker universe with time-symmetric expansion, Nuclear Phys. B, 312:673.

Panangaden, P. and Wald, R.M., 1977, Probability distribution for radiation from a black hole in the presence of incoming radiation, Phys. Rev. D, 16:929.

Papastamatiou, N.J., and Parker, L., 1979, Asymmetric creation of matter and antimatter in the expanding universe, Phys. Rev.D, 19:2283.

Parker, L., 1969, Quantized fields and particle creation in expanding universes I, Phys. Rev., 183:1057.

Parker, L., 1971, Quantized fields and particle creation in expanding universes II, Phys. Rev. D, 3:346.

Unruh, W.G., and Weiss, N., 1984, Acceleration radiation in interacting field theories, Phys. Rev. D, 29:1656.

Wald, R.M., 1976, Stimulated-emission effects of particle creation near black holes, Phys. Rev. D, 13:3176.

Wald, R.M., 1979, Existence of the S-matrix in quantum field theory in curved space-time, Ann. Phys., 118:490.

QUANTUM STRINGS IN CURVED SPACE TIMES

N. SANCHEZ

DEMIRM, Observatoire de Paris, Section de Meudon
92195 Meudon Principal Cédex, France

INTRODUCTION

We discuss string theory in the context of quantum gravity, with specific treatment of string quantization in curved space times. The contents of these lectures is the following. In lecture I we discuss thermal and ground state effects of quantum strings : (i) (first) string quantization in Rindler space-time and its applications to the Scharzschild geometry (ii) the Hawking-Unruh effect in string theory and its description in terms of analytic mappings. In Lecture II we give a new approach to string quantization in curved space time in order to take into account strong curvature effects of the geometry. We apply our method to (i) de-Sitter space-time (ii) the quantum string dynamics and scattering in black hole space times. In lecture III we treat the case of boosted black-holes in the ultra relativist limit (gravitational schock wave or Aichelburg Sexl geometry) in the context of (i) the particle scattering at the Planck energy scale and (ii) quantum string dynamics and scattering in this geometry, (in this case, the problem can be solved exactly).

CONTEXTUAL BACKGROUND

Perhaps the main challenge in theoretical physics today is the unification of all interactions including gravity. At present, string theories appear as the best candidates to achieve such an unification. However, many technical and conceptual problems remain and a quantum theory of gravity is still non available. Continuous effort over the last quarter of a century has demonstrated the many difficulties encountered in repeated attempts to construct such a theory and have also indicated some of the particular properties which an eventual complete theory will have to possess. The amount of work in that direction can be by now presented in two different sets which have most evolved (and remain) separated : (i) conceptual unification (introduction of the uncertainty principle in general relativity, the interpretation problem, Q.F.T. in curved space time and by accelerated observers, Hawking radiation and its consequences, "wave function of the universe"...) (ii) grand unification (the unification of all interactions including gravity from the particle physics point of view. Gravity is considered as a massless spin two particle (the graviton) : supergravities, Kaluza-Klein theories and the more successful superstrings).

Quantum Mechanics in Curved Space-Time
Edited by J. Audretsch and V. de Sabbata
Plenum Press, New York, 1990

Whatever the final theory of the world will be if it is to be a theory of everything, we would like to know what new understanding it theory of everything, we would like to know what new understanding it will give us about the singularities of classical general relativity. If string theory would provide a theory of quantum gravity, it should give us a proper theory (not yet existent) for describing the ultimate state of quantum black holes and the initial (say, very early) state of the Universe. That is, a theory describing the physics (and the geometry) at Planck energies and lengths.

Till now, gravity has not completely been incorporated in string theory : strings are most frequently formulated in a flat space-time. Gravity appears through massless spin two particles (graviton). One disposes only of partial results for strings in curved backgrounds ; these mainly concern the problem of consistency (validity of quantum conformal invariance) through the vanishing of the beta functions. The non-linear quantum string dynamics in curved space-time have only been studied in the slowly-varying approximation for the goemetry (background field method) where the field propagator is essentially taken as the flat space Feynman's propagator. Clearly, such approximations are useless for the study of strings in strong curvature regimes where quantum gravity effects are important. Our aim is to properly understand strings in the context of quantum gravity. As a first step in this program we propose ourselves to study Q.S.T. (Quantum string Theory) in curved spacetimes. (Of course the main goal should be to extract the particle spectrum and the space time itself from string theory but, as it is known, one is very far from doing that explicitely). There are different kind of effects to be considered here : ground state and thermal effects associated to the fact that in General Relativity there are no prefered reference frames, and to the possibility of having different choices of time and curvature effects which as we will see will modify the mass formula, critical dimension and scattering amplitudes of strings. There are also **conceptual aspects** which appear for instance when strings are restricted to live in a bounded region of the space-time, ie in the presence of event horizons, and which imply quantum fluctuations of the event horizon (and of the light-cone itself).

THERMAL AND GROUND STATE EFFECTS

The formulation of Quantum Field Theory in non trivial (curved or flat) space times has given new fundamental features with respect to the usual understanding of Q.F.T. in trivial (Minkowski flat) space time, i.e. : i) the possibility for a given field theory to have different alternative well defined Focks spaces (different "sectors" of the theory) ; (ii) the presence of "intrinsic" statistical features (temperature, entropy) arising from the non-trivial structure (geometry, topology) of the space time and not from a superimposed statistical description of the quantum matter fields. Relevant examples are Q.F.T. on the Rindler manifold and its analytic mapping extensions (1-2), black-holes and cosmological (de Sitter) space-times.

Quantum field theory developed for curvilinear (accelerated) coordinates in flat space in a way which can be directly generalized to curved space-time is an useful step for a physical and mathematical discussion of the full theory. The genuine coordinate independence which is so familiar in the classical theory of general relativity is not a particular property of gravity but a fundamental principle prevalent in all descriptions of physical laws. On the other hand, the apparent difference which results from the treatment of a quantum field theory in a variety of coordinate systems (in either curved or flat space-time) is not a coordinate effect at all, but is a consequence of the fact that physically different quantum states are correctly described by the quantum theory as being physically distinct. "Canonical" states for different coordinate systems are physically different (each timelike vector field leads to a separate indication of what constitutes a definition of positive frequency).

Strings in Rindler space

Much of the present interest in string theories comes from the hope that they may provide a finite theory of quantum grivity. It seems natural in this context to investigate the quantization of strings in a curved space-time. As a first step in this programme we study the quantization of a string in a uniformly accelerated space (D-dimensional Rindler space). Although this is a flat space-time, it possesses a space-time structure including an event horizon, similar to a black-hole manifold. In order to quantize the string properly in this manifolds, a horizon regularization is needed. This regularization can be introduced through the definition of the Rindler co-ordinates as follows :

$$x_1 \pm x_0 + \varepsilon = e^{\alpha(X^1 \pm X^0)}, \qquad x^i = X^i \quad , \quad i = 2, \ldots, D \tag{1}$$

where (x^1 , x^0) and (X^1 , X^0) are respectively Minkowskian (Kruskal-like) and Rindler (Schwarzschild-like) co-ordinates. α is a constant defining the proper acceleration of the Rindler observers (is equal to the surface gravity in the black hole case) and ε is an infinite simal parameter of the order of the Planck length. The horizon is now at a finite distance $|X^1 \pm X^0| \sim 1/\alpha \ln 1/\varepsilon$. This regularization reflects the fact that a classical description of the geometry is no longer valid at distances of the order of the Planck length.

We quantize the string in a light-cone gauge where the light-cone variables are ($x^1 \pm x^0$) and ($X^1 \pm X^0$). This choice is particularly convenient here since the acceleration points in the x^1 direction. We recall that for Q.F.T. in accelerated frames (in flat and curved space-times), different time-like vector fields lead to inequivalent positive frequency modes 1)-3). In the light cone gauge, positive frequency modes with respect to the time-like variable (τ) on the string world sheet are physical states when is identified with the appropriate time variable in the physical space-time. We associate inertial and accelerated particle states of the string to positive frequency modes with respect to x^0 and X^0 respectively. We find that the accelerated frequencies of the accelerated modes differ in a large factor.

$$\lambda_0 = \frac{2\pi\alpha}{\ln\left(\frac{2\pi}{\varepsilon} + 1\right)} \quad , \tag{2}$$

from the inertial ones. Physically this factor reflects the indefinite increasing of the string length when it approaches the event horizon. We develop the string dynamics in Rindler space and write the corresponding constraints. The longitudinal co-ordinates X^0 and X^1 obey non linear equations of motion but, as in the inertial case, they can be eliminated in terms of the transverse co-ordinates which are the independent dynamical variables. Two possible situations appear depending on whether the centre of mass has a uniform speed or a uniform acceleration. The mass formulae are derived in each case. The Poincaré invariance of the flat Minkowski space-time has a non-linear realization in terms of the Rindler co-ordinates. We prove that the passage from the inertial to accelerated modes of the string is a canonical transformation both at the classical and quantum levels. We also check that an explicit realization of the Poincaré algebra can be constructed in terms of the accelerated modes, provided (at the quantum level) one sets D = 26 and uses symmetric ordering of the operators. The explicit expression of the Bogoliubov coefficients relating the inertial and Rindler modes of the string is found. The expectation value of the Rindler number mode operator in the ground state (tachyon) of the string is computed. This follows a thermal distribution [Eq. (62)] with temperature

$$T_s = \frac{\alpha}{2\pi} \tag{3}$$

This is the same Hawking-Unruh value that appears in the fiel-theoretical context. However, if one measures the frequency in dimensionless units (1,2,....) instead of multiples of λ_0 (eq. (2)), the temperature of the ground state is a (very large) pure number

$$T_0 = \frac{T_s}{\lambda_0} = \frac{1}{4\pi^2} \ln\left(\frac{2\pi}{\varepsilon} + 1\right)$$

we find the expectation value of the accelerated mass M'operator in the (tachyon) ground state ; it turns out to be a large positive number

$$M' = 2\pi T_0 \tag{5}$$

The major features of the string quantization in Rindler space also holds for a string in a Schwarzschild space-time. An appropriate light-cone gauge can be introduced for left or right movers in the null Kruskal and Schwarzschild co-ordinates. Now, Eq. (2) for λ_0 gives the relation between the Kruskal and Schwarzschild frequencies and T_S and T_0 Eqs. (3)-(4) give the temperatures characterising the ground state of the string in the black hole manifold with

$$\varepsilon \sim \left(\frac{M_{Pl}}{M}\right)^{(D-2)/(D-3)} \quad , \quad \alpha \equiv K \sim \left(\frac{M_{Pl}}{M}\right)^{\frac{D-2}{D-3}}$$

Here Mpl and M stand for the Planck mass and the blak hole mass respectively. It can be noticed that the Hagedorn temperature in this context $(1/\sqrt{\alpha'} \sim M_{Pl})$ is always $\gtrsim$ Ts $\sim$ Mpl $(\mathrm{Mpl}/M)^{1/(D-3)}$ since a basic requirement for the present semiclassical treatment is M $\gtrsim$ Mpl. The investigations presented here can also be extended to the case of fermionic strings and to more general (non uniform) accelerations described by analytic (holomorphic) mappings as in the approach of Ref.2).

Horizon regularization in Rindler-space - Introduction of the εparameter

Let us discuss now some features of Rindler space which are important for our study of strings in this space. The Rindler transformation maps the right-hand wedge $x^1 \gg |x^0|$ of Minkowski space onto the whole Rindler space $-\infty \ll X^1, X^0 \lesssim \infty$(the whole Minkowski space can be covered using four different Rindler patches. As is known, a quantum field in Rindler space is in a thermal state with temperature $T = \alpha/(2\pi)$. In addition, ultra-violet divergences arise in the free energy and entropy of quantum fields from the existence of a horizon at $x^1 = |x^0|$, $(X^1 = -\infty)$in the space-time. The same problem appears in the case of a four-dimensional black hole. Let us illustrate this phenomenon by considering a free massive scalar field . In D-dimensional Rindler space, the positive frequency modes are

$$\Psi = \frac{1}{(2\pi)^{(D-2)/2}} \, e^{i(-\lambda X^0 + k^i X^i)} \, \Phi(X^1) \, , \quad \lambda > 0 \tag{6}$$
$$k^i \in \mathbb{R}$$

where $\Phi(X^1)$ satisfies

$$\left[\frac{d^2}{dX^{1\,2}} + \lambda^2 - (m^2 + k^{i\,2})\alpha^2 e^{2\alpha X^1}\right] \Phi(X^1) = 0$$

the total number $(\mathcal{N}_\lambda)$ of wave modes with fequency less than λ can be computed in the semi-classical approximation for $\lambda \gtrsim \alpha$. This is enough to study the ultra-violet behaviour of the quantities interesting us. In the W.K.B. approximation $\mathcal{N}_\lambda$ is given by

$$\pi \mathcal{N}_\lambda = \int_{-H}^{a(\lambda)} dX^1 \int \prod_{i=1}^{D-2} \frac{dk^i}{2\pi} \sqrt{\lambda^2 - (m^2 + k^{i\,2})\alpha^2 e^{2\alpha X^1}} \tag{7}$$

The integration is taken over the values of K^i and X^1 for which the argument of the square root is positive. Here a (λ) is the classical turning point

$$e^{\alpha a(\lambda)} = \lambda / m\alpha \quad ,$$

and H is a large cut-off (H $\gtrsim$ $1/\alpha$) on the negative Rindler co-ordinate X^1 . This shifts the horizon by replacing the light-cone $x^1 = |x_0|$ as a boundary of Rindler space-time by the hyperbola

$$(x^1)^2 - (x^0)^2 = e^{-2\alpha H} \tag{8}$$

This regularization takes into account the fact that a classical description of the geometry is no longer valid at distances of order of the Planck length. Thus,

$$e^{-2\alpha H} \sim l_{Pl}$$

Evaluating $\mathcal{N}_\lambda$ for large H it yields

$$\pi \mathcal{N}_\lambda = \frac{1}{4} \frac{(\lambda/\alpha)^{D-1}}{(4\pi)^{(D-3)/2}(D-2)} e^{\alpha H(D-2)} \left[1 + O(e^{-2\alpha H})\right] \tag{9}$$

Then the free energy and the entropy at temperature T :

$$F = -\int_0^\infty \mathcal{N}_\lambda \, d\lambda \, (e^{\lambda/T} - 1)^{-1} \quad , \qquad S = -\partial F/\partial T \, ,$$

are equal to

$$F = \frac{-\Gamma(D)\,\zeta(D)\,T\,\alpha^{1-D}\,e^{\alpha H(D-2)}}{(4\pi)^{(D-1)/2}(D-2)} \tag{10}$$

$$S = -DF/T > 0$$

In Rindler space, $T = \alpha/2\pi$ and so

$$F = -\Gamma(D)\,\zeta(D)\,\alpha\,(4\pi)^{1/2-D}\,\pi^{-D}\,e^{\alpha H(D-2)}$$

$$S = \Gamma(D)\,\zeta(D)\,2\pi D\,(4\pi)^{1/2-D}\,\pi^{-D}\,2\pi D\,e^{\alpha H(D-2)} \tag{11}$$

We explicitly see that F and S need the ultra-violet cut-off H to be finite. This is equivalent to considering the following mapping defining the accelerated co-ordinates

$$x^1 - x^0 + \varepsilon = e^{\alpha(X^1 - X^0)}$$

$$x^1 + x^0 + \varepsilon = e^{\alpha(X^1 + X^0)} \tag{13}$$

where $\varepsilon \simeq e^{-\alpha H}$, $-H \leq X^1 \leq +\infty$.

<u>Analytic mappings and the Hawking Unrugh effect in String Theory</u>

As it is known, the string action in a D-dimensional (generically curved) space-time is given by

$$S = \frac{1}{2\pi\alpha'} \int d\sigma d\tau \sqrt{g}\, g^{\alpha\beta} G_{AB}(X)\, \partial_\alpha X^A \partial_\beta X^B \tag{14}$$

$G_{AB}(X)$ and $g_{\alpha\beta}(\sigma,\tau)$ stand for the space-time and the world-sheet metrics respectively, namely

$$ds^2 = G_{AB}(X)\, dX^A dX^B \quad ; \qquad 0 \leq A,B \leq D-1 \tag{15}$$

$(2\pi\alpha')^{-1}$ is the string tension (we will take here $\alpha' = 1$).

The equations of motion and the constraints are

$$\Box^2 X^A + \Gamma^A_{BC}(X)\, \partial_\alpha X^B \partial_\beta X^C g^{\alpha\beta} = 0 \tag{16}$$

$$T_{\alpha\beta} = G_{AB}\, \partial_\alpha X^A \partial_\beta X^B - \frac{1}{2} g_{\alpha\beta}\, \partial_\sigma X^A \partial^\sigma X^B = 0 \tag{17}$$

$\Gamma^A_{BC}(X)$ stand for the Christoffel connections of the metric G_{AB}. As is well known, the action (14) is invariant under reparametrizations of the world-sheet co-ordinates

$$\delta X^\mu = \varepsilon^\alpha(\xi)\, \partial_\alpha X^\mu(\xi) \quad , \quad \xi^\alpha = (\sigma, \tau)$$

under which the two-dimensional metric $g_{\alpha\beta}$ goes to

$$\delta g_{\alpha\beta}(\xi) = \varepsilon^\gamma \partial_\gamma g_{\alpha\beta} + (\partial_\alpha \varepsilon^\alpha)\, g_{\gamma\beta} + (\partial_\beta \varepsilon^\gamma) g_{\alpha\beta} .$$

This reparametrization invariance allows us to choose $g_{\alpha\beta}$ in the conformal form (so called "conformal gauge"), namely,

$$\begin{aligned} g_{\alpha\beta} &= \Lambda(\xi)\, \eta_{\alpha\beta} , \\ ds^2 &= \Lambda(\sigma,\tau)\, (d\sigma^2 - d\tau^2) \end{aligned} \tag{18}$$

The choice of the conformal gauge still allows the reparametrizations

$$\begin{aligned} x_+ + \varepsilon &= f(x'_+ + \ell) \\ x_- + \varepsilon &= g(x'_- + \ell) \end{aligned} \tag{19}$$

(where $x_\pm = \sigma \pm \tau$, $x'_\pm = \sigma' \pm \tau'$ and we have introduced the constants ε, ℓ whose meaning will be clear in the sequel). In this way, we can choose the light-cone gauge:

$$V \equiv X_0 + X_1 = p_+ x_+ \tag{20}$$

where the proportionality constant $p_+ > 0$ is the momentum of the centre-of-mass of the string.

An important step in field as well as string quantization is the definition of positive frequency states and its associated ground state. It is by now well known, in QFT (in either flat or curved space-times) that "canonical states" for different co-ordinate systems are physically different (each timelike vector field leads to a separate indication of what constitutes a positive frequency)[1),2)]. There are different ways (and there is thus an ambiguity) in choosing such a basis. This makes it possible for a given field theory to have different alternative well-defined Fock spaces (different "sectors" of the theory). To illustrate this feature in string theory, let us consider the simplest case $G_{AB} = \eta_{AB}$, i.e.,

$$dS^2 = -dX^{0^2} + dX^{1^2} + \ldots\ldots + dX^{D-1^2} , \tag{21}$$

which means both that the space-time is flat and that the string is described in an inertial frame. Usually the string is described in this frame where

$$\partial_{x_+} \partial_{x_-} X^A = 0 , \qquad A = 0, 1, \ldots, D-1 \tag{22}$$

and then positive frequency modes are defined with respect to the inertial time X°. In the light-cone gauge, X° is proportional to τ and therefore the modes

$$\overleftarrow{\varphi}_n = \frac{1}{2\sqrt{\pi|n|}} e^{-inx_+} , \qquad \overrightarrow{\varphi}_n = \frac{1}{2\sqrt{\pi|n|}} e^{inx_-} \tag{23}$$

define the (inertial) particle states of the string.

Obviously, if we consider the (σ',τ') parametrization of the string world sheet [Eq. (19)], i.e.,

$$\sigma' + l = \frac{1}{2}\left[G(\sigma+\tau+\varepsilon) + F(\sigma-\tau+\varepsilon)\right] , \quad F \equiv f^{-1}$$
$$\tau' = \frac{1}{2}\left[G(\sigma+\tau+\varepsilon) - F(\sigma-\tau+\varepsilon)\right] , \quad G \equiv g^{-1} \tag{24}$$

for which we have

$$\partial_{x'_+} \partial_{x'_-} X^A = 0 , \tag{25}$$

positive frequency modes with respect to τ', namely

$$\overleftarrow{\phi}_n = \frac{1}{2\sqrt{\pi|n|}} e^{-i\lambda_n x'_+} , \qquad \overrightarrow{\phi}_n = \frac{1}{2\sqrt{\pi|n|}} e^{i\lambda_n x'_-} \tag{26}$$

do not define positive frequency modes with respect to X°, i.e., are not inertial particle states of the string. However, they are positive frequency modes with respect to another (accelerated) time $X^{\circ\prime}$ defined by

$$X_1 - X_0 = p_- f\,(X_1' - X_0')$$
$$X_1 + X_0 = p_+ g\,(X_1' + X_0') \tag{27}$$

where f is the same as in Eq. (19). This corresponds to the description of the string in an accelerated reference frame $\{X_0', X_1', \ldots, X_{D-1}'\}$, with acceleration

$$\tilde{\alpha} = \frac{1}{[f'g']^{1/2}}\ \partial_{X_1'}\,[\ln(f'g')]$$

and metric

$$ds^2 = p_+ p_- f'(X_1' - X_0')\,g'(X_1' + X_0')\,[-dX_0'^2 + dX_1'^2] + dX_2'^2 + \cdots + dX_{D-1}'^2 \tag{28}$$

where, for simplicity, we have taken

$$X_i' = X_i \qquad , \qquad (i = 2, \ldots, D-1)$$

For the description of the string in this frame we can always choose

$$V \equiv X_1' + X_0' = x_+' + \ell \tag{29}$$

as it follows from Eqs. (19), (20) and (27). Thus, Eq. (25) defines the accelerated particle states of the string. [The choice (29) is particularly useful here because of the acceleration points in the X^1 direction]. In this frame, the equations of motion of the string are

$$\partial_{x_+'}\partial_{x_-'}U' + \frac{f''}{f'}\partial_{x_+'}U'\partial_{x_-'}U' + \left(\frac{f''}{f'} - \frac{g''}{g'}\right)\partial_{x_+'}U'\partial_{x_-'}V' = 0 \tag{30}$$

$$\partial_{x_+'}\partial_{x_-'}V' + \frac{g''}{g'}\partial_{x_+'}V'\partial_{x_-'}V' - \left(\frac{f''}{f'} - \frac{g''}{g'}\right)\partial_{x_+'}U'\partial_{x_-'}V' = 0$$

$$\partial_{x_+'}\partial_{x_-'}X'^i = 0 \tag{31}$$

The equations of motion of the physical (transverse degrees) of freedom are the same in both frames. The equations of motion of the longitudinal co-ordinates (U',V') are different, but it can be shown[3] that, as in the inertial case[4], they can be eliminated in terms of the transverse co-ordinates by using the constraints

$$T'_{\pm\pm} = f'(U')\, g'(V')\, \partial_{\pm'} U\, \partial_{\pm'} V + \partial_{\pm'} X^i \partial_{\pm'} X^i \approx 0$$

$$T'_{+-} = 0 \tag{32}$$

The Poincaré invariance of flat space-time has a non-linear realization in terms of the accelerated co-ordinates, but an explicit realization of the Poincaré algebra can be constructed in terms of the accelerated modes provided (at the quantum level) one sets D = 26 and uses symmetric ordering of the operators[3].

The manifold (σ',τ') defined by Eq. (24) is the convenient world sheet parametrization for an accelerated string or a string in an accelerated space-time. This is true in either flat or curved space-time [although in the last case, Eq. (25) can only be valid asymptotically for $x'_\pm \to \pm\infty$].

Let us describe the string in the reference frames (21) and (28) referred in the sequel as (I) and (A) respectively. We consider here a closed string and take for simplicity f ≡ g. The boundary conditions are

$$X^\mu(0,\tau) = X^\mu(\pi,\tau)$$

$$X^\mu(0,\tau') = X^\mu(L_\varepsilon,\tau') \tag{33}$$

Notice that

$$0 < \sigma < \pi \qquad , \qquad -\infty < X^0 < +\infty \quad .$$

$$-\infty < \tau < +\infty \tag{34}$$

In order to have

$$0 < \sigma' < L_\varepsilon \qquad , \qquad -\infty < X'^0 < +\infty \, ,$$

$$-\infty < \tau' < +\infty \tag{35}$$

appropriate boundary conditions on the mappings must be imposed. Condition (33) is automatically satisfied from Eqs. (19) with

$$L_\varepsilon = F(\varepsilon + 2\pi) - F(\varepsilon) \qquad F \equiv f^{-1}$$
$$l = F(\varepsilon) \quad , \qquad (36)$$

Condition (35) on τ' and X'^0 is not particular to strings. It is required to get a consistent quantization (of fields or strings) in accelerated manifolds and to have a complete in (out) basis[2]. On the other hand, the manifold A can cover only a (bounded) region (namely $|X^1| > |X^0|$) of the original (I) one. Similarly, the manifold (σ', τ') can cover only a domain ($|\sigma| > \tau$) of the inertial or global world sheet. All these considerations are satisfied by taking mappings such that

$$u_\pm = f(\pm\infty) \, , \qquad (37)$$

where $u_+(u_-)$ are constants which can take independently finite or infinite values and $u_+ > u_-$. This means that the inverse mapping $F \equiv f^{-1}$ has singularities at $u = u_\pm$, i.e.,

$$F(u_\pm) = \pm\infty \qquad (38)$$

Singularities of these mappings describe the asymptotic regions of the space-time. Critical points of f, i.e.,

$$f'(\pm\infty) = 0 \qquad (39)$$

describe event horizons at $u = u_-$ and $u = u_+$. In this case $u_-(u_+)$ are finite and the manifold A cover the region $u_- < |X^1 \pm X^0| < u_-$ of the global (I) space-time. If $u_\pm = \pm\infty$, there are no horizons. In this approach, the well-known Rindler's space corresponds to the mapping

$$f = e^{\alpha U'} \quad , \quad \alpha = \text{const.} \qquad (40)$$

In this case, $u_- = 0$ and $u_+ = +\infty$; $u = u_- = 0$ is an event horizon and the manifold covers the right-hand wedge $|U| > 0$. (In order to cover the whole plane, four Rindler's patches are needed.) For the string world sheet we have

$$x_\pm + \varepsilon = e^{\alpha(x'_\pm + l)} \qquad (41)$$

and we see how the presence of the constant ε is necessary in order to have a finite string period in the manifold (σ', τ'). For the Rindler's mapping (41) we have

$$
\begin{aligned}
l &= \frac{1}{\alpha} \log \varepsilon \\
L_\varepsilon &= \frac{1}{\alpha} \log \left(\frac{2\pi}{\varepsilon} + 1 \right)
\end{aligned}
\tag{42}
$$

For $\varepsilon \to 0$, there is a stretching effect of the string due to the presence of an event horizon in the world sheet.

Let us now consider the string quantization in the frames (I) and (A). The string co-ordinates can be split into left and right movers as

$$
X^A(\xi) = \overleftarrow{X}^A(\xi_+) + \overrightarrow{X}^A(\xi_-) \tag{43}
$$

where $\xi_\pm$ stands for $\sigma \pm \tau$ or $\sigma' \pm \tau'$. As usual, left and right modes of the closed string decouple both in the classical and quantum theories. We consider first the left movers. The wave modes are periodic functions of x_+ and x_- with period 2π. The left modes which are only functions of x_+ will be periodic functions of $x_+' = \sigma' + \tau'$ with period L_ε. In the inertial frame (I) we have

$$
\overleftarrow{X}^i(x_+) = \frac{q^i}{2} + \frac{p^i}{2} x_+ + \frac{1}{2} \sum_{n \neq 0} \frac{\overleftarrow{a}_n^i}{\sqrt{|n|}} e^{-inx_+}
$$

$$
\overleftarrow{U}(x_+) = q_- + p_- x_+ + \frac{i}{2} \sum_{n \neq 0} \frac{\overleftarrow{\mu}_n}{n} e^{-inx_+} \qquad p_- = \frac{\mu_0}{2} \tag{44}
$$

The independent dynamical variables $\overleftarrow{a}_n^i (n \in z)$, q_i, q_+ and p_+ obey the canonical commutation relations

$$
\begin{aligned}
[\overleftarrow{\alpha}_n^i, \overleftarrow{\alpha}_m^j] &= n\,\delta_{n+m}\,\delta^{ij}, & \overleftarrow{\alpha}_n^i &= -i\sqrt{|n|}\,\mathrm{sgn}\,\overleftarrow{a}_n^i, \; n \neq 0 \\
[q^i, \overleftarrow{\alpha}_0^j] &= i\,\delta^{ij}, & \overleftarrow{\alpha}_0^i &= p^i \\
[q_-, p_-] &= i
\end{aligned}
\tag{45}
$$

Then

$$
[\overleftarrow{X}^i(x_+), \overleftarrow{p}^i(y_+)] = \frac{\delta^{ij}}{2} \delta(x_+ - y_+)
$$

The energy-momentum tensor, the conformal generators and the mass operator are

$$T_{++}(x_+) = \sum_{n\in Z} \frac{L_n}{2\pi} e^{-inx_+} ,$$

$$L_n = \frac{1}{2}\sum_m \overleftarrow{\alpha}_m^i \overleftarrow{\alpha}_{n-m}^i + p_+ \overleftarrow{u}_n \tag{46}$$

$$L_0 = \frac{1}{2}\sum_m \overleftarrow{\alpha}_m^i \overleftarrow{\alpha}_{-m}^i + 2p_+p_-$$

and

$$M^2 = p_+p_- - p_i^2 = \sum_m \frac{n}{2}(a_n^{i+} a_n^i + a_n^i a_n^{i+})$$

$$= \sum_m n\, a_n^{i+} a_n^i - \frac{(D-2)}{24} \tag{47}$$

In the accelerated frame (A) we have

$$\overleftarrow{X}^i(x'_+) = \frac{Q^i}{2} + \frac{P^i}{2}x'_+ + \sum_{n\neq 0} \overleftarrow{C}_n^i \overleftarrow{\Phi}_n(x'_+)$$

$$\overleftarrow{U}(x'_+) = u_0 + \frac{P_-}{2}x'_+ + \frac{i}{2}\sum_{n\neq 0}\frac{\gamma_n}{n} e^{-i\lambda_n x'_+} \tag{48}$$

with

$$\overleftarrow{X}(x'_+ + L_\varepsilon) - \overleftarrow{X}(x'_+) = \frac{P^i}{2\pi} L_\varepsilon \tag{49}$$

$$\overleftarrow{U}(x'_+ + L_\varepsilon) - \overleftarrow{U}(x'_+) = \frac{P_-}{2} L_\varepsilon$$

and

$$P^i = \frac{2\pi}{L_\varepsilon} p_i \quad , \quad P_- = \frac{2\pi}{L_\varepsilon} p_-$$

The left modes provide the T'_{++} component of the energy-momentum tensor. The accelerated conformal generators (L'_n) follow from the Fourier transform of T'_{++}:

$$T'_{++}(x'_+) = \sum_{-\infty}^{\infty} \frac{L'_n}{L_\varepsilon} e^{-i\lambda_n x'_+}$$

$$L'_n = L_n'^{\perp} + L_n'^{\parallel} \simeq 0 \tag{50}$$

$$L_n^{\leftarrow\perp} = -\frac{1}{L_\varepsilon} \sum_m \overleftarrow{\beta}_m{}^i \overleftarrow{\beta}_{n-m}^i$$

$$\overleftarrow{C}_n{}^{i\,+} = i\,\mathrm{sgn}\, \overleftarrow{\beta}_n{}^i/\sqrt{|n|} \;, \quad \overleftarrow{\beta}_n{}^i = \overleftarrow{\beta}_{-n}^i \;, \quad \beta_0 = p^i, \quad \gamma_0 = p_- .$$

We find the following Fock representation

$$\begin{aligned} [\overleftarrow{C}_m{}^i, \overleftarrow{C}_m{}^j] &= \delta_{nm}\,\delta^{ij} \;, \qquad [\overleftarrow{\beta}_n{}^i, \overleftarrow{\beta}_m{}^j] = n\,\delta^{ij}\delta_{n+m,0} \\ [P^i, Q_0{}^j] &= -\left(\frac{2\pi}{L_\varepsilon}\right) i\,\delta^{ij} \qquad (51) \\ [p_+, u_0] &= -i \end{aligned}$$

which implies

$$[\overleftarrow{X}^i(\xi), \overleftarrow{P}^j(\xi')] = \frac{i}{2}\,\delta(\xi-\xi')\,\delta^{ij} \qquad (52)$$

For a string whose centre-of-mass follows an accelerated world line in the longitudinal direction X^1, of the type

$$x^1 \pm x^0 \pm a_\pm = A_\pm\, e^{\pm\alpha\tau'} \;, \quad a_\pm, A_\pm = \text{const.}$$

it is convenient to use the expansion[3)]

$$U = U_0 + \frac{1}{f(x'_+)}\left[\tilde{p}_- + \frac{i}{2}\sum_{n\neq 0} r_n\, e^{-i\lambda_n x'_+}\right] \qquad (53)$$

for which the mass formula is given by

$$\begin{aligned} M'^2 = (\alpha L_\varepsilon)^2 p_+ \tilde{p}_- - p_i^2 &= \sum_1 \frac{n}{2}\left(C_n^{i\,+} C_n^i + C_n^i C_n^{i\,+}\right) \qquad (54) \\ &= \sum_1 n\, C_n^{i\,+} C_n^i - \frac{(D-2)}{24} \end{aligned}$$

The operators M^2 and M'^2 [Eqs. (47) and (54)] are different but both have the same eigenvalues.

The same treatment applies to the right movers $\vec{X}^A$ in a gauge defined by

$$\vec{U}' = p_-\, x_-$$

and with $\vec{U}' = x'_- + \ell$,

ie $x_- + \varepsilon = f(x'_- + \ell)$

The (I) and (A) Fock representations are related by a Bogoliubov transformation

$$C_m^i = p^i O_m + \sum_{n=1}^{\infty} (A_{mn} a_n^i + B_{mn} a_n^{i+}), \quad m \neq 0 \qquad (55)$$

with Bogoliubov coefficients

$$A_{mn} = \langle \phi_{\lambda m}, \varphi_n \rangle = \frac{1}{2\pi} \sqrt{\frac{n}{m}} \int_0^{2\pi} e^{in\sigma - i\lambda_m F(\sigma+\varepsilon)} d\sigma$$

$$B_{mn} = \langle \phi_{\lambda m}, \varphi_n^* \rangle = \frac{1}{2\pi} \sqrt{\frac{n}{m}} \int_0^{2\pi} e^{in\sigma + i\lambda_m F(\sigma+\varepsilon)} d\sigma \qquad (56)$$

and zero-mode contribution

$$O_m = \langle \frac{1}{2} (x_+ - \frac{2\pi}{L_\varepsilon} x'_+), \phi_{\lambda m} \rangle$$

The ground state of the string is defined by

$$\begin{aligned} a_n^i |0\rangle &= 0 \qquad \forall n > 1 \\ p^i |0\rangle &= 0 \end{aligned} \qquad (57)$$

We also have

$$\begin{aligned} C_m |0'\rangle &= 0 \qquad \forall m > 1 \\ p^i |0'\rangle &= 0 \end{aligned} \qquad (58)$$

but $C_m |0\rangle \neq 0$. The expectation value of the accelerated number operator in the ground state $|0\rangle$ is equal to (not sum over i)

$$N^i(\lambda_m) \equiv \langle 0 | C_m^i \, C_m^{i+} | 0 \rangle = N(\lambda_m) = \sum_{n=1}^{\infty} | B_{mn} |^2 \qquad (59)$$

which is i-independent. The expectation value of the accelerated M'^2 operator is equal to

$$M'^2 = 24 \sum_1^{\infty} n \, N(n) - 1 \qquad (60)$$

where N(n) is given by Eq. (59) with $\lambda_n = (2\pi/L_\varepsilon)n$. If $f(x'_+ + \ell) = e^{\alpha(x'_+ + \ell)}$, then:

$$B_{mn}(n \gg 1) = \mathcal{C}(\lambda_m) \frac{n^{-i\lambda_m/\alpha}}{\sqrt{n}} + \frac{b}{\sqrt{n}} + O\left(\frac{1}{n^{3/2}}\right)$$

$$\mathcal{C} = \frac{i}{2\pi} \frac{e^{-\frac{\pi\lambda_m}{2\alpha}}}{\sqrt{m}} \Gamma\left(1 + \frac{i\lambda_m}{\alpha}\right) , \qquad (61)$$

$$b = -\frac{i}{2\pi} \left(\frac{2\pi}{\sqrt{m}}\right)^{i\lambda_m/\alpha}$$

and

$$N(\lambda_m) = \frac{1}{\left(e^{2\pi\lambda_m/\alpha} - 1\right)} + \frac{\alpha}{2\pi\lambda_m} \qquad (62)$$

$$M'^2 = 24\left[\sum_{n=1}^{\infty} \frac{n}{(e^{n/T_0} - 1)} + T_0\right] - 1$$

$$M'^2_{\varepsilon \ll 1} = \frac{1}{2\pi} \ln \frac{2\pi}{\varepsilon} \qquad (63)$$

[Recall that in field theory, $B_{\lambda k}$ $(k \gg 1) = \mathcal{C}(\lambda)\ (k/\sqrt{k})^{-i\lambda/\alpha}$, with $\lambda \in R$.] $N(\lambda_n)$ [Eq. (62)] is equal to a Planckian spectrum at the Hawking-Unruh temperature

$$T_s = \frac{\alpha}{2\pi} \qquad (64)$$

In N(n), the temperature is

$$T_0 = \left(\frac{L_\varepsilon}{2\pi}\right) T_s = \frac{a}{2\pi} \underset{\varepsilon \ll 1}{=} \frac{1}{(2\pi)^2} \ln\left(\frac{2\pi}{\varepsilon}\right) \qquad (65)$$

The expectation value of M'^2 in the ground state $|0\rangle$ is <u>positive</u>. The tachyon level of this state is filled up with accelerated modes following a thermal distribution. For any other mapping f different from the exponential one [Eq. (41)], B_{nm} $(n \gg 1) \to e^{-n}$ and $N(\lambda_n)$ and N(n) are non-thermal.

It can be noted that the acceleration as well as the temperature T_s appearing in N(n) are re-scaled by the factor $(L_\varepsilon/2\pi)$, i.e., the ratio of the string period in the inertial and in the accelerated frame. Thus, the dimensionless temperature T_0 does not depend on the acceleration parameter α of Rindler's space, but on the parameter

$$a = \left(\frac{L_\varepsilon}{2\pi}\right)\alpha = \frac{1}{2\pi} \ln\left(\frac{2\pi}{\varepsilon} + 1\right) \underset{\varepsilon \ll 1}{=} -\frac{\ell\, \alpha}{2\pi} \qquad (66)$$

We see that the parameter ε [$\ell \equiv \ell(\varepsilon)$] introduced in the mappings on the world sheet, plays a fundamental rôle in the string context. ε acts as a regulator or cut-off to avoid the presence of an event horizon in the world sheet and to get a finite string period. Its magnitude is of the order of the Planck length.

TRANSFORMATION BETWEEN THE STATES: DERIVATION OF THE PARAMETER ε FROM THE CONSTRAINT EQUATIONS

We see that the transformations (19) are not without consequences but change the ground state and in general, the quantum states, except for mappings belonging to the O(2,1) group. Such invariance mappings are the Möbius or bilinear transformations. Under the transformations (19) and (27), the states transform as

$$| \rangle \rightarrow |'\rangle = e^{i\hat{G}} | \rangle \tag{67}$$

where $a_n | \rangle = 0$, $C_n |'\rangle = 0$, $\forall_n$ and

$$\hat{G} = \sum_n \theta_n \left(C_n C_n - C_n^+ C_n^+ \right)$$

[this is an operatorial representation for the Bogoliubov transformation, Eq. (55), with Bogoliubov coefficients $\cosh\theta_n$ and $\sinh\theta_n$].

The vacuum expectation value of $T_{\mu\nu}$ transforms as

$$\langle T_{\mu\nu} \rangle = \langle T_{\mu\nu}' \rangle + \oplus_{\mu\nu} + P_{\mu\nu} \tag{68}$$

$P_{\mu\nu}$ is any conserved traceless tensor taking into account the dependence of $\langle T_{\mu\nu} \rangle$ on the quantum state. It represents the non-local part of $T_{\mu\nu}$. $\oplus_{\mu\nu}$ depends on the mapping and represents the local part. In the conformal gauge we have

$$P_{--} = \mathcal{U}_-(x'_-) \quad , \quad P_{++} = \mathcal{U}_+(x'_+) \quad , \quad P_{+-} = P_{-+} = 0 \tag{69}$$

$$\oplus_{--} = -\frac{1}{12\pi} \sqrt{f'}\, d^2_{x'_-} \left(\frac{1}{\sqrt{f'}} \right)$$

$$\Theta_{++} = -\frac{1}{12\pi}\sqrt{g'}\, d^2_{x'_+}\left(\frac{1}{\sqrt{g'}}\right) \tag{70}$$

$$\Theta_{+-} = \Theta_{-+} = 0$$

$\mathcal{U}_-$ and $\mathcal{U}_+$ are arbitrary functions of the indicated variables. Equations (69) derive from the conditions $\nabla^\nu P_{\mu\nu} = 0$ and $P_\nu{}^\nu = 0$. Therefore, the constraints

$$\langle T_{\mu\nu} \rangle = 0 \quad , \quad \langle T_{\mu\nu}' \rangle = 0$$

yield to the equations[5)]

$$\sqrt{f'}\, d^2_{x'_-}\left(\frac{1}{\sqrt{f'}}\right) - 12\pi\, \mathcal{U}_-(x'_-) = 0 \tag{71}$$

$$\sqrt{g'}\, d^2_{x'_+}\left(\frac{1}{\sqrt{g'}}\right) - 12\pi\, \mathcal{U}_+(x'_+) = 0$$

These are zero-energy Schrödinger equations

$$\frac{d^2}{dx'^2_\pm}\Psi_\pm - 12\pi\, \mathcal{U}_\pm(x'_\pm)\,\Psi_\pm = 0 \tag{72}$$

for the mappings f and g. By giving the potentials $\mathcal{U}_\pm$, Eqs. (72) determine the wave functions

$$\Psi_- = \frac{1}{\sqrt{f'}} \quad , \quad \Psi_+ = \frac{1}{\sqrt{g'}} \tag{73}$$

Because $\mathcal{U}_\pm$ are arbitrary functions [compatible with the boundary conditions (37)], Eqs. (72) do not yield additional constraints on the mappings f and g, but a way of connecting the mapping to a potential problem.

The first term of Eq. (71) is the Schwarzian derivative of f:

$$D[f] = \frac{f'''}{f'} - \frac{3}{2}\left(\frac{f''}{f'}\right)^2 , \tag{74}$$

which is invariant under the Möbius or bilinear transformations. Under these transformations, f becomes a new function but D[f] is invariant determining the same ground state of the string. In particular, $\mathcal{U}_+ = \mathcal{U}_- = 0$ determine f(g) as

$$f = \frac{\alpha x'_- + \beta}{\gamma x'_- + \delta} \quad , \quad (\alpha\delta - \beta\gamma) = 1 \qquad (75)$$

The ground state defined by this mapping can be considered as a reference or "minimal" state at zero temperature with respect to which other ground states corresponding to non-zero potentials $\mathcal{U}_\pm$ appear as excited or thermal ones. If $\mathcal{U}_+ = \mathcal{U}_- = \mathcal{U}_0$ = constant > 0, then

$$\psi_\pm = A\, e^{-K x'_\pm} \quad , \quad f = \frac{1}{\alpha A^2} e^{\alpha x'_+} - \varepsilon \qquad (76)$$

where A is a normalizing constant (we choose $A^2 = e^{-\alpha \ell}/\alpha$) and K is the zero-energy transmission coefficient

$$K \equiv \frac{\alpha}{2} = \sqrt{12\pi\, \mathcal{U}_0} \, ,$$

For $\varepsilon \to 0$, the mapping (76) defines an event horizon at $x_\pm = 0$ ($x'_\pm = -\infty$) and carries an intrinsic temperature $T_s = \alpha/2\pi$, as can be seen by putting $t = i\tau$ ($x_+ = x + i\tau$) and then $0 \leqslant \tau' \leqslant 2\pi/\alpha$. The temperature appears related to the height of the potential, namely

$$T_s = \sqrt{\frac{12\, \mathcal{U}_0}{\pi}}$$

and the parameter ε arises naturally as an integration constant. This temperature T_s characterizes the spectrum $N(\lambda_n)$, Eq. (62).

Similar equations to (71) also appear in the so-called "back reaction problem" in two dimensions[6)] [as a consequence of the (±±) components of the semi-classical Einstein equations, the (+−) component giving rise to the Liouville equation for the geometry because of the conformal anomaly]. Equations (71) can also be derived in the context of conformal field theories on higher genus Riemann surfaces (the potential playing the rôle of the zero-point energy) in connection with the approach of Ref. 7).

We have shown that with appropriate boundary conditions, the holomorphic mappings of reparametrization invariance of string theory can be interpreted as a change of co-ordinate frame in the space-time in which strings are embedded. These mappings change the ground state in the quantized theory except for transformations belonging to the O(2,1) group. This allowed us to discuss in a systematic way the Hawking-Unruh effect in string theory. The transformations describing the world sheet of an accelerated string need the introduction of an additional parameter ε with respect to those describing the trajectories of accelerated point particles.

The results found here apply also to curved space-time. For the most important metrics in general relativity, the presence of isometry groups allows the maximal analytic extension of the (D-dimensional) manifold to be performed through the extension of a relevant two-dimensional manifold containing the time axis and a suitable spatial co-ordinate. This maximal analytic extension is performed by mappings like [Eq. (27)], where X_1, X_0 are Kruskal (maximal) type co-ordinates and X_1', X_0' are of the Schwarzschild type. [For the role played by these mappings in the context of Q.F.T., see Refs. 8) and 9).]

STRINGS NEAR BLACK HOLES

Our investigation of strings in Rindler space-time can be applied to the case of strings in a black hole background. Black hole solutions of Einstein equations exist in D-space-time dimensions (D ⩾ 4)[10]. These solutions are asymptotically flat and generalize the Schwarzschild space-time of four dimensions; they have the metric

$$ds^2 = -\left(1 - \frac{C}{R^{D-3}}\right) dT^2 + \frac{dR^2}{\left(1 - \frac{C}{R^{D-3}}\right)} + R^2 d\Omega^2_{D-2} \qquad (77)$$

R is the radial co-ordinate, $d\Omega^2_D$ is the line element on the unit D-sphere and the constant C is > 0. The surface

$$R = C^{1/D-3} \equiv R_S$$

is an event horizon (there are both past and future event horizons) and R = 0 is a space-like singularity. The horizon radius R_S is related to the black hole mass M by

$$C = 16\pi G \frac{M}{(D-2) A_{D-2}}$$

where

$$A_{D-2} = \frac{2\pi^{(D-1)/2}}{\Gamma(\frac{D-1}{2})}$$

is the area of a unit (D-2) sphere and G has dimensions of length^{D-2}.

The mass and the surface gravity K of the black hole are related by

$$K = \left(\frac{D-3}{2R_s}\right) = \frac{(D-3)}{2}\left[\frac{(D-2)\,A_{D-2}}{16\pi G M}\right]^{\frac{1}{D-3}} \tag{78}$$

For D = 4 this yields the standard relations $R_S = 2GM$ and $K = 1/(4GM)$.

The Kruskal extension of this Schwarzschild manifold is given by the mapping

$$r_K \pm t_K = e^{K(R^* \pm T)} \tag{79}$$

where

$$R^* = R + R_s^{D-3} \int \frac{dR}{(R^{D-3} - R_s^{D-3})}$$

$$-\infty \leq R^*, T \leq +\infty \quad .$$

This is the same exponential mapping as the Rindler transform with K instead of α. The Rindler co-ordinates are similar to the Schwarzschild (R^*, T) ones and the Minkowskian co-ordinates are similar to the Kruskal (global) co-ordinates (r_K, t_K). The event horizon $R = R_S$ corresponds to $R^* = -\infty$. As discussed above, a large cut-off ($H \gtrsim 1/K$) is needed in the negative Schwarzschild co-ordinate R^*. This shifting of the horizon is equivalent to considering a shifting ε in the mapping

$$r_K \pm t_K + \varepsilon = e^{K(R^* \pm T)} \tag{80}$$

with

$$\varepsilon = e^{-KH} \sim \ell_p^{\,2}$$

and thus

$$-H \leq R^* \leq +\infty$$

reflecting the fact that a classical description of the geometry is no longer valid at distances of order of the Planck length. We will take

$$\varepsilon = \frac{\lambda_c}{R_S}, \quad \text{where} \quad \lambda_c = \frac{1}{M}$$

is the Compton length of the black hole (here $\hbar = c = 1$). Thus the shifting of the horizon is $H = (1/K) \ln\varepsilon$, with

$$\varepsilon = \pi^{1/2} \left[\frac{(D-2)}{8\,\Gamma\left(\frac{D-1}{2}\right)} \right]^{\frac{1}{D-3}} \left(\frac{M_{Pl}}{M} \right)^{\frac{D-2}{D-3}} \tag{81}$$

and

$$K = \frac{\pi^{1/2}(D-3)}{2} \left[\frac{(D-2)}{8\,\Gamma\left(\frac{D-1}{2}\right)} \right] \left(\frac{M_{Pl}}{M} \right)^{\frac{1}{D-3}} M_{Pl} \tag{82}$$

Following on the same lines of argument discussed above, for the choice of gauge and parametrizations of the string world sheet and considering only left movers, we have

$$\overleftarrow{v} \equiv r_K + t_K = p_+ x_+ \tag{83}$$

$$V \equiv R^* + T = x'_+ + \frac{1}{K} \log p_+ \tag{84}$$

and

$$x_+ + \varepsilon = e^{K x'_+}$$

(Here the longitudinal direction of the string is in the radial direction.)

Positive frequency modes, $\overleftarrow{\varphi}_n$ with respect to the Kruskal time t_K and $\overleftarrow{\phi}_n$ with respect to the Schwarzschild time T can be defined. The Schwarzschild frequency is equal to

$$\lambda_n = \frac{2\pi}{L_\varepsilon} n, \qquad n = 1, 2, \ldots \tag{85}$$

where $$L_\varepsilon = \frac{1}{K} \ln \left(\frac{2\pi}{\varepsilon} + 1 \right)$$

For $\varepsilon \ll 1$, the frequency spacing tends to zero reflecting the stretching effect of the string near the horizon as seen by a Schwarzschild external observer. Associated to the modes $\overleftarrow{\varphi}_n$ and $\overleftarrow{\phi}_n$ we will have Kruskal and Schwarzschild operators a_n and C_n respectively and a vacuum state defined by

$$a_n | 0_K \rangle = 0 , \qquad \forall n > 1$$

On the other hand, in order to have a smooth Euclidean manifold from a black hole space-time with topology $R^2 \times S^{D-2}$, the Schwarzschild imaginary time iT must be identified with a period

$$\beta = \frac{2\pi}{K}$$

Then the same periodicity in the imaginary time appears in the correlation functions of string co-ordinates, indicating that the string is in equilibrium with a heat bath at the Hawking temperature

$$T_s = \frac{K}{2\pi} \qquad (86)$$

The same temperature T is recovered in the function $N(\lambda_m)$, i.e.,

$$N(\lambda_n) = \langle 0_K | C_{\lambda_n}^+ C_{\lambda_n} | 0_K \rangle ,$$

which gives a Planckian distribution for the Schwarzschild modes but with a "filter" $|g(\lambda_n)|^2$ equal to the absorption cross-section of the black hole. In the spectrum N(n) in which frequency is measured in dimensionless units 1,2,..., the temperature of the Planckian distribution is equal to

$$T_0 = \frac{1}{4\pi^2} \ln \left(\frac{2\pi}{\varepsilon} \right) \gg 1 \qquad (87)$$

One can consider different higher dimensional black hole space-times, namely a 26-dimensional or a four-dimensional black hole with the extra 22 dimensions compactified in a torus[11]. Intermediary situations can also be envisaged but it must be noticed that the qualitative properties of the string quantization will be the same since they depend upon the horizon structure in the two variables R, T (or X^0, X^1, for Rindler space). We hope to come back to this problem elsewhere.

It can be noticed that the Hagedorn temperature (T_m) in this context is

$$T_m = \frac{1}{\sqrt{\alpha'}} \sim M_{Pl}$$

Then,

$$\frac{T_s}{T_m} \sim \left(\frac{M_{Pl}}{M}\right)^{\frac{1}{D-3}}$$

and we have

$$T_s \lesssim T_m$$

since the basic requirement of the present semiclassical treatment is $M \gtrsim M_{Pl}$.

NEW APPROACH TO STRING QUANTIZATION IN CURVED SPACE-TIMES

The main feature of strings propagating in a curved space-time is that the equations of motion [Eq. (16)] are non-linear in X_A, so right and left movers interact with each other and also with themselves. It must be noticed that purely left modes (or right modes) are exact solutions of Eq. (16), namely

$$X_A = q_A(\sigma+\tau) \qquad \text{or} \qquad X_A = q_A(\sigma-\tau) \tag{88}$$

When G_{AB} is the metric of a symmetric space, the equations of motion possess an associated linear system and exact solutions can be constructed, by using the inverse scattering method. In Ref. 12), we propose a general perturbative scheme to solve the equations of motion and constraints both classically and quantum mechanically. We start from an exact given solution of Eqs. (16) and develop in perturbations around. A possible starting point is a solution for the centre-of-mass motion of the string $q_A(\tau)$ where

$$\ddot{q}^A(\tau) + \Gamma^A_{BC}(q)\,\dot{q}^B(\tau)\,\dot{q}^C(\tau) = 0 \qquad (89)$$

The world-sheet time variable τ is identified here with the proper time of the centre-of-mass trajectory. Another possibility is to take pure left(right) mover solutions. Then we set

$$X_A(\sigma,\tau) = q_A(\sigma,\tau) + \eta_A(\sigma,\tau) + \xi_A(\sigma,\tau) + \cdots \qquad (90)$$

Here $q_A(\sigma,\tau)$ is an exact solution of Eq. (16) and η_A is a solution of the linearized perturbation around q_A. That is

$$\partial^2\eta^A + \Gamma^A_{BC}(q)\left(\partial_- q^B\partial_+ + \partial_+ q^B\partial_-\right)\eta^C + \\ + \partial_C\Gamma^A_{BD}(q)\,\partial_+ q^B\,\partial_- q^D\,\eta^C = 0 \qquad (91)$$

$\xi_A(\sigma,\tau)$ fulfil the second order perturbation equation around q_A:

$$\partial^2\xi^A + \Gamma^A_{BC}\,\partial_+\eta^B\,\partial_-\eta^C + \frac{1}{2}\eta^D\eta^E\left(\partial^2_{DE}\Gamma^A_{BC}\right)\partial_+ q^B\,\partial_- q^C + \\ + \eta^D\left(\partial_D\Gamma^A_{BC}\right)\left(\partial_+ q^B\partial_- + \partial_- q^B\partial_+\right)\eta^C + \\ + \Gamma^A_{BC}\left(\partial_+ q^B\partial_- + \partial_- q^B\partial_+\right)\xi^C + \left(\partial_D\Gamma^A_{BC}\right)\partial_+ q^B\,\partial_- q^C\,\xi^D = 0 \qquad (92)$$

One can consider the higher-order perturbations but in this note we will restrict ourselves to first and second orders.

It must be noticed that we are treating the space-time metric exactly and taking the string oscillations around its centre-of-mass $q_A(\tau)$ as perturbation. So, our expansion corresponds to low excitations of the string, as compared with the energy scales of the metric G_{AB}. For example, this method is exact in flat space-time. In the Schwarzschild geometry, it will correspond to an expansion in ω/M where ω is the frequency mode and M the black hole mass. Thus, our approximation applies to black hole masses larger than the string energy. In other words, this corresponds to a strong gravitational field expansion. This can be equivalently considered as an expansion in powers of $\sqrt{\alpha'}$. Actually, since $\alpha' = \ell^2_{p\ell}$, ($\ell_{p\ell}$ is the Planck length), the expansion parameter turns out to be the dimensionless constant

$$g \equiv \sqrt{\pi}\,\frac{l_P}{R_c} = \frac{1}{l_P M} \simeq \frac{\omega}{M} ,$$

where the length R_c characterizes the curvature radius of the space-time under consideration and M its associated mass (the black hole mass in the Schwarzschild geometry, the mass of the Universe in a cosmological model). In most of the interesting situations, one clearly has $g \ll 1$.

It must be noticed that even for small g, the metric and its derivatives may be very large in some regions of the space-time. This shows that our method has a larger domain of applicability than the background field method where one must have everywhere

$$\left| \sqrt{\alpha'}\, \partial_C G_{AB}(X) \right| \ll 1 .$$

The first order equation describes the interaction between the string modes and the curved space-time geometry. The interactions between the string modes themselves start to appear from the second order perturbation (ξ^A) equation.

The constraint equations must also be expanded in perturbations. We find up to terms higher than the second order:

$$\begin{aligned} T_{\pm\pm} &= G_{AB}(q)\,\partial_\pm q^A \partial_\pm q^B + 2\,G_{AB}\,\partial_\pm q^A \partial_\pm \eta^B + \\ &+ \eta^C \partial_C G_{AB}(q)\,\partial_\pm q^A \partial_\pm q^B + \frac{1}{2}\eta^C \eta^D \partial_{CD} G_{AB}(q)\,\partial_\pm q^A \partial_\pm q^B + \\ &+ 2\eta^C \partial_C G_{AB}(q)\,\partial_\pm q^A \partial_\pm \eta^B + \xi^C \partial_C G_{AB}(q)\,\partial_\pm q^A \partial_\pm q^B + \\ &+ 2\,G_{AB}(q)\,\partial_\pm q^A \partial_\pm \xi^B + G_{AB}(q)\,\partial_\pm \eta^A \partial_\pm \eta^B \simeq 0 . \end{aligned} \tag{93}$$

See Ref. 12) where we have applied our method to the case in which the exact solution q_A describes the centre-of-mass motion.

This defines the Ln generators as

$$T_{\pm\pm} = \frac{1}{2\pi} \sum_{n \in Z} L_n\, e^{in(\sigma \pm \tau)} \tag{94}$$

In this second order approximation the constant operators Ln are bilinear in the fields η^A. One must solve the linear equation (92) for ξ^A expressing them as a bilinear functional of η^A. Inserting this expression for ξ^A in Eq. (93) give the generator Ln as operators bilinear in the η^A. Closed solutions of ξ^A, η^A andLn are given below for the de Sitter universe.

Let us now consider the case where the exact solution describes the centre of mass motion. It fulfils besides Eq. (89)

$$m^2 = G_{AB}(q)\, \dot{q}^A(\tau)\, \dot{q}^B(\tau) \qquad (95)$$

This defines the (mass)2 of the string. The fluctuation equation (9) and (10) simplify in this case. One can expand in Fourier series

$$\eta^A(\sigma,\tau) = \sum_{n\in\mathbb{Z}} e^{in\sigma}\, C_n^A(\tau) \qquad (96)$$

and the C_n^A satisfy <u>decoupled</u> ordinary differential equations

$$\ddot{C}_n^A + n^2 C_n^A + 2\,\Gamma^A_{BC}(q)\,\dot{q}^B \dot{C}_n^C + \dot{q}^B \dot{q}^C\, \partial_C \Gamma^A_{BD}(q)\, C_n^D = 0 \qquad (97)$$

In this case, the constraint equation read to second order :

$$\begin{aligned} T_{\pm\pm} &= m^2 \pm 2\,G_{AB}\,\dot{q}^A(\partial_\pm \eta^B + \partial_\pm \xi^B) + G_{AB}\,\partial_\pm\eta^A \partial_\pm \eta^B + \\ &+ (\eta^C + \xi^C)\,\partial_C G_{AB}(q)\,\dot{q}^A \dot{q}^B + \frac{1}{2}\eta^C \eta^D \partial_{CD} G_{AB}(q)\,\dot{q}^A\dot{q}^B + \\ &+ 2\,\partial_C G_{AB}(q)\,\eta^C \dot{q}^A\, \partial_\pm \eta^B \simeq 0 \end{aligned} \qquad (98)$$

here ξ^A is to be expressed in terms of the η^A through the solution of the linear equation (10). This procedure can be iterated to higher orders. One has only to solve <u>ordinary linear</u> differential equations at any step.

Non-physical degrees of freedom are to be eliminated by imposing the constraints (Eq. (98). In addition, the mass spectrum follows from the component of these constraints. it must be noticed that one must use the solution of the equation of motion of second order (ξ^A) in order to construct physically interesting $T_{\pm\pm}$. It is easy to check that this $T_{\pm\pm}$ is conserved order by order in the perturbation expansion. This perturbative scheme can be consistently extended to any order of perturbation and around any exact solution of the equations of motion.

<u>Quantum strings in de Sitter space-time</u>

As a first application of this general method we considered the de Sitter space-time. It is defined by the metric

$$ds^2 = \left(\frac{R_0}{X^0}\right)^2 \left[(dX^0)^2 - (dX^i)^2 \right], \quad 1 \leq i \leq D-1 \qquad (99)$$

Here $R_0 = H^{-1} = 3/\sqrt{\Lambda}$ and the curvature scalar $R = HD(D-1)$ is constant. In this case the center of mass motion (eq. (89) admits the solution

$$\begin{aligned} q^0(\tau) &= \frac{gm}{p\,\sinh(gm\tau)} \qquad\qquad q^1(\tau) = r^1 - \frac{gm}{p}\coth(gm\tau) \\ q^i(\tau) &= r^i \;, \quad 2 \leq i \leq D-1 \;, \qquad p^2 = \sum_{i=1}^{D-1} p_i^2 \end{aligned} \qquad (100)$$

Here r^i and p^i are constants ($1 \leq i \leq D-1$) and we choose the X^1-axis in the direction of the motion.

The first order perturbation equations (91) read in this case

$$\partial^2 \eta^0 + \frac{2}{q^0}(\dot q^0 \dot\eta^0 + \dot q^1 \dot\eta^1) - [(\dot q^0)^2 + (\dot q^1)^2]\eta^0/(q^0)^2 = 0$$

$$\partial^2 \eta^1 + \frac{2}{q^0}(\dot q^0 \dot\eta^1 + \dot q^1 \dot\eta^0) - 2\dot q^0 \dot q^1 \eta^0/(q^0)^2 = 0$$

$$\partial^2 \eta^i + 2\dot q^0 \dot\eta^i/q^0 = 0 \quad , \quad 2 \leq i \leq D-1 \tag{101}$$

This set of equations can be solved exactly in closed form (12). The general solution is

$$\eta^0(\sigma,\tau) = \frac{1}{p\sinh^2(mg\tau)}\left[A_1(\sigma,\tau) + B(\sigma,\tau)\cosh(mg\tau)\right]$$

$$\eta^1(\sigma,\tau) = -\frac{1}{p\sinh^2(mg\tau)}\left[B(\sigma,\tau) + A_1(\sigma,\tau)\cosh(mg\tau)\right]$$

$$\eta^i(\sigma,\tau) = \frac{A_i(\sigma,\tau)}{p\sinh(mg\tau)} \quad , \quad 2 \leq i \leq D-1 \tag{102}$$

Where A_α and B fulfil the free field equations

$$(\partial_\tau^2 - \partial_\sigma^2)\, B(\sigma,\tau) = 0$$

$$(\partial_\tau^2 - \partial_\sigma^2 - m^2g^2)\, A_\alpha(\sigma,\tau) = 0 \quad , \quad 1 \leq \alpha \leq D-1 \tag{103}$$

Therefore, they admit the Fourier expansions

$$A_\alpha(\sigma,\tau) = \sum_{n\in Z}\left[\gamma_n^\alpha\, e^{i(n\sigma-\omega_n\tau)} + \gamma_n^{\alpha +}\, e^{-i(n\sigma-\omega_n\tau)}\right]$$

$$B(\sigma,\tau) = \sum_{n\in Z}\left[\beta_n\, e^{in(\sigma-\tau)} + \tilde\beta_n\, e^{-in(\sigma+\tau)}\right] \tag{104}$$

$$\beta_n = \beta_{-n}^+ \quad , \quad \tilde\beta_n^+ = \tilde\beta_{-n} \quad , \quad \omega_n \equiv \sqrt{n^2 - m^2g^2}$$

The existence of imaginary frequencies for

$$|n| < gm \tag{105}$$

exhibits an instability of de Sitter geometry under the presence of strings. As it is shown in ref. (12) eqs. (102) - (104) define a canonical representation of the commutation rules provided we impose

$$[\gamma_n^\alpha, \gamma_\ell^{\beta +}] = \frac{m^2g^2}{4\pi\omega_n}\,\delta_{n\ell}\,\delta^{\alpha\beta}$$

$$[\beta_n, \beta_\ell] = \frac{m^2g^2}{4\pi n}\,\delta_{n+\ell,0} \tag{106}$$

In summary we can solve all constraints (up to this order) by the gauge choice (12)

$$B(\sigma,\tau) = \alpha m \tau \tag{107}$$

where α is a constant. Furthermore all zero modes can be and must be discarted since they just represent deformations of the centre of mass solution (100). Finally the $L_0 \simeq 0$ constraint yields the mass spectrum. This is no more a linear one. We find (12).

$$\frac{\pi}{2} m^2 = (D-1)\,\mathcal{E}(m^2) + \sum_{n\neq 0}\Omega_n(m^2)\sum_{\alpha=1}^{D-1} a_n^{\alpha +} a_n^\alpha + O(g^2)$$

where
$$\Omega_n(m^2) = \sqrt{n^2 - m^2g^2} + \frac{m^2g^2}{2\sqrt{n^2-m^2g^2}} \tag{108}$$

$$\mathcal{E}(m^2) = \sum_{n=1}^{\infty}\Omega_n(m^2) \tag{109}$$

and
$$a_n^\alpha = 2\sqrt{\pi\omega_n}\,\gamma_n^\alpha/(mg) \quad , \quad [a_n^\alpha, a_\ell^{\beta +}] = \delta_{n\ell}\,\delta^{\alpha\beta} \tag{110}$$

There are (D-1) physical string excitations for each n ($\neq 0$) all of them associated with positive norm states in the Fock space. Recall that in flat space-time one has (D-2) components. Eqs (109)-(110) permit to obtain the mass spectrum in powers of g. Besides the tachyon scalar particle associated to the ground state, we find two quanta states with mass

$$m^2 = -\frac{D-1}{12} + 2 + O(g^2) \tag{111}$$

Therefore (almost) massless spin-two and zero particles appear at D= 25 They can be identified with the graviton and the dilaton respectively. It is a general feature of de Sitter space-time to lower the critical dimensions in one unit. It must be remarked that the zero-point fluctuation energy (109) is finite upon regularizing the sum.

In addition, we found that the Regge trajectories deviate significatively from linear ones. Moreover we find that only states with $m^2 \lesssim 0$ (1) and spin $\lesssim 0.9/g2$ are stable at tree level. Beyond these values the particle masses become complex. Therefore, one does not need to go to the one-handle level to find a width for these states. We refer to more details to ref. (12).

Quantum string dynamics and scattering in black-hole space-times

We study the bosonic string in a Schwarzschild geometry. The D-dimensional generalization of the Schwarzschild metric is given by

$$ds^2 = -\left[1-\left(\frac{R_s}{R}\right)^{D-3}\right](dX^0)^2 + \frac{dR^2}{\left[1-\left(\frac{R_s}{R}\right)^{D-3}\right]} + R^2\, d\Omega^2_{D-2} \tag{112}$$

Here R is the radial coordinate. $d\Omega^2_D$ is the line element of the unit D-sphere and the Schwarzschild radius R=Rs defines the event horizon surface

$$R_s = \left[\frac{16\,GM}{(D-2)\,A_{D-2}}\right]^{1/(D-3)}, \qquad A_D = \frac{2\pi^{(D+1)/2}}{\Gamma((D+1)/2)} \tag{113}$$

The gravitational constant G has dimensions of $(\text{length})^{D-2}$.

The string equations of motion read in this geometry

$$\partial^2 X^0 + \frac{K}{a(R)}\left(R_s/R\right)^{D-2} \partial_\mu X^0\, \partial^\mu R = 0$$

$$\partial^2 R - \frac{K}{a(R)}\left(R_s/R\right)^{D-2}(\partial_\mu R)^2 - a\left[R(\partial_\mu \Omega^i)^2 - K\left(\frac{R_s}{R}\right)^{D-2}(\partial_\mu X^0)^2\right] = 0$$

$$\partial^2 \Omega^i + \frac{2}{R}\partial^\mu R\, \partial_\mu \Omega^i - \Omega^i(\Omega^j\, \partial^2 \Omega^j) = 0$$

here

$$(\Omega^i)^2 = 1, \qquad 1 \le i \le D-2, \qquad a \equiv 1-\left(\frac{R_s}{R}\right)^{D-3} \tag{114}$$

$$K = \frac{D-3}{2R_s}$$

The constraint equations write here

$$T_{\pm\pm} = \frac{1}{a}(\partial_\pm R)^2 - a(\partial_\pm X^0)^2 + R^2(\partial_\pm \Omega^i)^2 \simeq 0 \tag{115}$$

Following our general method we set

$$X^A(\sigma,\tau) = q^A(\tau) + \eta^A(\sigma,\tau) + \xi^A(\sigma,\tau) + \cdots \tag{116}$$

where $q^A(\tau)$ obeys the center of mass equation (89). That is

$$m^2 + \frac{1}{\alpha'^2 a}(\dot q^R)^2 - \frac{E^2}{a^2} + \frac{L^2}{q_R^2} = 0 \tag{117}$$

where E is the energy and L the angular momentum. Eq. (117) describes the motion in the plane defined by the spherical coordinates $\Theta^i = \pi/2$. That is to an angular motion only in the azimutal angle φ. We have

$$q^1 = \cos\varphi \qquad q^2 = \sin\varphi \qquad q^a = 0, \quad a > 2$$

and
$$\dot{\varphi} = \frac{\alpha' L}{q_R^{\,2}} \tag{118}$$

Eqs. (117)-(118) are solvable by quadratures
$$\tau = \int^{q_R} \frac{ds}{\sqrt{E^2 - a(s)(m^2 + L^2/s^2)}} \tag{119}$$
We recognize here an effective potential
$$V_{eff}(q) = \left(m^2 + \frac{L^2}{q^2}\right)\left[1 - \left(\frac{R_s}{q}\right)^{D-3}\right] \tag{120}$$
depending on the initial energy and momentum the particle will be absorbed or elastically scattered by the black-hole.

The first order fluctuations $\eta^A(\sigma,\tau)$ obey here the equations :
$$\partial^2\eta^* + \left[\frac{a''}{2a^2}\left(\dot{q}^{R\,2} + \alpha'^2 E^2\right) - \frac{L^2\alpha'^2}{(q^R)^4}\right] a\eta^* +$$
$$+ \frac{a'}{a}\left(\dot{q}^R \dot{\eta}^* + \alpha' E \dot{\eta}^0\right) - 2 q^R \dot{q}^i \dot{\eta}^i = 0$$
$$\partial^2\eta^i - 2\frac{\dot{q}^R}{q^R}\dot{\eta}^i - 2\dot{q}^i \partial_\tau\left(\frac{a\eta^*}{q^R}\right) - \frac{L^2\alpha'^2\eta^i}{(q^R)^4} - 2 q^i \dot{q}^j \dot{\eta}^j = 0 \tag{121}$$
Here we use the R* coordinate defined by
$$R^* = R + R_s^{D-3}\int^R \frac{dR'}{R'^{D-3} - R_s^{D-3}} \tag{122}$$

Eqs. (121) can be easily solved in the asymptotic region $\tau \to \pm\infty$. There, the center of mass is very far from the center of forces and the space-time is practically flat. We set
$$\eta^A(\sigma,\tau) = \frac{\alpha^A(\sigma,\tau)}{q^R(\tau)} \tag{123}$$
and we expand
$$\alpha^A(\sigma,\tau) = \sqrt{\alpha'}\sum_{n\in Z}\sum_{B=0}^{D-1}\frac{f_{n,\pm}^{A,B}(\tau)}{n}\left\{\alpha_{n,\pm}^B e^{-in\sigma} + \tilde{\alpha}_{n,\pm}^B e^{in\sigma}\right\}_A \tag{124}$$
here the functions $f_{n,\pm}(\tau)$ fulfil the fluctuation equations (121) with the boundary conditions
$$\lim_{\tau\to\pm\infty} f_{n,\pm}^{A,B}(\tau) = e^{-in\tau}\delta_{AB} \tag{125}$$
The choice of a positive frequency factor $e^{-in\tau}$ in eq. (125) corresponds to "in" or "out" particles states for $\tau \longrightarrow -\infty$ or $\tau \longrightarrow +\infty$ respectively. Since we have free oscillators in both asymptotic limits, we get
$$\lim_{\tau\to\mp\infty} f_{n,\pm}^{A,B}(\tau) = A_{n,\pm\,A,B}^{A,B}\, e^{-in\tau} + B_{n,\pm\,A,B}^{A,B}\, e^{in\tau} \tag{126}$$

where the constant coefficients $A_{n,\pm}$ and $B_{n,\pm}'$ depend on the detailed form of the eqs. (121) for all τ. As a consequence of eqs. (124)-(126) we find that outgoing and ingoing operator modes are related by the Bogoliubov transformation

$$\alpha^A_{n,+} = \sum_B \left(A^{A,B}_{n,+} \; \alpha^B_{n,-} + B^{A,B}_{n+} \; \tilde{\alpha}^B_{-n,-} \right)$$
$$\tilde{\alpha}^A_{n,+} = \sum_B \left(A^{A,B}_{n,+} \; \tilde{\alpha}^B_{n,-} + B^{A,B}_{n,+} \; \alpha^B_{-n,-} \right) \qquad (127)$$

We see two main effects in the transitions between the string modes produced by the black-hole :

a) Polarization changes in the modes (without changing their left or right character)

b) Pair mode creation. Each pair formed by modes of opposite chirality.

For the modes orthogonal to the scattering plane, we find (13,14)

$$B^{ij}_{n,+} = \delta^{ij} \; B_{n,+} \quad , \qquad 2 < i,j \leq D-1 \qquad (128)$$

Therefore, we find for the pair creation amplitude

$$\langle \overleftarrow{n}_{out} \, , \, \overrightarrow{n}_{out} \mid 0_{in} \rangle = B_{n,+} \qquad (129)$$

where $|\overleftarrow{n}_{out}, \overrightarrow{n}_{out}\rangle$ stands for an outgoing state with the left and right n^{th} modes occupied. For an excited initial state, we find

$$\langle \overrightarrow{m}_{out} \, , \, \overleftarrow{n}_{out} \, , \, \overrightarrow{n}_{out} \mid \overleftarrow{l}_{in} \rangle = \delta_{ml} \, B_n \, , \qquad l \neq n$$

The coefficients An and Bn have been derived in ref. (14) for large impact parameters explicitely. We found for b>>Rs

$$A_{n \; b \gg R_s} = i\alpha' \left(\frac{R_s}{b}\right)^{D-3} \frac{p\sqrt{\pi}}{b} \, \Gamma\left(\frac{D}{2}+1\right) \left[1 + \frac{D-1}{D} \frac{m^2}{p^2} \right]$$
$$B_{n \; b \gg R_s} = p\alpha' \left(R_s^{D-3} / b^{D-2} \right) F\left(m/p \, , \, nb/\alpha' p \right) \qquad (130)$$

where the function (F(x,y) is defined in ref (14) as a double integral. It must be noticed that Bn is appreciable different from zero when the characteristic interaction time is of the same order of magnitude as the vibration time of the mode. That is

$$\frac{b}{\alpha' p} \simeq \frac{2\pi}{n} \qquad \text{or} \qquad n \simeq \frac{\alpha' p}{b} \qquad (131)$$

when n>>α'p/b, Bn is very small (see eq.130)

The string black hole cross section is computed in ref (14) following our method.

At zeroth order we find for the center of mass cross section

$$\left(\frac{d\sigma}{d\Omega}\right)_{cm} = \left(\frac{b}{\sin\theta}\right)^{D-3} \frac{db}{d\theta}_{\theta \to 0} = \left[\frac{4GM\,\Gamma\left(\frac{D}{2}-1\right)\pi^{2-D/2}\left(1 + \frac{D-3}{D-2}\frac{m^2}{p^2}\right)}{(D-3) \qquad \theta^{D-1+1/D-3}} \right]^{\frac{D-2}{D-3}} \qquad (132)$$

This generalizes the Rutherford formula valid for D = 4. Taking into account the first quantum correction yields

$$\left(\frac{d\sigma}{d\Omega}\right)_{elastic} = \left(\frac{d\sigma}{d\Omega}\right)_{cm} \left[1 - \sum_{n' \neq n} |B_{n'}|^2 \right] \qquad (133)$$

By elastic cross section we mean when both the initial state and the final state are the nth mode. The inelastic cross section is proportional to Bn^2.

In conclusion our method allows to explicitely compute physical quantities for strings in curved space-time in powers of l_{Pl}/Ro (Ro = characteristic radius of the geometry). This is possible even for singular metrics. Our method is more powerful than the background field method that only provides explicitely the renormalization constants and beta functions.

Although we have limited ourselves to bosonic strings, fermionic strings can be treated analogously. In addition one can also treat curved space-times where some dimensions have been compactified.

III - PARTICLE AND STRING SCATTERING AT THE PLANCK ENERGY SCALE AND THE AICHELBURG-SEXL GEOMETRY

In order to improve our present (still fragmentary) understanding of quantum gravity, physical insights at the energies of the Planck scale need to be explored (be that in a theory of fields or strings). When the curvature associated to the space time geometry is too strong, the usual description of particle fields or strings in flat space time is no longer valid, the dynamics of the quantum fields (or strings) are governed by their equations of motion in the classical background geometry. Relevant examples are the Hawking effect and the description of quantum fields in curved space times , string quantization in curved space times and more recently 15 the particle scattering in the Aichelburg-Sexl (AS) geometry. As stressed in ref. 15, the AS metric 16 is relevant to the particle scattering at high energy $(s \gtrsim m_{pl}^2)$. In this paper we investigate the quantum dynamics and scattering of a closed string in the AS geometry. This describes the following physical process : the scattering of a string (or one of its possible particle states) by a particle with energy of the order or larger than the Planck mass (m_{pl}). At this energy scale it is necessary to take into account the gravitational field created by such highly energetic particle according the General Relativity. (This is also true even when the rest mass of the particle is small as compared with $m_{pl} = 10^{-5}$ gr). The gravitational field of an ultrarelativistic (neutral and spinless) particle is precisely described by the AS geometry. The motion of the particle is taken here along the x-axis with speed + 1. The interpretation of the AS geometry as a schock wave has been considered in ref (17). We give the D-dimensional generalization of this metric (this generalization has also been given recently in ref. (18). The string is considered here as a test string, that is its energy is much smaller than the energy associated to the AS space time. A general feature of strings propagating in curved space time is that the equations of motion are non-linear, so right and left movers interact with each other and also with themselves. A remarquable feature here is that the string equations of motion in the D-dimensional AS geometry are exactly solvable.

The procedure of ref.17 to derive a shock wave metric starting from a solution of the Einstein equations can be easily extended to D dimensions. Let be a solution of the Einstein equations given by

$$dS^2 = 2\,A\,(u,\,v')\,du\,dv' + g\,(u,v')\,(dx^{i'})^2 \qquad (134)$$

$$u = t - X \qquad 1 \leqslant i \leqslant D-2$$
$$v = t + X$$

Let us make the shift

$$v \rightarrow v' + f(X^i) \qquad \text{for } u > 0$$

Then, the metric eq. (134) reads

$$ds^2 = 2\,A\,(u,v)\,du\,dv + g\,(u,v)\,(dx^i)^2$$
$$- 2\,A\,(u,v)\,f\,(X^i)\,\delta(u)\,du^2 \qquad (135)$$

where $v = v' + f(X^i)\, H(u)$

$H(u)$ being the step function

Eq. (135) fulfils the Einstein equations provided

$$\partial_v A(u,v) = \partial_v g(u,v)$$

$$\frac{A}{g} \nabla_i^2 f - \frac{\partial^2_{uv} g}{g} f = 32\pi G p\, A^2 \delta^{D-2}(\rho) \qquad (136)$$

where the right and side corresponds to a point source moving at the speed of light ($T_{uu} = p\, \delta^{D-2}(\rho)\, \delta(u)$)

If the starting metric is flat space time :
$A = 1/2$, $g = 1$

and we have

$$\nabla_i^2 f = 16\pi G p\, \delta^{D-2}(\rho) , \quad \rho^2 = \sum_{i=1}^{D-2} X^{i\,2} \qquad (137)$$

whose solution is given by

$$f = \begin{cases} K/\rho^{D-4} , & K = 16\pi G p/(D-4)\,\Omega_{D-2} , \; D>4 \\ 8Gp \ln \rho & D = 4 \end{cases}$$

Recall that the scalar Green function in D dimensions satisfies

$$\nabla^2 G = -\Omega_{D-2} , \qquad \Omega_D = \frac{2\pi^{D/2}}{\Gamma(D/2)}$$

$$G = \frac{1}{(D-4)\,\rho^{D-4}} \qquad (138)$$

String quantization in the Aichelburg-Sexl Geometry

The Aichelburg-Sexl metric in D dimensions is given by

$$ds^2 = dU\,dV - (dX^i)^2 + f_D(\rho)\,\delta(U)\,dU^2 \qquad (139)$$

where U,V are null (longitudinal) coordinates

$$U = T - X , \quad V = T + X \qquad (140)$$

X^i are the transverse coordinates,

$$\rho = \sqrt{(X^i)^2} , \qquad 2 \leq i \leq D-2$$

and

$$f_D(\rho) = \begin{cases} K/\rho^{D-4} & K = \frac{8\pi^{2-D/2}}{(D-4)} G p\, \Gamma(\frac{D}{2}-1) \quad D>4 \\ 8Gp \ln \rho & D = 4 \end{cases} \qquad (141)$$

Here we are interested in D>4 dimensions. This metric describes the gravitational field of a ultra relativistic particle moving in the V-direction with momentum p . This represents a gravitational impulsive (or "shock") wave located at U = 0 superimposed to Minkowski space time. The only non-vanishing components of the Christoffel connections are :

$$\Gamma^V_{UU} = \partial_U G_{UU} = f_D(\rho)\ \delta'(U)$$

$$\Gamma^V_{Ui} = \partial_i G_{UU} = \partial_i f_D(\rho)\ \delta(U) = - \frac{K(D-4)\, X^i\, \delta(U)}{\rho^{D-2}} \tag{142}$$

$$\Gamma^i_{UU} = \frac{1}{2}\Gamma^V_{Ui}$$

The equations of motion of the string

$$\Box^2 X^A + \Gamma^A_{BC}\ \partial_\mu X^B\ \partial^\mu X^C = 0 ,$$

read in this metric

$$(\partial^2_\sigma - \partial^2_\tau)\, U = 0$$

$$(\partial^2_\sigma - \partial^2_\tau)\, V + \delta'(U)\, f_D(\rho)\, [(\partial_\sigma U)^2 - (\partial_\tau U)^2] + $$
$$+ 2\,\delta(U)\,\partial_i f_D(\rho)\,[\partial_\sigma U\, \partial_\sigma X^i - \partial_\tau U\, \partial_\tau X^i] = 0$$

$$(\partial^2_\sigma - \partial^2_\tau)\, X^i - \frac{1}{2}\,\delta(U)\,\partial_i f_D(\rho)\,[(\partial_\sigma U)^2 - (\partial_\tau U)^2] = 0 \tag{143}$$

where we have taken the world sheet metric $ds^2 = \Lambda(\sigma,\tau)(d\sigma^2 - d\tau^2)$ in the conformal gauge. Without loss of generality we can fix the gauge by setting $U = \alpha' p_U \tau$. Therefore we have

$$U = \alpha' p_U \tau$$

$$(\partial^2_\sigma - \partial^2_\tau)\, V - \delta'(\tau)\, f_D(\rho(\sigma,\tau)) - 2\,\delta(\tau)\,\partial_\tau f_D(\rho(\sigma,\tau)) = 0$$

$$(\partial^2_\sigma - \partial^2_\tau)\, X^i - \frac{p_U \alpha'}{2}\,\delta(\tau)\,\partial_i f_D(\rho(\sigma,\tau)) = 0 \tag{144}$$

These equations can be recasted as

$$U = p_U \alpha' \tau$$

$$(\partial^2_\sigma - \partial^2_\tau)\, V - \delta'(\tau)\, f_D(\rho(\sigma,\tau=0)) - \delta(\tau)\,\partial_\tau f^{\cdot}_D(\rho(\sigma,\tau)) = 0$$

$$(\partial^2_\sigma - \partial^2_\tau)\, X^i - \frac{p_U \alpha'}{2}\,\delta(\tau)\,\partial_i f_D(\rho(\sigma,\tau)) = 0 \tag{145}$$

where we have used the property

$$f(\tau)\,\delta'(\tau) = f(0)\,\delta'(\tau) - \dot f(0)\,\delta(\tau)$$

Eqs (145) can be exactly solved. The transverse coordinates X satisfy the free string equations in flat space-time except at $\tau = 0$, where their derivatives are discontinuous that is

$$(\partial_\sigma^2 - \partial_\tau^2)\, X^i_{\gtrless} = 0 \qquad \forall\, (\sigma, \tau \neq 0)$$

and

$$\left[\partial_\tau (X^i_> - X^i_<)\right]_{\tau=0} + \frac{P_v \alpha'}{2}\left[\partial_i f_D(\rho)\right]_{\tau=0} = 0 \qquad (146)$$

$$X^i_>(\sigma, 0) = X^i_<(\sigma, 0) \qquad (147)$$

The exact solution of eq. (145)- (147) is given by

$$U = P_v \alpha' \tau$$

$$V = F(\sigma-\tau) + G(\sigma+\tau) - \frac{1}{2} H(\tau)\left[f_D(\rho(\sigma+\tau)) + f_D(\rho(\sigma-\tau))\right] - $$

$$+ \frac{1}{2} H(\tau)\left[\partial_\tau f_D(\rho(\sigma+\tau)) + \partial_\tau f_D(\rho(\sigma-\tau))\right] \qquad (148)$$

$$X^i_{\gtrless} = \overrightarrow{X}^i_{\gtrless}(\sigma-\tau) + \overleftarrow{X}^i_{\gtrless}(\sigma+\tau)$$

where

$$\rho(\sigma,\tau) = \sqrt{X^{i\,2}(\sigma,\tau)}$$

$$f_D(\rho(x)) = K/\rho_{(x)}^{D-4} \quad , \quad \partial_\tau f_D(\rho(x)) = - \frac{K(D-4) X^i \dot X^i}{\rho^{D-2}}$$

$H(\tau)$ is the step function and

$$\overrightarrow{X}^i_{\gtrless}(x) = \frac{C^i(x)}{2} + \frac{E^i(x)}{2} \mp \frac{\tilde K}{4} P_v \alpha' \int_{\sigma_i}^{x} \frac{d\sigma'\, C^i(\sigma')}{\left[\sqrt{C^{j\,2}(\sigma')}\right]^{D-2}}$$

$$\overleftarrow{X}^i_{\gtrless}(x) = \frac{C^i(x)}{2} - \frac{E^i(x)}{2} \pm \frac{\tilde K}{4} P_v \alpha' \int_{\sigma_i}^{x} \frac{d\sigma'\, C^i(\sigma')}{\left[\sqrt{C^{j\,2}(\sigma')}\right]^{D-2}} \qquad (149)$$

here $C^i = X^i(\sigma,\tau=0)$, E^i, F and G are arbitrary functions of the indicated variables, the integration bounds σ_i will be obtained below and

$$\tilde K = \frac{(D-4)}{2} K = 4\pi^{2-D/2}\, G\mu\, \Gamma\left(\frac{D}{2} - 1\right) \qquad (150)$$

K is given by eq. (142). The symbol > (<) labels the solution in the $\tau > 0$ ($\tau < 0$) region. In the last term of eqs. (149), the upper sign corresponds to the (>) - solution and the lower sign to the (<) - solution.

Since the X^i obey the wave equation both for >o and < o, we can expand them in harmonic oscillators

$$\overrightarrow{X}^i_>(\sigma-\tau) = \frac{q^i_>}{2} - \frac{\alpha' p^i_>}{2}(\sigma-\tau) + \sqrt{\alpha'} \sum_{n\neq 0} \frac{\tilde\alpha^i_{n>}}{n} e^{in(\sigma-\tau)} \qquad (151a)$$

$$\overleftarrow{X}^i_>(\sigma+\tau) = \frac{q^i_>}{2} + \frac{\alpha' p^i_>}{2}(\sigma+\tau) + \sqrt{\alpha'} \sum_{n\neq 0} \frac{\alpha^i_{n>}}{n} e^{-in(\sigma+\tau)} \qquad (151b)$$

$\overrightarrow{X}^i_<(\sigma-\tau)$ = the same expression as eq. (151.a) with $> \rightarrow <$

$$\overleftarrow{X}^{i}_{<}(\sigma+\tau) = \text{the same expression as eq. (151.b) with } > \to < \quad (152)$$

The continuity constraint eq. (147) yields

$$\overrightarrow{X}^{i}_{>}(\sigma) - \overrightarrow{X}^{i}_{<}(\sigma) = -\left[\overleftarrow{X}^{i}_{>}(\sigma) - \overleftarrow{X}^{i}_{<}(\sigma)\right]$$

$$\alpha^{i}_{n<} - \tilde{\alpha}^{i}_{-n<} = \alpha^{i}_{n>} - \tilde{\alpha}^{i}_{-n>}, \quad n \neq 0 \qquad (153)$$

$$q^{i}_{<} = q^{i}_{>}$$

In summary, we have found the exact solution to the string equations of motion in the D-dimensional AS Geometry. The longitudinal coordinate V expresses in terms of the transverse ones through eq (148). The transverse coordinates admit flat space expansions for $\tau > 0$ and $\tau < 0$ eqs (151) - (152) . The Fourier coefficients in both regions are related by the matching relations eqs. (II.19) - (II.22). They can be rewritten as

$$p^{i}_{>} - p^{i}_{<} = \frac{\tilde{K}}{2\pi\alpha'} \int_0^{2\pi} d\sigma\, D^{i}(\sigma)$$

$$\alpha^{i}_{n>} - \alpha^{i}_{n<} = \frac{i\tilde{K}}{4\pi\sqrt{\alpha'}} \int_0^{2\pi} d\sigma\, e^{in\sigma} D^{i}(\sigma) \qquad (154)$$

$$\tilde{\alpha}^{i}_{n>} - \tilde{\alpha}^{i}_{n<} = \frac{i\tilde{K}}{4\pi\sqrt{\alpha'}} \int_0^{2\pi} d\sigma\, e^{-in\sigma} D^{i}(\sigma)$$

This is the exact non-linear transformation relating the creation and annhilation operators for $\tau>0$ and $\tau<0$. Notice here that D^{i} is a non-linear functionnal of the operators $(\alpha^{j}_{n<} - \tilde{\alpha}^{j}_{-n<})$

$$D^{i}(\sigma) = \frac{P_U \alpha' C^{i}(\sigma)}{\left[\sqrt{C^{j2}(\sigma)}\right]^{D-2}}, \qquad C^{i}(\sigma) = X^{i}(\tau=0) = q^{i} + \sqrt{\alpha'} \sum_{n\neq 0} \frac{e^{-in\sigma}}{n}(\alpha^{i}_{n<} - \tilde{\alpha}^{i}_{-n<})$$

The energy-momentum tensor

$$T_{\pm\pm}(\sigma,\tau) = G_{AB}(X)\, \partial_{\pm} X^{A}\, \partial_{\pm} X^{B} \qquad (155)$$

reads in this metric

$$T_{\pm\pm}(\sigma,\tau) = \partial_{\pm} U\, \partial_{\pm} V - (\partial_{\pm} X^{i})^2 + \delta(U)\, f_D(\rho)(\partial_{\pm} U)^2$$

where

$$x_{\pm} = (\sigma \pm \tau), \qquad \partial_{\pm} = \frac{1}{2}(\partial_{\sigma} \pm \partial_{\tau})$$

From eqs (148) we find

$$(T_{\pm\pm})_{<} = (T_{\pm\pm})_{<\,\text{flat}} \qquad \tau<0 \qquad (156)$$

$$(T_{\pm\pm})_{>} = (T_{\pm\pm})_{>\,\text{flat}} + \frac{P_U\alpha'}{2} H(\tau)\, \partial_{\pm}\left[f_D(\sigma\pm\tau) - \dot{f}_D(\sigma\pm\tau)\right] \qquad \tau>0$$

where

$$(T_{\pm\pm})_{\gtrless\,\text{flat}} = \pm P_U \alpha'\, \partial_{\pm} V_{\text{flat}} - (\partial_{\pm} X^{i}_{\gtrless})^2 \qquad (157)$$

$f_D(\sigma,\tau)$ and $\dot{f}_D(\sigma\pm\tau)$ are given by eqs. (148) and V_{flat} is just the free part of the solution for the V-coordinate (eq. 148)

$$V_{flat}\ F(\sigma-\tau)+G(\sigma+\tau)$$

In the light-cone gauge where we are working, these constraints completely determine the V coordinate of the string

$$\pm\partial_\pm V_{flat} = \frac{1}{P_v\alpha'}(\partial_\pm X^i_<)^2 \quad , \quad \tau<0 \quad (158)$$

$$\pm\partial_\pm V_{flat} = \frac{1}{P_v\alpha'}(\partial_\pm X^i_>)^2 - \frac{1}{2}\partial_\pm\left[f_D(\sigma\pm\tau)-\dot{f}_D(\sigma\pm\tau)\right], \quad \tau>0 \quad (159)$$

This formula gives the solution of the constraints. It can be checked from the transverse solution of the equations of motion that eq. (159) is a consequence of eq. (158) as it should be.

The conformal generators

$$L_n = \int_0^{2\pi} d\sigma\, e^{in\sigma}\, T_{++}(\sigma,\tau=0)$$

can be computed in terms of the $\tau<0$ basis or in terms of the $\tau>0$ basis. As T_{++} is conserved, the L_n generators computed in the < and > - basis are the same. We have

$$L_n = L_n\text{flat}(\alpha_<,\tilde{\alpha}_<) = \qquad (160)$$
$$= L_n\text{ flat}(\alpha_>,\tilde{\alpha}_>) + \Delta L_n(\alpha_>)$$

where L_n flat is the standard free expression

$$L_n\text{ flat}(\alpha,\tilde{\alpha}) = \sum_m (\alpha^i_m\,\alpha^i_{n-m} + \tilde{\alpha}^i_m\,\tilde{\alpha}^i_{n-m}) \qquad (161)$$

and ΔL_n stand for the Fourier components of eq. (156)

In particular,

$$L_0 = L_{0<}\text{ flat}$$
$$= L_{0>}\text{ flat} + \frac{P_v\alpha'}{2}\left[f_D(2\pi)-f_D(0)\right] \mp \frac{P_v\alpha'}{2}\left[\dot{f}(2\pi)-\dot{f}(0)\right] \qquad (162)$$

Here the argument of f_D indicates the value of σ (τ = o is omitted) and

$$L_0\text{ flat} = \frac{1}{2}\sum_m(\alpha^i_{-m}\,\alpha^i_m + \tilde{\alpha}^i_m\,\tilde{\alpha}^i_{-m}) + 2P_+P_- \quad (163)$$

The mass formula follows from the $L_0 \simeq$ o constraint

$$M^2 = M^2_{flat} = \frac{1}{\alpha'}\sum_{n=1}^{\infty}(\alpha^{i+}_{n<}\,\alpha^i_{n<} + \tilde{\alpha}^{i+}_{n<}\,\tilde{\alpha}^i_{n<}) =$$
$$= \frac{1}{\alpha'}\sum_{n=1}^{\infty} n\,(a^{i+}_n\,a^i_n + \tilde{a}^{i+}_n\,\tilde{a}^i_n) \qquad (164)$$

where

$$\alpha^i_n = -i\sqrt{|n|}\,(\text{sgn } n)\,a_n\,.$$

The mass spectrum is the same as in flat space time but there is an scattering effect on the string due to the shock wave geometry. The term ΔL_n describes the excitations between the internal (particle) states of the string due to this scattering. The expression of L_o in terms of $\alpha_>$ is different from the flat space expression but the spectrum is the same. The critical dimension at which massless states appear is D=26, the same as in flat space .

Scattering amplitudes and pair creation

The transformation eq. (154) mixes non-linearly the particle and antiparticle modes and the right and left movers. This transformation contains exactly all the information concerning the interaction of the string with the AS geometry, the elastic and inelastic scattering and pair creation process. In order to get more explicit expressions for these processes we will express the exact solution as

$$X^i(\sigma,\tau) = q^i(\tau) + \eta^i(\sigma,\tau) , \tag{165}$$

that is, an in refs (2,3) we expand X^i around the center of mass solution plus fluctuations around it.Here eq.(165) is the exact solution of the equations of motion. At the first order in η^i we have

$$D^i(\sigma) = \frac{P_\upsilon \alpha'}{(q_j^2)^{\frac{D}{2}-1}} \left[q^i + \Delta_{ij}\, \eta^j + O(\eta^2) \right] \tag{166}$$

where

$$\Delta_{ij} = \delta_{ij} - (D-2)\frac{q_i q_j}{(q_\ell^2)} \tag{167}$$

All quantities here are taken at $\tau = o$. (Δ_{ij} identically satisfies

$\mathrm{Tr}\,\Delta_{ij} = 0$

From eqs. (154) and (167) we get the linearized transformation

$$P_>^i - P_<^i = \frac{\tilde{K} P_\upsilon}{(q_\ell^2)^{\frac{D}{2}-1}}\, q^i + O(\eta^2) \tag{168}$$

Notice that

$$\alpha_{-n} = \alpha_n^+ \quad , \quad \tilde{\alpha}_{-n} = \tilde{\alpha}_n^+$$

For the oscillators we get the linearized transformation

$$\begin{aligned} \alpha^i_{n\,>} &= (\delta^i_j + \mathcal{B}_n{}^i{}_j)\, \alpha^j_{n\,<} - \mathcal{B}_n{}^i{}_j\, \tilde{\alpha}^{j\,+}_{n\,<} \\ \tilde{\alpha}^i_{n\,>} &= (\delta^i_j + \mathcal{B}_n{}^i{}_j)\, \tilde{\alpha}^j_{n\,<} - \mathcal{B}_n{}^i{}_j\, \alpha^{j\,+}_{n\,<} \end{aligned} \tag{169}$$

with

$$\mathcal{B}_n{}^i{}_j = \frac{i}{2n}\frac{\tilde{K} P_\upsilon \alpha'\, \Delta^i{}_j}{(q_\ell^2)^{\frac{D}{2}-1}} = i\,\frac{\Delta^i_j}{n}\,\frac{2\pi^{2-D/2}}{(q_\ell^2)^{\frac{D}{2}-1}}\, P_\upsilon \alpha' G_D\, \Gamma(\tfrac{D}{2}-1) \tag{170}$$

This is a Bogoliubov transformation expressing the particle operators at $\tau > 0$ as a linear combination of the particle and antiparticle operators at $\tau < 0$.

For fixed n, transitions take place between the internal oscillatory modes of the string. Notice that the effect of mixing particle and antiparticle modes changes at the same time their right or left character. There is also an effect on the polarization of the modes (without changing their right or left character). The number of left modes minus the number of right modes stays equal to zero in both processes. In other words, if for $\tau < o$, the string has a right (or left) excited mode with a given polarization j, then for $\tau > o$ there will be : (i) an amplitude ($\delta_i^j + \beta_n{}^i{}_j$) for a right (or left) mode polarized in the i-direction and (ii) an amplitude $\beta_n{}^i{}_j$ for a left (or right) anti-particle mode polarized in the i-direction ($2 \leq i \leq D-2$)

Let us define the ingoing vacuum state $|0_<\rangle$ at $\tau < o$:

$$\alpha^i_{n<} |0_<\rangle = 0 \text{ for all } n \tag{171}$$

Then we find

$$\begin{aligned} \langle 0_< | N^i_{n>} | 0_< \rangle &= \langle 0_< | \alpha^{i\,+}_{n>} \alpha^i_{n>} | 0_< \rangle = \\ &= \frac{1}{2} | \beta^i_{jn} |^2 \end{aligned} \tag{172}$$

$$\langle 0_< | \tilde{N}^i_{n>} | 0_< \rangle = \langle 0_< | \tilde{\alpha}^{i\,+}_{n>} \tilde{\alpha}^i_{n>} | 0_< \rangle = \frac{3}{2} | \beta^i_{jn} |^2$$

that is,

$$N^i_n = \langle 0_< | N^i_{n>} + \tilde{N}^i_{n>} | 0_< \rangle = | \beta^i_{jn} |^2$$

where

$$| \beta^i_{jn} |^2 = \frac{1}{4n^2} \frac{\tilde{K}^2 (P_v \alpha')^2}{(q_e^2)^{D-2}} (D^2 - 5D + 8) \tag{173}$$

$\tilde{K}$ is given by eq.

This represents the number of modes for $\tau > o$ at the nth-level created from the $\tau < o$ vacuum. That is, pair creation out of the $\tau < 0$ vacuum state takes place for $\tau > o$ and in <u>all</u> the modes as a consequence of the scattering by the shock wave geometry. Each pair here is formed by a right and left mode. The total number over all n is

$$\mathcal{N} = \sum_n N(n) = \left[P_v \alpha' \frac{(D-4)}{4} \frac{\tilde{K}}{\rho_o^{D-2}} \right]^2 (D^2 - 5D + 8) \frac{\pi^2}{6} \tag{174}$$

where

$$\rho_o = \sqrt{q^{i\,2}(0)}$$

We see that the excitation of the string by the geometry is inversely proportional to the center of mass impact parameter ρ_o, for large ρ_o.

We can also write

$$\alpha_{n>}^{i} = e^{G}\, \alpha_{n>}^{i}\, e^{-G}$$

where G = $\sum_n \theta_n \left(\alpha_{n<}^{+} \tilde{\alpha}_{n<}^{+} - \alpha_{n<} \tilde{\alpha}_{n<} \right)$

and

$$\text{th}\, \theta_n = \frac{\beta_n}{(1 + \beta_n)}$$

Therefore,

$$|0_{>}\rangle = e^{G}\, |0_{<}\rangle$$

where

$$\alpha_{n>}^{i}\, |0_{>}\rangle = 0 \qquad \forall n$$

<u>The center of mass motion and string cross section</u>

We use lower case letters for the center of mass (c.m.) motion and keep capital ones for the string solutions. The center of mass trajectory is just the motion of a test particle in curved space time, that is

$$\ddot{q}^{A}(\tau) + \Gamma^{A}_{BC}(q)\, \dot{q}^{B} \dot{q}^{C} = 0 \tag{174}$$

The (c.m) solution always admits the constant of motion

$$\alpha'^2 m^2 = - G_{AB}(q)\, \dot{q}^{A} \dot{q}^{B} \,, \tag{175}$$

which defines the (classical) mass of the string in the curved geometry. The goedesic equations (174) read

$$\frac{d^2 u}{d\tau^2} = 0$$

$$\frac{d^2 v}{d\tau^2} + f_D(\rho)\, \delta'(u) \left(\frac{du}{d\tau}\right)^2 + 2\, \partial_\tau f_D(\rho)\, \delta(u) \left(\frac{du}{d\tau}\right) = 0 \tag{176}$$

$$\frac{d^2 q^i}{d\tau^2} + \frac{1}{2}\, \partial_i f_D(\rho)\, \delta(u) \left(\frac{du}{d\tau}\right)^2 = 0$$

which can be written as

$$u = p_u \alpha' \tau$$

$$\frac{d^2 v}{d\tau^2} + f_D(\rho_0)\, \delta'(\tau) + \dot{f}_D(\rho_0)\, \delta(\tau) = 0$$

$$\frac{d^2 q^i}{d\tau^2} + \frac{1}{2}\, \partial_i f_D(\rho_0)\, p_u \alpha'\, \delta(\tau) = 0 \,, \quad \rho_0 = \sqrt{q^{i2}(0)} \tag{177}$$

The general solution is given by

$$u = p_u \alpha' \tau$$

$$v = p_v \alpha' \tau - \dot{f}_D(\rho_0)\, \tau H(\tau) - f_D(\rho_0)\, H(\tau)$$

$$q^i = q_0^i + p_0^i \alpha' \tau - \frac{1}{2} \partial_i f_D(\rho_0)\, p_u \alpha' \tau\, H(\tau) \qquad (178)$$

here

$$q_0^i \equiv q^i(0)$$
$$p_0^i = \frac{1}{\alpha'} \left[\dot{q}^i(\tau)\right]_{\tau=0^-} \equiv p_{0<}^i \qquad (179)$$

The c.m. trajectory follows continuous straight lines at $\tau = 0$ in the transverse $(1 \leq i \leq D-2)$ directions with discontinuous momenta

$$p_{0>}^i - p_{0<}^i = -\frac{p_u}{2}\, \partial_i f_D(\rho_0) \;.$$
$$(q_{0>}^i = q_{0<}^i) \qquad (180)$$

The motion is discontinuous at $\tau = 0$ in the v-direction.

Notice that

$$\dot{q}^i(0) = \alpha' p_0^i - \frac{\alpha' p_u}{4}\, \partial_i f_D(\rho_0)$$

and

$$\dot{f}_D(\rho_0) = \dot{q}^i(0)\, \partial_i f_D(\rho_0) = \alpha' p_0^i\, \partial_i f_D(\rho_0) - \frac{\alpha' p_u}{4}\left[\partial_i f_D\right]^2 \qquad (181)$$

The mass of the c.m. solution can be easily computed from eq.(175), that is

$$\alpha'^2 m^2 = \left(\frac{dq^i}{d\tau}\right)^2 - \left(\frac{dv}{d\tau}\right) p_u \alpha' - f_D(\rho_0)\, \delta(\tau)\, p_u \alpha' \qquad (182)$$

By replacing the solution given by eqs. (178) and by using eq. (181), the H (τ) and $\delta(\tau)$ dependences cancel out as it should be and we have

$$m^2 = p_0^{i\,2} - p_u p_v \qquad (183)$$

Now, we can write the trajectories as

$$v = \frac{(p_0^{i\,2} - m^2)\, u}{p_u^2} + \left\{\frac{1}{4}\left[\partial_i f_D(\rho_0)\right]^2 - \frac{p_0^i}{p_u}\, \partial_i f_D(\rho_0)\right\} u\, H(u) -$$
$$- f_D(\rho_0)\, H(u) \;. \qquad (184)$$
$$q^i = q_0^i + \frac{p_0^i}{p_u}\, u - \frac{1}{2}\, \partial_i f_D(\rho_0)\, u\, H(u).$$

We see that the particular solutions given in eq.(9) of ref (18) correspond to take $m = 0$ and $p_0^i = 0$.

Let us consider now the scattering angle of the particle as a function of its impact parameter. ρ_0 is the distance from the origin to the point where the particle trajectory intersects the $u = o$ $(x = t)$ plane (eq. 178). If we call φ the angle between $\vec{\rho}_0$ and the ingoing velocity we have

$$\sin\varphi = \frac{p_{x<}}{|\vec{p}_<|} = \frac{p_{x<}}{\sqrt{p_x^2 + p_0^{i\,2}}} \qquad (185)$$

where

$$p_x = \frac{1}{2}\left(p_v - p_u\right)$$

Therefore

$$b = \rho_0 \sin \varphi$$

Now the scattering angle Θ is just the angle between the ingoing and outgoing momenta, that is

$$\cos \Theta = \frac{\vec{P}_> \cdot \vec{P}_<}{|\vec{P}_>|\,|\vec{P}_<|} = \frac{1 + \dfrac{\alpha' \rho_0^i \left(\rho_0^i - \frac{P_v}{2} \partial_i f_D\right)}{P_x \left(P_x \alpha' - f_D/2\right)}}{\sqrt{1 + \dfrac{\rho_0^{i\,2}}{P_x^2}} \;\sqrt{1 + \dfrac{\alpha'^2 \left(\rho_0^i - \frac{P_v}{2} \partial_i f_D\right)^2}{\left(P_x \alpha' - \frac{1}{2} f_D\right)^2}}} \tag{186}$$

For $\rho_0 \to \infty$ $(\Theta \ll 1)$ we get

$$\Theta_{\rho_0 \to \infty} = \frac{\tilde{K}}{\rho_0^{D-3}\, \vec{p}^{\,2}} \sqrt{(P_v P_x)^2 + \left(\rho_0^{i\,2} - 2 P_v P_x\right) \frac{\left(q_0^j \rho_0^j\right)^2}{\rho_0^2}}$$

$$= \frac{4 G \mu}{\rho_0^{D-3}}\, \pi^{2 - D/2}\, \Gamma\left(\frac{D}{2} - 1\right) R(\varphi) \tag{187}$$

$$R(\varphi) = \frac{1}{\vec{p}^{\,2}} \sqrt{(P_v P_x)^2 + \left(\rho_0^{i\,2} - 2 P_v P_x\right) p_c^{i\,2} \cos^2 \varphi}$$

The D dimensional c.m. cross section is given by

$$\frac{d\sigma_D}{d\Omega_{D-1}} = \left(\frac{b(\Theta, \Phi)}{\sin \Theta}\right)^{D-3} \frac{1}{\left|\dfrac{d\Theta}{db}\right|_{\Phi}}, \qquad d\sigma_D = b^{D-3}\, db\, d\Omega_{D-2} \tag{188}$$

where the azymutal variable Φ is the angle between the planes $(P_>^i, P_<^i)$ $(q_{rc}^i, P_<^i)$:

$$\cos \Phi = \cot g\, \varphi\, \operatorname{ctg} \Theta - \frac{q_0^i P_>^i}{\sin \varphi \sin \Theta\, |q_0|\, |P_>|} \tag{189}$$

For large b, from eqs.(187) and (188) we find

$$\left(\frac{d\sigma}{d\Omega}\right)_{\substack{cm \\ \Theta \ll 1}} = \frac{\left[4 G \mu\, \Gamma\left(\frac{D}{2} - 1\right)\right]^{\left(\frac{D-2}{D-3}\right)}}{(D-3)}\, \Theta^{-\left(D - 1 + \frac{1}{D-3}\right)} \left[\pi^{2 - D/2} R(\varphi)\right]^{\left(\frac{D-2}{D-3}\right)} \tag{190}$$

Let us now consider the string in an initial state described by the nth-mode. The string elasti c cross section is given by

$$\frac{d\sigma}{d\Omega} = \left(\frac{d\sigma}{d\Omega}\right)_{cm} \left| \langle \overleftarrow{n}_> | \overleftarrow{n}_< \rangle \right|^2$$

$$= \left(\frac{d\sigma}{d\Omega}\right)_{cm} \left[1 - \sum_{n' \neq n} |B_{n'}|^2\right. \tag{191}$$

From eqs. (174) and (191) we get

$$\left(\frac{d\sigma}{d\Omega}\right) = \left(\frac{d\sigma}{d\Omega}\right)_{cm} \left[1 - \left(\frac{P_0 \alpha' G p}{\rho_0}\right)^2 \left(\frac{1}{\rho_0}\right)^{2(D-3)} C_D \left(1 - \frac{6}{\pi^2 n^2}\right) \right] \tag{192}$$

where $C_D = (2/3)\pi^{6-D} \Gamma^2(D/2-1)\,(D^2-5D+8)$

and $(d\sigma/d\Omega)_{cm}$ is given by eq.(190)

Compar ison with the black hole case

It is interesting to compare these results with the corresponding ones for a string in the Schwarzschild geometry. In the Schwarzschild's geometry we have found

$$\begin{aligned} \tilde{\alpha}^i_{n\,out} &= (1 - A_n)\, \tilde{\alpha}^i_{n\,in} + B_n\, \alpha^{i\,+}_{n\,in} \\ \alpha^i_{n\,out} &= (1 - A_n)\, \alpha^i_{n\,in} + B_n\, \tilde{\alpha}^{i\,+}_{n\,in} \end{aligned} \tag{193}$$

where

$$A_n = \left(\frac{R_s}{b}\right)^{D-3} \left(\frac{i p \alpha'}{b}\right) \sqrt{\pi}\, \frac{\Gamma\left(\frac{D}{2}+1\right)}{\Gamma\left(\frac{D+1}{2}\right)} \tag{194}$$

$$B_n = \left(\frac{R_s}{b}\right)^{D-3} \left(\frac{i p \alpha'}{b}\right) \frac{\pi}{2} \left|\frac{2nb}{\alpha' p}\right|^{D/2} e^{-\left|\frac{2nb}{\alpha' p}\right|} B_D \left[1 + O\left(\frac{\alpha' p}{n}\right)\right]$$

These results hold at first order in α' and at first order in $(R_s/b)^{D-3}$. Here b, p and m are the impact parameter, momentum and mass of the center of mass respectively and

$$R_s^{D-3} = \frac{8GM}{(D-2)}\, \pi^{-\frac{(D-3)}{2}}\, \Gamma\left(\frac{D-1}{2}\right)$$

$$B_D = \left[D + 3 - i \left(\frac{D-1}{2}\right)^2 \right] \frac{1}{(D-5)}\, \frac{1}{2^{(P+1)/2}}\, \frac{1}{\Gamma\left(\frac{D+1}{2}\right)}$$

Thus, we have

$$A_n = \frac{1}{b^{D-2}}\, 4i\alpha' p\, GM\, \pi^{D/2} \frac{D}{(D-1)}\, \Gamma\left(\frac{D}{2}-1\right) \left[1 + \frac{(D-1)}{D} m^2/p^2 \right]$$

$$B_n = \frac{1}{b^{D-2}}\, 4i\alpha' p\, GM\, \pi^{D/2} \frac{\sqrt{\pi}}{(D-2)}\, \Gamma\left(\frac{D-1}{2}\right) B_D \left|\frac{2nb}{\alpha' p}\right|^{D/2} e^{-\left|\frac{2nb}{\alpha' p}\right|} \tag{194}$$

whereas in the AS metric we have

$$B^i_{n\,j\,AS} = \frac{1}{\rho_0^{D-2}}\, 2i\alpha' P_0\, G p\, \pi^{1-D/2}\, \Gamma\left(\frac{D}{2}-1\right) \frac{\Delta_{ij}}{n} \tag{195}$$

Notice that for the black hole geometry, $\mathcal{A}_n$ is independent of n and vanishes exponentially at high n, In the AS geometry, $\mathcal{A}_{n\,j}^{\,i}$ and $\mathcal{B}_{n\,j}^{\,i}$ differ only in δ^{i}_{j} and are proportionnal to 1/n for all n.

It is also interesting to compare the elastic scattering cross sections. In the black hole geometry we have found

$$\left(\frac{d\sigma}{d\Omega}\right)_{BH} = \left(\frac{d\sigma}{d\Omega}\right)_{\substack{cm\\BH}} \left[1 - \left(\frac{\pi\alpha' p}{b}\right)^2 \left(\frac{R_s}{b}\right)^{2(D-3)} \sum_{n'\neq n} \left|F\left(\frac{m}{p}, \frac{n'b}{\alpha' p}\right)\right|^2\right]$$

$$= \left(\frac{d\sigma}{d\Omega}\right)_{\substack{cm\\BH}} \left[1 - \left(\frac{\alpha' p GM}{b}\right)^2 \left(\frac{1}{b}\right)^{2(D-3)} a_D \sum_{n'\neq n} \left|F\left(m/p, \frac{n'b}{\alpha' p}\right)\right|^2\right] \quad (196)$$

Where $\sum|F|^2$ is dimension less and a_D a factor only depending on D .

$$a_D = \left[\frac{8}{D-2}\,\Gamma\left(\frac{D-1}{2}\right)\right]^2 \pi^{D-3} .$$

In the AS geometry we have

$$\left(\frac{d\sigma}{d\Omega}\right)_{AS} = \left(\frac{d\sigma}{d\Omega}\right)_{\substack{cm\\AS}} \left[1 - \left(\frac{\alpha' p\, G f}{\rho_0}\right)^2 \left(\frac{1}{\rho_0}\right) C_D \left(1 - \frac{6}{\pi^2 n^2}\right)\right]$$

$$\text{where}\quad C_D = (2/3)\,\pi^{6-D}\, \Gamma(D/2-1)\,(D^2-5D+8) \quad (197)$$

We see that the classical result $(d\sigma/d\Omega)_{cm}$ gets corrections of order α'^2 in the black hole case and in the AS geometry. The dependence on b and on the metric parameters of the corrections are the same in the both geometries, the dependence on n is different. The deflection angles at first order in $(1/b)^{D-3}$ are respectively given by

$$\Theta_{BH} = \frac{4GM}{b^{D-3}}\, \pi^{2-D/2}\, \Gamma\left(\frac{D}{2}-1\right) \left[1 - \left(\frac{D-3}{D-2}\right)\frac{m^2}{p^2}\right] \quad (198)$$

and

$$\Theta_{AS} = \frac{4Gf}{\rho_0^{D-3}}\, \pi^{2-D/2}\, \Gamma\left(\frac{D}{2}-1\right) \frac{1}{\vec{p}^{\,2}} \sqrt{(P_0 P_x)^2 + (P_0^{i\,2} - 2P_0P_x)\, P_0^{i\,2}\cos^2\varphi} \quad (199)$$

The small angle c.m. cross sections are

$$\left(\frac{d\sigma}{d\Omega}\right)_{\substack{cm\\BH}} = \frac{\left[4GM\,\Gamma\left(\frac{D}{2}-1\right)\right]^{\frac{D-2}{D-3}}}{(D-3)}\, \Theta^{-\left(D-1+\frac{1}{D-3}\right)} \left\{\pi^{2-D/2}\left[1 + \left(\frac{D-3}{D-2}\right)\frac{m^2}{p^2}\right]\right\}^{(D-2)/D-3}$$

$$\left(\frac{d\sigma}{d\Omega}\right)_{\substack{cm\\AS}} = \frac{\left[4Gf\,\Gamma\left(\frac{D}{2}-1\right)\right]^{\frac{D-2}{D-3}}}{(D-3)}\, \Theta^{-\left(D-1+\frac{1}{D-3}\right)} \left\{\pi^{2-D/2} . R(\varphi)\right\}^{\frac{D-2}{D-3}}$$

For D = 4 :

$$\Theta_{BH} = \frac{4GM}{b}\left(1+\frac{m^2}{2p^2}\right)$$

$$\Theta_{AS} = \frac{4G\hbar}{\rho_0}\frac{1}{\vec{p}^{\,2}}\sqrt{(P_U P_x)^2 + (P_0^{i\,2} - 2P_U P_x)P_0^{i\,2}\cos^2\varphi} \qquad (200)$$

and

$$\left(\frac{d\sigma}{d\Omega}\right)_{cm\,BH} = \frac{(4GM)^2}{\Theta^4}\left(1+\frac{m^2}{2p^2}\right)^2$$

$$\left(\frac{d\sigma}{d\Omega}\right)_{cm\,AS} = \frac{(4G\hbar)^2}{\Theta^4}\frac{1}{p^4}\left[(P_U P_x)^2 + (P_0^{i\,2} - 2P_U P_x)P_0^{i\,2}\cos^2\varphi\right] \qquad (201)$$

When the initial momentum is collinear with the shock wave propagation (x) we have

$$\varphi = \pi/2 , \qquad p_0^{i} = 0 , \qquad b = \rho_0 ,$$

and eq.(199) yields

$$\Theta_{AS} = \frac{8G\hbar}{b^{D-3}}\,\pi^{2-D/2}\,\Gamma\left(\frac{D}{2}-1\right) \qquad (202)$$

for a massless particle ($p_x = - p_0$) .
Comparing Θ_{BH} eqs.(198,200) with Θ_{AS} eq.(202) we see that the scattering angle for two massless particles is <u>twice</u> the scattering angle for a masssive and a massless particle in the large b and small Θ limit .

Further perspectives and discussion

It is interesting to study the scattering of the scalar particle corresponding to the string ground state in the AS geometry. The scalar vertex operator in curved space time fulfils the Klein-Gordon equation

$$(\Box^2 - \mu^2)\, T(X) = 0, \qquad \Box^2 = \frac{1}{G}\partial_A\left(\sqrt{G}\,G^{AB}\partial_B\right) \qquad (203)$$

The generalization of the flat space scattering amplitude to curved space time is given by

$$A(k_1, k_2) = \int d^2z_1\, d^2z_2 \,\langle 0_<| : \Psi^*_{out}(k_2, X(z_2)) : \; : \Psi_{in}(k_1, X(z_1)) : |0_<\rangle \qquad (204)$$

here (k_1, k_2) are the on-shell ingoing and outgoing momenta of the scalar particle. $\Psi_{in}(k, X)$ and $\Psi_{out}(k,X)$ are solutions of eq. (203) with the boundary conditions

$$\Psi_{in}(k,X) = e^{i\,k.X} \qquad X^0 \to -\infty \qquad (205)$$

$$\Psi_{out}(k,X) \underset{X^\circ \to +\infty}{=} e^{ik.X}$$

X(z) in eq. (204) is the string coordinate operator and the normal ordering :: is taken with respect to the ingoing ground state $|0_<\rangle$

$$\alpha_{n<}^{i}\, |0_<\rangle = 0 \quad , \quad n > 0$$
$$P_<\, |0_<\rangle = 0 \qquad (206)$$

z_1, z_2 stand for the world sheet coordinates. In flat space time the amplitude eq. (204) is trivial :

$$A\,(k_1,\, k_2) = \text{const.}\ \delta^D\,(k_1 - k_2)$$

In the D-dimensional AS metric, the Klein-Gordon equation can be solved in closed form [20] . (The in-solution with the boundary condition eq.(205) is given by [20] .

$$\Psi_{in}(k, X) = \begin{cases} e^{i[\vec{k}_\perp . \vec{X}_\perp - \frac{\omega V}{4} - \frac{U}{\omega}(k_\perp^2 + m^2)]} & U<0 \\ e^{-\frac{i\omega V}{4}} \int dp_\perp^{D-2}\, e^{i[\vec{p}_\perp . \vec{X}_\perp - \frac{U}{\omega}(m^2+p_\perp^2)]}\, S(\vec{k}_\perp - \vec{p}_\perp) & U>0 \end{cases} \qquad (207)$$

where

$$S(\vec{k}_\perp - \vec{p}_\perp) = \int \frac{dy_\perp^{D-2}}{(2\pi)^{D-2}}\, e^{i\vec{Y}_\perp . (\vec{k}_\perp - \vec{p}_\perp) - i\varphi_D(|\vec{Y}_\perp|)} \qquad (208)$$

and $\varphi_D(|\vec{Y}_\perp|)$ is the D dimensional generalization of the 't Hooft phase-shift [15] .

$$\varphi_D(\rho) = 4\omega K / \rho^{4-D}$$

The integral in eq. (204) covers the both regions τ<o and τ>o. For τ> o one must use the string operators $X_>^\mu$ (z) eqs. (151, 152). Notice that they express in terms of the $\alpha_{n<}$- oscillators through the non-linear transformation eq.(154) .

It should be pointed out that the effect of the geometry manifests in the scattering amplitude eq.(204) in two combined ways : (i) in the wave function through the non-trivial S matrix eq.(208) already present in the quantum point particle theory and (ii) in the α_n^i operators related to $\alpha_{n<}^i$ through the non linear transformation eq.(154) specific of the string. If one ignores the difference between the oscillators $\alpha_{n>}$ and $\alpha_{n<}$, the amplitude (A (k_1, k_2) gives the quantum point particles S matrix eq.(208). The full computation of the string amplitude A (k_1, k_2) will be the subject of a forthcoming paper.

Recently the graviton-graviton scattering has been studied in the context of superstring theory in flat space time [21] (Regge-Gribov techniques or multiple Reggeized graviton exchange). The eikonal expression given in refs.21 and 22 correspond to the resommation of graviton exchanges in flat space time which yields the relativistic Coulombian amplitude (23,24). For example the dominant term of large b in eq. (6.20) of ref 21 (also in eq.12 of ref.22 corresponds to the relativistic Coulombian amplitude. For D = 4 the leading term in the small angle regime of ref. (21). gives the cross section :

$$\left(\frac{d\sigma}{d\Omega}\right)_{\Theta \ll 1} = \frac{4\, G^2 E^2}{(\Theta/2)^4} \tag{209}$$

This coincides with the scattering cross section of two gravitons in the graviton-exchange approximation (25). The factor 4 in eq. (209) is characteristic of ultrarelativistic particles. Notice that eq. (209) is also the scattering cross section for two identical scalar particles through one graviton exchange in the ultrarelativistic limit. Also notice that the scattering of (unpolarized) gravitons by a scalar particle of mass M in the Ultrarelativistic limit is given by

$$\left(\frac{d\sigma}{d\Omega}\right)_{UR} \underset{\Theta \ll 1}{=} 4\,\frac{G^2 M^2}{(\Theta/2)^4} \tag{210}$$

whereas that in the non-relativistic limit it is given by

$$\left(\frac{d\sigma}{d\Omega}\right)_{NR} \underset{\Theta \ll 1}{=} \frac{G^2 M^2}{(\Theta/2)^4} \tag{211}$$

Finally, recall that for large b and small Θ, the scattering cross section of massless neutral particles by any static gravitational field of mass M is equal to eq. (211). This describes only the weak (large distance) behaviour of the metric.

The resommation of string (or field theory) amplitudes in flat space time as performed in the litterature 21, 22 reproduces the linear effects (weak or large distance behaviour) of the gravitational field. Although further classical and quantum corrections are indeed present in higher gravi-Reggeon loops, their explicit evaluation and resommation does not seem simple. For high curvature one must consider the curved space time metric associated to such an enormous energy to start. Such strong gravitational effects are appropriately described by exactly solving the field equations in the curved backgrounds (the AS metric in the present case).

In ref. [27] we study the space time geometry of ultrarelativistic charged particles and their scattering. First at all, we start from the Reissner - Nordstrom solution describing the gravitational field of a particle of mass m and charge e. We apply to it a Lorentz boost, and take the ultrarelativistic limit $\gamma \to \infty$, $m = (\gamma^{-1} p) \to 0$ and $e = (\gamma^{-1/2} p_e) \to 0$, keeping the kinetic momentum p and the electromagnetic momentum p_e fixed, (here $\gamma = (1 - v^2)^{-1/2}$). We do this in four and in D dimensions and find in this way the following metric

$$ds^2 = du\,dv + dx_{\perp}^{i\,2} - f_D(\rho)\,\delta(u)\,du^2 \quad , \quad \rho = \sqrt{x_{\perp}^{i\,2}} \qquad (212)$$

where

$$f_D(\rho) = \begin{cases} \dfrac{K}{\rho^{D-4}} + \dfrac{B}{\rho^{2(D-4)+1}} \quad , & D > 4 \\[2ex] K = -8\pi^{2-D/2}\,G p\,\Gamma\left(\frac{D}{2}-1\right) , & B = \dfrac{-6\pi G p_e^2}{(D-2)(2D-7)} \displaystyle\prod_{j=0}^{D-4} \frac{2D-7-2j}{2(D-3-j)} \end{cases}$$

$$f(\rho) = -8 G p \ln \rho - \frac{3}{2}\pi \frac{G p_e^2}{\rho} \quad , \quad D = 4 \qquad (213)$$

We also obtain the same result by generating from flat space time, the gravitational shock wave of a boosted point particle of mass m and charge e in the ultrarelativistic limit $v \to c$, $m = (\gamma^{-1} p) \to 0$ and $e = (\gamma^{-1/2} p_e) \to 0$. The metric eq. (212) generalizes the AS metric to the case where the ultrarelativistic particle has a non - zero electromagnetic momentum p_e, besides the kinetic momentum p. It should be also noticed that this limit is a peculiar configuration for the electromagnetic field: the field is itself zero in this limit, but " its square " (the energy - momentum tensor) which generates curvature is non - zero and proportional to $\delta(u)$. On the other hand, it should be pointed out, that the boosted Reissner - Nordstrom geometry (with γ finite), is well defined in the three cases (i) $e^2 > m^2$, (ii) $e^2 < m^2$ and (iii) $e^2 = m^2$. However the ultrarelativistic limit only can be taken from the solution with $e^2 > m^2$, that is that corresponding to a point particle of mass m and charge e at the irremovable singularity $r = Q$. This is in agreement with the interpretation and the elementary particle context in which we are using this geometry:the ratio Ge^2 / m^2 is always bigger than one ($G=10^{19}$Gev). (This is not the case, of course, for astrophysical objects). The solutions representing black holes with $e^2 < m^2$ do not exist in the ultrarelativistic limit.And the case $e^2 = m^2$ (extreme black hole) leads in the ultrarelativistic limit to the original AS metric without any modification ($p_e = 0$).

We also study the scattering of relativistic charged particles at the Planck scale. In a frame where only one of the particles (particle 2 say) has an energy of the order of m_{pl} and both particles have masses m_1 , $m_2 << m_{pl}$, particle 2 is ultrarelativistic and the space time around it is described by our geometry, eq. (212). Particle 1 is a test particle since its energy is much smaller than m_{pl}. We first consider the case in which $e_1 = 0$, thus this scattering problem reduces to solve the Klein - Gordon equation in the curved space of eq. (212). This type of equation is not a well defined mathematical problem since it is a first order equation in u with a $\delta(u)$ potential. (The wave function Φ is discontinuous at u=0 and the product $\delta(u)\Phi$ is undefined).In reference [20] two rigorous procedures have been used to give sense to it: continuous and lattice regularisations. Here we are most concerned with the continuous regularisation which leads to the same phase shift as that found in reference [15] :

$$\varphi_{D\,cont}(\rho) = \frac{\omega}{4} f_D(\rho)$$

here in our problem $f_D(\rho)$ is given by eq.(210).
In D = 4 we find

$$S_{cont}(s,t) = \frac{2^{-2iG(s-m_1^2)}}{\pi |t|^{1-iG(s-m_1^2)}} \sum_{n=0}^{\infty} \frac{1}{n!}\left[\frac{-i\alpha G(s-m_1^2)\sqrt{|t|}}{4p}\right]^n$$

$$\times \frac{\Gamma\left[1-iG(s-m_1^2)-n/2\right]}{\Gamma\left[iG(s-m_1^2)+n/2\right]} \qquad (214)$$

where $\alpha = 3/2\,\pi\, p_e^2$

When $p_e = 0$, we recover the scatering matrix S_{cont} of the standard AS geometry [15] :

$$S_{cont}(s,t)\Big|_{p_e=0} = \frac{1}{\pi}\left(\frac{4}{|t|}\right)^{iG(s-m_1^2)} \frac{\Gamma\left[1-iG(s-m_1^2)\right]}{\Gamma\left[iG(s-m_1^2)\right]} \qquad (214)$$

The S - matrix for $p_e \neq 0$ is an infinite superposition of the S - matrices for $p_e = 0$ and thus becomes an infinite series in (s, t) . S_{cont} for $p_e \neq 0$ exhibits an infinite sequence of imaginary poles at the values $Gs = i\,m/2$, ($m \in \mathbb{Z}$). (Recall that for $p_e = 0$, S_{cont} exhibits poles only in the negative imaginary axis $Gs = -i(n+1)$, $n = 0,1,2,\ldots$).

Thus, the effect of the electromagnetic momentum p_e increases by a factor four the number of poles (or " resonances ").In ref.[27] we also obtain $S_{lattice}$ which exhibits cuts in both s and t variables and generalizes for $p_e \neq 0$ the expression given in reference [20]. We find

$$S_{cont}(s \to m_1^2, t) = S_{lattice}(s \to m_1^2, t) =$$

$$= -\frac{i}{\pi t} G(s - m_1^2) \tag{215}$$

that is the one graviton exchange amplitude (as in the $p_e = 0$ case) and

$$S_{cont}(s, t \to 0) = -\frac{i}{\pi t} Gs ,$$

$$S_{lattice}(s, t \to 0) = -\frac{i}{\pi t} \frac{1}{G(s-m_1^2)\ln^2(-t/4)} \left[1 - (i\alpha G/8\mu)(s - m_1^2)\right] \tag{216}$$

In ref.[27] we also compute the behaviour of S_{cont} for $s \to \infty$ with t fixed, which has a different t - dependence from that of the $p_e = 0$ case, and for the $s \to \infty$ and $t \to \infty$ regime with s / t fixed.

The scattering in which both particles 1 and 2 are charged is also discussed. In the context discussed above, the dynamics of this process is described by the Klein - Gordon equation for the charged field (particle 1 with charge e_1) in the metric of particle 2 with electromagnetic momentum p_e. Since at the ultrarelativistic limit the electromagnetic potential A_μ is a singular configuration and the $g_{\mu\nu}$ itself contains a $\delta(u)$ term, the terms $A^\mu A_\mu$ diverge. We derive the Klein - Gordon equation for this problem using the metric and the electromagnetic field for γ large but not infinite (that is, v almost but not 1) and (m, e) small (but not zero). This is the correct equation describing the dynamics of this scattering process. Since for g large but not infinite, the metric coefficient g_{uu} does not factorizes as a function of ρ times $\delta(u)$ but it is

$$g_{uu} = \frac{1}{\sqrt{(x - vt)^2 + \rho^2(1 - v^2)}} - \frac{1}{\sqrt{(x - vt)^2 + (1 - v^2)}} ,$$

the Klein - Gordon equation is very complicated to analize (even for $p_e = 0$).

It should be noticed that the effect of the charge in the scattering of ultra - high energetic particles is highly non - trivial. The effect of charge is not just to cause a shifting $G s \to G s + e_1 e_2 / 4\pi$ in the scattering amplitude as has been suggested in reference [15]. As we show in this paper, the shock wave metric itself is modified by the effect of the charge, the contribution to $f(\rho)$ of the terms with p and p_e are different. Of course, the effect found here will not always overshadow the contributions of ordinary electromagnetic interactions proportional to $e_1 e_2$. It is not hard to see in which limit any of these effects dominates over the other. The gravitational field of the electromagnetic field is negligible as long as the impact parameter b satisfies

$$(e^2)/b \ll m ,$$

or

$$(e^4)\, t \ll m^2, \quad \text{in Planck units,}$$

Actually ,we are dealing here with just a special case of "corrections" due to a deviation from the pointlike mass distribution of the original particles.
Particles with finite charge (e) cannot move at the speed of light. At $\gamma=\infty$ the charge should vanish as $e(\gamma) = \gamma^{-1/2} p_e$, with finite constant p_e (the mass vanishes as $m(\gamma) = \gamma^{-1} p$ with constant p). An ultrarelativistic particle of momenta p and p_e can interact with a particle of mass m and zero charge, but not with a finite charge (the interaction is infinite). For γ large but finite , the interaction does not diverge . The interaction of two ultrarelativistic particles of momenta p_{e1} and p_{e2} should be also a well defined, although a very complicated non - linear collision problem. The inclusion of spin is also an interesting non- trivial problem and it is now under investigation [28].Others generalizations are given in ref.[29].

The interaction of a neutral test string with the shock wave curved geometry discussed here is well defined at the $\gamma=\infty$ limit and it is non linear.

The non linear equations of motion of a neutral test string in the curved shock wave geometry discussed here are exactly solvable. The exact general solution of the equations of motion and constraints both at the classical and quantum level, have been found in reference [19]. This solution can be easily generalized to the present charged metric just using our $f_D(\rho)$ [eq.(212)]. There is a non linear transformation relating the $\tau < 0$ oscillator operators $a^i_{n<}$ to the $\tau > 0$ operators $a^j_{n>}$ and $a^j_{n>}{}^+$. The explicit expressions for the deflection angle $\Theta(\rho_o)$, (ρ_o is the impact parameter of the center of mass of the string), hold here at the leading behaviour since the contribution to $f_D(\rho)$ of the term depending on p_e is proportional to $\rho^{-2(D-4)-1}$.

More recently generalizations of the gravitational shock wave space times have been given [see ref.29].
This problem has raised much interest recently. For other approachs see refs[30-33].

REFERENCES

1) See, for example : N.D. Birrell and P.C.W. Davies, "Quantum Fields in Curved Space", Cambridge University Press, Cambridge, England, 1982, and references therein.

2) N. Sanchez, Phys. Rev. 24D (1981) 2100
N. Sanchez and B.F. Whiting, Phys. Rev. 34D (1986) 1056.

3) H.J. de Vega and N. Sanchez, Nucl. Phys. 299B, 818 (1987)

4) P. Goddard, J. Goldstone, C. Rebbi and C.B. Thorn - Nucl. Phys. 56B (1973) 109.

5) N. Sanchez , Phys. Lett 195B, 160 (1987).

6) N. Sanchez, Nucl. Phys. 266B q(1986) 487

7) M. Martellini and N. Sanchez, Phys. Lett. B, 192B, n°3, 4 (1987) 361.

8) N. Sanchez, in the Proceedings of the Second Marcel Grossmann Meeting, Ed, R. Ruffini, (North Holland, 1982) pp.501.

9) N. Sanchez and B.F. Whiting, in "Quantum concepts in space and time" Eds. C. Isham and R. Penrose, (Oxford University Press, 1985), pp. 319-324

10) R.C. Myers and M.J. Perry, Ann. Phys. (N.Y.) 172 (1986) 304

11) R.C. Myers, Phys. Rev. 35D (1987) 455.

12) H.J. de Vega and N. Sanchez, phys. Lett. 197B , (1987) 455.

13) H.J. de Vega and N. Sanchez, Nucl. Phys. Phys. 309B, 552 (1988)

14) H.J. de Vega and N. Sanchez, Nucl. Phys. 309B, 577 (1988)

15) G. t Hooft, Phys. Lett. B. 198, 61 (1987) and Duke Univ.

16) P.C. Aichelburg dand R.U. Sexl, Gen. Rel. Grav. 2, 303, (1971)

17) T. Dray and G.t'Hooft, Nucl. Phys. B 253, 173 (1985)

18) V. Ferrari, P. Pendenza, G. Veneziano CERN-TH preprint 4973/88 (to appear in Gen. Rel. Gra.)

(19) H.J. de Vega and N. Sanchez, "Quantum string scattering in the Aichelburg-Sexl geometry", LPTHE-Paris/DEMIRM Meudon preprints 88-25 and 88095 (to appear in Nucl. Phys. B).

20) H.J. de Vega and N. Sanchez, "Paticle scattering at the Planck Scale and the Aichelburg-Sexl Gometry", LPTHE-Paris/DEMIRM Meudon preprints 88-11 and 88096 (to appear in Nucl. Phys. B).

21) D. Amati, M. Ciafaloni and G. Veneziano, Phys. Lett. B. 197, 81 (1987) and CERN-TH preprint 4886/87 (to appear in Int. Jour. Mod. Phys.)

22) I.J. Muzinich and M. Soldate, Phys. Rev. D37, 359 (1988) M. Soldate, Phys. Lett. B186, 321 (1987)

23) H.DI. Abarbanel and C. Itzykson, Phys. Rev. Lett. 23, 53 (1969)

24) M. Lévy and J. Sucher, Phys. Rev. 186, 1656 (1969)

25) B. de Witt, Phys. Rev. 162, 1195 (1967)

26) D. Amati and C. Klimčik, Phys. lett. B, (1988)

27) C.O. Lousto and N. Sanchez, "The curved schock wave space time of ultrahigh relativistic particles and their scattering", DEMIRM-Meudon preprint (1989)

28) C.O. Lousto and N. Sanchez, Manuscript in preparation

29) C.O. Lousto and N. Sanchez, Phys. Lett. 220B, 55 (1989)

30) G. Veneziano, CERN - TH preprint 5019, April 1988

31) G. Klimčik, Phys. Lett. B208, 373 (1988)

32) D. Amati and C. Klimčik, PHys. Lett. B210, 92 (1988)

33) D. Gross and Mende, Nucl. Phys. B303, 407 (1988)

The probabilistic time and the semiclassical approximation of quantum gravity

Mario A. Castagnino

Instituto de Astronomía y Física del Espacio
Casilla de Correo 67 - Sucursal 28, 1428 Buenos Aires - Argentina
and
Departamento de Física
Facultad de Ciencias Exactas y Naturales, Universidad de Buenos Aires
Ciudad Universitaria - Pabellón I, 1428 Buenos Aires, Argentina

ABSTRACT: The notion of "Probabilistic Time" is studied as a way to obtain Quantum Field Theory in curved space-time from Quantum Gravity. It is shown how ideal and real clocks measure the probabilistic time and how the universe can be used a a real clock. The problem of the arrow of time is studied because it is related with the definition of probabilistic time.

1. Introduction

As it is well known Quantum Field Theory in Curved Space can be considered the semiclassical limit of Quantum Gravity [1], [2]. Therefore it is very interesting to study the connection between these two subjects and see how we can solve the problem or understand the concept of one of them with the techniques and ideas of the other and viceversa. Nevertheless to perform this program we have a major obstacle. Quantum Gravity, at least in its canonical version, works only with spatial variables, as it is well known there is no time coodinate in this theory. On the other hand Quantum Field Theory in curved space is formulated in a four dimensional manifold which, of course, has time coodinates. Thus a major goal of the program is the reconstruction of the time notion from Quantum Gravity. This can be done in two different, but complementary ways.

i) Introducing classical time as a notion that can be defined in the classical limit only ([1], [3], [4]).

Quantum Mechanics in Curved Space-Time
Edited by J. Audretsch and V. de Sabbata
Plenum Press, New York, 1990

ii) Introducing the probabilistic time, a notion that can be exactly defined in the quantum period and that coincides with classical time in the classical limit ([5] to [9]).

We wil study the second approach, that we consider more complete and profound, and see how this problem is related with many features of the theory: the arow of time, the notion of Herakleitian time, the colapse of the wave function, ideal and real clocks, etc. Even if our theory can not be consideed as a final answer to al these problems it seeds light on all of them and their relation.

Let us begin giving the definition of probabilistic time in the simplest case. Let us conside an homogeneous universe with metric

$$dS^2 = \sigma^2(-N^2 d\tau^2 + a^2 d\Omega_3^2), \tag{1.1}$$

where a is the radius of the universe, $d\Omega_3^2$ is the metric of the three-dimensional unit sphee, S_3, N is the lapse, τ a time coordinate and $\sigma^2 = l_{pl}^2/24\pi^2$, $m_{pl}^{-2} = l_{pl}^2 = 16\pi G$. Using Quantum Gravity we can compute $\Psi(a)$ the "wave function of the universe" $|\Psi(a)|^2$ is proportional to the "unnormalized" probability to find a metric of radius a. Then what we have now is a colection of metrics labeled by a and a way to know the "unnormalized" probability density of each of them $|\Psi(a)|^2$, and we want to reconstruct the notion of time.

But let us first stress that Quantum Gravity is formulated in such a way that the notion of time is not used and that this fact stems naturally from the whole philosophy of this theory. In fact, time in General Relativity is only a coordinate, not an observable like a, thus the results that the theory gives, like $\Psi(a)$, must only be a function of the observables, like a or other geometrical parameters if the metric is not isotropic, and not of the coordinates like t or x^μ. That is completely true but it is also true that we can use coordinates in the intermediate steps if they disappear from the final results. This will be our philossophy, we can define and use a physical coordinate time, that will be absent from the final quantum gravity results but will be present if we give semiclasical results, because in this case the time will be an observable that can be measued with a physical device, the real clock.

Then let us try to reconstruct the notion of time, at the quantum gravity level, in a heuristic way. Let us first order all the metrics by the growing of a, small a first and big a latter, from 0 to infinite (fig. 1.a) (i.e. let us adopt the cosmological arrow of time, we will return to this point latter on). Now the ordered spheres are three dimensional manifolds S^3 (fig. 1.a) and we want to reconstruct a four dimensional manifold where time can be defined. Then let us give to all S^3 a "time thickness" $\Delta\theta$ equal to all of them (fig. 1.b) and let us "glue" all the S^3 together (fig. 1.c), they will form a four dimensional manifold V^4 that we can consider the spacetime.

The time thickness $\Delta\theta$ of the slices added together will give a time parameter θ of V^4, let us compute this parameter. In the interval a, $a + da$ there are a number of spheres proportional to $|\Psi(a)|^2\mu(a)da$, where $\mu(a)$ is the measure of a, then the corresponding $d\theta$ will be

$$d\theta = \theta_0|\Psi(a)|^2\mu(a)da, \tag{1.2}$$

where θ_0 is an arbitrary constant that we shall choose latter on. Then θ is

$$\theta = \theta_0 \int_0^a |\Psi(a)|^2\mu(a)da. \tag{1.3}$$

We shall call this parameter the "probabilistic time".

This procedure gives a "frequency" interpretation to $|\Psi(a)|^2$ because in the manifold V^4, that we have constructed, if we want to know "how many" metrics there are in the interval a, $a + da$ we can cut the manifold in n slices of "time thickness" $\Delta\theta$ and count them, and the result will be proportional to $|\Psi(a)|^2\mu(a)$.

Therefore we have introduced the notion of time in Quantum Gravity in an exact way because there is no approximation neither in this procedure nor in eq. (1-3).

It is remarkable that the time parameter introduced in this heuristic, but exact way, has several very important properties in the classical and semiclassical limits:

i) It turns out to be the classical time in the classical limit if we fix the constant θ_0 properly (cf. [4], [6] and [7]).

ii) It plays the role of time in the Schrödinger equation in the semiclassical limit ([6], [7], [10] and [11]).

iii) From (1-3) we can obtain $a = a(\theta)$. It is the classica soution of the Einstein equation in the classical limit and the solution of the Einstein equation with the back reaction term in the semiclassical limit [9].

Thus θ is an exact notion with a logical classical limit behaviour but with properties (namely **iii**) that go beyond the classical domain. But certainly time is a very rich physicall structue and we would like to see if the just introduced "probabilistic time" has all the properties that normal time has, and how this new notion can be used to solve some of the deep problems of Quantum Gravity.

Perhaps the most fundamental problem of Quantum Gravity is to combine the apparently antagonics Einstein's ideas, on one side, with the ones of Born and Schrödinger, on the other, in a harmonic scenario, because Quantum Gravity consists, at least in its Hamiltonian version, in the study using ordinary Quantum Mechanics of a physical system gravity, whose classical model is General Relativity. The collision of these two formalisms, Quantum Mechanics (QM) and General Relativity (GR) produces the welll known set of problems that make Quantum Gravity

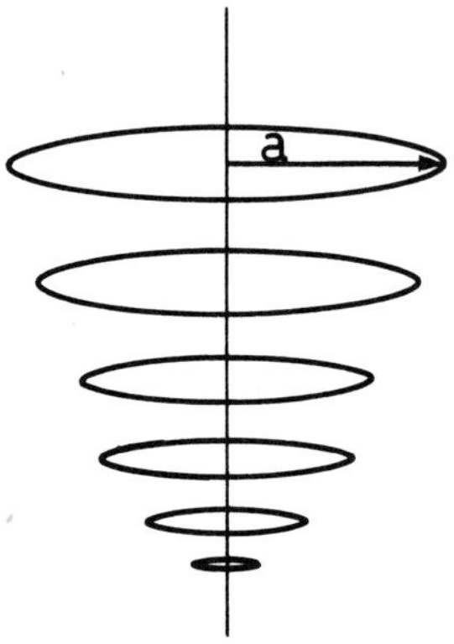

Fig.1a

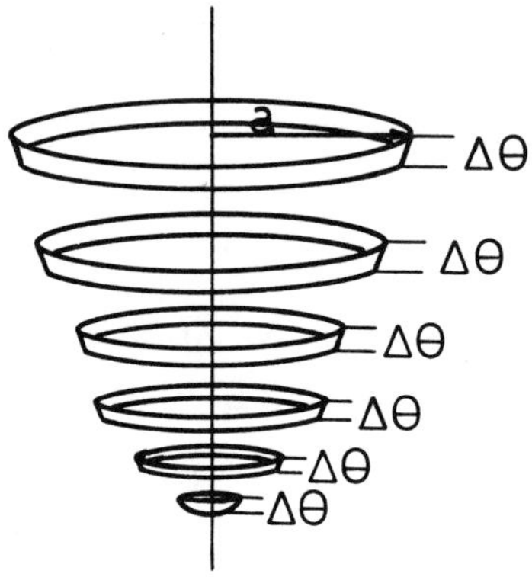

Fig.1b

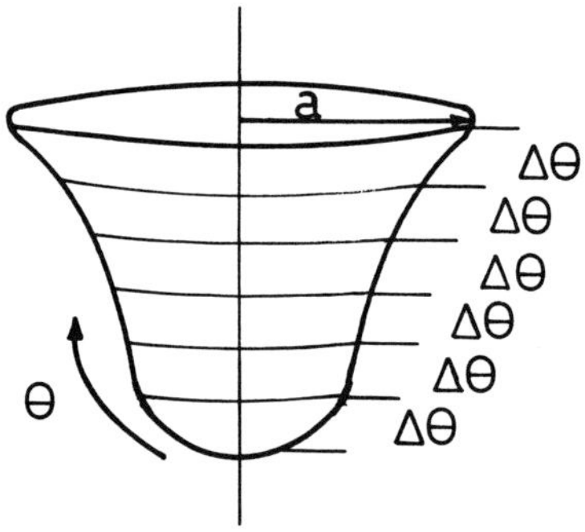

Fig. 1c

so difficult. In fact, the apparent incompatibility of these formalisms, QM and GR, becomes critical when the role played by the time notion is analized.

-In QM, time is an absolute parameter, a c-numbe and therefore different to all other coordinates that turn out to be observables.

-In GR, time is just a coordinate, completely eqivalent to any other coordinate. Thus no physical property makes time a privilegiated variable.

Of course we must emphasize that we cannot perform a canonical quantization if this privileged absolute time, that we shall call Hamiltonian time, is missing. An ordinary and undifferenciated GR time coordinate certainly cannot play that role (our candidate for Hamiltonian time is of course Probabilistic Time).

Following paper [12] we can say that Hamiltonian time has two fundamental properties.

a. It is an ordering parameter. As we know measurements make the wave functions colapse (at least in the small subsystems inside the universe because a measurement of all the universe is a controversial subject). The wave function "before" the measurement is different than the wave function "after" the measurement. The Hamiltonian time, that in this case is not a dynamical variable but just an ordering parameter, tells what the quantum state is that corresponds to the period "before" the measurement and what the other one is that corresponds to the period after the meassurement. This feature of time is not dynamical but it is essential to deal with the colapse phenomenon. In a Quantum theory with no time we cannot even think about colapse.

b. It has a "Herakleitian" property. Herakleitus stated that "Time is that which allows contradictory things to occur". That a particle would be in a point of coordinate x is, in principle contradictory with the fact that the same particle would be at $x' = x$ because the particle cannot be simultaneously at x and at x'. Nevertheless it can be at x at a certain time t and at x' at another time t'. Thus time makes possible contradictory situations.

But for a given time the particle must have one and only one position. We shall call Herakleitian any time parameter with this property, namely that for a given value of the time parameter the system under study can have one and only one configuration. Hamiltonian time is Herakleitian by definition but a generic time parameter could not be Herakeitian in general. This fact can be seen in fig. 2. In fig. 2.a we have drawn the space-time path, or history, of a partice in plane $t - x$, where t is the Hamiltonian ordinary time. Every horizontal straight line $t = const.$ intersects the history at one and only one point. In fig. 2.b we have drawn an arbitrary time parameter $\tau = \tau(t, x)$. As we can see the curves $\tau = const.$ intersect the history in an arbitrary number of points: zero, one, two,...

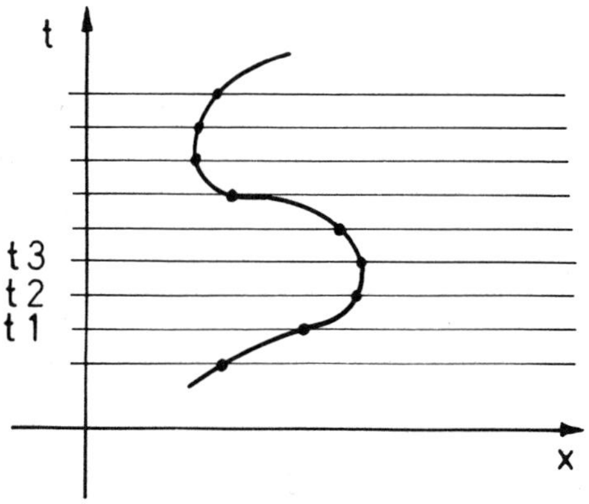

Fig. 2a

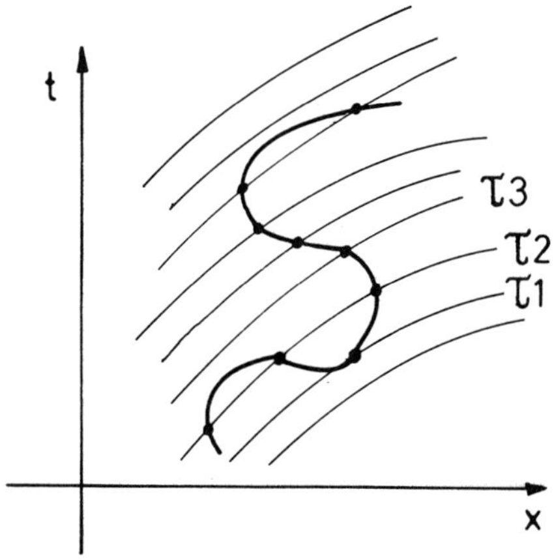

Fig. 2b

It is precisely the Herakleitian property of time that allows the normalization of the wave functions, that promotes the state space to a normed space and that yields the probabilistic intepretation. In fact, to introduce the notion of probability we need to have a **complete** set of **exclusive** events. Returning to the example of the particle moving in the plane $t - x$ we can say that at a certain Herakleitian time t, the particle cannot be at more than one point, the position of the particle at x and at $x' = x$ are **exclusive.** As all the possible histories intersect the horizontal straight line $t = const$ and we must consider all sort of histories, every point of this line is intersected by one or several histories. Thus the set of positions $x \epsilon R$ is complete. Now as we have a complete and exclusive set of particle positions we can nomalize the probability $|\Psi(x)|^2$ as

$$\int |\Psi(x)|^2 dx = 1 \tag{1.4}$$

and define a norm in the state space. If we do not have a Herakleitian time parameter all this process is impossible.

In QM we just **postulate** the existence of a time parameter endowed with properties **a** and **b**, i.e. the Hamiltonian time. Let us emphasize this fact, we just **postulate** that such a time exists, because no real clock can measure this parameter exactly. In fact only ideal clocks measure Hamiltonian time in an exact way but we do not have this kind of clocks in our laboratory, they are just ideal. We only have real clocks that measure Hamiltonian time in an approximate way. Really real clocks are not even Herakleitian because they can go "backward" in time. Even so we do not reject the notion of Hamiltonian time, we just postulate its existence and content ourselves to measure this time approximately.

On the other hand in GR we only have an infinite set of time parameters and we cannot say if some of them have properties **a** and/or **b**.

Our recipe to solve the problem will be that in Quantum Gravity, as in QM, we must postulate that a Hamiltonian time exists endowed with properties **a** and **b** (and this time will turn out to be our probabilistic time). We wil see how we can define and measure this time.

The lectures will be organized as follows.

Section 2: We shall study a toy model with the same patology as GR and find the natural way to quantize it. Then we shall see that if time is introduced in the model under some reasonable hypotheses it must be the probabilistic time.

Section 3: We shall introduce the Ehrenfest theorem.

Section 4: We shall study and ideal clock and see that it measures the probabilistic time in an exact way.

Section 5: We shall introduce a real clock and see that it measures the probabilistic time only in an approximated way.

Section 6: We shall use all what we have learnt from the toy model with real universe, first with Friedmann homogeneous and unperturbated universe and later on with a perturbated one.

Section 7: We shall study small subsystems of the universe and see that for them we obtain the usual physical laws.

Section 8: We shall study the problem of the arrow of time.

Section9: We shall obtain our main conclusions and list the main lines of research opened by these lectures.

2. A Toy Model

Following always paper [12] let us study our problem using an extremely simple toy model but with the same kind of patology of GR. Furthermore this model will be extremelly useful in all the lectures. Let us study the physical system with dynamical variables N, a, x and Lagrangian

$$\begin{aligned} &\mathcal{L}(\check{N}, N, \check{a}, a, \check{x}, x, \tau) = \\ &= NL\left(\frac{\check{a}}{N}, a, \frac{\check{x}}{N}, x\right) \end{aligned} \tag{2.1}$$

where "ˇ" symbolyzes the derivative with respect to the time parameter τ.

Let us observe that the Lagrangian has not an explicit time dependence, thus the corresponding Hamiltonian is a constant. Furthermore there is not an explicit $\check{N}$ dependence. Thus it is impossible to define the momentum P_N so we cannot use N as a variable in Hamilton's equations, N will only be a Lagrange multiplier.

Finally let us remark that there is an invariance group because the action is invariant under the transformation

$$\tau \to \tau'(\tau), \tag{2.2}$$

$$N \to N' = N\frac{d\tau}{d\tau'}. \tag{2.3}$$

Therefore neither τ nor N are physical observables because they can be substituted by τ' and N'. What kind of objects are τ and N exactly? τ is a coordinate in a one dimensional manifold M_1 and N is the only component of a covariant tangent vector of M_1. Thus Lagrangian (2.1) is similar to the GR Lagrangian because both Lagrangians are functions of coordinates, or components of vectors or tensors, defined on a manifold. Nevertheless lagrangian (2.1) can be studied using the ordinary methods of Cassical or Quantum Mechanics. The canonical momenta are

$$
\begin{aligned}
P_a &= \frac{\partial \mathcal{L}}{\partial \breve{a}} = L(1), \\
P_x &= \frac{\partial \mathcal{L}}{\partial \breve{x}} = L(3),
\end{aligned}
\tag{2.4}
$$

where $L(i)$ symbolizes the partial derivative of L with respect to its ith variable. The Hamiltonian reads

$$
\begin{aligned}
\mathcal{H} &= P_a \breve{a} + P_x \breve{x} - NL = \\
&= L(1)\breve{a} + L(3)\breve{x} - NL = NH,
\end{aligned}
\tag{2.5}
$$

where

$$
H = \left[L(1)\frac{\breve{a}}{N} + L(3)\frac{\breve{x}}{N} - L \right]. \tag{2.6}
$$

Let us see how we can interpretate H. From eq. (2.3) we can see that from the couples (τ, N), (τ', N'), (τ'', N''), ... we can construct an invariant because

$$
N d\tau = N' d\tau' = N'' d\tau'' \cdots \tag{2.7}
$$

We will call this invariant dt. t belongs to the special couple $(t, 1)$ and for this couple the Lagrangian (2.1) is simply $L(\dot{a}, a, \dot{x}, x)$ (where now the over dot symbolizes the derivative with respect to t); and H is the corresponding Hamiltonian. As in H there is not an explicit dependence of t, H is also a constant.

Finally if we make the variation of the action with respect to the Lagrange multiplier N we obtain

$$
\frac{\partial \mathcal{L}}{\partial N} = L - \frac{1}{N}\left[L(1)\dot{a} + L(3)\dot{x}\right] = -H = 0. \tag{2.8}
$$

Thus "on shell" both $H = 0$ and $\mathcal{H} = 0$, these equations are the constraint of the system.

Let us now go to the quantum study of the system. Let us first observe that if we treat Lagrangian (2.1) as a Quantum Gravity Lagrangian we must erase every notion of time and as N is a gauge variable the wave function must be $\phi(a, x)$, i.e. a function of the observables of the system a and x. With this philosophy the field equation would be

$$
H\phi(a, x) = 0, \tag{2.9}
$$

because $H = 0$ is the classical constraint. This is what we actually do in Quantum Gravity and eq. (2.9) would be the corresponding Wheeler-De Witt equation. If one of the time parameters is privileged because is Herakleitian the symmetry is broken and the constraint desappears. The natural procedure would be to use the postulates of QM and write the Schrödinger equation (cf. [23])

$$i\frac{\partial\Psi}{\partial\tau} = \mathcal{H}\Psi = NH\Psi, \tag{2.10}$$

where now $\Psi = \Psi(a, x, N, \tau)$ and we shall suppose that $\mathcal{H}$ or H are hermitian operators. But as we said in the introduction as N and τ are not observables but coordinates, a physical result like the probability $|\Psi|^2$ cannot be a function of these variables. Coordinates can be used in the intermediate calculation but must desappear in the final results. By the way, this is the state of affairs in GR that we would like to extend also to Quantum Gravity. But as $dt = Nd\tau$ we can write eq. (2.10) as

$$i\frac{\partial\Psi}{\partial t} = H\Psi, \tag{2.11}$$

where now Ψ is only a function of three variables:

$$\Psi = \Psi(t, a, x) = \Psi(\int Nd\tau, a, x). \tag{2.12}$$

The three variables are gauge invariant and the coordinates N and τ appear now only through the invariant combination $\int Nd\tau$. Now as H is a constant we can compute the eigen equation

$$H\phi_n(a, x) = E_n\phi_n(a, x) \tag{2.13}$$

and the general solution of the problem will be

$$\begin{aligned}\Psi(t, a, x) &= \sum_n e^{-iE_n(t-t_0)}\phi_n(a, x) = \\ &= \sum_n e^{-iE_n\int_{\tau_0}^{\tau} Nd\tau}\phi_n(a, x).\end{aligned} \tag{2.14}$$

If $E_0 = 0$ is one of the eigen values we obtain the Wheeler-De Witt equation

$$H\phi_0(a, x) = 0 \tag{2.15}$$

as the time independent Schrödinger equation for $E_0 = 0$ (as it was already proposed in work [13]).

Now we must remark that not all the times τ can be Herakleitian because they can change in an arbitrary way and τ can go backwards ($N < 0$). Thus we cannot say that these time parameters have properties **a** and **b**. We can only postulate that the time parameter endowed with these properties is t, that we can consider the Hamiltonian time of our system. Of course we cannot verify this postulate because we are only studying a toy model. But if time $t = \int N d\tau$ has a distinct physical role we must somehow measure it. According to the ordinary interpretation of Quantum Gravity based in Wheeler-De Witt equation (2.9) we do not have a time, so we do not need to measure it. On the contrary if we take as a field equation the Schrödinger equation (2.10) or (2.11), we cannot say that only the time independent equation (2.15) has a physical meaning. Thus we must somehow measure the time t and therefore we must find a "clock" that measures t. We have only two dynamical variables a and x that can be used as clock hands. Among them we must choose the one that would have the most Herakleitian behaviour, i.e. the one we believe that oscilates, the less or the more massive one. Let us suppose that a is this variable (when we will study the Quantum Gravity case this concept will be clearer: a is the radius of the universe, the more "massive" variable). Thus we would like to find a function $\theta(a)$ such that

$$t = \theta(a). \tag{2.16}$$

Now let us suppose that a solution of the Schrödinger equation (2.11) exists such that

$$\Psi(t, a, x) = D(t, a)\phi_0(a, x), \tag{2.17}$$

where $D(t, a)$ is a function that peaks sharply at $t = \theta(a)$ and it is almost zero if $t \neq \theta(a)$ and $\phi_0(a, x)$ is the ground state of eq. (2.15). This supposition can be either right or wrong depending on the system Lagrangian L. We will see that in the relevant examples (ideal and real clocks, and the homogeneous and inhomogeneous universes) this supposition is always right in an exact or approximated way. Furthermore, let us remark that equation (2.17) means that we can separate our system in two non- interacting subsystems, the clock with wave function $D(t, a)$ and the rest of the system (the "universe") with the wave function $\phi_0(a, x)$ corresponding to its ground state. It seems that this separation is always possible at least in an approximated way.

As H is hermitian we can define the ordinary Schrödinger inner product that turns out to be t-invariant:

$$< \Psi, \Psi' >= \int\int \Psi^* \Psi' \mu(a) da dx, \tag{2,18}$$

where $\mu(a)$ is the measure of variable a, and we take the measure of variable x equal to one for simplicity (this inner product can be eventually divergent but we know how to solve this problem [9]).

Now $D(t, a)$ is the wave function of the clock. Therefore it must be something like

$$D(t, a) = \triangle(t, t - \theta(a)), \tag{2.19}$$

where $|\triangle(t,\theta)|$ is a function like the one of fig. 3, where $A(t)$ and $B(t)$ are functions of t and $B(t) \ll A(t)$. In this way, in almost all the cases, $t \simeq \theta(a)$ and t will be measured by a via the function $a(t)$. Furthermore we will suppose $\triangle$ is normalized in the inner product

$$((\triangle,\triangle')) = \int \triangle^* \triangle' d\theta, \tag{2.20}$$

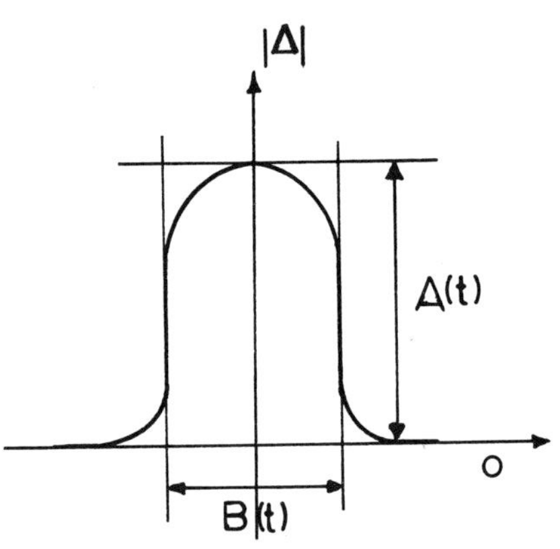

Fig.3

thus

$$\int |\triangle(t-\theta)|^2 d\theta \simeq B(t)A(t)^2 = 1. \tag{2.21}$$

Of course we cannot prove this property if we do not know the Lagrangian L, but we will see that this property can be proved in the examples studied below.

Now if F(θ) is an arbitrary but well behaved function we have

$$\begin{aligned}\int F(\theta)|\triangle(t-\theta)|^2 d\theta \simeq F(t)\int |\triangle(t-\theta)|^2 d\theta \simeq \\ \simeq A(t)^2 B(t) F(t) \simeq F(t)*\end{aligned} \tag{2.22}$$

*Thus we can say that $|\triangle(t-\theta)|^2 \simeq \delta(t-\theta)$ but for the moment we prefer our formalism.

Then the norm corresponding to the inner product (2.18) reads

$$
\begin{aligned}
< \Psi, \Psi > &= \int\int |\triangle(t, t-\theta)|^2 |\phi_0(a, x)|^2 \mu(a) da dx = \\
&= \int\int |\triangle(t, t-\theta)|^2 |\phi_0[a(\theta, x)]|^2 \frac{da(\theta)}{d\theta} \mu[a(\theta)] d\theta dx = \\
&= \int |\phi_0[a(t), x]|^2 \frac{da(t)}{d\theta} \mu[a(t)] dx = \\
&= \mu[a(t)] \frac{da(t)}{d\theta} \int |\phi_0[a(t), x]|^2 dx = \\
&= const.
\end{aligned}
\tag{2.23}
$$

Now if we define an x-inner product as:

$$(\phi, \phi') = \int \phi^* \phi' dx \tag{2.24}$$

and if all the suppositions we have made make eq. (2.23) valid, this equation reads

$$\mu[a(t)] \frac{da(t)}{d\theta} (\phi_0, \phi_0) = \theta_0^{-1} = const., \tag{2.25}$$

where θ_0 is an arbitrary constant. We can now calculate function $\theta = \theta(a)$, it is

$$\theta = \theta_0 \int_0^a (\phi_0, \phi_0) \mu(a) da, \tag{2.26}$$

i.e., it is the probabilistic time introduced in eq. (1.3) for our toy model. In fact in a general case we can have not one but n variables $x, x^1, x^2, ..., x^n$, and all will be the same with the substitutions $x \to x^i, dx \to dx^n$. Thus eq. (1.3) corresponds to the simplest case $n = 0$, but equations similar to (2.25) with one x can be found in papers [5] to [8] (x is the scalar field φ) and with many $x's$ in paper [9].

Then if all hypotheses introduced are fulfilled i.e.:

i) We can separate our system into a clock and the "universe", as in eq. (2.17).

ii) $D(t, a)$ is the wave function of a clock, namely a function like the one of eq. (2.19) and fig. 3.

iii) $\triangle$ are normalized with the product (2.20).

we can say that Hamiltonian time t is measured by the hand a via the function $t = \theta(a)$ and this measure is more (less) precise if function $\triangle(t, t-\theta)$ is more (less) sharp. The wave function $\Psi(t, a, x)$ can be written as

$$
\begin{aligned}
\Psi(t, a, x) &= \triangle(t, t-\theta) \phi_0(a, x) = \\
&= \triangle(t, t-\theta) \phi_0[a(\theta), x] = \psi(t, \theta, x),
\end{aligned}
\tag{2.27}
$$

where in the last expression we have changed a by θ using $a = a(\theta)$. For t = 0 the wave functional would be

$$\psi(0,\theta,x) = \triangle(0,-\theta)\phi_0[a(\theta),x]. \tag{2.28}$$

We can take as an initial state at t = 0 an eingenstate of variable θ with eigenvalue 0, i.e. to make $\triangle(0,-\theta)$ extremely sharp such that

$$\theta\Psi(0,\theta,x) = 0. \tag{2.29}$$

In this way we have synchronized t and θ because when $t = 0$ it is also $\theta = 0$. Depending on Lagrangian $\mathcal{L}$, when time passes, the function $\triangle$ will keep its shape or, in general, will be smeared and the measurement of t will become less exact. Anyhow in this way the Hamiltonian time can be measured and this measure yields probabilistic time. We only need to verify that L is such that the three introduced hypotheses would be fulfilled in all the relevant cases as we will see below.

But first, we will see that there is some ambiguity in the reasoning of this section and we will see how to eliminate it.

3.Ehrenfest Theorem

If we review the demostration of the last section we can realize that, even if the hypothesis **(ii)** is sufficient to obtain the probabilistic time definition (2.26), it is not completely necessay. In fact let us suppose that hypothesis **(i)** is valid and also that we define probabilistic time as in eq.(2.26), then eq. (2.25) is valid. So from hypothesis **(i)** and normalization condition $<\Psi,\Psi>= const.$, we can obtain

$$\begin{aligned}
&\int\int |D(t,a)|^2|\phi_0(a,x)|^2\mu(a)dxda = \\
&= \int\int |D[t,a(\theta)]|^2|\phi_0[a(\theta),x]|^2\mu(a)\frac{da}{d\theta}d\theta dx = \\
&= \int |D[t,a(\theta)]|^2\mu(a)\frac{da}{d\theta}(\phi_0,\phi_0)d\theta.
\end{aligned} \tag{3.1}$$

Then from eq. (2.25)

$$\int |D[t,a(\theta)]|^2 d\theta = const. \tag{3.2}$$

Then only **(iii)** is necessary and not **(ii)** to obtain definition (2.26), i.e. function D must be only a normalized one and must not be a peaked function when $t = \theta(a)$. Even so the idea of the measurement of t using a clock with wave function $D(t,a)$ yields necessarily to a peaked wave function and not only a normalized one.

Furthermore if **(ii)** is valid we can prove Ehenfest theorem as follows.

Let us compute the mean value of a for the time variable $\Psi(t,a,x)$ We have

$$<\Psi|a|\Psi>= \int\int a|\triangle|^2|\phi_0|^2\mu(a)dadx =$$

$$= \int a(\theta)|\triangle(t,t-\theta)|^2(\phi_0,\phi_0)\mu(a)\frac{da}{d\theta}d\theta = \tag{3.3}$$
$$= \theta_0^{-1}\int a(\theta)|\triangle(t,t-\theta)|^2 d\theta = \theta_0^{-1}a(t),$$

where we use the fact that $\triangle$ is a peaked function according to hypothesis **(ii)**.

As it is explained in papers [7] and [9] the real physical object is $\theta_0^{1/2}\Psi$ and not Ψ, thus the real mean value is

$$\theta_0 < \Psi|a|\Psi >= a(t), \tag{3.4}$$

i.e., Ehrenfest theorem for a. This fact shows that $a(t)$ can be considered as the mean value of operator a contrary to the opinion in paper [1].

We can now compute the mean vaue of x:

$$< \Psi|x|\Psi >= \int\int x|\triangle|^2|\phi_0|^2\mu(a)dadx. \tag{3.5}$$

But we can define

$$\frac{\int x|\phi_0(a,x)|^2}{\int|\phi_0(a,x)|^2dx} = \tilde{x}(a), \tag{3.6}$$

i.e. the mean value of x after a measurement of a or, in other words, the mean value computed in the eigenspace of the eigenvalue a. Then

$$< \Psi|x|\Psi > = \int|\triangle(t,t-\theta)|^2\tilde{x}[a(\theta)]\mu(a)\frac{da}{d\theta}d\theta(\phi_0,\phi_0) =$$
$$= \theta_0^{-1}\int|\triangle(t,t-\theta)|^2\tilde{x}[a(\theta)]d\theta = \theta_0^{-1}\tilde{x}[a(t)], \tag{3.7}$$

thus

$$\theta_0 < \Psi|x|\Psi >= \tilde{x}[a(t)], \tag{3.8}$$

i.e. Ehrenfest theorem for x. In both demonstrations we have used the fact that $\triangle$ is a peaked function, thus hypothesis **(ii)** is necessary not only to use the clock idea but to obtain the classical limit using Ehrenfest theorem.

4.The Ideal Clock

Let us now see how the three hypotheses of section **3** are fulfilled in several examples. Let us begin with a well known example, the ideal clock. In this case the Hamiltonian H of eq. (2,.11) is

$$H = -i\frac{\partial}{\partial a} + h(a,x), \tag{4.1}$$

where $h(a,x)$ is a selfadjoint operator that is only a c-number with respect to a (i.e. it has not a-derivatives). Then in this case eq. (2.11) reads

$$i\frac{\partial\Psi}{\partial t} = -i\frac{\partial\Psi}{\partial a} + h(a,x)\Psi. \tag{4.2}$$

Then, let us introduce the decomposition (2.17) in order to see if hypothesis **(i)** is fulfilled. We obtain

$$i\phi_0 \frac{\partial D}{\partial t} = -iD\frac{\partial \phi_0}{\partial a} - i\phi_0 \frac{\partial\ D}{\partial a} + Dh\phi_0. \tag{4.3}$$

But $H\phi_0(a,x) = 0$, then the equation for D reads

$$\frac{\partial D}{\partial t} + \frac{\partial D}{\partial a} = 0. \tag{4.4}$$

This equation can be satisfied by a function $D(t,a)$ thus hypothesis **(i)** is satisfied. Moreover $D(t,a)$ must be

$$D(t,a) = \triangle(t-a), \tag{4.5}$$

where $\triangle$ is a one variable function. If we take at t $= 0$, $D(0,a) = \delta(a)/[\delta(0)]^{1/2}$ we obtain for all times

$$D(t,a) = \frac{\delta(t-a)}{[\delta(0)]^{1/2}}. \tag{4.6}$$

This function in fact satisfies hypotheses **(ii)** and **(iii)** thus the three hypotheses are satisfied and

$$\Psi(t,a,x) = \frac{\delta(t-a)\phi_0(a,x)}{[\delta(0)]^{1/2}} \simeq \phi_0(t,x). \tag{4.7}$$

In this case $t = \theta(a)$ is simply $t = a$ and the ideal clock measure time t in a completely exact way. As $t = a$, $\phi_0(t,x)$ satisfies the ordinary Schrödinger equation

$$i\frac{\partial \phi_0}{\partial t} = h(t,x)\phi_0. \tag{4.8}$$

Therefore in this case the ideal clock with "hand" a works perfectly and measures Hamiltonian time t precisely. If we have and ideal clock from the toy model we can obtain the ordinary Schrödinger equation (4.8) and ordinary QM.

There is however a small problem: h is a selfadjoint operator but H is not. Thus in principle $< \Psi'; \Psi >$ is not a $t - constant$. Precisely

$$i\frac{\partial}{\partial t} < \Psi'; \Psi > = < \Psi'; H\Psi > - < H\Psi'; \Psi > \neq 0. \tag{4.9}$$

Nevertheless *

$$\begin{aligned} < \Psi'; \Psi > &= \int\int \Psi'^{*} \Psi da dx = \\ &= \int\int \frac{\delta(t-a)}{\delta(0)} \delta(t-a) \phi'^{*} \phi da dx \simeq \\ &\simeq \int \phi'^{*} \phi dx = (\phi', \phi). \end{aligned} \tag{4.10}$$

And as $t = a$

$$\frac{\partial < \Psi'; \Psi >}{\partial t} \simeq \frac{\partial (\phi', \phi)}{\partial a} = 0, \tag{4.11}$$

because h is a selfadjoint operator. As in this case $(\phi_0, \phi_0) = const.$, eq. (2.26) reads

$$\theta = \theta_0 a \simeq a \tag{4.12}$$

and a is also the probabilistic time. Thus probabilistic time turns out to be an exact definition when the clock is ideal.

Another, much more serious problem is that there are not ideal clocks in nature because the kinetic component of Hamniltonian (4.1), p_a, is not bounded from below. What we can find in nature are kinetic terms like the one of a particle of mass m, $p_a^2/2m$, that are bounded from below. To measure the Hamiltonian time exactly is only an idealization that can not be performed in a real experimental world. Therefore we turn now to the problem of measuring Hamiltonian time in a approximated way with real clocks.

5. Real Clock

Let us now study the most interesting example: the real clock. In fact from this example it will inmediately follow from the Quantum Gravity case. Let H be the following equation

$$H = -\frac{f(a)}{\mu'_a(a)}\frac{\partial}{\partial a}\frac{1}{\mu'(a)}\frac{\partial}{\partial a} + h(a, x), \tag{5.1}$$

where again $h(a, x)$ is a selfadjoint operator, but a c-number for a, $\mu'_a(a)$ is a "measure" of a considered as a variable in M_1, a one dimensional manifold and $f(a)$ and arbitrary function with a role that will be explained in the next section.

The field equation with the decomposition (2.17) reads

$$\begin{aligned} i\phi_0\frac{\partial D}{\partial t} = &-\frac{D}{\mu'_a(a)}f(a)\frac{\partial}{\partial a}\frac{1}{\mu'_a(a)}\frac{\partial \phi_0}{\partial a} - 2\frac{f(a)}{\mu'^2_a(a)}\frac{\partial D}{\partial a}\frac{\partial \phi_0}{\partial a} - \\ &- \phi_0\frac{f(a)}{\mu'_a(a)}\frac{\partial}{\partial a}\frac{1}{\mu'_a(a)}\frac{\partial D}{\partial a} + Dh\phi_0 \end{aligned} \tag{5.2}$$

But $H\phi_0 = 0$, i.e.

$$-\frac{f(a)}{\mu'_a(a)}\frac{\partial}{\partial a}\frac{1}{\mu_a(a)}\frac{\partial \phi_0}{\partial a} + h\phi_0 = 0. \tag{5.3}$$

Therefore the equation for D reads

$$\begin{aligned} i\phi_0\frac{\partial D}{\partial t} = &-\phi_0\frac{f(a)}{\mu'_a(a)}\frac{\partial}{\partial a}\frac{1}{\mu'_a(a)}\frac{\partial D}{\partial a} - \\ &- 2\frac{f(a)}{\mu'_a(a)}\frac{\partial \phi_0}{\partial a}\frac{1}{\mu'_a(a)}\frac{\partial D}{\partial a}. \end{aligned} \tag{5.4}$$

* Let $\mu(a) = 1$ in this case

Now if we devide this equation by ϕ_0 we obtain that

$$\frac{1}{\phi_0(a,x)}\frac{\partial\phi_0(a,x)}{\partial a} \tag{5.5}$$

must be a function of a only and not of x because $D = D(t,a)$ and there are only t and a derivatives in eq. (5.4). Therefore we cannot satisfy in general eq. (5.4) or, what is equivalent, hypothesis **(i)** cannot be fulfilled in general, thus all the procedure stated in section **2** fails in this case. Nevertheless this procedure can be followed in an approximated way if we compute ϕ_0 using a Born-Oppenheimer approximation (cfr. [1]), i.e. we write ϕ_0 as

$$\phi_0(a,x) = e^{iK(a)}J(a,x) \tag{5.6}$$

and we suppose that

$$\frac{1}{J(a,x)}\frac{\partial J(a,x)}{\partial a} \ll \frac{\partial K(a)}{\partial a}. \tag{5.7}$$

Then

$$\frac{\partial\phi_0}{\partial a} \simeq ie^{iK(a)}K(a)J(a,x) \tag{5.8}$$

and in fact

$$\frac{1}{\phi_0(a,x)}\frac{\partial\phi_0(a,x)}{\partial a} \simeq iK(a). \tag{5.9}$$

Thus real clocks only work in a approximated way because hypothesis **(i)** is only satisfied approximately.

As H is selfadjoint from $H\phi_0 = 0$ we obtain $H\phi^* = 0$. Thus if zero is a non degenerate eigenvalue, ϕ_0 is real. We will follow the computation in this case but the case 0 complex is not difficult as can be seen in appendix **I**. If ϕ_0 is real and we multily eq. (5.4) by ϕ_0 and integrate on dx using the inner product of eq. (2.24) we get

$$i(\phi_0,\phi_0)\frac{\partial D}{\partial t} = -\frac{f(a)}{\mu_a'(a)}\frac{\partial}{\partial a}\left[\frac{(\phi_0,\phi_0)}{\mu_a'(a)}\right]\frac{\partial D}{\partial a}. \tag{5.10}$$

But we can change variable a to b and, as under a change of coordinate $\mu'(a)da = \mu'(b)db = \ldots$ is invariant, we can choose a variable b such that

$$\frac{(\phi_0,\phi_0)}{\mu_b'(b)} = B = const. \tag{5.11}$$

Thus equation (5.10) reads

$$i\frac{\partial D}{\partial t} = -\frac{f(b)}{[\mu_b'(b)]^2}\frac{\partial^2 D}{\partial b^2}. \tag{5.12}$$

Therefore the equation that D must satisfy is a Schrödinger equation with only a kinetic term with variable mass (if $f(b) > 0$ as it will be in the examples). Then hypothesis **(ii)** is verified because this equation has a peaked solution, if we give a peaked initial condition we will have a peaked solution that will spread on the course of time. Also this solution can be normalized in the variable b and in another variable $\theta = \theta(b)$ too if $d\theta/db$ is suficiently regular when $|b| \to \infty$ as will be the case. Thus hypothesis **(iii)** is also verfied (see Apendix **II**). Then we can say, using the conclusions of section **2**, that t is measured by the probabilistic time given by eq. (2.26).

But let us try to obtain this result in a different way. Now we know that a function $\triangle$ exists such that

$$D(t,b) = \triangle[t, t - F(b)] \tag{5.13}$$

that sharply peaks when $t = F(b)$. This last equality can be written as $b = F^{-1}(t)$. Let us call $\tau = F^{-1}(t)$. Eq. (5.13) can be written as:

$$D(t,b) = \triangle(\tau, \tau - b). \tag{5.14}$$

But $\partial^2\triangle/\partial b^2$ is also a function that sharply peaks at $\tau - b$, then eq. (5.12) reads

$$i\frac{\partial\triangle}{\partial t} = -\left[\frac{f(b)}{\mu_b'(b)^2}\right]\frac{\partial^2\triangle}{\partial b^2} = -\frac{f(\tau)}{[\mu_b'(\tau)]^2}\frac{\partial^2\triangle}{\partial b^2} \tag{5.15}$$

because when $\tau \neq b$ the equation is satisfied because both members are more or less equal to zero, thus we must only be woried by the case $\tau = b$. Now we can compute the parameter τ. It turns out to be

$$\tau = \int \frac{f(t)}{[\mu_b'(t)]^2}dt. \tag{5.16}$$

In fact, with this definition of τ, eq. (5.15) reads

$$i\frac{\partial\triangle}{\partial\tau} = -\frac{\partial^2\triangle}{\partial b^2}, \tag{5.17}$$

which has the solution

$$\triangle(\tau, b) \simeq exp\left[\frac{-(b-\tau)^2}{(\delta_0^2 + 4i\tau) - (i/2)(b-\tau)}\right] \tag{5.18}$$

that is, in fact, sharply peaked when $\tau = b$. Now the relation between parameter a and b is

$$\frac{da}{db} = \frac{\mu'_b}{\mu'_a} = \frac{(\phi_0, \phi_0)_{(a)}}{B\mu'_a(a)}. \tag{5.19}$$

Then

$$b = B\int_0^a \frac{\mu'_a(a)}{(\phi_0, \phi_0)_{(a)}} da. \tag{5.20}$$

Now from eqs. (5.11) and (5.16) we get

$$\tau = B^2 \int_0^t \frac{f(t)}{(\phi_0, \phi_0)^2_{(t)}} dt. \tag{5.21}$$

As function $\triangle(\tau, \tau - b)$ has its maximum when $\tau = b$ or

$$\int_0^a \frac{\mu'_a(a)}{(\phi_0, \phi_0)_{(a)}} da = B\int_0^t \frac{f(t)}{(\phi_0, \phi_0)^2_{(t)}} dt \tag{5.22}$$

if we derivate with respect to a we obtain

$$\begin{aligned} &\frac{d}{da}\int_0^a \frac{\mu'_a(a)}{(\phi_0, \phi_0)_{(a)}} da = \\ &= B\frac{dt}{da}\frac{d}{dt}\int_0^t \frac{f(t)}{(\phi_0, \phi_0)^2_{(t)}} dt, \end{aligned} \tag{5.23}$$

i.e.

$$\frac{\mu'_a(a)}{(\phi_0, \phi_0)_{(a)}} = \frac{dt(a)}{da} B \frac{f[t(a)]}{(\phi_0, \phi_0)^2_{[t(a)]}}. \tag{5.24}$$

Thus t is a function of a, precisely

$$t \simeq \theta = \frac{1}{B}\int_0^a \frac{(\phi_0, \phi_0)_{(a)}}{f(a)} \mu'_a(a) da. \tag{5.25}$$

Thus we have again obtained eq. (2.26) making $\theta_0 = B^{-1}$, an arbitrary normalization constant, but we have the new factor $f(a)$. We can understand why this factor is there, the real measure of a, not considered as a variable in M_1, a one dimensional manifold, but a variable in superspace, is

$$\mu(a) = \frac{\mu'_{(a)}(a)}{f(a)}. \tag{5.26}$$

In fact, in Quantum Gravity the operators in the r.h.s. of eq. (5.1) will be the D'Alembertian in superspace:

$$\frac{1}{\sqrt{-G(a)}}\frac{\partial}{\partial a}\frac{\sqrt{-G(a)}}{a}\frac{\partial}{\partial a}. \tag{5.27}$$

Thus $\mu(a) = \sqrt{-G(a)}$.* Thus eq. (5.25) coincides with eq. (2.26).

Then we have again obtained a function $\triangle$ that peakes at $t = \theta(a)$ and verifies the three hypotheses of section **2**, but only approximately. Anynow the identification $t = \theta(a)$ stays, but it is not exact as in section **4** because function $\triangle(\tau, b)$ of eq. (5.18) is "spread" in the course of time and its "width" increases as

$$\delta^2 = \delta_0^2 + 4i\tau \simeq \delta_0^2 + 4i\tau(\theta). \tag{5.28}$$

Then what we have now is a "real clock" that allows us to identifie the hamiltonian time with the probabilistic time, but only as an approximation (cfr. Appendix **III**).

The standard real clock is just a usual particle with $\mu'_{(a)} = \sqrt{2m}$ and $f(a) = 0$. Thus the kinetic term reads

$$-\frac{f(a)}{\mu_{(a)}(a)}\frac{\partial}{\partial a}\frac{1}{\mu_{(a)}(a)}\frac{\partial}{\partial a} = -\frac{1}{2m}\frac{\partial^2}{\partial a^2} = \frac{p_a^2}{2m} \tag{5.29}$$

i.e., the kinetic energy of an ordinary particle with mass m.

6. Quantum Gravity

Now let us use the ideas developed in the above section in our main problem, Quantum Gravity. We will use the notation of paper [10] that we will now review. The metric reads

$$dS^2 - (N^2 - N_i N^i)d\tau^2 + 2N_i dx^i d\tau + h_{ij}dx^i dx^j, \tag{6.1}$$

where $i, j = 1, 2, 3$ are the indices corresponding to the tridimensional sections of space time, that we will consider only as compact. We can see that metric (6.1) is invariant under coordinate transformations in the three-dimensional sections, i.e.:

$$\begin{array}{ll} \tau \to \tau' = \tau, & x^i \to x'^i = x'^i(x^i), \\ N \to N' = N, & N^i \to N'^i = \frac{\partial x'^i}{\partial x^i}N_i, \end{array} \quad h_{ij} \to h'_{ij} = \frac{\partial x^i}{\partial x'^i}\frac{\partial x^j}{\partial x'^j}h_{ij}. \tag{6.2}$$

---

* $\mu'(a) = \frac{a}{\sqrt{-G(a)}}$; $f(a) = -a/G(a)$.

It is also invariant under changes of the parameter τ i.e.

$$\begin{aligned}\tau \to \tau' = \tau'(\tau), \qquad & x^i \to x'^i = x^i, \\ N \to N' = \tfrac{d\tau}{d\tau'}N, \quad & N^i \to N'^i = \tfrac{d\tau}{d\tau'}N^i, \quad h_{ij} \to h'_{ij} = h_{ij}.\end{aligned} \tag{6.3}$$

These are the invariances that we must keep in all the formalism. We can inmediately see that we have two invariants

$$\begin{aligned} "t(\tau, x^i)" &= \int N(\tau, x^i)d\tau, \\ t^i(\tau, x^j) &= \int N^i(\tau, x^j)d\tau. \end{aligned} \tag{6.4}$$

None of them can be used as Hamiltonian time because they are functions of x^i. Thus if we want to define a satisfactory Hamiltonian time we must use a mean value of $N(\tau, x^i)$ as we will see. Anyhow we do not have any physical role for $t^i(\tau, x^j)$, so it is reasonable to eliminate this quantities from our scenario, considering only the coordinates where $N^i = 0$. We shall follow this line, even so some alternative lines might exist. The action reads

$$S = \int (L_g + L_m)d^3x dt, \tag{6.5}$$

where the gravitational Lagrangian is

$$L_g = -\frac{m_p^2}{16\pi}N\left[G^{ijlk}K_{ij}K_{kl} + h^{1/2}R^{(3)}\right], \tag{6.6}$$

where $h = det(h_{ij})$ and

$$K_{ij} = \frac{1}{2N}\left[-\frac{\partial h_{ij}}{\partial \tau} + 2N_{(i,j)}\right], \tag{6.7}$$

$$G^{ijkl} = \frac{1}{2}h^{1/2}(h^{ik}h^{jl} + h^{il}h^{jk} - 2h^{ij}h^{kl}). \tag{6.8}$$

We adopt the following matter Lagrangian

$$\begin{aligned} L_m = &-\frac{1}{2}Nh^{1/2}\Big[\frac{1}{N^2}\left(\frac{\partial\varphi}{\partial\tau}\right)^2 - 2\frac{N^i}{N^2}\frac{\partial\varphi}{\partial\tau}\frac{\partial\varphi}{\partial x^i} - \\ &-\left(h^{ij} - \frac{N^iN^j}{N^2}\right)\frac{\partial\varphi}{\partial x^i}\frac{\partial\varphi}{\partial x^j} - m^2\varphi^2\Big]. \end{aligned} \tag{6.9}$$

We can see that the action introduced is invariant under transformations (6.2) and (6.3). The conjugated momenta are ($\dot{} = d/d\tau$):

$$\pi^{ij} = \frac{\partial L_g}{\partial \dot{h}_{ij}} = h^{1/2}\frac{m_p^2}{16\pi}\left(K^{ij} - h^{ij}K\right), \tag{6.10}$$

$$\pi_\phi = \frac{\partial L_m}{\partial \dot{\varphi}} = -\frac{h^{1/2}}{N^{1/2}}\left(\dot{\varphi} - N^i\frac{\partial\varphi}{\partial x^i}\right). \tag{6.11}$$

The Hamiltonian reads

$$H = \int \left(\pi^{ij}\dot{h}_{ij} + \pi_\phi \dot{\varphi} - L_g - L - m)\right) d^3x = \\ = \int (NH_0 + N_i H^i) d^3x, \tag{6.12}$$

where

$$H_0 = -\frac{16}{m_p^2} G_{ijkl}\pi^{ij}\pi^{kl} - \left(\frac{m_p^2}{16\pi}\right) h^{1/2} R^{(3)} + \\ + \frac{1}{2} h^{1/2} \left(\frac{\pi_\varphi^2}{h} + h^{ij}\frac{\partial\varphi}{\partial x^i}\frac{\partial\varphi}{\partial x^i} + m^2\varphi^2\right) \tag{6.13}$$

$$H^i = 2\pi^{ij}_{/j} + h^{ij}\frac{\partial\varphi}{\partial x^j}\pi_\varphi, \tag{6.14}$$

where the "/" symbolizes the covariant derivative in the 3-dimensional sections. The quantities N and N^i can be considered Lagrange multipliers. Thus the constraints are

$$H_0 = 0, H^i = 0. \tag{6.15}$$

To obtain the quantum theory we must substitute

$$\pi^{ij} \to -i\frac{\delta}{\delta h_{ij}(x)}, \pi_\varphi \to -i\frac{\delta}{\delta\varphi}. \tag{6.16}$$

Then H_0 and H^i become operators and we obtain an ordinary QM version for Quantum Gravity with a wave function $\phi(h_{ij}, \varphi)$ and the constraint $H_0 = 0$ becomes the main field equation of the theory, the Wheeler-De Witt equation:

$$H_0\phi = 0, \tag{6.17}$$

while operator H_i is the generator of the diffeomorphisms in the 3-dimensional spaces thus the corresponding quantum constraint

$$H_i\phi = 0 \tag{6.18}$$

means only that the theory is invariant under transformation (6.2).

We will now study the problem in two important examples.

6.1. Unperturbated Friedmann Model

In this case we will take the 3-spaces and the field φ homogeneous and isotropic. Thus metric (6.1) will be metric (1.1), i.e.

$$dS^2 = \sigma^2(-N^2 d\tau + a^2 d\Omega_3^2), \tag{6.1.1}$$

where a is the radius of the universe, $d\Omega_3^2$ the metric of the unit sphere and $\sigma^2 = (2/3)\pi m_{pl}^2$. Now N is only a function of a (and $N^i = 0$), therefore in this case we have only one invariant independent of x^i (cf. eq. (6.4)) that we can consider our Hamiltonian time (as in eq. (2.7)):

$$t(\tau) = \int N(\tau) d\tau. \tag{6.1.2}$$

Now we can rescale the mass and the potencial as $m \to \sigma^{-1} m$, $\varphi \to (2^{1/2}\pi\sigma)^{-1}\varphi$ and the new action reads

$$S = -\frac{1}{2}\int d\tau N a^3 \Big[\frac{1}{N^2 a^2}\left(\frac{da}{d\tau}\right)^2 - \frac{1}{a^2} - \frac{1}{N^2}\left(\frac{d\varphi}{d\tau}\right)^2 + m^2\varphi^2\Big], \tag{6.1.3}$$

where we have performed the x-integration. As we can again verify the new action is independent under the transformation (6.3) and the classical Hamiltonian reads

$$H = \frac{N}{2}\left(\frac{\pi_a^2}{a} - \frac{\pi^2\varphi}{a^3} - a^2 + a^3 m^2 \varphi^2\right). \tag{6.1.4}$$

The momenta are

$$\begin{aligned}\pi_a &= a\frac{da}{N d\tau}\\ \pi_\phi &= a^3\frac{d\varphi}{N d\tau},\end{aligned} \tag{6.1.5}$$

and the Wheeler-De Witt equation reads

$$H\phi = \frac{1}{2}Ne^{-3\alpha}\left(-\frac{\partial^2}{\partial\alpha^2} + \frac{\partial^2}{\partial\varphi^2} + 2V\right)\phi(a,\varphi), \tag{6.1.6}$$

where

$$\begin{aligned}\alpha &= \ln a,\\ V &= \frac{1}{2}\left(e^{6a}m^2\varphi^2 - e^{4a}\right).\end{aligned} \tag{6.1.7}$$

Now we can write the Schrödinger equation for the new interpretation

$$i\frac{d\Psi}{d\tau} = H\Psi = \frac{1}{2}Ne^{-3\alpha}\left(-\frac{\partial^2}{\partial\alpha^2} + \frac{\partial^2}{\partial\varphi^2} + 2V\right)\Psi, \tag{6.1.8}$$

if we introduce the Hamiltonian time this equation reads

$$i\frac{d\Psi}{dt} = \frac{1}{2}e^{-3\alpha}\left(-\frac{\partial^2}{\partial\alpha^2} + \frac{\partial^2}{\partial\varphi^2} + 2V\right)\Psi. \tag{6.1.9}$$

Then we can repeat the procedure of section **5**, the real clock technique, and decompose function Ψ in two factors

$$\Psi(t,a,\varphi) = \triangle(t - \theta(a))\phi_0(a,\varphi), \tag{6.1.10}$$

where ϕ_0 satisfies de Wheeler-De Witt equation (6.1.6) and $\triangle(t-\theta(a))$ is a function sharply peaked at $t = \theta(a)$. Then if we suppose that the three hypotheses of section

2 are satisfied, the probabilistic time (that can be computed from eq. (5.25) with $B^{-1} \to \theta'_0, f \to e^{-3\alpha}, \mu'_a \to 1, a \to \alpha$, where we take as a hand of the clock the radius of the universe) is:

$$\theta = \theta_0 \int_{-\infty}^{\alpha} (\phi_0, \phi_0)_{(\alpha)} e^{3\alpha} d\alpha = \theta_0 \int_{0}^{a} (\phi_0, \phi_0)_{(a)} a^2 da, \tag{6.1.11}$$

where

$$(\phi_0, \phi_0)_{(a)} = \int_{-\infty}^{+\infty} |\phi_0(a, \varphi)|^2 d\varphi. \tag{6.1.12}$$

In fact, in this case the supermetric reads

$$G_{AB} = \begin{pmatrix} a & 0 \\ 0 & -a^3 \end{pmatrix}, \tag{6.1.13}$$

where $A, B = 0, 1$ and 0 corresponds to a while 1 corresponds to φ. Then

$$t \simeq \theta = \theta_0 \int_{0}^{a} (\phi_0, \phi_0)_{(a)} \sqrt{-G} da = \theta_0 \int_{0}^{a} da \int_{-\infty}^{+\infty} d\varphi |\phi_0(a, \varphi)|^2 \sqrt{-G}. \tag{6.1.14}$$

We have obtained the probabilistic time for the Quantum Gravity case, introduced in papers [5] to [9], that can also be written as

$$t \simeq \theta = \theta_0 \int_{0}^{a} \mu_a da \int_{-\infty}^{+\infty} |\phi_0(a, \varphi)|^2 \mu_\varphi d\varphi, \tag{6.1.15}$$

with $\mu_a = \sqrt{a}$ and $\mu_\varphi = \sqrt{a^3}$ as it can be obtained from metric (6.1.13).

At this point two observations are in order.

First. That we have chosen a as the clock hand. Can we choose other hands? The only other candidate in our model is φ. But the three hypotheses of section **2** may be valid in order to arrive to eq. (6.1.15) (at least in this kind of treatmente). And we know that hypothesis **(i)** is only approximately satisfied because and equation similar to (5.6) is valid in this case, i.e.

$$\phi_0(a, \varphi) = e^{iK(a)} J(a, \varphi), \tag{6.1.16}$$

with

$$\frac{1}{J(a, \varphi)} \frac{\partial J(a, \varphi)}{\partial a} \ll \frac{\partial K(a)}{\partial a}. \tag{6.1.17}$$

This is a reasonable hypothesis for the field a because this variable has not an oscilatory nature, but it is not a natural hypothesis for the field φ that surely oscillates rapidly. Therefore φ cannot be a clock hand. Only a in this example, or other non oscilatory variables in other more complex examples, can be considered as logical candidate for hand. The oscilatory fields are ruled out.

Second. As $|\phi_0(a,\varphi)|^2\mu_a$ and μ_φ are positive, eq. (6.1.15) yields $d\theta/da > 0$, namely that variable θ follows the cosmological "arrow of time". But θ has its own arrow given by eq. (5.28) with $\tau = \tau(\theta)$ (the growing function (5.21)), the "arrow of spreading". In eq. (5.25) we are making the hypothesis that both arrows coincide and this hypothesis has not a physical base. We will discuss this problem further on.

6.2. The Perturbated Friedmann Model

If we like to go beyond the minisuperspace model studied in the preceding seccion to a more general case, we find the typical problem of Quantum Gravity: we do not know how to solve functional differential equations. Thus it is better to study the perturbated Friedmann model where we have an infinite set of modes but a discrete infinite set. Then, let us consider the metric

$$h_{ij} = a^2(\Omega_{ij} + \epsilon_{ij}), \tag{6.2.1}$$

where Ω_{ij} is the metric of a unit three-sphere and ϵ_{ij} can be considered as a perturbation that we can expand in spherical harmonics as

$$\epsilon_{ij}(x) = \sum_{A,n,l,m} a^{(A)}_{nlm} Q^{(A)n}_{ijlm}(x), \tag{6.2.2}$$

where $Q^{(A)n}_{ijlm}$ are spherical harmonic functions of different types on the three-sphere for the two-rank tensors (A = 1,2, ..., 6) corresponding to the six types of harmonic function: scalar, vector, tensor types (cfr. [10]);n, l, m are the typical quantum number of these functions and $a^{(A)}_{nlm}$ are coefficientes. N, N_i, and φ can also be expanded in spherical harmonics as

$$\begin{aligned} N &= N_0(\tau)\left(1 + \sum_{n,l,m} g_{nlm} Q^n_{lm}\right), \\ N_i &= a(\tau) \sum_{B,m,l,n} K^{(B)}_{nlm} Q^{(B)n}_{ilm}, \\ \varphi &= \sigma^{-1}\left(\frac{1}{2^{1/2}\pi}\varphi(\tau) + \sum_{n,l,m} f_{nlm} Q^n_{lm}\right), \end{aligned} \tag{6.2.3}$$

where $N_0(\tau)$, $a(\tau)$ and $\varphi(\tau)$ are only functions of the time parameter τ, and therefore, constant on each three dimensional surface S; Q^n_{lm} are scalar spherical harmonics;

$Q_{il}^{(B)n}$ are vectorial spherical harmonics ($B = 1, 2$) and $g_{nlm}, K_{nlm}^{(B)}$ and f_{nlm} are coefficients. Then the wave function is:

$$\Psi(\tau, N, N_i, h_{ij}, \varphi) = \Psi(\tau, N_0, g_{nlm}, a, K_{nlm}^{(B)}, a_{nlm}^{(A)}, \varphi, f_{nlm}), \tag{6.2.4}$$

i.e. not a functional of function of τ and x^i but a functional of an infinite set of functions of τ.

We will symbolize the three indices n, l, m just with one index n. With all these prescriptions the action turns out to be

$$S = S_0 + \sum_n S_n, \tag{6.2.5}$$

where S_0 is the action of the unperturbated model (eq. (6.1.3) and S_n are

$$S_n = \int N_0 d\tau L_n, \tag{6.2.6}$$

where the Lagrangian L_n are

$$L_n = L_n\left(g_n, a, \frac{K_n^{(B)}}{N_0}, a_n^{(A)}, \varphi, f_n, \frac{\dot{a}}{N_0}, \frac{\dot{a}_n^{(A)}}{N_0}, \frac{\dot{\phi}}{N_0}, \frac{\dot{f}}{N_0}\right). \tag{6.2.7}$$

Then we can see that transformation (6.3) in this case reads:

$$\begin{array}{ll} \tau \to \tau' = \tau'(\tau), & x^i \to x'^i = x^i, \\ N_0 \to N_0' = \frac{d\tau}{d\tau'} N_0, & K_n^{(B)} \to K_n'^{(B)} = \frac{d\tau}{d\tau'} K_n^{(B)} \end{array} \tag{6.2.8}$$

and $g, a, a_n^{(A)}, \varphi$ and f_k remain invariant. In this case we have not the problem stated after eq. (6.4) and we can define the time parameter

$$t(\tau) = \int N_0(\tau) d\tau, \tag{6.2.9}$$

which is invariant under transformation (6.2.8) and which can be considered as the Hamiltonian time.

The Hamiltonian reads

$$H = N_0\left(H_0 + \sum_n H_2^n + \sum_n g_n H_1^n + \sum_{n,B} K_n^{(B)} H_1^{n(B)}\right), \tag{6.2.10}$$

where H_0 is the Hamiltonian of the unperturbated model (6.1.4) with $N = 1$ and the indices $0, 1$ and 2 give the order of the corresponding H in the perturbation ϵ_{ij}. The corresponding formulae can be found in ref. [10].

The constraints can be found making the variation with respect to N and N_i, or what is now the same thing, making the variation with respect to the coefficients of these quantities: N_0, g_n and $K_n^{(B)}$, we get

$$H_0 + \sum_n H_2^n = H_1^n = H_1^{n(B)} = 0. \tag{6.2.11}$$

The traditional wave function is

$$\phi = \phi(a, a_n^{(A)}, \varphi, f_n). \tag{6.2.12}$$

The supermomenta equations are

$$H_1^{n(B)} \phi = 0 \tag{6.2.13}$$

and the superhamiltonian equation yields

$$H_1^n \phi = 0 \tag{6.2.14}$$

and the Wheeler-De Witt equation is

$$\left(H_0 + \sum_n H_2^n \right) \phi = 0. \tag{6.2.15}$$

Using the new formalism the wave function is

$$\Psi = \Psi \left(\tau, N_0, g_n, K_n^{(B)}, a, a_n^{(A)}, \varphi, f_n \right) \tag{6.2.16}$$

and the Schrödinger equation reads

$$i \frac{\partial \Psi}{\partial \tau} = H\Psi = N_0 \left(H_0 + \sum_n H_2^n + \sum_n g_n H_1^n + \sum_{n,B} K_n^{(B)} H_1^{n(B)} \right) \Psi \tag{6.2.17}$$

that inmediately can be written as

$$i \frac{\partial \Psi}{\partial t} = \left(H_0 + \sum_n H_2^n + \sum_n g_n H_1^n + \sum_{n,B} K_n^{(B)} H_1^{n(B)} \right) \Psi. \tag{6.2.18}$$

Now we can try to find the solution with a wave function like

$$\begin{aligned} \Psi = \triangle(t - \theta(a)) \prod_n \triangle^{(0)}(g_n) \triangle^{(1)}(K_n^{(1)}) \triangle_n^{(2)} \phi_0(a, \varphi) \times \\ \times \prod_n \phi_n(a, \varphi, a_n^{(A)}, f_n) \end{aligned} \tag{6.2.19}$$

where functions $\triangle^{(i)}$ are very sharply peaked functions at zero. As $x\triangle^{(i)}(x) \simeq 0$, and $\triangle^{(i)}(0) = const.$ we can define a new wave function

$$\tilde{\Psi} = \triangle(t - \theta(a)) \phi_0(a, \varphi) \times \prod_n \phi_n(a, \varphi, a_n^{(A)}, f_n) \tag{6.2.19'}$$

that satisfies

$$i\frac{\partial\tilde{\Psi}}{\partial t} = \left(H_0 + \sum_n H_2^n\right)\tilde{\Psi}. \tag{6.2.20}$$

Now, as the operator H_0 and H_2^n contain only second derivatives with respect to the variable $a, \varphi, a_n^{(A)}, f_n$ and ordinary c-number functions of these variables, we can write

$$H_0 + \sum_n H_2^n = g^{\alpha\beta}(h_\gamma)\partial_\alpha\partial_\beta + U(h_\gamma), \tag{6.2.21}$$

where

$$h_0 = \alpha; h_i = \varphi, a_n^{(A)}, f_n; i \neq 0; \partial_\beta = \frac{\partial}{\partial h_\beta} \tag{6.2.22}$$

and where $g^{\alpha\beta}(h_\gamma)$ can be considered the metric of the minisperspace of the h_γ and $U(h_\gamma)$ a potencial, i.e., a simple function of the h_γ. Of course we can introduce a different ordering in the operator $H_0 + \sum H_2^n$. Let s choose the ordering that makes the operator invariant under the change of coordinates h_γ in the minisuperspace, i.e.

$$H_0 + \sum_n H_2^n = \frac{1}{\sqrt{g}}\partial_\alpha\left(\sqrt{g}g^{\alpha\beta}\partial_\beta\right) + U, \tag{6.2.23}$$

where $g = |det(g_{\alpha\beta})|$ (cfr. [14]). We can now make an arbitrary change of variables in such a way that

$$\begin{aligned} h^0 &= \alpha \to h^{0'} = \tilde{\alpha} = \alpha + F(h_i), \\ h^i &\to h^{i'} = h^{i'}(h^i). \end{aligned} \tag{6.2.24}$$

Then we can calculate the change of coordinate coefficients

$$\begin{aligned} \theta_0^{0'} &= \frac{\partial\tilde{\alpha}}{\partial\alpha} = 1, \quad \theta_0^{i'} = \frac{\partial h^{i'}}{\partial\alpha} = 0, \\ \theta_i^{0'} &= \frac{\partial F(h^i)}{\partial h^i}, \quad \theta_{j'}^{i'} = \frac{\partial h^{i'}}{\partial h^j} \neq 0. \end{aligned} \tag{6.2.25}$$

The Jacobian matrix of the transformation is

$$J = \begin{pmatrix} 1 & \frac{\partial F}{\partial h^i} \\ 0 & \frac{\partial h^i}{\partial h^j} \end{pmatrix} \tag{6.2.26}$$

and the Jacobian, det $J \neq 0$ if $det(\partial h^i/\partial h^j)$, is chosen different from zero. The inverse matrix is

$$J^{-1} = \begin{pmatrix} \frac{\partial h^0}{\partial h^{0'}} & \frac{\partial h^0}{\partial h^{i'}} \\ \frac{\partial h^i}{\partial h^{0'}} & \frac{\partial h^i}{\partial h^{i'}} \end{pmatrix} = \begin{pmatrix} 1 & \triangle_{i'} \\ 0 & \left(\frac{\partial h^{i'}}{\partial h^i}\right)^{-1} \end{pmatrix}, \tag{6.2.27}$$

where the quantities $\triangle_{i'}$ must satisfy the equation

$$\frac{\partial F}{\partial h^i} + \frac{\partial h^{i'}}{\partial h^i}\triangle_{i'} = 0. \tag{6.2.28}$$

Under this change of coordinates the metric changes as

$$g^{0'0'} = g^{00} + 2\frac{\partial F}{\partial h^i} g^{0i} + \frac{\partial F}{\partial h^i}\frac{\partial F}{\partial h^j} g^{ij}$$

$$g^{0'i'} = \frac{\partial h^{i'}}{\partial a} g^{00} + \frac{\partial F}{\partial h^j}\frac{\partial h^{i'}}{\partial a} g^{j0} + \frac{\partial h^{i'}}{\partial h^j} g^{0j} + \frac{\partial F}{\partial h^i}\frac{\partial h^{i'}}{\partial h^j} g^{ij}. \tag{6.2.29}$$

But the minisupermetric of our problem is (cfr. [10])

$$g^{\alpha\beta} = a^{-3}\gamma^{\alpha\beta}(h^i), \tag{6.2.30}$$

where $a = e^{\alpha}$ is only a function of a. Then from equations (6.2.29) we can impose the following conditions

$$\begin{aligned} g^{0'0'} &= a^{-3}\left[\gamma^{00} + 2\frac{\partial F}{\partial h^i}\gamma^{0i} + \frac{\partial F}{\partial h^i}\frac{\partial F}{\partial h^j}\gamma^{ij}\right] = -a^{-3}e^{3F}, \\ g^{0'i'} &= a^{-3}\left[\frac{\partial h^{i'}}{\partial a}\gamma^{00} + \frac{\partial h^{i'}}{\partial h^j}\gamma^{0j} + \frac{\partial F}{\partial h^i}\frac{\partial F}{\partial h^j}\gamma^{ij}\right] = 0. \end{aligned} \tag{6.2.31}$$

The expressions under the brackets are only functions of the $h^i$and not of α or a. Thus if ν is the dimension of the minisuperspace (really $\nu \to \infty$ at the end of the calculations) we have ν equations to fix ν variables. Precisely, the first equation fixes F and the other $\nu - 1$ equations fix the $\nu - 1$ $h^{i'}(h^i)$. This computation can be seen in detail in some particular cases (cfr. [2]). Under these conditions we have

$$\begin{aligned} \partial_{\tilde{\alpha}}\sqrt{g} &= \frac{\partial}{\partial\tilde{\alpha}} a^{-(3/2)\nu}\sqrt{|\gamma(h^i)|} = \frac{\partial\alpha}{\partial\tilde{\alpha}}\frac{\partial}{\partial\alpha} a^{-(3/2)\nu}\sqrt{|\gamma(h^i)|} + \\ &+ \frac{\partial h^i}{\partial\tilde{\alpha}}\frac{\partial}{\partial h^i} a^{-(3/2)\nu}\sqrt{|\gamma(h^i)|} = \frac{\partial}{\partial\alpha} a^{-(3/2)\nu}\sqrt{|\gamma(h^i)|} = \\ &= \sqrt{|(\gamma(h^i)|}\frac{\partial}{\partial\alpha} a^{-(3/2)\nu} = \sqrt{|\gamma(h^i)|}\frac{\partial}{\partial\tilde{\alpha}} a^{-(3/2)\nu}, \end{aligned} \tag{6.2.32}$$

where $\tilde{a} = e^{\tilde{\alpha}}$ ando also $\tilde{a} = ae^{-F}$. Then we can see that, in the coordinate system we have chosen, $\sqrt{\gamma(h^i)}$ can be commuted with $\partial/\partial\tilde{\alpha}$ and the same happens, of course, with e^F. Then in the new coordinates the operator of eq. (6.2.32) reads

$$\begin{aligned} H_0 + \sum_n H_2^n &= -\frac{1}{\sqrt{g}}\partial_{\tilde{\alpha}}\left(\sqrt{g}\tilde{a}^{-3}\partial_{\tilde{\alpha}}\right) + \frac{1}{\sqrt{g}}\partial_{i'}\left(\sqrt{g}g^{i'j'}\partial_{j'}\right) + U = \\ &= -\frac{1}{\tilde{a}^{-(3/2)\nu}}\partial_{\tilde{\alpha}}\tilde{a}^{\frac{-3(\nu+2)}{2}}\partial_{\tilde{\alpha}} + \frac{1}{\sqrt{g}}\partial_{i'}\left(\sqrt{g}g^{i'j'}\partial_{j'}\right) + U. \end{aligned} \tag{6.2.33}$$

Now we can use the theory of the real clock of section **5**, because the Schrödinger equation (6.2.20), in the new coordinate system reads

$$\frac{i\partial\tilde{\Psi}}{\partial t} = \left[-\frac{1}{\tilde{a}^{-(3/2)\nu}}\partial_{\tilde{\alpha}}\tilde{a}^{\frac{-3(\nu+2)}{2}}\partial_{\tilde{\alpha}} + h(\tilde{a}, h^i)\right]\tilde{\Psi}, \tag{6.2.34}$$

where $h(\tilde{a}, h^i)$ is an operator with no $\tilde{\alpha}$ derivatives, thus we have found an operator like the one of equation (5.1) where

$$\mu'_a = \tilde{a}^{\frac{3(\nu+2)}{2}},$$
$$f(a) = \tilde{a}^3. \tag{6.2.35}$$

Then the probabilistic time turn out to be

$$\theta(\tilde{a}) = \theta_0 \int \left(\phi_0 \prod_n \phi_n, \phi_0 \prod_n \phi_n \right) \tilde{a}^{\frac{3\nu}{2}} d\tilde{a}, \tag{6.2.36}$$

where in fact, $\tilde{a}^{\frac{3\nu}{2}} \sqrt{|\gamma(h^i)|}$ is the measure that corresponds to the minisupermetric (6.2.3) and $\sqrt{|\gamma(h^i)|}$ is the measure that we must use in the inner product $(\phi_0 \prod_n \phi_n, \phi_0 \prod_n \phi_n)$.

In the case $\varphi = 0$, i.e. there is no matter field, the explicit calculation of $\tilde{a}$ can be found in ref. [2] (but with another factor ordering). It is

$$\tilde{\alpha} = \alpha + \frac{1}{6} \sum_n (n^2 - 4)(a_n + b_n)^2 + \frac{1}{2} \sum_n a_n^2 - 2 \sum_n \frac{(n^2-4)}{(n^2-1)} b_n^2$$
$$- 2 \sum_n (n^2 - 4) c_n^2 - 2 \sum_n d_n^2, \tag{6.2.37}$$

where a_n, b_n, c_n and d_n symbolize the coefficients $a_n^{(A)}$ (there is an even and an odd case for c_n and d_n, thus we obtain the six types of harmonic functions).

The Wheeler - De Witt equation takes, in fact, the form (5.1), it is eq. (6.2.15) with

$$H_2^n = \frac{1}{2} \left[-\frac{\partial^2}{\partial d_n^2} + (n^2 - 1)\tilde{a}^4 d_n^2 \right], \tag{6.2.38}$$

i.e. an operator with no $\tilde{\alpha}$ derivatives.

As we can see to finish with the whole program we must change the hand, from the radius of the universe a (or α), to a combination of this radius and a quadratic form of the fluctuations (cfr. eq. 6.2.37). Thus the choice of the hands is not so simple as in section **6.1** because in this section, to implement the demonstration, we need the "conformal symmetry" of eq. (6.2.30) to find the correct hand. Nevertheless it is possible that similar constructions can be found using another kind of hand.

Finally we see that the examples of section **6.1** and 6.2 follow the program proposed in paper [12], that now we believe is almost complete because the last example is quite general.

7. The Correspondence Principle

It is very important to verify, as we will do in this section, that the introduced Hamiltonian time is just our ordinary time, the one we use when we study a small subystem of the universe. We will call this necessary physical fact the "correspondence principle" (cfr. [14]). Therefore let us study a small subsystem of the universe with Hamiltonian $H_x(a,x)$ where a symbolizes the radius of the universe (or eventually the whole set of cosmological variables) and x is the characteristic variable of the small subystem (eventually a set of observables) interacting with the universe, whose Hamiltonian is $H_a(a)$. Furthermore $H_x(a,x)$ has no a-derivative. Then our choice takes into account the following logical physical facts:

a. The small subsystem has no influence on the universe as a whole.

b. On the contrary the universe influences the small subystem but only in a "potential" way, not in a kinematic one. In fact we can imagine that the cosmological variables have some influence in the small subsystem but there are no examples of a derivative interaction.

The total Hamiltonian is

$$H = H_a(a) + H_x(a,x). \tag{7.1}$$

The Schrödinger equation is

$$i\frac{d\phi}{dt} = H\phi. \tag{7.2}$$

where t is our Hamiltonian time. This equation can be solved if we make the factorization

$$\phi(t,a,x) = A(t,a)X(t,a,x). \tag{7.3}$$

Then we obtain

$$i\frac{dA}{dt} = [H_a - f(a,t)]\,A, \tag{7.4}$$

$$i\frac{dX}{dt} = [H_x + f(a,t)]\,X, \tag{7.5}$$

where $f(a,t)$ is the "separation constant". Equation (7.5) is almost a Schrödinger equation for the small subsystem, the only difference is that the factorization (7.3) is ambiguous because we can take an arbitrary factor function of a and t, from A into X and the factorization remains the same. In the same way we can eliminate function $f(a,t)$ from eqs. (7.4) and (7.5) defining

$$S' = Xexp\left[i\int f(a,t)dt\right], \tag{7.6}$$

$$A' = Aexp\left[-i\int f(a,t)dt\right]. \tag{7.7}$$

In fact

$$
\begin{aligned}
i\frac{dX'}{dt} &= ie^{+i\int f dt}\frac{dX}{dt} - fX' = \\
&= e^{+i\int f dt}\left[H_x + f\right]X - fX' = H_x X'.
\end{aligned}
\tag{7.8}
$$

Thus between X and X' we only have a phase difference and the new factor X' satisfies the ordinary Schrödinger equation, using our Hamiltonian time, as we anticipate. Thus Hamiltonian time is just the ordinary time that we use in everyday Schrödinger equation.

Everything we have said seems just a small exercise and a generalization of the factorization (2.17) of section **2**. Nevertheless we can obtain from this exercise very important conclusions.

First of all we know that ordinary Schrödinger time is endowed with the two properties stated in section **1**: it is an ordering parameter for colapse and it is Herakleitian, thus our Hamiltonian time, which we have proved coincides with an ordinary Schrödinger time, also has the two properties. Of course it can be said that these two properties can only be tested in a small subsystem and not in the universe as a whole. This is, in fact, true but the two properties were stated thinking on small experimental subystem, and not on the universe as a whole, because we cannot make any experiment of this kind. Therefore our Hamiltonian time is an ordering parameter for colapse and Herakleitian with the full meanning that we gave to this definition in section **1**.

Then we realize that, as our Hamiltonian time and also the probabilistic time, i.e. its aproximate real clock version, are ordering time parameter, they have their own arrow of time "built in". What we have done is to adapt ordinary Q M to the problem of the whole universe, thus what we have found is the typical arrow of time of quantum mechanics that points out the order of colapse, the direction of the smearing or spreading of a packet (eq. 5-27), etc. We will call this arrow the quantum arrow of time. As this definition is based only in QM principles, the quantum arrow is, in principle, independent of the cosmological and thermodinamical arrow of time (cfr. [15]) that are based in other kind of phenomena. We will follow this discussion in the next section.

A last question: where is the GR general change of coordinates in all this formalism? Because all our Schrödinger equations are not "relativistic invariants". Of course we can answer that we have worked with a particular foliation, i.e. a particular $1+3$ decomposition, and that the whole theory is invariant under the transformations (6.2) and (6.8). But this is not the whole answer we are looking for: what happens when we change the foliation? This is a very difficult mathematical (and probably physical) problem, treated e.g. in papers [16] and [17] which we will not discuss here. But of course this is a problem, not because of our treatment, but of the Hamiltonian formalism that we are using. What we have done is to use this formalism to obtain new conclusions and concepts. We hope that this very deep problem could be solved in the future.

8. The Arrow of Time

At the end of section **6.1** and in section **7** we have mentioned the problem of the "arrow of time" that is closely related to the definition of probabilistic time, because in eq. (1.3), (5.25), (6.1.15) or (6.2.36) we have avoided the problem postulating that $d\theta/da > 0$ i.e. that the sense of growing of the probabilistic time ($d\theta$) coincides with the cosmological arrow of time (da). On the contrary in section **7** we have introduced the quantum arrow, that would be the natural arrow of the probabilistic time. Then we would correct eqs. (1.3), (5.25), (6.1.15) or (6.2.36) introducing a $+1(-1)$ when the cosmological arrow coincides (does not coincide) with the quantum arrow. Then it is better to study the whole problem in this section.

Let us begin reviewing the problem of the reversibility and irreversibility of time, in classical and quantum mechanics. As it is clearly stated in paper [18] the time-inversion transformation in classical mechanics does not mean inversion of the time flow, but rather the inversion of the dependence of all other coordinates on time, i.e. the invariance under time inversion is not the invariance of the system under the transformation $t \to -t$ but rather under the transformation

$$q^k(t) \to q^k(-t) \quad t \to t. \tag{8.1}$$

In fact, the equations of classical mechanics are not, in general, invariant under transformation $t \to -t$ (e.g. the Hamilton equations are not invariant), but almost all classical models (the non dissipative ones) have Lagrangians and Hamiltonians invariant under transformation (8.1). Analogously, at the quantum level, the reversibility of the theory is not the invariance under $t \to -t$ (e.g. Schrödinger equation is not invariant under this transformation while Klein Gordon equation is invariant). At the quantum level the theory istime invariant if the probabilities do not change under the substitution t by $-t$, i.e.:

$$|\phi(t,x)|^2 = |\phi(-t,x)|^2; t \to t, \tag{8.2}$$

as it is the case with Schrödinger equation (a not reversible one under the change $t \to -t$) if we complete the T inversion with the C inversion. In fact this equation is invariant under

$$\phi(t,x) \to \phi^*(-t,x); t \to t. \tag{8.3}$$

From this equation (8.2) follows inmediately. For the Klein Gordon equation things are even simpler because this equation is invariant under the transformation

$$t \rightarrow -t.$$

Therefore the classical mechanics of non dissipative processes and the quantum and relativistic mechanics (if we do not use non symmetrical interactions under CP or T as the weak interaction) are time-reversible according to the above definitions. Then the problem of the "arrow of time" can be stated in two different ways:

a - All microscopical processes are reversible (with exception of the processes related to the weak interaction). How can we explain the existence of dissipative and irreversible macroscopical processes? (cfr. [19]).

b - There are different arrow of time:

* The psychological arrow (cfr. e.g. [15] and [20])

* The thermodinamical arrow (the one defined by the entropy growing)

* The electrodynamical arrow (we choose the retarded solution of the field equation and not the advanced one)

* The cosmological arrow (the one defined by the expansion of the universe)

* The quantum arrow (that we have introduced in section **7**)

The problem is to know if all these arrows have the same direction.

It is shown in paper [15] that, in fact, the three first arrows of the list have the same direction. The problem, as stated in papers [15] and [20] is to know if the fourth arrow coincides with the first three ones or if it only coincides in certain period of the universe evolution (i.e. if we take the three first arrows as the fidutial ones, to know if the universe always expands or if it bounces or oscilates in some way).

We can see that both ways to formulate the problem are related, because the thermodynamical arrow of **b** (and therefore the psychological and electromagnetical arrows) is the macroscopic arrow of **a**, and the microscopical arrow of **a** could be considered the quantum arrow of the list in **b**, i.e. the arrow of spreading of wave packets and the collapse of wave functions.

Thus, the problem can be stated in the following way:

a.- To define the microscopic arrow because in principle microscopic equation would have no arrow since these microscopic equations are reversible (the quantum arrow will be our candidate to play this role).

b.- See if this arrow coincides or not with the cosmological arrow.

c. Finally see if these arrows coincide or not with the thermodynamical macroscopic arrow.

We shall study the two first steps using probabilistic time method. As we have seen universe can be considered as a clock and the radius of the universe a as its hand. Let us first see if this clock is endowed with an arrow of time. Let us begin supposing that somehow the universe is an ideal clock.

a. Ideal Clock. At first sight it seems that when a clock is working it defines an arrow of time. In fact it is not so because it can work backwards, if it is a reversible device, and it defines the opposite arrow of time. When we use an ideal clock the arrow of time we are using is our psychological arrow of time that tells us if the clock is working forwards or backwards (i.e. its motion coincides or not with our psychological arrow of time). Can we obtain somehow an arrow of time from an ideal clock? In principle the answer is no, because the wave function of the clock $D(t, a)$ has always the same shape and it only changes places (i.e. $D(t, a) \simeq \delta(t - a)$ something like fig. 4). Thus there is no way to obtain a privilegiated direction, fig. 4 is symmetric and homogeneous.

b. Real clock. The behaviour of the real clock is completely different. If we give a very peaked shape to function $D(t, a)$ at a time t_0 we know that the shape of the function is smeared towards the past and the future (fig. 5). Then the real clock gives two privilegiated directions from t_0 because the shape of the functions $D(t,_a)$ are no longer the same as in the case of the ideal clock. But fig. 5 being non homogeneous is still symmetric, then the real clock gives two directions but does not tell which is the one pointing towards the future of towards the past. In fact, the "reversibility" of the Schrödinger equation yields this symmetry.

c. The Universe. As we proved in section **6**, the universe can be considered as a real clock, but as a physical structure it has also something else. In the real clock all the initial times t_0 are equivalent, we can take any instant of time to put the very peaked shape of the wave function $D(t, a)$ and begin the smearing. On the contrary, the great majority of cosmological models has two very particular times: t_B, the "Big-Bang", i.e. the time when $a = 0$ and t_p, i.e. the time of today or present time. Thus if we put t_0, the time where the smearing begins it coincidence with t_B, the two directions are different because t_p belongs to one of them but not to the other (fig. 6). The direction from t_B to t_p gives the good arrow of time pointing towards the future, and the real clock can tell this direction at any time because it is one of

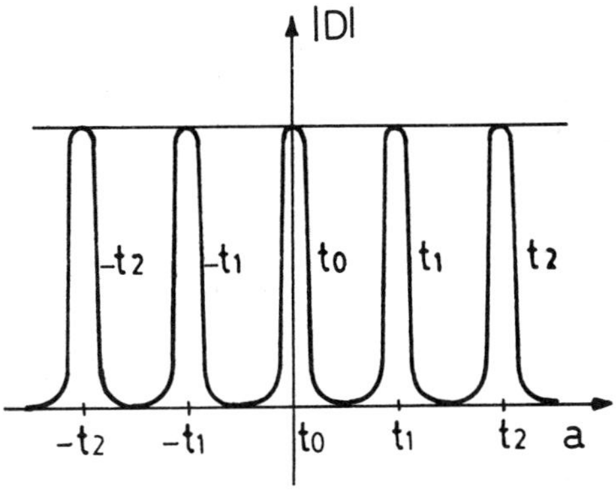

Fig.4

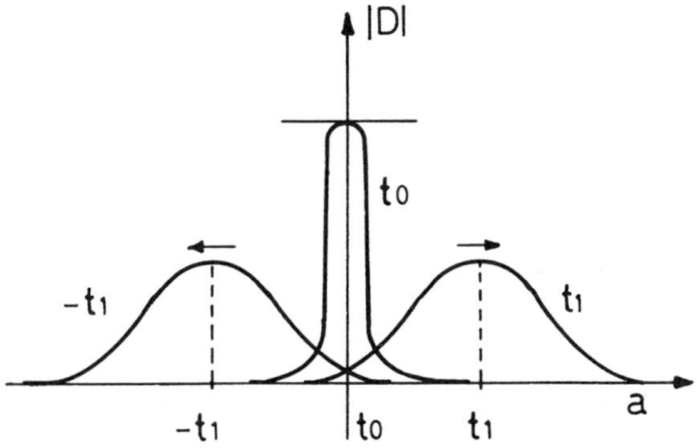

Fig.5

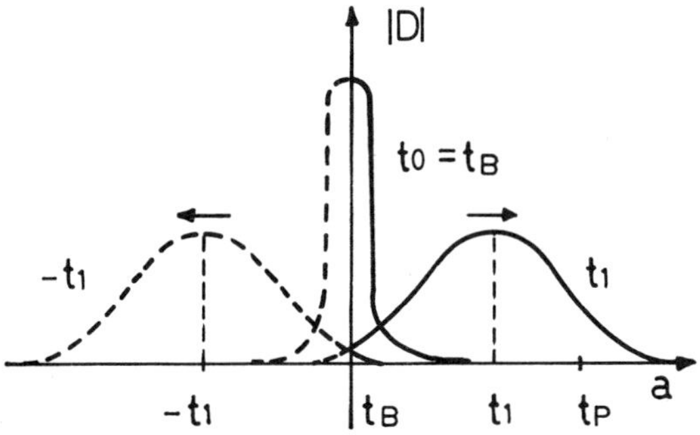

Fig.6

its two smearing directions. On the contrary if it would be an ideal clock it would not tell anything because it has not smearing directions.

Furthermore the universe begins at t_B, therefore the past of the universe "before" t_B does not exist (that is why we do not draw in solid lines this part in fig. 6). Then we can see that figures 4 and 5 are symmetric while fig. 6 is not and it gives us an arrow of time. Perhaps the question if we are in the solid line zone or in the dashed line zonehas not even a physical sense either because the dashed line zone does not exist or because we can mathematically identify both zones in an "antipodal" way. Of course we can put the peaking point of the wave function at t_p, then the past direction is defined as the one that has time t_B etc., etc..

Therefore we can consider as microscopical arrow of time the smearing of spreading arrow because this arrow is the one based in microscopic and quantum phenomena. In this way we can say that it coincides with the arrow of collapse because both arrows are based in the same QM set of postulates, and their coincidence can be verified in small subsystems of the universe in the frame of QM theory. On the contrary the macroscopic-thermodynamical arrow and the cosmological arrow are based on a completely different physical theories and therefore can be considered independent arrows.

Thus we have completed step **a**, we can identify the microscopical arrow with what we have called the quantum arrow (i.e. the spreading and collapse arrow). The following steps (**b** and **c**) are to see if this arrow coincides with the two others.

But let us first precise which variables have the direction of the different arrows:

* The direction of growing of the entropy S is the thermodynamical macroscopic arrow.

*The direction of growing of the radius of the universe is the cosmological arrow.

*The direction of growing of $|\delta^2|$ the width of the wave packet $D(t, a)$ (cfr. eq. (5.27)) is the quantum arrow.

But

$$|\delta^2| = |\delta_0^2 + 4i\tau|, \tag{8.4}$$

thus the direction of growing of τ is also the quantum arrow. But from eq. (5.21) the direction of growing of τ and t are the same because $B^2 > 0$ and $(\phi_0, \phi_0)_t^2 > 0$ and from section **5**, t is measured by θ thus the direction of growing of the probabilistic time computed inverting eq. (5.21) as $t = \theta = f(\tau)$ is also the quantum arrow. That is why we have said at the end of section **8** that the probabilistic time has its arrow of time "built in". Of course from (8.4) we see that there are two smearing directions, nevertheless we know how to choose the right one.

Now we can see what the real physical meaning of probabilistic time is. In the introduction we defined this parameter "glueing" together the S^3 geometries from the "Big-Bang" ($a = 0$) to the present ($a = a$) (cfr. eq. 1.3) endowed with a "time thickness", then we realize that the probabilistic time is the time-lenght between the Big-Bang and the present. But in the introduction we ordered all the S^3 using the cosmological arrow but now we realize that what we must do is to order the S^3 using the quantum arrow, that is to say using the arrow of $|\delta^2|, \tau, t$ or θ (obtained from 5.21). But eqs. (1.3), (5.25), (6.15) or (6.2.36) give θ as a function of a and a has another independent arrow. Thus we must correct these formulae introducing a factor -1 in the periods when the quantum arrow does not coincide with the cosmological arrow. These alterations of sign can be obtained changing the sign of B in eq. (5.20) or (5.25) in different periods when variables b or θ change their direction. On the contrary in eq. (5.21) we have $B^2 = 0$ instead of B, therefore t and τ have always the same arrow: the quantum or smearing arrow.

These changes of sign of the constant B can be done in a continuous way. In fact if we would like to pass from $d\theta/da > 0$, where the variable θ has the cosmological arrow to a period $d\theta/da < 0$ where θ has this arrow reversed we must necessarily pass through a point where $d\theta/da = 0$ or $(\phi_0, \phi_0)_a = 0$ and there we can change the sign of B in eq. (5.25) in a continuous way.

Then θ can show eventual oscilations of the radius of the universe, that were impossible to see if we use the cosmological arrow of time only. With this correction θ measures the time elapsed from the Big-Bang to the present.

In this way we have computed step **b**, we know how to compare the quantum arrow with the cosmological arrow. We do not have, by now, a definite answer to step **c** because we do not have a well defined notion of entropy (cfr. [13]) but we believe that most likely the quantum arrow and the thermodynamical arrow coincide because this coincidence takes place in the small subsystem of the universe.

Today it seems observationally that the three arrows, quantum, thermodynamical and cosmological coincide. Then it looks like that this is a necessary condition to impose to the wave function of the universe and a way to single out the right wave function as we will see in the next section.

Let us finally observe that in the period where the universe oscilates the parameter a is not Herakleitian, there are more that one universe configuration for each a. The correction using the quantum arrow of time in eqs. (1.3), (5.25), (6.1.5) and (6.2.23) makes that the probabilistic time θ turns out to be Herakleitian anyhow.

9. Conclusion

In this section we will review the main conclusions and the main line of research that we think this paper opens.

The first conclusion is that we have characterized the time-coordinate that we must add to the three dimensions of quantum gravity to obtain the four dimensional manifold of classical or semiclassical theories, it is the probabilistic time defined in this lectures. We can show that this time defined here in an exact way, coincides with the one defined via the WKB or Born-Oppenheimer approximations that can be found in refs. [7] and [9]. We have also studied the properties of this time parameter as an ordering parameter and its Herakleitian property.

A very important ingredient to define probabilistic time is the arrow of time. This problem was neglected in the preceding papers [5] to [9]. Now it is solved by the introduction of the quantum arrow of time, the natural arrow of time for probabilistic time.

Finally perhaps the most important conclusion is that the existence of Hamiltonian time was only postulated but we have seen how we can approximately measure this time with probabilistic time. This time measurement is similar to the one made when we measure ordinary classical time using a real clock in QM. Therefore the Hamiltonian time and classical time are based in the same kind of reasoning and facts, both are postulated and both cannot be exactly measured because we do not have ideal clocks, and both can only be measured approximately by real clocks. Thus both notions of time are equally based and if we accept classical time in QM we must, as welll, accept Hamiltonian time in Quantum Gravity.

We believe that the main line of research originated by these lectures is the following. We think that continuing our line of research it would possibly give rigorous bases to the choice of the wave functions of the universe and find a set of conditions that would single out the right one. This reasoning would be more or less as follows. The wave function of the universe must be chosen in the set of wave functions with a good classical behaviour in the WKB approximation, i.e. it must have the form $J(a, \phi)e^{\pm iK(a)}$, because with this behaviour we know that the universe satisfies the classical Einstein equation in the classical limit and the "back reaction" equation in the semiclassical approximation (cfr. [9]). Then we woul have a choice of the sign in $J(a, \phi)e^{\pm iK(a)}$. But the relation between probabilisti time and a in the classical limit is:

$$\frac{da}{d\theta} = -\frac{1}{a}\frac{dK}{da}. \tag{9.1}$$

Then if we want the quantum arrow to coincide with the cosmological arrow as it happens nowadays, as $d|K|/da > 0$ in our model, we must choose the minus sign and the outgoing "wave function". Thus we would arrive to Vilenkin's wave function of the universe (cfr. [21]) and to a happy end because this wave function has the right behaviour to yield new inflation and chaotic inflation. Perhaps with a good entropy definition for the whole universe we would understand why WKB approximation plays such an important role in this line of reasoning.

Appendix I

Let us suppose that

$$\phi_0(a,x) \neq \phi_0^*(a,x). \tag{A1.1}$$

Then let us observe that in Quantum Gravity a is the radius of the universe or a scale factor, that appears in the metric

$$dS^2 = -N^2 dt^2 + a^2 d\Omega^2_{(3)}. \tag{A1.2}$$

Therefore a is always found under the form a^2. In fact we usually consider the universe in the interval $0 \leq a < +\infty$, so if in some equation a appears it can be substituted by $|a|$, e.g. $\sqrt{-g} \simeq Na^3$ is really $N|a|^3$. Thus Quantum Gravity is invariant under the change $a \to -a$. Thus function $\phi_0(a,x)$ is even in the variable a.

$$\phi_0(-a,x) = \phi_0(a,x). \tag{A1.3}$$

On the other hand function $\triangle$ of eq. (5.18) has the property

$$\triangle^*(\tau,b) = \triangle(-\tau,-b). \tag{A1.4}$$

Thus under the change of variables (5.20) and (5.21) we also have

$$\triangle^*(t,a) = \triangle(-t,-a). \tag{A1.5}$$

Let us conjugate eq. (5.4) *

$$-i\phi_0^* \frac{\partial D^*}{\partial t} = -f \frac{\phi_0^*}{\mu'_a} \left(\frac{\partial}{\partial a}\right) \frac{1}{\mu'_a} \frac{\partial D^*}{\partial a} - 2 \frac{f}{\mu'_a} \frac{\partial \phi_0^*}{\partial a} \frac{1}{\mu'_a} \frac{\partial D^*}{\partial a}. \tag{A1.6}$$

As $\triangle = D$, using eq. (1.3) and (1.5) we get

$$i\phi_0^* \frac{\partial D}{\partial t} = -f \frac{\phi_0^*}{\mu'_a} \left(\frac{\partial}{\partial a}\right) \frac{1}{\mu'_a} \frac{\partial D}{\partial a} - 2 \left[\frac{f}{\mu'_a}\right] \frac{\partial \phi_0^*}{\partial a} \frac{1}{\mu'_a} \frac{\partial D}{\partial a}. \tag{A1.7}$$

Using the product (2.4) we can "premultiply" eq. (5.4) by ϕ_0^* and integrate, "post-multiply" eq. (1.7) by ϕ_0 and integrate in x so we get

$$i(\phi_0,\phi_0) \frac{\partial D}{\partial t} = -f \frac{(\phi_0,\phi_0)}{\mu'_a} \left[\frac{\partial}{\partial a}\right] \frac{1}{\mu'_a} \frac{\partial D}{\partial a} - 2 \frac{f}{\mu'_a} (\phi_0, \frac{\partial \phi_0}{\partial a}) \frac{1}{\mu'_a} \frac{\partial D}{\partial a}, \tag{A1.8}$$

$$i(\phi_0,\phi_0) \frac{\partial D}{\partial t} = f \frac{(\phi_0,\phi_0)}{\mu'_a} \left[\frac{\partial}{\partial a}\right] \frac{1}{\mu'_a} \frac{\partial D}{\partial a} - 2 \frac{f}{\mu'_a} \left(\frac{\partial |phi_0}{\partial a}, \phi_0\right) \frac{1}{\mu'_a} \frac{\partial D}{\partial a}. \tag{A1.9}$$

Adding these last equations we get eq. (5.5) and from this point demonstration of section **5** can be continued even in the case of complex wave functions.

Appendix II

It is interesting to show that the inner product related with eq. (5.12) is really the one that was defined in eq. (2.20). In fact the t-invariant product that makes the r.h.s. of eq. (5.12) a selfadjoint operator is

$$\int D_1^* D_2 \frac{(\mu_b'(b))^2}{f(b)} db, \tag{A2.1}$$

but from eqs. (5.20) and (5.25) we have

$$db = \frac{f(a)}{\mu_a'(a)} d\theta. \tag{A2.2}$$

Therefore

$$\int D_1^* D_2 \frac{(\mu_b'(b))^2}{f(b)} db = \int D_1^* D_2 d\theta. \tag{A2.3}$$

This fact shows that, really, hypothesis **(iii)** of section **2** is satisfied since the solution of eq. (5.12) is normalized using the inner product (2.20).

Appendix III

Let us compute the precision of the real clock. If the initial precision is δ, i.e. the width of the wave packet, after a time $\delta(t)$ will be (cfr. [22] p. 64 eq. 20)

$$\delta(t) = \frac{\delta}{2}\sqrt{1 + \frac{4t^2}{\delta^4}}. \tag{A3.1}$$

Let us restablish the ordinary units

$$\frac{\delta(t)}{t_{pl}} = \frac{\delta}{2t_{pl}}\sqrt{1 + \frac{4(t/t_{pl})^2}{(\delta/t_{pl})^4}}, \tag{A3.2}$$

i.e.

$$\delta(t) = \frac{\delta}{2}\sqrt{1 + \frac{4t^2 t_{pl}^2}{\delta^4}}. \tag{A3.3}$$

then the clock will have a reasonable precision if

$$\frac{4t^2 t_{pl}^2}{\delta^4} < 1, \tag{A3.4}$$

or

$$t < \frac{\delta^2}{2t_{pl}} = \frac{1}{2}\frac{\delta}{t_{pl}}\delta. \tag{A3.5}$$

* f and μ_a' are real functions.

Then if e.g. $\delta = picosec = 10^{-12} sec$ is

$$\frac{\delta}{t_{pl}} = 10^{32} \tag{A3.6}$$

and

$$t < 10^{19} sec. \tag{A3.7}$$

This upper bound is longer than the age of the universe. Then the real clock has a very good precision for all the duration of the universi if it is considered as a normal laboratory clock.

Aknowledgements

The author is grateful for the hospitality of the Free University of Brussels where this work was begun and to E. Gunzig, F. D. Mazzitelli and P. Nardone for stimulating discussions and criticisms

References

[1] - Hartle J.B., "Predictions in Quantum Gravity". NATO Advanced Summer Institute, Cargése (1986).

[2] - Wada S., Nucl. Phys. B **276**, 729 (1986).

[3] - Banks T., Nucl Phys. B **249**, 332 (1985).

[4] - Hawking S. W., page D. N., Nucl Phys B **264**, 185 (1986).

[5] - Castagnino M., "The appearence of time in Quantum Gravity", IV Quantum Gravity Seminar, Moscow 1987, Ed. M.A. Markov et al. World Scientific, Singapore (1988).

[6] - Castagnino M., "Time and probability in Quantum Gravity", SILARG VI, Rio de Janeiro, 1987, Ed. M. Novello. World Scientific, Singapore (1988).

[7] - Castagnino M., "Probabilistic time in Quantum Gravity", Phys. Rev. D, in press (1988).

[8] - Castagnino M., D'Negri C., "On the interpretation of Quantum Gravity", A. Friedmann Coomemoration Conference, Leningrad, 1988. To be published by World Scientific, Singapore.

[9] - Castagino M., Mazzitelli F.D., "Probabilistic time and the interpretation of Quantum Gravity". Jour. of Theo. Phys. In press (1989).

[10] - Halliwell J.J., Hawking S.W., Phys. Rev. D bld31, 1777 (1985).

[11] - D'Eath P.D., Halliwell J.J., Cambridge University report (1986), unpublished.

[12]- Unruh W. G., "Time and Quantum Gravity", IV Quantum Gravity Seminar, Moscow 1987. Ed. M.A. Markov et al. World Scientific, Singapore (1988).

[13] - Kandrup H.E., Class. and Quant. Grav., **5**, 903 (1988).

[14] - Vilenkin A., "The interpretation of the wave function of the universe", Tuft University, preprint (1988).

[15] - Hawking S. W., Phys. Rev. D. **32**, 2489 (1985).

[16] - Ashtekar A., Magnon A., Proc. Roy. Soc. (Lond.) **A 346**, 375 (1975) and C. R. Acad. Sci. (Paris) **281**, 875 (1975) and **286**, 531 (1978).

[17] - Kay B., Comm. Math. Phys., **62**, 55 (1978).

[18] - Hajicek P., Phys Rev. D, **34**, 1040 (1981).

[19] - Prigogine I., George C., Henin F., Rosenfeld L., Chemica Scripta, **4**, **5**, 32 (1973).

[20] - Hawking S. W., "The direction of time", Univ. of Cambridge, preprint (1988).

[21] - Vilenkin A., Phys. Rev. D, **37**, 888 (1988).

[22] - Cohen-Tannoudji C. et al., "Quantum Mechanics", John Wiley and sons. New York (1977).

[23] - After finishing this paper I learnt about the work: Unruh W., Wald R. M., "Time and the Interpretation of Canonical Quantum Gravity" (Institute of Theo. Phys., Univ. of California, preprint NSF-ITP-88-190, Pys. Rev. D **40**, 2598 (1989)) which is essential to understand eq. (2.10).

QUANTUM AND STATISTICAL EFFECTS IN SUPERSPACE COSMOLOGY*

B. L. Hu

Newman Laboratory of Nuclear Studies
Cornell University
Ithaca, New York 14853+

and

Department of Physics and Astronomy
University of Maryland
College Park, Maryland 20742

- Dedicated to the martyrs of June 4, 1989,
whose sacrifice presages a truly people's China,
whose unflinching spirit inspires the whole humanity.

ABSTRACT

The general aim of these lectures is twofold:

1) To develop a viewpoint for understanding the kinematics of quantum cosmology in the minisuperspace approximation: i.e., superspace as the space of collective excitation modes of spacetime. Emphasis here is put on the spectrum of the Lifshitz operator in spacetime and that of the Wheeler-DeWitt operator in superspace. Some established theorems on the zero-mode structure and asymptotic distribution of eigenvalues of invariant operators in curved space are invoked.

2) To apply some recent results in quantum field theory in curved spacetime to problems in quantum cosmology. Specifically, the analysis of a) infrared behavior and b) dissipative behavior of quantum fields in curved spacetimes are used to elicit the views that a) minisuperspace cosmology is the infrared limit of quantum gravity, and b) dissipative behavior appears in the minisuperspace modes which can lead to memory lost of the initial conditions.

---

* These lectures expound some personal viewpoints and approaches towards current problems in quantum cosmology presented in a semipedagogical way. They are not reviews. Many results quoted here are also of a preliminary nature.

Quantum Mechanics in Curved Space-Time
Edited by J. Audretsch and V. de Sabbata
Plenum Press, New York, 1990

I. THE MINISUPERSPACE OF HOMOGENEOUS COSMOLOGY

A. From Spacetime to Superspace

We begin with the class of homogeneous (Bianchi) cosmology[4,5] and see how its quantum dynamics can be described via the ADM formalism[16] and the minisuperspace approximation.[21,22,23] The line element of Bianchi universes are given by

$$ds^2 = -dt^2 + \sum_{a,b=1}^{3} \gamma_{ab}(t)\, \sigma^a(x)\, \sigma^b(x) \tag{1.1}$$

The metric $\gamma_{ab}(t)$ on the homogeneous 3-space Γ depends only on the cosmic time t. It can be written as $\gamma_{ab}(t) = e^{2\alpha}(e^{2\beta})_{ab}$, where $a = e^{\alpha}$ is the scale factor ($\mathrm{Vol} = a^3$ is the spatial volume) and β_{ij} ($\mathrm{tr}\, \beta_{ij} = 0$) is the anisotropy factor. The σ^a are the basis 1-forms on Γ obeying the differential relation

$$d\sigma^a = \frac{1}{2} C^a{}_{bc}\, \sigma^b \wedge \sigma^c \tag{1.2}$$

where $C^a{}_{bc}$ are the structure constants. In terms of coordinate differentials dx^i, $\sigma^a = S^a{}_i(x)\, dx^i$, and the metric tensor ${}^{(3)}g_{ij}$ is

$$g_{ij}(x,t) = \sum_{a,b} \gamma_{ab}(t)\, S^a{}_i(x)\, S^b{}_j(x) \tag{1.3}$$

In the Hamiltonian (ADM) formulation of general relativity the Einstein action is given by

$$S_E = \frac{(4\pi)^2}{16\pi G} \int \pi^{ij}\, dg_{ij} \tag{1.4}$$

where π^{ij} is the canonical momenta conjugate to g_{ij}. The spatial volume $\mathrm{Vol} = \int d^3x = (4\pi)^2$ is here chosen to match with that of a closed universe. $\ell_{p\ell} = (16\pi G)^{-1/2}$ is the Planck length. The accompanying energy constraint is given by

$$C^0 = 0 = \sqrt{-g}\, \left\{ g^{-1}\left[\frac{1}{2} (\pi^k{}_k)^2 - \pi^i{}_k\, \pi^k{}_i\right] + {}^3R \right\} \tag{1.5}$$

where g is the determinant and ${}^3R$ the scalar curvature formed from the 3-metric g_{ij}. The momentum constraint

$$C^i = -2\pi^{ij}{}_{|j} = 0 \tag{1.6}$$

is satisfied identically owing to spatial homogeneity.

For the class of homogeneous cosmologies with diagonal metric, β_{ij} = diag $(\beta_1, \beta_2, \beta_3)$. Since β_{ij} is traceless, one can replace these by the deformation parameters β_+, β_- at constant "volume" α, where $\beta_{1,2} = \beta \pm_+ \sqrt{3}\,\beta_-$ and $\beta_3 = -2\beta_+$. Equivalently, one can use the polar coordinates β, γ ($\beta_+ = \beta \cos\gamma$ and $\beta_- = \beta \sin\gamma$), known in nuclear physics[28,29] as the deformation and shape parameters. This is an example of a 2-dimensional minisuperspace. Writing

$$\pi^i{}_k = (2\pi)^{-1} p^i{}_k + \frac{1}{3}\, \delta^i{}_k\, (\pi^l{}_l) \tag{1.7}$$

as the sum of a traceless component $p^i{}_k$ and the trace part, the Einstein Lagrangian S_E becomes

$$S_E = \int (p_+\, d\beta_+ + p_-\, d\beta_- - H d\Omega) \tag{1.8}$$

where $\Omega = -\alpha$ plays the role of "time" for Hamiltonian cosmology. The energy constraint appears in the form

$$H^2 = (2\pi)^2\, (\pi^k{}_k)^2 = p_+^2 + p_-^2 - 24\pi^2 g\, {}^{(3)}R = K + U \tag{1.9}$$

which consists of a kinetic energy term

$$K_{tran} = p_+^2 + p_-^2$$

and a potential energy term

$$U = -\, 24\pi\, g^2\, {}^{(3)}R \tag{1.11}$$

The dynamics of the universe governed by Einstein's equations is described by the motion of a point ("universe point") particle of unit mass in a 2-dim minisuperspace (mnss) with coordinates β_+, β_-. The metric of the mnss can be derived from the quadratic kinetic energy. For the class of diagonal homogeneous cosmology,

$$ds^2 = -\, d\Omega^2 + d\beta_+^2 + d\beta_-^2 \tag{1.12}$$

For example, the Bianchi Type I universe has

$$C^a_{bc} = 0\ , \quad {}^{(3)}R = 0\ , \quad H = \sqrt{p_+^2 + p_-^2}\ . \tag{1.13}$$

Since there is no spatial curvature, the potential energy is zero and the velocity of the universe point is unity with respect to Ω "time". For Bianchi Type IX universes,

$$C^a_{bc} = \varepsilon_{abc}, \quad {}^{(3)}R = \frac{6}{a^2}\,(1-V), \tag{1.14}$$

where the anisotropy potential

$$V(\beta) = \frac{1}{3}\, tr(e^{4\beta} - 2\, e^{-2\beta} + 1) \tag{1.15}$$

has a 3-fold symmetry in the (β_+, β_-) plane. $V(\beta)$ is the same for the diagonal and the general homogeneous cosmologies. The velocity of the universe point differs from unity by the amount of this time-dependent potential. Different Bianchi types with different C^a_{bc} differ in $V(\beta)$ but has the same K in Hamiltonian cosmology. In quantum cosmology, the classical trajactory in mnss is replaced by a wave function ψ which is a functional of the coordinates β_a and momenta p_a $(p_\pm = \frac{\hbar}{i}\frac{\partial}{\partial\beta_\pm})$ satisfying the canonical quantization condition $[\beta_a, p_b] = i\,\delta_{ab}$. The Hamiltonian constraint $H = i\hbar\, \partial/\partial\Omega$ yields the Wheeler-DeWitt equation for the wave function ψ

$$-\frac{\partial^2\Psi}{\partial\Omega^2} + \frac{\partial^2\Psi}{\partial\beta_+^2} + \frac{\partial^2\Psi}{\partial\beta_-^2} + \mathcal{R}\Psi = 0 \tag{1.16}$$

where $\mathcal{R} = g\ {}^{(3)}R$. This is in the form of a Klein-Gordon equation in a mnss with metric (1.12),

$$\Box\psi_{ss} + \mathcal{R}\psi = 0 \quad , \quad (\Box_{ss} + \mathcal{R})\ \Psi = 0 \tag{1.17}$$

the Laplace-Beltrami operator being

$$\Box_{ss} = \frac{1}{\sqrt{-G}}\ \frac{\partial}{\partial x^A}\ (\sqrt{-G}\ \ G^{AB}\, \frac{\partial}{\partial x^B}\ ) \tag{1.18}$$

where x^A (A = 1, 2, 3 for the metric (1.12) are the coordinates and G_{AB} the metric in mnss. $(G = \det G_{AB})$.

To conclude this section, note the following correspondance between spacetime and superspace, and the role played by the extrinsic and intrinsic curvatures.

B. General (non-diagonal) Homogeneous Cosmologies

For the class of general homogeneous cosmologies, the metric $\gamma_{ab}(t)$ is nondiagonal and has six components. The 3x3 symmetric traceless tensor β_{ij} can, however, be diagonalized by a rotation matrix. $\hat{R}(\phi,\theta,\psi)$, i.e.

$$\hat{\beta} = \hat{R}^{-1}\ \hat{\beta}_d\ \hat{R}$$

where

$$\hat{\beta}_d = \mathrm{diag}\ (\beta_1, \beta_2, \beta_3) = (\alpha, \beta_+, \beta_-) \tag{1.19}$$

The three Euler angles (φ, θ, ψ) which parametrize $\hat{R}$ can then play the role of three off-diagonal components of γ_{ab}. If one uses the scale factor $\alpha = -\Omega$ as the "time" coordinate, then the other 5 "canonical coordinates" are

$$q^A = (\beta_+, \beta_-, \varphi, \theta, \psi) \qquad (A = 1, ..5) \tag{1.20}$$

Table I

<u>Metric</u>: diag $\gamma_{ab}(t) = (\beta_1, \beta_2, \beta_3)$ ↓	<u>Spatial geometry</u>
<u>Coordinates</u> in mnss: $q^A = (\alpha, \beta, \gamma)$	Basis 1-forms σ^a ↓
("time", deformation & shape parameters)	structure constant $C^a_{\ bc}$ describes <u>symmetry</u> ↓
<u>Momenta</u> $p^A = (H, p_+, p_-)$ extrinsic curvature ↓	intrinsic curvature ^3R ↓
<u>kinetic energy</u> $K = p_+^2 + p_-^2$	<u>Potential energy</u> $U = 24\pi^2 g\ ^3R$
translational (diag.homo.) only, (1.10), + rotational (non-diag. homo.) (1.30)	(different for different Bianchi Types.)
"Total energy" $H^2 = K + U$	

Their differentials for β_{ij} can be expressed as[27] (cf. Ref. 28)

$$d\hat{\beta}_{(ij)} = \hat{R}^{-1}\{\hat{\alpha}_1\, d\beta_+ + \hat{\alpha}_2\, d\beta_- + $$

$$+ \hat{\alpha}_3\, \sigma_3 \sinh \delta_3 + \hat{\alpha}_4\, \sigma_2 \sinh \delta_2 - \hat{\alpha}_5\, \sigma_1 \sinh \delta_1\}\, \hat{R} \tag{1.21}$$

where

$$\delta_i = 2(\beta_j - \beta_k) \qquad \text{(i, j, k cyclic)} \tag{1.22}$$

and

$$\alpha_1 = \begin{bmatrix} 1 & & \\ & 1 & \\ & & -2 \end{bmatrix}, \quad \alpha_2 = \begin{bmatrix} \sqrt{3} & & \\ & -\sqrt{3} & \\ & & 0 \end{bmatrix}$$

$$\alpha_3 = \begin{bmatrix} 0 & 1 & 0 \\ 1 & 0 & 0 \\ 0 & 0 & 0 \end{bmatrix}, \quad \alpha_4 = \begin{bmatrix} 0 & 0 & 0 \\ 0 & 0 & 1 \\ 0 & 1 & 0 \end{bmatrix}, \quad \alpha_5 = \begin{bmatrix} 0 & 0 & 0 \\ 0 & 0 & 1 \\ 0 & 1 & 0 \end{bmatrix}. \tag{1.23}$$

α_A form a basis for the set of traceless symmetric 3x3 matrices. Here σ^a are the invariant basis 1-forms of SO_3. We can view them as the canonical "coordinates" (frames) in a symmetric space replacing the Euler angles. The canonical momenta conjugate to q^A are

$$p_A = (p_+, p_-, p_\varphi, p_\theta, p_\psi) \qquad (A = 1, \ldots 5) \tag{1.24}$$

The canonical form $p_A\, dq^A$ requires that the momenta conjugate to β_{ij} be

$$p_{ij} = \hat{R}^{-1}(p_A\, \alpha_A)\hat{R}\ . \tag{1.25}$$

Or, explicitly,

$$6\, p_{(ij)} = \hat{R}^{-1}\left\{\hat{\alpha}_1\, p_+ + \hat{\alpha}_2\, p_- + \frac{3\,\hat{\alpha}_3\, L_3}{\sinh^2\delta_3} + \frac{3\,\hat{\alpha}_4\, L_2}{\sinh^2\delta_2} + \frac{3\,\hat{\alpha}_2\, L_1}{\sinh^2\delta_1}\right\}\hat{R} \tag{1.26}$$

where L_a are the invariant vectors of SO_3-symmetric space dual to σ^a. See Ref. 5,40 for details. The Einstein action is then cast in the canonical form:

$$I = \int (p_{ij}\, d\beta_{ij} - H d\Omega) \tag{1.27}$$

The Hamiltonian constraint is

$$H^2 = 6\,\mathrm{Tr}(p^2) - 24\pi^2 g\, {}^3R - 24\,\pi^2 \sqrt{g}\, \mathcal{L}_M \tag{1.28}$$

where $\mathcal{L}_M$ is the matter Lagrangian. The presence of matter is necessary for rotation in homogeneous universes. The momentum constraint $\pi^{ij}{}_{|j} = \frac{1}{2} \mathcal{L}^i_M$ is related to the vector components of the Lagrangian. (We ignore the factor-ordering problem here, see Refs. 12, 18, 21 for discussions)

The Hamiltonian constraint (1.28) can be written as a sum of kinetic, potential and matter parts respectively.

$$H^2 = K_{kin} + U + M \tag{1.29}$$

The potential energy term related to ^3R is the same as that in the diagonal case. The kinetic energy for this class of metric has a translation and a rotation component,

$$K = p_+^2 + p_-^2 + \sum_{a=1}^{3} \frac{L_a^2}{2\mathscr{I}_a} = K_{translation} + K_{rotation}, \tag{1.30}$$

where

$$\mathscr{I}_a\,(\beta_+, \beta_-) = \frac{1}{6}\sinh^2 \delta_a \tag{1.31}$$

plays the role of "moments of inertia". We will return to this model in the next lecture.

Superspace Metric. Associated with this kinetic energy quadratic form, one can write down the superspace metric for the class of general (nondiagonal) homogeneous cosmology as

$$ds^2 = 24[-d\Omega^2 + d\beta_+^2 + d\beta_-^2 + \sum_{a=1}^{3} \ell_a^2\,(\sigma^a)^2] \tag{1.32}$$

where

$$\ell_a^2\,(\beta_+, \beta_-) = 2\,\mathscr{I}_a = \frac{1}{3}\sinh^2 \delta_a \tag{1.33}$$

This was shown by DeWitt[13,22] to be a space of constant curvature (^6R = constant). For the case where there is only one non-zero nondiagonal metric component (this is called the "symmetric" rotating universe by Ryan[27]), $\sigma^3 = d\varphi$, then

$$ds^2 = 24(-\,d\Omega^2 + d\beta_+^2 + d\rho^2) \tag{1.34}$$

where

$$d\rho^2 = d\beta_-^2 + \frac{1}{3}\sinh\,(2\sqrt{3}\;\beta_-)d\varphi^2 \tag{1.35}$$

is the metric for a 2-dim space of constant negative curvature H^2. This space has the topology of $M_o(\Omega) \times M_o(\beta_+) \times SL(2)/SO(2)$, where the quotient space is a symmetric space of type AI of rank 1. The superspace for the general case has topology $M_o(\Omega) \times SL(3)/SO(3)$, where the quotient is a symmetric space of type AI of rank 2.

C. Superspace Dynamics and Nuclear Collective Model

After this brief review, let me digress for a moment to expound the physical meaning of superspace dynamics. Perhaps the best way is to bring in the analog of nuclear collective model[28,29]. Recall that in the metric functions γ_{ab} the diagonal components α and $\beta_{\pm}$ measure the scale (volume) and deformation of the geometry from the isotropic configuration. For spatially closed universes, $\beta_{\pm} = 0$ gives the Robertson-Walker universe, $\beta_{-} = 0$ gives the Taub universe. The off-diagonal metric components can be replaced by the three Euler angles θ, ϕ, ψ parametrizing the rotation matrix which diagonalizes γ_{ab}. Similarly, in nuclear collective model, the dynamical variables which describe the quadrupole (J=2) deformation of a spherical nucleus can be characterized by five amplitude functions.

$$\alpha_{2\mu} = \sum_{\nu} a_{2\nu} D^{2}_{\mu\nu}(g) \quad , \quad \mu = \pm 2, \pm 1, 0 \quad , \tag{1.36}$$

where $D_{\mu\nu}(g)$ is the characteristic function and $g = (\phi, \theta, \psi)$ is an element of the rotation group SO_3. From the symmetry of deformation the five coefficients $a_{2\nu}$ reduce to two real independent variables

$$a_{22} = a_{2,-2} = \frac{1}{\sqrt{2}} \beta \sin\gamma = \frac{1}{\sqrt{2}} \beta_{-} \ , \tag{1.37}$$

and $a_{20} = \beta \cos\gamma = \beta_{+}$,

where $\beta = \sqrt{\beta_{+}^{2} + \beta_{-}^{2}}$ is the deformation parameter and $\gamma = \tan^{-1} (\beta_{-}/\beta_{+})$ is the shape parameter. These two parameters, together with the three Euler angles (ϕ, θ, ψ), give a complete description of the quadrupole excitations of a spherical nucleus. The kinetic energy of nuclear collective motion is given by

$$K = \frac{1}{2} D \sum_{\mu} |\dot{\alpha}_{2\mu}|^{2} = K_{\text{translation}} + K_{\text{rotation}} \tag{1.38}$$

It consists of a translational part

$$K_{\text{translation}} = \frac{1}{2} D_{\beta\beta} \dot{\beta}^{2} + D_{\beta\gamma} \dot{\beta} \dot{\gamma} + \frac{1}{2} D_{\gamma\gamma} \dot{\gamma}^{2} \tag{1.29}$$

and a rotational part

$$K_{\text{rotation}} = \sum_{a} \frac{L_{a}^{2}}{2\mathcal{I}_{a}} \tag{1.40}$$

where $D(\beta, \gamma)$ are the variable "mass" tensor, and $\mathcal{I}_{a}(\beta, \gamma)$ of nuclear deformation is assumed to be dependent only on the shape parameters (β, γ) or (β_{+}, β_{-}),, but independent of the orientation. For small oscillations $U(\beta_{+}, \gamma)$ assumes a harmonic form $U = \frac{1}{2} C \sum_{\mu} |\alpha_{2\mu}|^{2}$. The Hamiltonian in the harmonic approximation is then given by (the Bohr Hamiltonian)

$$H = K + U = K_{vib} + K_{rot} + U_{vib} \tag{1.41}$$

where

$$U_{vib} = \frac{1}{2} C \beta^2, \tag{1.42}$$

$$K_{vib} = \frac{1}{2} D (\dot{\beta}^2 + \beta^2\dot{\gamma}^2) \tag{1.43}$$

and

$$\mathscr{I}_a = 4 D \beta^2 \sin^2(\gamma - \frac{2\pi}{3} a), \; a = 1,2,3 \tag{1.44}$$

For fixed deformations, β and γ = const, K_{rot} is the kinetic energy of a rotor with moments of inertia $\mathscr{I}_a$. For varying β, γ, the $\mathscr{I}_a$ depend on deformation, and the rotation and vibrational modes become coupled, forming the vib-rot spectra.

We see that this form bears close resemblance with the Hamiltonian of the general homogeneous cosmology. The resemblance is more than just formal. As we will see: From the pont of view of gravitational perturbation theory this general form can be obtained from perturbations of spacetimes with larger symmetry, e.g. RW, Type I or Type IX diagonal cases. This analogy suggests that superspace can be thought of as the space of all excitation modes (configurations) of spacetime. There are also differences:

a) The super-Hamiltonian (1.28) is of a quadratic form, leading to a Klein-Gordon equation (1.17) whereas the nuclear Hamiltonian is of linear order in time, giving rise to a Schrodinger equation.

b) The moments of inertia $\mathscr{I}_a$ for the nuclei is of the circular functional form (1.44), whereas for cosmology they are the hyperbolic form (1.31).

c) One can make a small oscillation approximation in the nuclear potential, but in cosmology the translational kinetic energy term always dominates over the potential term near the singularity for the class of velocity-dominated solutions.[6,7,8] However, there is a regime at late times where the time-dependent potential $V(\Omega, \beta_\pm)$ can be approximated by a near-harmonic (corners closing up, near-spherical potential) form.[7,25] This is the so-called quasi-oscillatory regime[7,25], where, in nuclear physics terminology, the "vibrational" mode begins to appear.

Thus, starting from a background spacetime (zero-mode) of highest symmetry (e.g. the RW universe with SO_4 symmetry) each deformation mode adds an extra dimension D to the minisuperspace. Hence as we add on anisotropic deformation modes successively to the 1-dim (closed) RW universe (α), we get the Taub (D=2), mixmaster (D=3), "symmetric" rotating Type IX (D=4) and the "general" rotating Type IX (D=6). When we add on inhomogeneous modes to each of these, the dimension increases to ∞, as there are infinitely (for compact background spaces, countably ∞) many inhomogeneous perturbations modes. The longer wavelength perturbations are deformation modes whereas the shorter wavelength are gravitational waves. We now describe this way to construct a larger superspace.

II. COSMIC SPECTROSCOPY AND INFRARED BEHAVIOR: DIMENSIONAL REDUCTION IN SUPERSPACE

A. Gravitational Perturbations and the Midisuperspace (mdss) of Inhomogeneous Cosmologies

Superspace is the ∞-dimensional space of all 3-geometries. Minisuperspace usually refers to the space of all homogeneous 3-geometries, which has highest dimension D=6. Whether physics in minisuperspace is a faithful representation of quantum gravity remains from the start an open question. We would like to address this and related questions from a spectral analytic viewpoint. Instead of tackling the full superspace corresponding to inhomogeneous spacetimes, we study the smaller but still infinite-dimensional superspace generated from the gravitational perturbation modes of homogeneous cosmologies. We can call this the midisuperspace. The advantage is that one can exploit the symmetry of the background spacetime and use some well-known theorems on the spectrum of invariant operators in curved space.

Let us consider as a concrete example perturbations on a diagonal type IX (mixmaster) universe.[41,41] The background (homogeneous) metric $g^{(0)}_{ij}(x,t)$ is given by (1.3). The perturbed (inhomogeneous) metric is

$$g_{ij}(x,t) = g^{(0)}_{ij}(x,t) + h_{ij}(x,t) \tag{2.1}$$

where the metric perturbations $h_{ij}(x,t)$ are expanded only to the linear order.(Many of our subsequent results are valid without the small amplitude restriction). It has a Fourier decomposition

$$h_{ij}(x,t) = \sum_{J=0}^{\infty}\sum_{K=-J}^{J}\sum_{M=-J}^{J} h^{JKM}_{ab}(t)\, D^{J}_{KM}(g)\, S^{a}_{i}(x)\, S^{b}_{j}(x) \tag{2.2}$$

where $D^{J}_{KM}(g)$ is the SO_3 representation function with quantum numbers (J,K,M). and $h^{JKM}_{ab}(t)$ are the amplitude functions of the (J,K,M)th mode. $h_{ij}(x,t)$ satisfies the Lifshitz equation[37] in the form

$$\Box_{st}\, h_{ij}(x,t) = 0 \tag{2.3}$$

In addition to the (spin 2) tensor modes corresponding to pure gravitational waves, there are also tensor modes constructed from applying the invariant operators on the scalar (spin 0) and vector (spin 1) harmonics. The scalar, vector and tensor perturbations are in general coupled to each other except for spaces of high symmetry (e.g. RW). Each member of the multiplet (J,K,M) adds one extra dimension to the superspace. Each Jth(= 0 to ∞) mode has $(2J+1)^2$ multiplicity. Hence the midisuperspace constructed from gravitational perturbations up to the Jth level has dimension $D = 3 \times 6(2J+1)^2$. The factor 3 accounts for the scalar, vector and tensor components. From earlier perturbation studies of homogeneous cosmology we take note of the following points:

1) The general (non-diagonal) homogeneous universes (e.g. Type IX) can be obtained from the lowest mode vector perturbation of the symmetric

(diagonal) universe.[41,41]

2) The diagonal homogeneous universes are themselves obtainable from anisotropic perturbations of the isotropic (RW) universe.

3) Gravitational perturbations on spatially-flat universe (e.g. Type I) with only extrinsic curvature can generate intrinsic curvature and contribute to the potential energy terms in the Hamiltonian (e.g. Type VII_o, IX from Type I).[44]

Gravitational perturbations on homogeneous cosmology offers an accessible way to construct higher-dimensional superspace (ss). In the perturbation framework, one can thus think of the

Dimension of ss ~ number of normal modes excited,
Metric of ss ~ kinetic energy of normal-modes, and
^3R of spacetime ~ potential energy of normal-modes.

In this picture, each point in superspace represents a particular excitation configuration of 3-geometry. This viewpoint also offers an easier physical interpretation of the quantum and statistical processes in superspace, as they can be thought of as the excitation and exchange of the normal-modes of spacetime. Similar to the phonons of lattice dynamics or the deformation modes of nuclear collective model discussed above. It is in this sense that we refer to superspace as cosmic spectroscopy. To examine the basic assumptions of minisuperspace quantum cosmology we will analyze the spectrum of the Lifshitz operator in spacetime and the spectrum of the Wheeler-DeWitt operator in superspace in what follows:

B. Minisuperspace Approximation and Zero-mode Dominance - Spectral Analysis of the Lifshitz Operator in Spacetime

After setting up the problem in this framework, we can now try to address one central issue in quantum cosmology, i.e., the minisuperspace approximation to quantum gravity.[21,50,51] In the framework of gravitational perturbation theory, the tower of (inhomogeneous) modes built up from a single mode of background spacetime (usually assumed to be homogeneous) with a given symmetry and topology encompasses a particular selection of superspace. Phrased in terms of the spectrum of the Lifshitz operator (2.3) the minisuperspace approximation is equivalent to a truncation of the higher modes. Whether the behavior of the lowest modes (homogeneous cosmology) is characteristic of the complete system(inhomogeneous cosmology), and whether the quantization of the lowest modes (quantum cosmology) is a faithful representation of the full theory (quantum gravity) depends on a number of factors:

1. The geometry and topology of background spacetime and the observation scale. The nature of the spectrum - discrete or continuous, the density of states-depends on these factors. Whether the lowest sector is a good representation of the physical world depends also on the energy scale of observation. Dimensional reduction in the Kaluza-Klein theory is a well-known example.[52] There, the topology of space (e.g. $M^4 x S^7$) is dictated by physical reasons (S^7 to admit the gauge groups of electroweak and strong interaction which are compact).[53] The dominance of the

zero-mode is apparent only at energies much lower than the Planck scale. In this sense we can say that 4-dimensional physics is the infrared limit of the more complete 11-dim theory. In gravitational perturbation theory, the inhomogeneous modes are of shorter wavelengths and thus lie at a higher energy. Thus at any one time (static spectrum) one can view the homogeneous mode as the infrared limit of the full theory. [Of course as the universe expands, all modes are red-shifted to a longer wavelength and lower energy. It is in this dynamical sense that we sometimes view the late (cosmic t) time regime as the infrared limit. This is not to be confused with the discussion of the next section when the approach towards singularity (as measured by $\Omega\to\infty$) is viewed as the infrared limit in superspace.] Whether a lowest mode at one time will stay as the lowest mode at a later time also depends on the particular dynamics and the coupling of modes.

2. Dynamics and Coupling. For spacetimes of lesser symmetry, e.g. mixmaster (but not RW or Taub), the perturbation modes $h^{JKM}_{ij}(t)$ are generally coupled. A glance at the spectrum as a function of deformation $E(\beta)$ (e.g. Fig. 5 of Ref. 40) shows that the lowest mode changes with the magnitude of deformation. Interestingly, when there is level-crossing, whether the system continues to occupy the same mode (which is of a lower energy before the level-crossing point) but now at a higher energy, or settles into a new mode of lower energy (which has higher energy before crossing) depends on how fast (elastic) or slow (plastic) the deformation is exerted (Wheeler, 1970 private communication). In cosmology, both deformation (β) and shape (γ) depend on the dynamics dictated by Einstein's equations. Hence in addition to mode-coupling due to the particular symmetry for certain classes of spacetimes, (e.g. mixmaster) there is always dynamical coupling between the modes at level-crossing.

3. Nonlinearity and Backreaction. In perturbation analysis one assumes that the amplitudes of the perturbation modes $g = g^{(0)} + \varepsilon h^{(1)} + \varepsilon^2 h^{(2)} + \ldots$ are small enough ($|\varepsilon| < 1$) so that the full non-linear equation can be approximated[55] by a sequence of coupled equations in increasing orders of ε. The Lifshitz equation is just the simplest one for the linear gravitational-wave modes $h^{(1)}_{\alpha\beta}$ in this hieararchy. To the second order in ε, the equation for $g^{(0)}$, the background metric,

$$G_{\mu\nu}(g^{o}) = \kappa \langle T_{\mu\nu}(h^{(1)}h^{(1)}\rangle \tag{2.4}$$

has a source term $T_{\mu\nu}$ proportional to the quadrature of $h^{(1)}_{\alpha\beta}$. To ensure that $T_{\mu\nu}$ shares this same symmetry as the background metric, one usually introduces some averaging scheme on $T_{\mu\nu}$. The average is performed over a range small compared to the curvature radius $a \sim R^{-1/2}$, but large compared with the wavelengths λ_n of h. For the high frequency waves the so-called Brill-Hartle-Isaacson[56] average yields an energy momentum tensor in the form of a radiation fluid. Minisuperspace cosmology assumes that the backreaction of the high-frequency modes on the lowest modes is negligible. This will no longer be a good assumption when the background field changes very fast compared to the natural frequency of the higher modes. (Backreaction of high-frequency gravitational waves on the background is discussed in Ref. 44, 48). The high-low mode separation in such cases becomes meaningless. Thus, adiabatic variation of the background (lower modes) is a precondition for such a distinction in general.

When the full non-linearity of the system is incorporated, this separation of background and perturbation can become blurry even at lower energy or late times. Soliton structure (an example is given by the Type VII universe, see Ref. 57,58) and chaotic behavior[59] can also arise. Nonadiabatic, nonlinear effects show up quantum mechanically as particle creation and stimulated emission at the Planck and Compton scales[60] which further undermines the validity of minisuperspace approximation.

4. Quantum and Statistical Effects

We will just make a few general remarks here on potential problems of the minisuperspace approximation arising from the quantization of some restricted degrees of freedom. Our next lecture will elaborate on the last point.

a) In the canonical quantization scheme, minisuperspace approximation amounts to simultaneously setting to zero both the amplitudes and momenta of the higher modes, which violates the uncertainty principle. Quantum fluctuations in these other degrees of freedom as well as their self-interaction and interaction with the lowest modes can influence the lowest mode behavior.[50]

b) In reducing a system from infinite to finite degrees of freedom, we are avoiding the questions of infinites, renormalization and the accompanying problems of anomalies and cancellations, which exist in the full theory (cf. Ref. 54).

c) Loss of coherence and correlation arise when only the dynamics of a subsystem is monitored. Dissipative behavior appear in the lower modes (treated as the system) when there is insufficient or inaccessible information about the high modes (treated as both). [Lecture III]

C. Infrared Behavior and Dimensional Reduction - Spectral Analysis of the Wheeler-DeWitt Operator in Superspace

The above discussion focuses on the spectrum (perturbation modes) of the Laplace-Beltrami (Lifshitz) operator on a background spacetime. There, infrared behavior refers to the lowest-lying modes which correspond to the homogeneous anisotropic cosmologies. Here we want to present another viewpoint to address the infrared problem, i.e., in terms of dimensional reduction. We shall adopt the methodology developed for studying the symmetry behavior of quantum fields in curved spacetime, and carry out an analysis based on properties of the Wheeler-DeWitt operator in minisuperspace. The wave function ψ is treated as an order parameter field which describes the quantum state of the universe.

In our previous work[61,62], we have discussed the symmetry behavior of an interacting scalar (order-parameter) field in product spaces with topology $R^d \times B^b$ where R is a d-dim noncompact space and B is a b-dim compact space. This includes many physically interesting systems such as

1) S^4, the Euclideanized de Sitter universe[65],

2) $R^1 \times S^3$ the Einstein, Taub and mixmaster universes[66],

3) $R^2 \times S^2$ the Einstein-Rosen[45] and axisymmetric solutions,

4) $R^3 \times S^1$ the imaginary-time finite temperature field theory

5) $M^4 \times S^1$ the Kaluza-Klein theory.

These systems are examples of what we define as "finite-size" systems. After a background-field splitting $\Phi = \hat{\phi} + \phi$, the (linearized) fluctuation field ϕ satisfies the wave equation

$$\hat{A}\phi = (\Box + M^2_{eff})\phi = 0 \quad , \tag{2.5}$$

$$M^2_{eff} = m^2 + \xi R + \frac{1}{2}\lambda\hat{\phi}^2 \; , \tag{2.6}$$

where $\Box$ is the Laplace-Beltrami operator on the 4-dim spacetime with 4-curvature R and the effective mass M_{eff} depends on the coupling ξ of field with R and self-coupling ($\lambda\phi^4$ assumed here, $\hat{\phi}$ being the background field). A number of observations can be made for this general problem:

1) <u>Zero-mode dominance:</u> For spacetimes with some compact spatial dimension or for invariant operators A of the fluctuation field with a discrete or band spectrum, the most important contribution to the infrared behavior comes from its zero mode or band.

2) <u>Dimensional reduction:</u> The decoupling of higher modes from the dynamics gives rise to dimensional reduction in the low energy limit. The symmetry behavior of these constrained systems is determined by the value of a parameter η equal to the ratio of the correlation length ξ ($\xi = M^{-1}_{eff}$, $M^2_{eff} = \partial^2 V/\partial\hat{\phi}^2|\hat{\phi}_{minima}$, where V_{eff} is the efffective potential) to the geometric scale L of the background spacetime. (For compact dimensions $L = 2\pi a$ where a is its curvature radius. For noncompact dimensions $L = \infty$). Dimensional reduction occurs for very large values of η.

3) <u>Effective infrared dimensions:</u> In the low energy limit the system behaves effectively as in a lower dimension. The effective infrared dimension (EIRD) for product spaces $R^d \times B^b$ (with d non-compact dimensions and b compact dimensions) is, to a first approximation, equal to d.

4) <u>Curvature dependence of correlation length:</u> The geometric and field parameters in the effective potential run with curvature and energy according to a set of renormalization group equations. The correlation length at different minimum energy states can thus decrease or increase with curvature, changing η accordingly. Therefore the EIRD can be different at states of different symmetry.

Let us now consider a similar problem on minisuperspace. The Wheeler-DeWitt equation (1.17) for the wave function of the universe ψ has the form of a Klein-Gordon equation in curved space (mnss). The interpretation of ψ as a field (3rd quantization - see remarks in Section II) is questionable, but is not absolutely necessary for our discussion. This is because even though the general results we quoted above were derived for quantum fields, the general framework is valid on the quantum mechanical level, (as applied to the Kaluza-Klein dimensional-reduction problem. For quantum mechanical problems see Refs. 61, 62). For phase transition considerations if one does not care about the full energetics of a system evolving from a metastable to a stable state, which requires the knowledge of the effective action, but is satisfied with finding out whether a state with certain symmetry is stable or not, then analysis based on the effective mass will be sufficient.[67-69] To this order (the mean field approximation) the effect of self-interaction is to modify the effective mass Eq. (2.6) through the $\lambda\hat{\phi}^2/2$ term. The wave equation (2.5) remains in the form of a free-field - which makes the interpretation of ψ as a "third-quantized" field unnecessary.

As a concrete example let us consider the minisuperspace of the general (non-diagonal) homogeneous cosmologies with metric (1.32). To simplify discussions let us also make the following assumptions:

1) <u>Velocity-dominated</u> approximation - This assumes that the extrinsic curvature terms in the Einstein equations dominate over the intrinsic curvature terms. This is equivalent to saying that the kinetic energy is greater than the potential energy K >> U in the Hamiltonian (1.28) or setting $\mathcal{R} = 0$ in (1.17). This assumption is valid near the singularity for the class of "generalized Kasner" solutions.[7,8]

2) <u>Adiabatic</u> <u>deformation</u> approximation - This assumes that the variation of ℓ_a in (1.33) [related to $\mathcal{I}_a$ via (1.31)] measured in Ω time is small compared to the characteristic time scale of rotation (determined by L_a). In nuclear collective model when the potential has a harmonic form, this assumes the $E_{vib} >> E_{rot}$, or that the moment of inertia is not influenced significantly by rotation during a period of vibration (irrotational approx.). For minisuperspace dynamics with velocity-dominated approximation, this assumes that $T_{transl} >> E_{rot}$, or that ℓ_a are approximately constants, taking on the adiabatic values of (β_+, β_-) in the Ω-time interval of interest.

Thus we will discuss the spectrum of $\Box_{ss}$ in (1.17) with $\mathcal{R}=0$ and metric given by (1.32). The "kinetic energy" term has a translational part and a rotational part. We want to see if we can use the general observations on infrared behavior described above to deduce a well-known result in quantum cosmology: the effect of rotation becomes less important compared to translation as the universe evolves towards the singularity.[27] Through the superspace representation we have transformed this problem of dynamics of spacetime into one about the geometry and topology of minisuperspace. We need to make these connections:

1) The low-energy [in this case, the kinetic energy K in (1.30)] infrared domain corresponds to the regime near the singularity, i.e. large Ω.

2) The 6-dim minisuperspace $(\Omega, \beta_+, \beta_-, \sigma^a)$ can in certain limits be reduced to a product space $R^1 \times R^2 \times B^b$, with the compact 3-space B^b describing rotation [metric $d\rho^2 = \sum \ell_a^2(\sigma^a)^2$, Eq. (1.32)].

3) The spectrum of the WDW operator in such limits yield a discrete or band structure, such that the behavior of the full theory (6-dim) at low energy is effectively that of a lower dimension [in this case, the 3- time (Ω) and space (β_+, β_-) dimensions].

Thus is the idea behind this way of seeing the dominance of the infrared behavior one needs to make the adiabatic approximation to make the reduction to a product space possible.[62] How realistic this assumption is is not clear.

A simpler example is the symmetric type IX with only one off-diagonal component in β_{ij} and $\theta = 0$, $\psi = 0$. The metric is given by (1.34). The Wheeler-DeWitt equation on this mnss can be written as

$$- \frac{\partial^2\Psi}{\partial\Omega^2} + \frac{\partial^2\Psi}{\partial\beta_+^2} + \frac{1}{\sinh(2\sqrt{3}\,\beta_-)} \frac{\partial}{\partial\beta_-}\left[\sinh(2\sqrt{3}\,\beta_-)\frac{\partial\Psi}{\partial\beta_-}\right]$$

$$+ \frac{3}{(\sinh 2\sqrt{3}\,\beta_-)^2}\frac{\partial^2\psi}{\partial\varphi^2} + e^{-4\Omega}(V-1)\Psi + \mu e^{-3\Omega}\left[1+4C^2e^{2\Omega}e^{4\beta_+}\right]^{1/2}\Psi = 0 \tag{2.7}$$

subject to the momentum constraint

$$p_\varphi = \mu C , \tag{2.8}$$

where μ and C are related to the matter content of the universe, which can be taken to be a perfect fluid. Near the singularity ($\Omega \to \infty$) the kinetic energy (extrinsic curvature) terms dominate over the potential (intrinsic curvature) and the matter terms. Thus we can ignore the last two terms in (2.7). The momentum constraint yields

$$\frac{\partial\Psi}{\partial\varphi} = i(\mu C)\Psi \tag{2.9}$$

There is an ambiguity in the Dirac method of treating the momentum constraint in that to satisfy

$$\psi(\beta_\pm, \varphi + \pi/2) = \Psi(\beta_\pm, \varphi) \tag{2.10}$$

we need to have $\mu C = 4n$, n integers, despite the fact that μ and C's are c-numbers. We will not address this issue here. Without any approximation, the topology of this space is $R^1(\Omega) \times R^1(\beta_+) \times H^2(\beta_-,\varphi)$, where H^2 is a 2-dim space of constant negative curvature. This does not qualitatively give the result we naively expect, but rather suggests a more intricate behavior somewhat similar to the reduced product space and "dynamical finite-size effect" we discussed in a related context.[61,67] (Infrared behavior of H^2 appears in a 2-dim inflationary universe model). It is only when $\sinh(2\sqrt{3}\beta_-)$ is slowly varying or approximately constant that this minisuperspace will effectively appear as $R^1(\Omega) \times R^1(\beta_+) \times S^1(\varphi)$. [A similar situation occurs in the reduced product space limit of the Taub universe (in spacetime) under extreme deformation leading to an "oblate" configuration $\ell_1 = \ell_2 >> \ell_3$. See Ref. 62]. In the "low energy" regime only the n=0 will contribute and the "effective infrared dimension" will be 3 rather than 4. This would then be in accord with the classical result that near the singularity "rotation" becomes less important compared to "translation" energy. This problem is under investigation by S. Sinha at Maryland. See Ref. (i).

D. Other Mechanisms of Dimensional Reduction

In the above we have presented two viewpoints for understanding minisuperspace (homogeneous) cosmology as the infrared limit of quantum gravity:

1) minisuperspace as the low-lying modes of the spectrum of the Lifshitz operator on a fixed background spacetime of high symmetry whose amplitude functions constitute the extra dimensions of the infinite-dimensional (midi) superspace of inhomogeneous cosmologies.

2) minisuperspace as the infrared sector of the Wheeler-DeWitt operator in superspace. The dynamics of spacetime (e.g. the relative importance of translational vs rotational degrees of freedom) is transformed via the superspace representation into a problem determined largely by the geometry and topology of minisuperspace (e.g. "finite-size" effect). Here, let me outline some additional mechanisms which can lead to dimensional reduction in superspace.

1) curvature and dynamics induced dimensional reduction. We refer here to the effect of the potential term [e.g. (1.11) (1.14) (1.15)] in the superhamiltonian, which, as we recall, arises from the intrinsic-curvature of the 3-geometries. For homogeneous cosmologies, the form of the potentials differs for each Bianchi Types (class A) but is the same for rotating and non-rotating universe of the same type. In the pictorial representation of Hamiltonian cosmology[24,25] one can for example view a deep "channel-run" solution in Type IX or VII as approximating a dimensional reduction. The special form of the potential confines the universe point to move deeper into the channel (where $\beta_+ >> \beta_-$) whence the three-dimensional (ℓ_1, ℓ_2, ℓ_3) mixmaster ("ellipsoidal" configuration) reduces dynamically to a two-dim $(\ell_1 = \ell_2 >> \ell_3)$ Taub universe ("spheroidal" or "pancake" configuration), or a 2 → 1 dim reduction in minisuperspace. Dimensional reduction of the Taub universe via finite size effect consideration discussed in Ref 62 & 64 is hereby aided by the channeling potential. In the fully dynamical picture, dimensional reduction is of-course never complete ("magnetic mirror" effect) nor even frequent.[25] How effective it is depends on how likely the universe point wanders into a channel and how long it oscillates in the deep channel, which in turn depends on the chaotic behavior[59] and the relative velocity of the universe point in comparison with the narrowing potential. Another example is the "quasi-oscillatory" solution where the universe point hovers in a small region near the origin (of β_+, β_- plane) bounded by the closed, near-spherical potentials. This depicts a 2-dim (mixmaster) to a zero-dimensional (near Friedmann) approximate reduction in the "spatial"-minisuperspace (β_+, β_-), which is possible only at very late times with the presence of matter.

2) Matter and dynamics induced dimensional reduction. What we are describing here is related to the problem of the approach of anisotropic homogeneous cosmologies to the isotropic state. Taking the cosmic spectroscopy viewpoint homogenization is dimensional reduction from midisuperspace to minisuperspace, while isotropization is dimensional reduction from minisuperspace to unisuperspace (FRW). These processes can be brought about by the different dynamics as influenced by the different potential forms in the Bianchi classes, as well as by the presence of matter - both classical and quantum. For classical matter, Collins and Hawking[9] and Doroshkevich, Zeldovich and Novikov[10] have studied the conditions upon which the different Bianchi Types can evolve to the FRW state at late times with fluid source. Misner and Matzner[11] studied the same problem with neutrino viscosity. For quantum matter, backreaction of particle creation can more effectively bring forth the isotropization[71] (more of shear anisotropy than curvature anisotropy) and possibly the homogenization of anisotropic, inhomogeneous cosmologies.[72]

3) Symmetry-breaking in group space and dimensional-reduction in superspace. The curvature-anisotropy which cannot easily be damped away to facilitate dimensional reduction is contained in the potential term in the superhamiltonian described above. The nine types of Bianchi cosmologies have larger or smaller symmetries depending on the dimension of the group of motion on the spatial-hypersurface. Types VIII, IX, (class A) VI_h and VII_h (class B) have the full 6-dim symmetry group, followed by the VI_o, VII_o and IV with dim = 5, and II, V with dim = 3 and Type I with dim = 0. The symmetry breaking patterns can be found in Refs. 4, 5, 9. Note that the dimension (d_g) of group space measures the homogeneity, whereas the dimension (d_s) in minisuperspace measures the anisotropy. There is no direct relationship between these dimensions, but only restrictive relationship. For example, for Bianchi Type IX universes, only the maximally symmetric isotropic FRW universe with $d_s = 1$ (α) can admit the full 6-parameter ($d_g=6$) symmetry group $SO_4 = SO_{3(L)} \times SO_{3(R)}$ where both left (space) and right (body) multiplication exist. The Taub universe with $d_s = 2$ (α, β_+) admits the 4-parameter $SO_{3(L)} \times U_{1(R)}$ group. The general Type IX with the 3-parameter symmetry group $SO_{3(L)}$ can have 3 (mixmaster), 4 ("symmetric" rotating) or 6 (general rotating) dimensional superspaces. A similar consideration on the relationship of symmetry breaking in the gauge group of the internal space and dimensional reduction in the higher-dimensional spacetime occurs in Kaluza-Klein theories.[53]

III. QUANTUM AND STATISTICAL PROPERTIES OF SUPERSPACE COSMOLOGY: Issues and Critiques

In this last lecture I would like to use the cosmic spectroscopy framework developed above (i.e. viewing minisuperspace as the low-lying modes of the Lifshitz operator in spacetime or the result of infrared dimensional reduction in superspace) to discuss some quantum statistical effects in quantum cosmology, with particular focus on the appearance of dissipative behavior of minisuperspace dynamics and its implications. The prototype we shall adopt is that of a system of coupled oscillators depicting the interaction of geometry and the matter fields. Or, in the perturbation scheme we introduced above, it could be the backreaction of the higher (inhomogeneous) modes on the zero (homogeneous cosmology) mode. In this way, from the statistical viewpoint we can once again address the question of the validity of the minisuperspace approximation. However, before I delve into statistical issues, I would like to add a few remarks on viewing the wave function in the Wheeler-DeWitt equation as a quantum field, and the related so-called "third quantization" program.

A. Quantum Field Theory in Superspace

There are three levels of treating minisuperspce dynamics i) the classical level[25] where Hamiltonian dynamics gives the trajectory $\beta(\Omega)$, ii) quantum mechanical level[21,31,36] where the wave function $\Psi(\beta,\Omega)$ gives the probability amplitude[21,51] of finding the universe in a certain configuration, and iii) quantum field level, where $\Psi(\beta,\Omega)$ is viewed as a field whose amplitude functions a_k in a normal mode decomposition become operators in this "third quantized" picture.[73-75] In the Fock representation, one constructs the "number operator $N_k=a_k^+ a_k$ which measures the number of quanta in mode k. The "vacuum" $|0\rangle$ is defined as the zero-quanta state, and so on. So far our discussion in superspace has been carefully confined to the first two levels. Owing to the similarity in form of the Wheeler-DeWitt equation in superspace with that of the Klein-Gordon equation in curved spacetime, it is tempting to construct a quantum field theory in superspace by treating Ψ as an operator. By imposing the canonical quantization rules and using the Bogolubov transformation to relate fields of in-out states, one can even begin to talk about the "Creation of Universe from Nothing",etc.[74,75] Admittedly one can formally carry out calculations in superspace like particle creation in curved spacetime, treating ψ as a scalar field and superspace as the background geometry. One can talk about expanding and contracting "virtual universe-pairs" or even a thermal distribution of such universes as in Hawking effect. But these descriptions do not make much physical sense because the very foundation of constructing field theory in superspace and using it to describe the creation and annihilation of universes is devoid of any operational meaning within the observational confines of this specific universe we live in. Before we indulge in these convenient generalizations, we have to ensure that the questions we try to ask are physically meaningful and the mathematical objects are operationally definable.[76]

Despite the inhibitions we put on the Wheeler-DeWitt equation as a field theory in superspace, one can nonetheless construct a field theory based on the Lifshitz operator in spacetime. This describes the quantized modes of excitation of a background spacetime - in the language we introduced in the first lecture. Gravitons are indeed such entities. They are the freely-propagating degrees of freedom in the excitation modes. Although we usually refer to the short-wavelength classical perturbations as gravitational waves, and the long-wavelength perturbations as deformation modes, they are one and the same thing, and gravitons are just the quantized modes. The amplitude functions [e.g., $h_{ab}^{JKM}(t)$ in Eq. (2.2)] can be quantized, in which case one can talk about the annihilation and creation of zero or many graviton states in a (perburvative) quantum theory of gravity. One can also regard, say, the five configuration parameters $b_\mu = (\beta_+, \beta_-, \theta, \phi, \psi)$ plus the scalar α as a set of second-quantized operators which can create or annihilate the quantum state describing a particular configuration of the universe. (These are, after all, the lowest homogeneous anisotropic perturbation modes of S^3.) The analogy we presented in Section II.C of Hamiltonian cosmology with nuclear collective model has a parallel in their quantum version. The second quantized version of nuclear collective model[28,29] is carried by the interacting boson model[30] (IBM), where the six so-called bosonic operators (α, b_μ) associated with the scalar-tensor deformation modes give a rather good description of the even-even nuclei spectra. To entertain another thought, the physics of quantized perturbation modes of spacetime as a theory of quantum cosmology (based on the Lifshitz operator on spacetimes of certain symmetry) can be viewed as the three-dimensional ("blob" , "jelly") extension of the two-dimensional membrane physics[77] and the one-dimensional string physics.[78] It is not difficult to see the similarity in structure between the Einstein-Rosen cylindrical wave solution (related to the Gowdy model mentioned in Section II) and the string solution.[79]

B. Statistical Effects of Quantum Fields and Quantum Cosmology

Many fundamental concepts of non-equilibrium statistical mechanics can be explained with the simple and well-studied model of a system of coupled oscillators.[80,81] One can use one mode to depict the system of interest and the remaining modes after some coarse-graining would constitute the environment, bath or reservoir. We have used this model to interpret the statistical meaning of dissipation in a dynamical quantum field system, as occurring in cosmological particle creation processes. Using a quantum mechanical analog of coupled oscillator system, Kandrup and I[108,109] applied the projection operator[92,93] techniques to separate the system of interest from the bath variables, and used a subdynamics analysis[94,95] to calculate the changes in the correlation of states due to interaction. Entropy generation from changes in the correlation of the system can be obtained in this manner. For non-interacting but dynamically excited systems, one can calculate the change of coherence in the system as measured by the variance in the coherent-state representation. If the initial state is an eigenstate of the number operator as in vacuum or thermal particle production processes, the number-phase uncertainty relation applied to an initial random-phase state would generate a net increase in number, which can be viewed as a measure of entropy generation.[110,111]

For interacting quantum field systems, changes of correlation due to interaction have also been studied via the Wigner function method.[98,99] In particular, field equations for the higher-order correlation functions in the BBGKY hierarchy has been derived.[113] Wigner function in curved space was discussed[102] and kinetic equations derived, which are useful for analyzing transport phenomena in curved space. Since Wigner function obeys an equation similar in form to that governing the classical distribution function and is directly related to results obtained from the coherent-state representation, it is often used for describing the semi-classical limit of a quantum theory. (However, as pointed out by Joos and Zeh, Kiefer, Halliwell, Padmanabhan, Unruh and, Zurek,[125-131], this is not enough. A necessary condition for the quantum to classical transition is that the wave function has to decohere.) Particle creation in strong or time-varying background fields and geometries can give rise to dissipative effects.[114,115] The statistical mechanics of these backreaction effects can also be understood in the system-bath paradigm. This was explained in my recent review,[112] One idea proposed there, i.e., that all effective theories should manifest dissipative behavior, is particularly relevant to minisuperspace cosmology. An effective theory can be the low-energy, long wavelength limit of a more complete theory of a larger symmetry, higher dimension etc. Our exclusion or ignorance (because we are confined to observations at a lower energy scale) of the complete details of the higher modes (bath) induces a dissipative behavior in the observable sector (system). This is a well-known phenomena in nonequilibrium statistical physics, but it has important significance for quantum cosmology.

The ideas and methods developed for studying the statistical properties of quantum systems can be applied to quantum cosmology. Wigner function was used to study the correlations in the wave function of the universe.[123-125] The system-environment paradigm of quantum measurement theory[90-94] was used to study decoherence[126-131] and quantum-to-classical transition in quantum cosmology. These are related to questions proposed in Sec. II.B.3 and II.B.4(c), i.e. the backreaction of higher modes on the low-lying modes. For example, on the question of the validity of minisuperspace approximation, one can say that it is at least incomplete, as it neglects the dissipative terms in the dynamical equation for the minisuperspace modes. Fluctuation will also appear in the system if the environment has a stochastic source. Existence of dissipation and memory lost also changes one's view on the issue of initial conditions in quantum cosmology. It means that there may exist a wider class of initial conditions than the Hartle-Hawking[31] or Vilenkin-Linde[34,35] choices which are compatible with the same later-time behavior. Philosophically this is similar to the chaotic cosmology proposal of Misner[11,24] where the outcome (late-time behavior) is determined more by the processes-dynamics and interaction, rather than the stipulation of specific initial conditions. See also the recent comments by C. H. Woo.[36]

1. Basic Issues of Quantum Cosmology and Statistical Effects

Before discussing dissipation in quantum cosmology, I should outline some general issues particular to quantum cosmology where statistical concepts could play an essential role. The central issues are: 1) the emergence of time, 2) the quantum to classical transition, and 3) choice of initial conditions. These issues are to varying degrees related to each other. Each one is by itself a fundamental subject to which many studies are devoted. Therefore I will only delineate the main points in each issue and show how they are related to the problem of dissipation in quantum cosmology.

a) The issue of time[93, 116-118, 129-131]

Time occupies an unusual role in quantum mechanics. It is, as Hartle describes it, the "sole observable which is not represented by an operator, but rather enters the theory as a parameter describing evolution".[116] Therefore, "unlike every other observable for which there are interfering alternatives (e.g. position and momentum) there is no observation which intereferes with the determination of an observation's time of occurrence".[116] The preferred time in quantum mechanics is associated with some fundamental consequences which include i) causality, ii) the notion of a complete description by a state on a spacelike surface, and iii) unitarity.[116]

It has been proposed that this peculiar feature of time in quantum mechanics might be approprite only to the late universe when spacetime behaves classically. This brings up the connection of the issue of time with quantum cosmology and the classical limit, the second issue named above. In quantum cosmology, there is no intrinsic time, but one can regard any one of the variables as playing the role of time and speak in terms of the conditional probabilities of observing the variables of interest in conjuction with a definite reading of this "time" variable.[116,121,122] For the purpose of our discussion here, the "time" variable could be the scale factor a of the universe and the direction of increasing a (expansion) defines an arrow of time against which all dynamical processes are measured (including decoherence and dissipation). According to Gell-Mann et al[121] and Hartle,[116] the existence of a late classical universe is "a consequence of the specific conditon in a more general sum-over-history framework of quantum prediction". In this view, in fact, only a limited set of initial conditions (e.g. the Hartle-Hawking condition) can possess classical limits. This is how the third issue named above enters.

b) Quantum to classical transition[90-94, 126-131]

Again this is a topic which appears in almost every discipline of theoretical physics. The view which appears to be most complete and fundational is that derived from the quantum theory of measurement,[90-94] which uses basic concepts in non-equilibrium statistical mechanics. [Quantum optics, which studies the quantum statistical properties of radiation interacting with a multilevel atom and the surrounding, is a good model where these concepts show]. Halliwell has applied this viewpoint to quantum cosmology.[126] According to him, at least two requirements must be satisfied before a system may be regarded as classical.

1) The system is in one of a number of definite states. This is possible only if the interference between different states is small, i.e. the system decoheres. This manifests as the

destruction of the off-diagonal elements of the density matrix.[119-120]

2) The wave function should be strongly peaked about some classical configuration. Since classical wave functions usually manifest strong correlation between coordinate and momenta, this latter property can be used as a partial indicator of the emergence of classical behavior.

The second criterion can be tested on a quantum system by the use of Wigner distributions, as Halliwell did for quantum cosmology.[123] The first problem, that of decoherence of the density matrix of the universe was first studied by Kiefer[129] and Zeh[130] and recently by Halliwell, Padmanabhan and others[126-128] by examining the interaction of the universe as a system with an environment of scalar fields or higher-mode gravitational perturbations. This is a basic problem in statistical physics. However, decoherence is only one consequence of this scheme which involves interaction with the environment. Dissipation in the system is another important consequence which appears in an almost ubiquitous manner. In particular, it is known that entropy generation from the change of correlations in such processes can provide a thermodynamic arrow of time, and dissipative dynamics of the system incurs memory lost of its initial conditions.

c) Initial conditions[31,116,121]

As explained in the first lecture, a physical description of the quantum dynamics of the universe dictated by the Wheeler-De Witt equation requires a reasonable theory of initial conditions. The attractiveness of the Hartle-Hawking[31] or Vilenkin[32] choices is that they are based on simple principles (e.g. compact 4-geometries) and they lead to reasonable and desirable consequences (e.g. inflation in the classical regime, although these conditions are not the only ones which can elicit these features. For questions on how well the HH or Vilenkin conditions engender inflation, see Ref. 35). One can even argue that the appearance of classical spactime which imparts the particular feature of time as we perceive it in quantum mechanics depends on special choices of the initial condition. We do not concur completely with this view. Objections to this viewpoint can be brought up based on chaotic-dynamical, statistical-mechanical and information theoretical reasonings (see, e.g. Ref. 36). I shall only discuss the statistical-mechanical reasoning involving the effects of dissipation and fluctuation here.

2. Quantum Theory of Dissipation: An Introduction

There are three commonly adopted ways to formulate a quantum theory of dissipation in statistical physics [see, e.g. Ref. 84-87].

a) The density operator method - based on a master equation for the reduced density matrix;

b) The noise operator method - based on a Langevin equation in the form of a damped harmonic oscillator driven by a stochastic source, or equivalently, the Fokker-Planck equation for the distribution function; or,

c) The influence functional method - based on an effective action for the system variables incorporating the averaged effects of the environment.

These method are very well established and have been applied to problems in quantum optics, atomic and molecular dynamics extensively. Recently we see more discussions of these issues and methods in problems of early universe quantum processes. Calzetta, Kandrup and I have used methods a) and c) for studying the dissipative nature of cosmological particle creation.[108-115] Graziani, Bruisma and Cornwall, Habib and Mijic, Hu and Zhang have used methods b), c) for the study of field dynamics in inflationary cosmology.[104-107] Recent discussion of decoherence and dissipation in quantum cosmology uses method a). I shall use this to illustrate the dissipation problem below as it is the least phenomenological. Application of methods b) and c) for statistical problems in quantum cosmology is currently under way.[134]

Let me outline the reduced density matrix method in general terms. I shall begin with the problem of two interacting systems treated on equal footing, paying special attention to the development of correlation between them. We can then make some simplifying assumptions about one system turning it into a bath, which is for our interest the "irrelevant" part, and derive the reduced density matrix for the remaining system, which is the relevant part we are interested in. This approach will enable us to examine the assumptions we put in, which is important for working in a new conceptual setting as quantum cosmology. Our approach here follows the method originally given by Lax. It gives equivalent result as the projection operator technique discussed in, say, Ref. 82. Details can be found in standard textbooks. Our discussion here follows Ref. 95.

Let two systems A and B interact with potential energy V(t). Their reduced density matrices ρ_A and ρ_B are given respectively by tracing over the complete sets of states $|A\rangle$, $|B\rangle$ of the other system, i.e.,

$$\rho_A(t) = \mathrm{Tr}_B\{\rho_{AB}(t)] = \sum_B \langle B|\rho_{AB}(t)B\rangle$$

$$\rho_B(t) = \mathrm{Tr}_A\{\rho_{AB}(t)\} = \sum_A \langle A|\rho_{AB}(t)|A\rangle \tag{3.1}$$

The equations of motion in the interaction picture are

$$i\hbar \frac{d}{dt}\rho_A(t) = \mathrm{Tr}_B\,[V(t),\, \rho_{AB}(t)]$$

$$i\hbar \frac{d}{dt}\rho_B(t) = \mathrm{Tr}_A[V(t),\, \rho_{AB}(t)] \tag{3.2}$$

At any time t, ρ_{AB} can be written in the form as a sum of a direct product term and a correlational term:

$$\rho_{AB}(t) = \rho_A(t) \otimes \rho_B(t) + \rho_C(t) \tag{3.3}$$

where $\rho_C(t)$ represents the correlation which develops in time between the systems from their interaction. If $\rho_C(t_o) = 0$ then initially the two systems are statistically independent. They will, of course, no longer be so for $t > t_o$. The correlational part is what determines both the degeneracy of coherence and the emergence ofdissipation in the system. It depends on the interaction (coupling) and the averaging (coarse-graining) measure.

After substituting (3.3) into (3.2) and some simplification we get

$$i\hbar \dot{\rho}_A(t) = [V_A(t), \rho_A(t)] + Tr_B[V(t), \rho_C(t)]$$

$$i\hbar \dot{\rho}_B(t) = [V_B(t), \rho_B(t)] + Tr_A[V(t), \rho_C(t)] \tag{3.4}$$

Here $V_A(t) = Tr_B\{V(t)\rho_B(t)\}$ is a "Hartree-type" self-consistent energy for system A, which measures the averaged interaction due to B. If B were a reservoir, this term usually produces a shift in the natural frequency Ω of A, which can be absorbed by a redefinition of Ω. Using (3.4) the equation for the correlation operator $\rho_C(t)$ is

$$\begin{aligned} i\hbar\dot{\rho}_C(t) &= [V(t), \rho_A \otimes \rho_B + \rho_C] \\ &\quad - \{[V_A(t), \rho_A]\rho_B + Tr_B[V(t), \rho_C]\}\rho_B \\ &\quad - \rho_A\{[V_B(t), \rho_B] + Tr_A[V(t), \rho_C]\} \end{aligned} \tag{3.5}$$

In problems for which the Hartree energies V_A and V_B are zero, if we also neglect the higher-order terms containing the commutator $[V, \rho_C]$, then we get

$$i\hbar \dot{\rho}_C(t) = [V(t), \rho_A(t) \otimes \rho_B(t)] \tag{3.6}$$

with the formal solution (assuming $\rho_C(t_o) = 0$)

$$\rho_C(t) = -\frac{i}{\hbar} \int_{t_o}^{t} dt' \; [V(t'), \rho_A(t') \otimes \rho_B(t')] \tag{3.7}$$

which for weak coupling $V << H_A$, H_B can be solved perturbatively by iteration.

The formalism developed so far accounts for the evolution of full backreaction (or rather, mutual interaction) in a self-consistent manner. If one of the system, say B, is a "reservoir", i.e., that it has a large number of degrees of freedom (frequency ω_k densely distributed) and that its characteristic period is much shorter than the Poincare recurrence time, then the combined system can be regarded as an <u>open system</u>, and dissipative behavior will emerge in the "system" A. (For a more mathematically rigorous discussion, see Ref. 85). The reservoir has the property that it is usually very "large" (so that it precludes full energy exchange with the system) and very "sluggish" (its overall statistical property changes little over Ω_k^{-1}). One can then assume

$$\dot{\rho}_B(t) \simeq 0, \text{ or } \rho_B(t') \simeq \rho_B(t_o) \tag{3.8}$$

and get the equation of motion for ρ_A (up to 2nd order in λ, the coupling constant in V)

$$\dot{\rho}_B(t) = -\frac{1}{\hbar^2}\int_{t_o}^{t} dt' \, Tr_B\Big[V(t), [V(t'), \rho_A(t') \otimes \rho_B(t_o)]\Big] \tag{3.9}$$

Equation (3.9) is one form of the master equation. As it stands, it is general and formal. Our purpose of sketching its derivation here is to call attention to its generic character and to make explicit the underlying assumptions. Explicit solutions can be obtained by working out some simple examples.

For instance, a prototype used in laser physics is[95-97]

a) system: single mode field of frequency Ω (lasing frequency) (described by annihilation operator a)

$$H_A = \hbar \, \Omega \, a^+ a \tag{3.10}$$

b) bath: simple harmonic oscillator with densely distributed frequencies ω_k (described by annihilation operator b)

$$H_B = \Sigma_k \, \hbar \, \omega_k \, b_k^+ \, b_k \tag{3.11}$$

c) interaction:

$$V(t) = \hbar \, \Sigma \, \lambda_k \, a \, b_k^+ \, e^{-i(\Omega-\omega_k)t} + \text{adjoint}. \tag{3.12}$$

where λ_k is the coupling constant of a, interacting linearly with the kth mode b_k. For lasers, the system consists of the atoms and the field of the lasing mode while the bath refers to the contribution of mirrors, scatterers, lattice vibrations, collisions and all the nonlasing modes. But one usually assumes that the atoms follow the motion of the field adiabatically and eliminates the atomic degrees of freedom. See, e.g., Ref. 97. From these studies some generic properties of quantum dissipative systems can be quoted, which are the same as those derived with influence integral or memory function methods. [See in particular Refs. 89, 92].

a) There is a <u>shift</u> in the system resonance frequency ($\frac{\gamma}{2}$ - $i\Delta\Omega$) as a result of its coupling to the reservoir. For linear coupling, the shift $\Delta\Omega \propto |\lambda|^2$ is proportional to the second order of the coupling constant.

b) <u>Dissipation</u> in the system occurs as energy is lost from the system to the reservoir (Ohmic if coupling is linear). This effect manifests as line-broadening. The dissipative constant γ determines the line width of the system's energy in the Breit-Wigner form.

c) Diffusion of fluctuations in the reservoir into the system. Usually the reservoir is assumed to have a white noise spectrum at finite temperature T. This formalism can also accomodate non-equilibrium stochastic processes, although in such cases it might be simpler to work with the Langevin equation.

d) Memory loss. We examine the conditions under which the damping is Markovian, i.e., that the system's future behavior is determined by the present but not the past. In equation (3.9), the reduced density matrix of the system depends on a double integral, one over the response frequencies ω of the bath, the other over the time interval τ when the correlation is strong. With the linear interaction of (3.12), $\gamma(t)$ is given by[95].

$$\gamma(t-t_o) \propto \int_o^{t-t_o} d\tau \int d\omega\, g^2(\omega)\, \mathscr{D}(\omega)\, \exp[i(\Omega-\omega)\tau] \qquad (3.13)$$

Observe that it depends on the coupling λ and the density of state $\mathscr{D}(\omega)$ of the bath oscillator states. Because of the exponential factor, the largest contribution of the reservoir modes comes from frequencies close to Ω. Only those ω with $|\omega-\Omega| < 1/\tau$ have significant effect on the dynamics of the system. If the reservoir spectrum is densely and broadly distributed around Ω, the frequency integral becomes negligible as τ increases because of destructive interference. The time interval τ for which the frequency integral has appreciable value is called the correlation time τ_c of the reservoir. This says that if little change occurs in the system A during this time (say, $\tau_c << \gamma^{-1}$, the damping time), the Markov approximation is valid.[95]

This last point is key to the memory lost property of dissipative systems. The description I have given above is of a general nature (it is in fact quoted from a standard text in laser physics, Ref. 95). What we need to do next is to consider how well the quantum cosmology problem fits these models and how much of these general features are preserved. The last point, if proven valid, would contest the current viewpoint that the initial conditions play a special and crucial role in quantum cosmology.

C. Dissipation in Quantum Cosmology: An example

The example I shall now use to illustrate the above features is that of a Robertson-Walker universe with scale factor a and a massive (m) minimally coupled scalar field Φ. This discussion is based on the recent work of Calzetta.[132] We are preparing a more comprehensive paper on this issue.[133] The "system" (A) in this problem is the RW universe, the scale factor a acting like the single mode field we introduced previously with "time"-dependent frequency. (Einstein's equation governing it is however not in a simple harmonic oscillator form). The "bath" (B) is the scalar field Φ with a "time" dependent multiple-mode frequency Ω_n. The coupling (C) has two components: a derivative coupling (C_1) whose magnitude depends on how fast the field changes with respect to the geometry, and a time-dependent, nonlinear contact coupling (C_2) proportional to ma^3. This model can also be used to depict RW universe with gravitational perturbations, as the tensor modes obey the Lifshitz equation which has the same form as the massless, minimally-coupled scalar wave equation. Our discussion in Lecture II says that one can view this problem as a midisuperspace formed with a tower of modes built on the RW universe, the latter now regarded as the homogeneous zero-mode. In this light our problem at hand would then address the problem of the backreaction of these higher modes on the zero-mode. As remarked before, this backreaction has many statistical effects: The coupling of these modes after coarse-graining acts like a bath, which decoheres the wave function of the universe and brings forth the emergence of classical behavior. It can also generate dissipation in the quantum dynamics of the universe (adds a viscous term in the Einstein equation) and incur memory lost of the initial conditions.

The classical action for this system is

$$S = \int d^4x \sqrt{-g} \left\{ \left[\frac{1}{2} \kappa R - \Lambda \right] + \frac{1}{2} \sum_{i=1}^{N} \left[\left(\partial \Phi_i \right)^2 - m_i^2 \Phi_i^2 \right] \right\} \tag{3.14}$$

where $\kappa = m_p^2/12$, m_p the Planck constant, $R = 6(\dot{a}/a^2 + \varepsilon/a^2)$is the scalar 4-curvature, $\varepsilon = +, 0, -1$, and Λ is the cosmological constant. We have introduced N scalar fields here although each one can act as a bath as it contains infinite number of modes. Upon setting $\pi = -m_p^2 \dot{a}$ and $p_i = a^2 \dot{\Phi}_i$ the momenta, we get the classical Hamiltonian for this combined system.

$$H = \pi\dot{a} + \sum p_i \dot{\Phi}_i - L \tag{3.15}$$

where L is the Lagrangian. The "kinetic energy" part of this Hamiltonian can be expressed as a metric on minisuperspace

$$d\mathscr{s}^2 = - m_p^2 da^2 + a^2 \sum_i (d\Phi_i)^2 \tag{3.16}$$

which is conformally flat. This has the same form as a (N+1)-dim Milne universe (or a RW universe with a=t) with spacetime metric

$$ds^2 = -dt^2 + t^2 \sum_i (dx_i)^2 \tag{3.17}$$

where one may draw the correspondence $t = m_p a$ and $x_i = m_p^{-1} \Phi_i$. While the RW scale factor plays the role of "time", the number of scalar fields plays the role of "spatial" dimensions. Writing $m_p da = a d\xi$, the metric (3.16) has a conformally-flat form

$$ds^2 = a^2[-d\xi^2 + \sum_i (d\Phi_i)^2] \tag{3.18}$$

where $a = e^{\xi/m_p}$ and ξ acts as a conformal "time" in mss. The metric in superspace has the scalar curvature

$$\mathcal{R} = - N(N-1)/(m_p a)^2 . \tag{3.19}$$

By choosing the "conformal" ordering[21], we can write down the Wheeler-DeWitt equation (1.17) $(\Box + \mathcal{R}) \tilde{\Psi} = 0$ for the mss metric (3.18). The potential term is given by the intrinsic curvature of the spacetime and that of the scalar field in the harmonic oscillator form (see Eq. (3.21) below).

Using the conformally-related wave function Ψ

$$\Psi(\xi, \Phi_i) = a^{(N-1)/2} \tilde{\Psi} \tag{3.20}$$

the Wheeler-DeWitt equation reads

$$\left\{\left[\frac{1}{2} \frac{\partial^2}{\partial \xi^2} + V(\xi)\right]_A + \left[\frac{1}{2} \sum_i \left(- \frac{\partial^2}{\partial \Phi_i^2} + m_\alpha^2 a^6 \Phi_i^2\right)\right]_B\right\}\Psi(\xi, \Phi_i) = 0 \tag{3.21}$$

where $V(\xi) = \Lambda a^6 - \frac{1}{2} \varepsilon m_p^2 a^4$. For massless fields the coupling between gravity and field simplifies greatly. We see that without the last term (3.21) is formally like a wave equation in spacetimes with homogeneous spatial sections. The $\frac{\partial^2}{\partial \Phi_i^2}$ term is the "spatial" Helmholtz operator ${}^{(3)}\Delta$ which has eigenvalues k_n determined by the symmetry of the underlying homogeneous "space". The spectral density of ${}^{(3)}\Delta$ has an effect on the decoherence and dissipation of the gravitational system a.

For massive fields, we can introduce new variables

$$\eta_i = m^{1/2} a^{3/2} \Phi_i \tag{3.22}$$

which simplifies the harmonic frequency of Φ_i to unity. The WDW equation now becomes

$$H\Psi(\xi, \eta) = \left\{\frac{1}{2} \frac{\partial^2}{\partial \xi^2} + \sum_i \frac{3}{2m_p} \eta_i \frac{\partial^2}{\partial \xi \partial \eta_i} + \frac{9}{8m_p^2} \sum_{ij} \eta_i \frac{\partial}{\partial \eta_i} \eta_j \frac{\partial}{\partial \eta_j}\right.$$

$$\left. + V(\xi) + \frac{1}{2} e^{3\xi/m_p} \sum_i m_i \left[- \frac{\partial^2}{\partial \eta_i^2} + \eta_i^2\right]\right\} \Psi (\xi, \eta_i) = 0 \tag{3.23}$$

The cross derivative terms arise from the fact that η_i contains also a. This is common to solutions to wave equations where there is no strict separation of variables by symmetry. [An example is the wave function $\Phi(x,t)$ in mixmaster universe expanded in terms of the asymmetric-top spatial wave function $w(\alpha(t), x)$, where time enters into the curvature radii implicitly. One finds that additional terms $R^{(1)}_{mn}$, $R^{(2)}_{mn}$ etc. appear in the wave equation which depict a form of dynamic coupling amongst the modes. [see Ref. 40 (1974)]. How important these terms are depends on how rapidly a changes relative to Φ_i. If a varies slowly, it can be treated as a parameter instead of a variable in the scalar field sector. We will make this assumption here (cf. Born-Oppenheimer approximation in atomic physics - here a plays the role of a nuclear variable, Φ_i the electronic variables). These cross terms become neglibigly small at large ξ, or in the classical regime [this can be seen by letting $m_p \to \infty$ in (3.23)].

We will begin by considering the effect of one scalar field (i=1) and extend it to N scalar field at the end. Note that except for the $\partial\eta_i\ \partial\eta_j$ term, each field Φ_i is coupled only to a but independent of others Φ_j. It is also more convenient to work with the conformally-scaled wave function $\hat{\Psi} = a^{-3/4}\ \hat{\Psi}$ satisfying the WDW equation $\hat{H}\hat{\Psi} = 0$ where,

$$\hat{H} = H + \frac{3}{4m_p}\frac{\partial}{\partial\xi} + \frac{9}{32m_p^2} \tag{3.24}$$

The statistical properties of this system is depicted by the density matrix

$$\rho(\xi,\eta_i,\ \xi',\ \eta_i') = \hat{\Psi}\ (\xi,\eta_i)\ \hat{\Psi}^*(\xi',\ \eta_i') \tag{3.25}$$

which obeys the corresponding equation

$$\hat{H}\ \hat{\rho} = 0. \tag{3.26}$$

$\hat{\rho}$ has the generic form

$$\hat{\rho} = \sum_{n,m} C_{nm}\ (\xi,\xi')\ f_n(\eta) f_m(\eta') \tag{3.27}$$

(now consider only one field i=1 and remember that η contains ξ as a parameter)

Here $f_n(\eta)$ are the Hermite polynomials

$$f_o(\eta) = \pi^{-1/4}\ e^{-\eta^2/2}$$

$$f_n(\eta) = (n!)^{-1/2}\ \frac{1}{\sqrt{2}}\ (\eta - \frac{\partial}{\partial\eta})\ f_{n-1}(\eta) \tag{3.28}$$

The WDW equation becomes

$$\hat{H}\ \hat{\rho} = \sum_{n,m} f_n(\eta')\Big\{ f_n(\eta)\ H_n\ C_{nm}$$

$$+ \frac{3}{4m_p}\left[\sqrt{n(n-1)}\ f_{n-2}(\eta) - \sqrt{(n+1)(n+2)}\ f_{n+2}(\eta)\right]\frac{\partial}{\partial\xi}\ C_{nm}$$

$$\frac{9}{32m_p}\left[\sqrt{n(n-1)(n-2)(n-3)}\; f_{n-4} + \sqrt{(n+1)(n+2)(n+3)(n+4)}f_{n+4}\right]C_{nm}\Big\} = 0 \quad (3.29)$$

where

$$H_n = \left\{\frac{1}{2}\frac{\partial^2}{\partial\xi^2} + V(\xi) + (n + \frac{1}{2})me^{3\xi/m_p} - \frac{9}{32m_p}\left[n(n-1)+(n+1)(n+2)\right]\right\} \quad (3.30)$$

For reasons given above, we will keep only the dominant n ± 2 mode couplings.

As discussed in the previous subsection, the object of interest to us is the reduced density matrix obtained by tracing out the field degrees of freedom

$$\hat{R}\;(\xi,\xi') = \int d\eta\; \hat{\rho}\;(\xi,\eta,\; \xi',\eta') \quad (3.31)$$

One can consider this as a result of applying some projection operator P on $\hat{\rho}$. Allowing the fact that the scalar field is dominated by the lowest state, one possible form of P can be

$$P = \delta(\xi-\tilde{\xi})\delta(\xi'-\tilde{\xi}')f_o(\eta)f_o(\eta')\delta(\tilde{\eta}-\tilde{\eta}') \quad (3.32)$$

which leads to

$$\hat{\rho}_R = P\;\hat{\rho} = f_o(\eta)f_o(\eta')\hat{R}(\xi,\xi') \quad (3.33)$$

Here $\hat{\rho}_R$ refers to the "relevant" part of the density matrix. The irrelevant part being $\hat{\rho}_I = \hat{\rho} - \hat{\rho}_R$ has $P\;\hat{\rho}_I = 0$. They satisfy the system (cf. Ref. 108).

$$P\hat{H}\;\hat{\rho}_R = -\;P\;\hat{H}\;\rho_I,$$

$$(1-P)\hat{H}\;\hat{\rho}_I = -\;(1-P)\;\hat{H}\;\rho_R \quad (3.34)$$

Tracing over the first equation gives

$$\left[\frac{1}{2}\frac{\partial^2}{\partial\xi^2} + V(\xi) + \frac{1}{2}\;m\;e^{3\xi/m_p} - \frac{9}{16m_p^{\;2}}\right]\hat{R}\;(\xi,\xi')$$

$$= -\int d\eta\;([P,\hat{H}]\;\hat{\rho}_I)\;(\xi,\eta,\xi',\eta) = F \quad (3.35)$$

Solving the equation for $\hat{\rho}_I$ with $\hat{H}$ given by (3.30), we get, after some work, the following equation for $\hat{R}\;(\xi,\xi')$

$$\left[\frac{\partial^2}{\partial \xi^2} + \gamma(\xi) \frac{\partial}{\partial \xi} + \Omega_o^2 (\xi) \right] \hat{R} (\xi, \xi') = F \tag{3.36}$$

where

$$\gamma(\xi) = \sqrt{\frac{2\pi\Omega_2(\xi)}{m_p}} \ \theta(\xi - \xi_r) \cos\left[S_o(\xi) - S_o(\xi_r) - \frac{\pi}{4}\right] , \tag{3.37}$$

Here

$$\Omega_o^2 = V(\xi) + \frac{1}{2} m e^{3\xi/m_p} - \frac{9}{16m_p^2} , \tag{3.38}$$

$$\Omega_2^2 = \Omega_o^2(\xi) + 2 m e^{3\xi/m_p} - \frac{27}{8m_p^2} ,$$

$$S_o(\xi) = \int^{\xi} d\xi' \ \Omega_o (\xi) , \tag{3.39}$$

and ξ_r is the resonance point where $\Omega_o^2 = \Omega_2^2$, i.e.

$$\xi_r = \frac{m_p}{3} \ln \left[\frac{27}{16m_p^2 m}\right] , \tag{3.40}$$

If N fields are present then

$$\gamma(\xi) = \int_o^{\infty} dm^2 \ \mathscr{D}(m^2) \ \theta(\xi - \xi_r(m)) \sqrt{\frac{2\pi}{m_p} \Omega_2(\xi, m)}$$
$$\cos\left[S_o(\xi, m) - S_o(\xi_r, m) - \frac{\pi}{4}\right] \tag{3.41}$$

where $\mathscr{D}(m^2)$ is the (mass) spectral function. Eq. (3.36) is the equation of motion for a driven damped harmonic oscillator with "velocity"-dependent viscosity function $\gamma(\xi)$. It has a similar structural form as the $\gamma(t)$ in (3.13), except the present case has derivative coupling. A similar approach to dissipation was discussed in our earlier work on dissipative properties of quantum fields.[112-115] The origin of dissipation in classical and quantum mechanics was illustrated by the simple two coupled-oscillator example in Ref. 113. Dissipation due to particle creation on background fields and background geometries using closed-time-path effective action method was treated in Refs. 113, 114 and 115, the former two being of contact (polynomial) coupling ($\lambda\phi^4$, $g\phi^3$ theories) while the last on anisotropy damping is dynamical (derivative) coupling. Since in these processes the viscosity function depends on a non-local (in "time") kernel and is history-dependent, strictly speaking the system evolves in a non-Markovian manner. But the overall consequence of dissipation is that the system's late time behavior depends insensitively on the initial conditions. The damping time is a measure of this sensitivity.

We have discussed in the above the conditions for the advent of dissipation and memory lost in quantum systems. As long as the reservoir spectrum is densely and broadly distributed around the natural frequency of the system and the coupling betweeen the fields and gravity is not too weak, the system will evolve in a way that its late time behavior will not depend sensitively on its initial conditions and the surrounding. There are many deeper issues one can discuss from the implication of this result. Let me just comment on one to end this lecture, that of the arrow of time. The dynamical time in our model is the scale factor. If we define expansion as the positive direction of dynamical time, then contraction is negative. Quantum physics is unitary with respect to dynamical time-reversal. Dissipation breaks this symmetry and defines a thermodynamic arrow of time as the direction of increasing entropy. From our earlier studies on the statistical effects of particle creation we learned that even in the contracting phase(negative dynamical time), there will still be dissipation (although the magnitude may differ because of difference in phase space volume for the created particles). The process discussed here is of a similar nature. This means that with dissipation the thermodynamic arrow of time will remain in the same direction in both the expanding and contracting phases. For earlier discussions of this issue in quantum cosmology see Hawking, Page[117] and the book by Zeh[93].

Epilogue: In these lectures I have outlined the basic issues in quantum cosmology, some of which - e.g., the minisuperspace approximation, the possibility of phase transitions, and the almost unavoidable dissipative effects - I feel should deserve better attention and more proper treatment. I have presented some viewpoints in addressing these issues and discussed different methods in solving problems related to these issues. As stated in the beginning these lectures are meant to complement the current views and approaches in the field, which emphasize the issue of initial conditions and use Euclidean path integral methods. (The reader should refer to many excellent reviews on these other aspects to get a more complete picture.) By bringing in insights from other (more mundane?) disciplines I hope to impart a richer physical meaning to a noble but somewhat barren field as quantum cosmology. By discussing the shared concerns and approaches in other related disciplines, I hope to stimulate new lines of thought and queries in this very intriguing yet fundamental topic in theoretical physics.

Acknowledgement: I thank my students and colleagues for discussions on many different topics covered in these lectures, especially Esteban Calzetta, Salman Habib, Henry Kandrup, Denjoe O'Connor, Kristin Schleich, Sukanya Sinha, Chris Stephens and Yuhong Zhang. Lectures II and III contain unpublished research results (References i) ii) and iii)) done in collaboration with E. Calzetta, S. Sinha and Y. Zhang. I enjoyed the hospitality of the particle and astrophysics theory groups at Cornell University while these lectures were prepared. This work is supported in part by NSF grant PHYS87-17155 to the University of Maryland.

REFERENCES AND GUIDE

The references below are given in conjunction with the topics discussed in these lectures. As these lectures are not general reviews, the list here is neither complete nor necessarily representative. The general and more balanced views are usually better represented when the references quoted in these works are included.

For details of these lectures, see:

i) B. L. Hu and S. Sinha, Minisuperspace Cosmology as Infrared Limit of Quantum Gravity. (Abstract in Proceedings of the 12th International Conference on General Relativity and Gravitation (GR12), Boulder, Colorado, July 1989); S. Sinha, Ph.D. thesis (University of Maryland 1990).

ii) E. Calzetta and B. L. Hu, Dissipation and Initial Conditions in Quantum Cosmology (1990); E. Calzetta, Cl. Quan. Grav. (1989); B. L. Hu, Physica A158, 399 (1989).

iii) B. L. Hu and Yuhong Zhang, Coarse-Grained Effective Action, (Abstract in GR12, July 1989); Yuhong Zhang, Ph.D. thesis (University of Maryland 1990).

Lecture I

A. Bianchi (Spatially-homogeneous) Cosmologies

1. L. Bianchi, Mem. Soc. It. Della Sci (DeiXL) (3) 11, 267 (1897)

2. E. Kasner, Am. J. Math. 43, 217 (1921)

3. A. H. Taub, Ann. Math. 53, 472 (1951)

4. G.F.R. Ellis and M.A.H. MacCallum, Comm. Math. Phys. 12, 108 (1969). M.A.H. MacCallum, in General Relativity: An Einstein Centenary Volume , ed. W. Israel and S. W. Hawking (Cambridge Univ. Press, 1979).

5. M. P. Ryan and L. C. Shepley, Homogeneous Relativistic Cosmology (Princeton Univ. Press. 1975).

Solutions near the cosmological singularity

6. E. M. Lifshitz and I. M. Khalatnikov, Adv. Phys. 12, 185 (1963)

7. V. A. Belinsky, I. M. Khalatnikov and E. M. Lifshitz, Adv. Phys. 19, 525 (1970); 31, 639 (1982).

8. D. M. Eardley, E.P.T. Liang and R. K. Sachs, J. Math. Phys. 13, 99 (1972).

Isotropization: Solutions with classical matter

9. C. B. Collins and S. W. Hawking, Ap. J. 180, 317 (1973)

10. A. G. Doroshkevich, V. N. Lukash and I. D. Novikov, Sov. Phys.-JETP 37, 739 (1973), Sov. Astron-AJ 18, 544 (1975).

11. C. W. Misner, Ap. J. 151, 431 (1968); R. A. Matzner and C. W. Misner, Ap. J. 171, 415 (1972) [neutrio viscosity].

B. Quantum Gravity and Quantum Cosmology

Classic papers:

12. B. S. DeWitt in Relativity, Groups and Topology, ed. B. DeWitt and C. DeWitt (Gordon & Breach, 1964). [1963 Les Houches Lectures].

13. B. S. DeWitt Phys. Rev. 160, 1113 (1967)

14. J. A. Wheeler, in 1963 Les Houches Lectures

15. J. A. Wheeler, in Battelle Recontres, ed. C. DeWitt and J. A. Wheeler (Benjamin, 1968) [Battelle Lectures].

Hamiltonian Quantization (work in the 70's)

16. R. Arnowitt, S. Deser and C. W. Misner, in Gravitation, ed. L. Witten (Wiley 1962).

17. M. P. Ryan, Hamiltonian Cosmology (Springer Lecture No. 13, 1972).

18. K. Kuchar, Phys. Rev. D4, 955 (1971) [cylindrical waves].

19. B. K. Berger, D. M. Chitre, V. E. Moncrief and Y. Nutku, Phys. Rev. D5, 2467 (1972) [spherically symmetric systems].

20. W. F. Blyth and C. J. Isham, Phys. Rev. D11, 768 (1975) [RW Universe with scalar field].

Minisuperspace:

21. C. W. Misner, in Magic Without Magic, ed. J. Klauder (Freeman, San Francisco, 1972).

22. B. S. DeWitt, in Relativity, M. Carmeli, S. K. Fickler and L. Witten (Plenum, New York, 1970) [Cincinnati Conference].

23. A. Fisher, in Cincinnati Conference (1969).

Mixmaster Universe:

24. C. W. Misner, Phys. Rev. 186, 1319 (1969)

25. C. W. Misner in Cincinnati Conference (1969)

26. R. A. Matzner, L. C. Shepley and J. B. Warren, Ann. Phys. (N.Y.) 57, 401 (1970).

27. M. Ryan, Ann. Phys. (N.Y.) 65, 506 (1971); 68, 541 (1971) [Symmetric and General Type IX].

Aside: Nuclear Collectic Model and Interacting Boson Model

28. A. Bohr, Mat. Fys. Medd. Dan. Vid. Selsk. 26 No. 14 (1952).
A. Bohr and B. R. Mottelson, ibid 27, No. 16 (1953); Nuclear Structure Vol. 2 (Benjamin, 1975).

29. D. L. Hill and J. A. Wheeler, Phys. Rev. 89, 1102 (1953).

30. F. Iachello and A. Arima, The Interacting Boson Model (Cambridge Univ. Press 1987).

C. Boundary Conditions in Quantum Cosmology: Euclidean Approach (Work in 80's)

31. J. B. Hartle & S. W. Hawking; Phys. Rev. D28, 2960 (1983);
S. W. Hawking; in Relativity, Groups & Topology II ed. B. DeWitt & R. Stora (North Holland 1984). [2nd Les Houches Lectures].
J. B. Hartle, in High Energy Physics, ed. M. J. Bowick and F. Gursey (World Scientific 1985). [Yale Summer School].

32. A. Vilenkin, Phys. Lett. 117B, 25 (1982); Phys. Rev. D27, 2848 (1983); D30, 509 (1984).

33. A. Linde, Sov. Phys. JETP 60, 211 (1984); L. P. Grischuk and Ya. B. Zel'dovich, in Quantum Structure of Space and Time, ed. M. J. Duff and C. J. Isham (Cambridge Univ. Press, 1982).

34. J. B. Hartle, Phys. Rev. D37, 2818 (1988); 38, 2985 (1988);
A. Vilenkin, ibid D39, 1116 (1989).

35. L. P. Grishchuk and L. V. Rozhansky, Caltech GRP-207 (1989).

36. C. H. Woo, Phys. Rev. D39, 3174 (1989)

For a list of work in euclidean quantum cosmology up to 1988 see, J. J. Halliwell, Quantum Cosmology: An Introductory Review and Bibliograph, NSF-ITP 88-131.

LECTURE II

A. Cosmic Spectroscopy - Gravitational Perturbation to Homogeneous Cosmology

Friedmann-Robertson-Walker:

37. E. M. Lifshitz, J. Phys. USSR 10, 116 (1946). (tensor hyperspherical harmonics)

38. J. M. Bardeen, Phys. Rev. D22, 1882 (1980). (gauge-invariant method)

39. J. J. Halliwell and S. W. Hawking, Phys. Rev. D31, 1777 (1985). (in midisuperspace, with Hartle-Hawking condition).

Mixmaster: (in spacetime, or rotating homogeneous cosmology in superspace)

40. B. L. Hu, Phys. Rev. D8, 1048 (1973); D9, 3263 (1974).

41. B. L. Hu & T. Regge, Phys. Rev. Lett. 29, 1616 (1972).

42. B. L. Hu, Phys. Rev. D12, 1551 (1975).

Type I:

43. T. E. Perko, R. A. Matzner & L. C. Shepley, Phys. Rev. D6, 969 (1972).

44. B. L. Hu, Phys. Rev. D18, 969 (1978).

Inhomogeneous:

45. A. Einstein and N. Rosen, J. Franklin Inst. 223, 43 (1937).

46. H. Bondi, F.A.E. Pirani and I. Robinson, Proc. R. Soc. Lond A251, 519 (1959).

47 R. Gowdy, Ann. Phys. (N.Y.) 83, 203 (1974), J. Math. Phys. 16, 224 (1975).

48. B. Berger, Ann. Phys. (N.Y.) 83, 458 (1974).

49. C. W. Misner, Phys. Rev. D8, 327 (1973).

B. Minisuperspace Quantization: Validity and Approximation

50. K. Kuchar and M. P. Ryan, Jr., in Proc. Yamada Conference XIV, ed. H. Sato and T. Nakamura (World Scientific, 1986).

51. K. Kuchar and M. P. Ryan, Univ. Utah preprint (July 1989).

Kaluza-Klein Theory: Dimensional reduction and renormalizability

52. e.g. T. Appelquist, A. Chodos and P.G.O. Freund, Modern Kaluza-Klein Theories (Addison-Wesley, 1987).

53. E. Witten, Nucl. Phys. B186, 412 (1981).
A. Salam and J. Strathdee, Ann. Phys. 131, 316 (1982).

54. M. Duff and D. J. Toms in Quantum Gravity, ed. M. A. Markov and P. C. West (Plenum, New York 1984).

Nonlinearity and backreaction

55. C. W. Misner, K. S. Thorne and J. A. Wheeler, Gravitation (Freeman 1972) Ch. 35.

56. D. Brill and J. B. Hartle, Phys. Rev. B135, 271 (1964).
R. Isaacson, Phys. Rev. 166, 1263, 1272 (1968).
Y. Choquet-Bruhat; Comm. Math. Phys. 12, 16 (1969).
M.A.H. MacCallum and A. H. Taub, ibib 30, 153 (1973).

Solitonic and chaotic behavior

57. V. Lukash, Sov. Phys. JETP 40, 792 (1975).

58. V. A. Belinsky and V. E. Zakharov, Proc. 2nd Marcel Grossmann Meeting, ed. R. Ruffini (North-Holland, 1982) p. 275.

59. See, e.g., J. D. Barrow and D. Chernikov, Phys. Rep. 85, 1 (1982);
I. M. Khalatnikov, E. M. Lifshitz, K. M. Khann, L. N. Shchur, Ya. G. Sinai, in General Relativity and Gravitation (GR10,

Padova, 1983) ed. B. Bertotti, F. de Felice and A. Pascolini, (Reidel, Dordrecht, 1984) p. 343.

60. For nonlinear quantum field effects see, e.g. B. L. Hu and Y. Zhang, Phys. Rev. D37, 2151 (1988).

C. Infrared Behavior of Quantum Fields in Curved Spacetime

Finite size effect in curved spacetimes:

61. B. L. Hu, in Proc. 5th Marcel Grossmann Meeting, Rome 1985, ed. R. Ruffini (N. Holland 1986); D. J. O'Connor, Ph.D. Thesis, Univ. of Maryland 1985; C. R. Stephens, Ph.D. Thesis, Univ. of Maryland 1986.

62. B. L. Hu and D. J. O'Connor, Phys. Rev. D36, 1701 (1987); D. J. O'Connor, C. R. Stephens and B. L. Hu, Ann. Phys. (NY) 190, 310 (1989)

63. D.J. O'Connor, B. L.Hu and T. C.Shen, Phys. Lett. 130B, 31 (1983). (Einstein Univ.).

64. T. C. Shen, B. L. Hu and D. J. O'Connor, Phys. Rev. D31, 2401 (1985). (Taub universe).

65. B. L. Hu & D. J. O'Connor, Phys. Rev. Lett. 56, 1613 (1986). (De Sitter universe).

66. B. L. Hu and D. J. O'Connor, Phys. Rev. D34, 2535 (1986) (Mixmaster universe).

Dynamical finite size effect and quasilocal fields:

67. B. L. Hu, in Proc. CAP-NSERC Summer Institute in Theoretical Physics, Edmonton, July 1987, ed. by K. Khanna et al (World Scientific 1988).

68. S. Sinha and B. L. Hu, Phys. Rev. D38, 2423 (1988).

69. A. Stylianopoulos and B. L. Hu, Phys. Rev. D39, 3647 (1989).

General decoupling theorem:

70. T. Appelquist and J. Carrozone, Phys. Rev. D11, 2856 (1975).

D. Isotropization and homogenization: Dimensional Reduction in Minisuperspace

i) due to classical matter source: Ref. 9, 10, 11.
ii) due to quantum field source: Ref. 71, 72.

For dissipation of anisotropy:

71. Ya. B. Zel'dovich and A. A. Starobinsky, JETP 34, 1159 (1972); B. L. Hu and L. Parker, Phys. Rev. D17, 933 (1978); J. B. Hartle and B. L. Hu, ibid D20, 1757 (1979); 20, 1772 (1979); P. R. Anderson, ibid D32, 1302 (1984); E. Calzetta and B. L. Hu, ibid D35, 495 (1987).

For dissipation of inhomogeneities:

72. Ya. B. Zel'dovich and A. A. Starobinsky, JETP Lett. 26, 252 (1977); B. L. Hu and L. Parker, Phys. Lett. 63A, 217 (1977); J. B. Hartle, Phys. Rev. Lett. 39, 1373 (1977), Phys. Rev. D23, 2121 (1981); G. Siemiemec-Ozieblo, Class. Quan. Grav. 1, 167 (1984); J. Frieman, Phys. Rev. D39, 389 (1989).

LECTURE III

A. Quantum Field Theory in Superspace

"Third quantization and universe creation"

73. S. Giddings and A. Strominger, Harvard preprint HUTP-88/A036.

74. A. Hosoya and M. Morikawa, Phys. Rev. D39, 1123 (1989).

75. M. McGuigan, Phys. D39, 2229 (1989).

76. W. G. Unruh and R. M. Wald are amongst those who strongly object to the third-quantization proposal in quantum cosmology (private communication at this workshop).

Strings, membranes and spacetime excitations

77. See, e.g. M. Duff and E. Sezgin (ed.) Proceedings of the Trieste Conference on Supermembranes and Physics in 2+1 Dimensions ICTP,Miramare, Trieste, July,1989. (World Scientific, Singapore, 1990)

78. See, e.g. M. B. Green, J. H. Schwarz and E. Witten, Superstring Theory, Vol. I, II (Cambridge Univ. Press 1987).

79. T. Matsuki and B. Berger, Phys. Rev. D39, 2893 (1989).

B. Quantum Statistical Effects: General Discussions

Irreversibility in coupled oscillator systems

80. G. W. Ford, M. Kac and P. Mazur, J. Math. Phys. 6, 504 (1963).

81. R. Rubin, J. Math. Phys. 1, 309 (1960); 2, 373 (1961).

Projection operator techniques

82. R. Zwanzig, in W. E. Britten, B. W. Downes and J. Downes (ed.) Lectures in Theoretical Physics III (Interscience, New York, 1961) pp. 106-141.

83. H. Mori, Prog. Theor. Phys. 33, 1338 (1965).

Subdynamics analysis and reduced density operator

84. R. Balescu, Equilibrium and Nonequilibrium Statistical Mechanics (Wiley, 1975).

85. E. B. Davies, Quantum Theory of Open Systems (Academic, London, 1976).

Noise-operator, Langevin and Fokker-Planck equations

86. N. G. van Kampen, Stochastic Processes in Physics and Chemistry (North-Holland 1981).

87. H. Riskin, Fokker-Planck Equation (Springer 1984).

88. R. P. Feynman and F. L. Vernon, Ann. Phys. 24, 118 (1963).

89. A. O. Caldeira and A. J. Leggett, Physica (Utrecht) 121A, 587 (1983); Ann. Phys. (N.Y.) 149, 374 (1983).

Quantum measurement theory and emergence of classical behavior

90. J. A. Wheeler and W. H. Zurek, (ed.) Quantum Theory and Measurement (Princeton Univ. Press 1983).

91. W. H. Zurek, Phys. Rev. D24, 1516 (1981); 26, 1862 (1982).

92. W. G. Unruh and W. H. Zurek, Phys. Rev. D40, 1071 (1989).

93. H. D. Zeh, The Physical Basis of the Direction of Time (Springer 1989).

94. E. Joos and H. D. Zeh, Z. Phys. B59, 223 (1985).

Quantum theory of dissipation: Examples from quantum optics

95. M. Sargent III, M. O. Scully and W. E. Lamb, Laser Physics (Addison-Wesley, 1974).

96. W. H. Louisell, Quantum Statistical Properties of Radiation (Wiley, 1974).

97. H. Haken, Laser Theory (Springer, 1983).

C. Non-equilibrium Quantum Fields in Curved Spacetime

Wigner function:

98. E. P. Wigner, Phys. Rev. 40, 749 (1932).

99. M. Hillery, R. F. O'Connell, M. O. Scully and E. P. Wigner, Phys. Rep. 106, 121 (1984).

Kinetic field theory:

100. P. Carruthers and F. Zachariasen, Rev. Mod. Phys. 55, 245 (1983).

In curved spacetime:

101. J. Winter, Phys. Rev. D32, 1871 (1985).

102. E. Calzetta, S. Habib and B. L. Hu, Phys. Rev. D37, 2901 (1988).

103. S. Habib and H. Kandrup, Ann. Phys. (N.Y.) July, 1989.

Inflationary cosmology:

104. F. R. Graziani, Phys. Rev. D38, 1122, 1131, 1802 (1988).

105. J. M. Cornwall and R. Bruinsma, Phys. Rev. D38, 3146 (1988).

106. S. Habib and M. Mijic (1989).

107. B. L. Hu and Y. Zhang (1989).

D. Entropy Generation in Cosmological Particle Creation and Interaction

Field entropy: correlation from interaction or dynamics

108. B. L. Hu and H. E. Kandrup, Phys. Rev. D36, 1776 (1987).

109. H. E. Kandrup, J. Math. Phys. 28, 1398 (1987).

"Coherence" entropy: particle number and phase duality

110. B. L. Hu and D. Pavon, Phys. Lett. B180, 329 (1986).

111. H. E. Kandrup, Phys. Rev. D37, 3505 (1988).

E. Dissipation in Quantum Fields & Semi-Classical Gravity

Review

112. B. L. Hu, Physica A158, 399 (1989) [Cleveland lecture].

BBGKY hiearchy for interacting fields:

113. E. Calzetta and B. L. Hu, Phys. Rev. D37, 2878 (1988).

Backreaction of particle creation:

114. E. Calzetta and B. L. Hu, Phys. Rev. D40, 656 (1989) [$g\phi^3$ theory].

115. E. Calzetta and B. L. Hu, Phys. Rev. D35, 495 (1987) [Type I and $\lambda\phi^4$ theory].

F. Statistical properties of Quantum Cosmology

Issue of Time:

116. J. B. Hartle in A. Ashtekar and J. Stachel (ed.) Proc. Second Osgood Hill Conference on Conceptual Problems in Quantum Gravity (Birkhäuser, Boston 1989); Phys. Rev. D37, 2818, 38, 2985 (1988).

117. S. W. Hawking, Phys. Rev. D32, 2489 (1985); D. Page, Phys. Rev. D32, 2496 ((1985).

118. C. J. Isham and K. Kuchar, Ann. Phys. (N.Y.) 164, 316 (1985).

Density matrix of the universe:

119. S. W. Hawking, Physica Scripta T15, 151 (1986)

120. D. Page, Phys. Rev. D34, 2267 (1986)

Conditional probability and prediction:

121. M. Gell-Mann, J. B. Hartle and V. Telegi (in preparation)

122. H. E. Kandrup, Cl. Quan. Grav. 5, 903 (1988).

Correlation and Wigner function in superspace:

123. J. J. Halliwell, Phys. Rev. D36, 3626 (1987).

124. H. Kodama, Kyoto University preprint (1988).

125. E. Calzetta and B. L. Hu, Phys. Rev. D40, 380 (1989).

Decoherence in quantum cosmology:

126. J. J. Halliwell, Phys. Rev. D39, 2912 (1989).

127. T. Padmanabhan, Phys. Rev. D39, 2924 (1989).

128. T. Fukuyama and M. Morikawa, Phys. Rev. D39, 462 (1989).

129. C. Kiefer, Cl. Qu. Grav. 4, 1369 (1987).

130. E. Joos, Phys. Lett. 116A; 6 (1986).

131. H. D. Zeh ibid 116A, 9 (1986); 126A, 311 (1988).

Dissipation and reduced density matrix in superspace:

132. E. Calzetta, Cl. Qu. Grav. (1989).

133. E. Calzetta & B. L. Hu, Dissipation in Quantum Cosmology (1990).

134. B. L. Hu and Y. Zhang, (1990).

On Quantum Gravity for Homogeneous Pure Radiation Universes

Mario A. Castagnino

Instituo de Astronomia y Fisica del Espacio
Buenos Aires, Argentina

Edgard Gunzig

and

Pascal Nardone*

Université Libre de Bruxelles,
Service Chimie-Physique
C.P. 231 - Campus Plaine
1050 Bruxelles

Abstract

Quantum Gravity is studied in the case of a homogeneous pure radiation universe. Several prescriptions to define the wave function of the universe are tested in this framework. A new idea to define this wave function is introduced.It is based on the final classical behaviour of the universe.

Introduction

The aim of these lectures is to discuss several "traditional" as well as new ideas concerning the definition of the wave function of the universe, in the framework of the simplest theoretical laboratory: the homogeneous pure

---

Supported partially by the "STICHTING FUND FOR SCIENCE, TECHNOLOGY AND RESEARCH"

* On leave from Konstanz University under CEE contract $n^o ST2 * 0449$

Quantum Mechanics in Curved Space-Time
Edited by J. Audretsch and V. de Sabbata
Plenum Press, New York, 1990

radiation universe. More precisely, we shall test the following ways to define the wave function of the universe:

a: Hartle-Hawking's wave function defined via their well known initial conditions [1].

b: Vilenkin's wave function defined via the prescription that"at singular boundaries of superspace, includes outgoing waves only" [2].

c: A wave packet definition: this corresponds, as will be fully described in what follows, to the final condition: the quantum evolution must lead to the present classical universe, or, in other words the wave function describing this universe must approach a wave packet characterizing the presently observed cosmological data.

The perspective of this discussion is to test this last idea, as we have the feeling that it is an appealing physical proposal . The wave function must be able to provide the whole range of physical phenomena,corresponding to the presently observed universe,and therefore acquire a wave packet structure delivering precisely this picture. This wave packet must be able to describe phenomena ranging from local particle interference properties,and,through fluctuations in the black-body radiation, up to the galaxies and the planets motion.

For example, let us suppose that the cosmological quantum system consists of a planet with its center located at X, an electron situated at x just undergoing a two-slices experiment, and the hand of a clock pointing at T (fig 1). The corresponding wave function will have the general form $\Phi(X, x, T)$. As the two slices experiment and the planet motion are practically independent, it follows that:

$$\Phi(X, x, T) = \Phi_P(X, T)\Phi_E(x, T) \qquad (1-1)$$

where $\Phi_P(X, T)$ and $\Phi_E(x, T)$ represent the wave functions of the planet and the electron respectively. For $\Phi_P(X, T)$ it is legitimate to choose a wave packet as we can measure the position X and the momentum Π_X with sufficient accuracy for astronomical purposes, so that Φ_P is of the form

$$\Phi_P(X, T) = \Delta(X - X(T)) \qquad (1-2)$$

where $X(T)$ is the classical trajectory of the planet (with time measured by the clock), and Δ is a very high peak function at zero.

In consequence of the electron's quantum behaviour, if $\Phi_1(x, T)$ is the solution of Schrodinger's equation corresponding to the second sealed slice of the screen, whereas $\Phi_2(x, T)$ represents the solution with the first one sealed, the wave function of the electron will take the form:

$$\Phi_E(x, T) = \Phi_1(x, T) + \Phi_2(x, T) \qquad (1-3)$$

which exhibits, as usual, the interference phenomenom. If we focus our attention on the planet only and disregard the electron slice experiment, the resulting wave function will obviously consist of that given in (1-2). As it appears legitimate to consider the universe,as a whole, as a macroscopical object, like the planet for example, its seems appealing to describe the universe quantum-mechanically with a wave packet, at least as a final condition. We

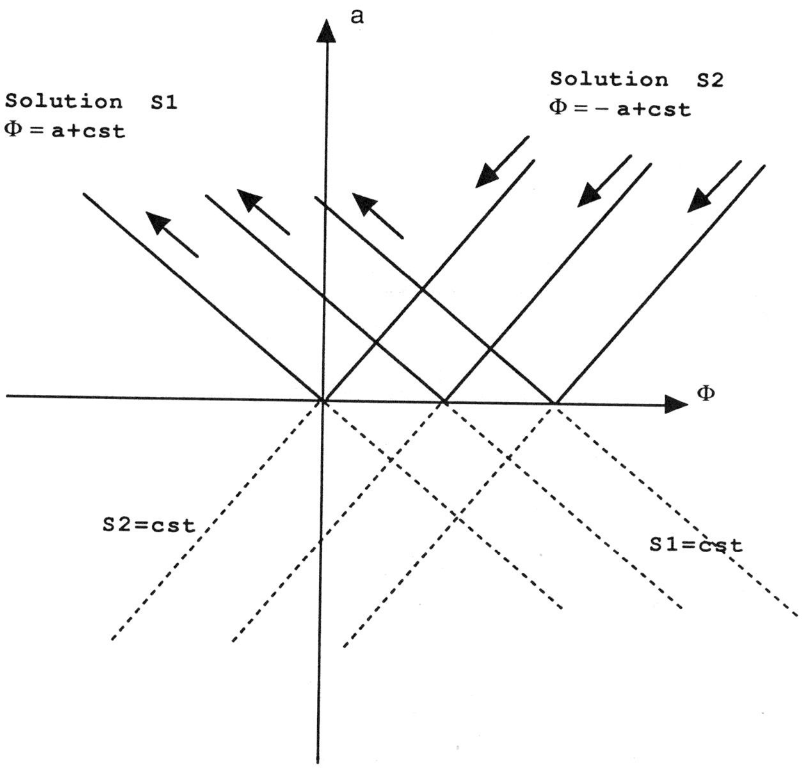

Fig 1

could of course follow a completely different approach such as the many-world interpretation for example [3], [4] (wherein living as well as dead cats could simultaneously coexist) . Nevertheless, we want to limit ourselves, in these lectures, to the following natural scenario: we live today in a classical macroscopic universe characterized by a set of (more or less) known cosmological parameters. As previously suggested, this universe can be quantum- mechanically described by a wave packet. This then plays the role of a final condition from which we will deduce the evolution of the universe backwards in time.

The wave function will be studied using two different methods related to the role played by time:

a. The "Orthodox" method,wherein there is no time, so that the wave function depends on the observables like the radius of the universe,or the matter field only(see [1],[6] [7].).

b. The Time Coordinate method, wherein the time coordinate remains in the formalism(as in paper [5]) and is explicitely used to perform all the computations; it desappears from the final physical results,thereby playing an intermediate coordinate role. We believe that this is the most powerfull method and moreover the best adapted to the spirit of General Relativity, in the sense that coordinates can be used in transitory steps ,while desappearing in the final coordinate-independent results. We will in fact conclude that the two methods yield similar results.

The lecture is organized as follows.

In section 2, we introduce the main formalism and compute the classical evolution of the model in the $\epsilon = 0, 1$ and -1 cases. Section 3 is devoted to the "orthodox formalism". More precisely, in section 3. 1 we study the case $\epsilon = 0$ and some logical analogies of Hartle-Hawwking and Vilenkin proposals. It will appear that the first of these fails to give the classical evolution, whereas the second will deliver an undefined wave function. It is precisely then that we introduce the wave packet concept together with an"exactly defined momentum" limit, which will be designate(for reasons explained in this part) as the "single-term Vilenkin's prescription". The merits of these concepts will be explicitely underlined.

Section 3.2 will be devoted to enlarging the reasonings to the cases $\epsilon = 1$ and $\epsilon = -1$ respectively. In section 4, we introduce the time-coordinate formalism. The Generalized Coherent States (GCS) are then introduced in section 4.1. It is shown in section 4.2 how the GCS lead to a satisfactory classical cosmological evolution. The conclusions are presented in section 5.

Formalism

We shall review in this section the main relations which will be used in the following developments.Consider a 4-dimensional metric with homogeneous space sections:

$$ds^2 = \sigma^2[-N^2(t)dt^2 + a^2(t)d\Omega_\epsilon^2] \qquad (2-1)$$

where $N(t)$ is an arbitrary lapse function,

$$\sigma^2 = \frac{l_{pl}^2}{24\pi^2}; \quad l_{pl}^2 = 16\pi G$$

and

$$d\Omega_\epsilon^2 = \frac{dr^2}{1-\epsilon r^2} + r^2 d\theta^2 + r^2 \sin^2\theta d\varphi^2 \qquad (2-2)$$

The "proper time version" of the metric(2.1) (i.e. $N = 1$)is

$$ds^2 = \sigma^2[-dt^2 + a^2(t)d\Omega_\epsilon^2] \qquad (2-3)$$

and the "conformal time version" (i.e. $N = a$) is

$$ds^2 = \sigma^2 a^2[-d\eta^2 + d\Omega_\epsilon^2] \qquad (2-4)$$

We will hereafter use the second version more often. Let us consider in addition an homogeneous matter field $\Phi(t)$ or $\Phi(\eta)$ with mass M(eventually$M = 0$). The corresponding action is:

$$S = l_{pl}^{-2}\int d^4x\sqrt{-g}(R^{(4)} - 2\Lambda)+$$

$$+\frac{1}{4\pi}\int d^4x\sqrt{-g}(g^{\mu\nu}\partial_\mu\Phi\partial_\nu\Phi + V(\Phi) + M^2\Phi^2 + \xi R^{(4)}\Phi^2) + \text{surfaceterms} \qquad (2-5)$$

Let us rescale all the parameters in the following way:

$$H^2 = \frac{\sigma^2\Lambda}{3} \qquad \varphi = (2\pi^2\sigma^2)^{1/2}\Phi a^{6\xi} \qquad (2-6a)$$

$$m = \sigma M \qquad V(\varphi) = \sigma^2 V(\Phi) \qquad (2-6b)$$

After a Legendre transformation, the action takes the form (with the metric (2.1) and a homogeneous field):

$$S = \int dt(\dot{\varphi}\Pi_\varphi + \dot{a}\Pi_a - N\mathcal{H}) \qquad (2-7)$$

where the superhamiltonian $\mathcal{H}$ is:

$$\mathcal{H} = \frac{1}{2a}[-\Pi_a^2 + a^{12\xi-2}\Pi_\varphi^2 + U(a,\varphi)] \qquad (2-8)$$

with the potential

$$U(a,\varphi) = -\epsilon a^2 + H^2a^4 + 6\epsilon\xi a^{2-12\xi}\varphi^2 + a^4(m^2\varphi^2 a^{-12\xi} + V(\varphi)) \qquad (2-9)$$

and the mini-supermetric is

$$G_{AB} = \begin{pmatrix} -a & 0 \\ 0 & a^{3-12\xi} \end{pmatrix} (A, B \equiv a, \varphi) \qquad (2-10)$$

The Hamiltonian constraint will be considered in the two important cases: $\xi = 0$,namely the minimal coupling

$$-\Pi_a^2 + a^{-2}\Pi_\varphi^2 - \epsilon a^2 + H^2 a^4 + a^4(m^2\varphi^2 + V(\varphi)) = 0 \qquad (2-11)$$

$\xi = 1/6$,namely the conformal coupling

$$-\Pi_a^2 + \Pi_\varphi^2 - \epsilon a^2 + H^2 a^4 + \epsilon\varphi^2 + a^4(m^2\varphi^2 a^{-2} + V(\varphi)) = 0 \qquad (2-12)$$

In the case $\xi = 0$ it is convenient to introduce a new variable

$$a = e^\alpha \qquad \alpha = \ln a \qquad (2-13)$$

Equation (2.11) then reads

$$-\Pi_\alpha^2 + \Pi_\varphi^2 - \epsilon e^{4\alpha} + H^2 e^{6\alpha} + e^{6\alpha}(m^2\varphi^2 + V(\varphi)) = 0 \qquad (2-14)$$

In this lecture we shall restrict ourselves to the simplest massless case, with neither cosmological constant nor matter field self-interaction. hence, $M = 0, H = 0, V = 0$ and the Hamiltonian constraints take the form (For $\xi = 0$)

$$-\Pi_\alpha^2 + \Pi_\varphi^2 - \epsilon e^{4\alpha} = 0 \qquad (2-15)$$

and for $\xi = 1/6$

$$-\Pi_a^2 + \Pi_\varphi^2 + \epsilon(-a^2 + \varphi^2) \qquad (2-16)$$

As usual, the conformal coupling turns out to be more symmetric in the massless case. Thus we will consider that equation (2.16) is the physical constraint, for the massless case, and we shall concentrate on this case only with some exeption as in section 3.1.

Let us now review the classical limit of our model. The Hamiltonian is $H = N\mathcal{H}$, so that it takes the form in conformal time

$$H = \frac{1}{2}[-\Pi_a^2 + \Pi_\varphi^2 + \epsilon(-a^2 + \varphi^2)] \qquad (2-17)$$

The Hamiltonian equations are

$$\frac{da}{d\eta} = \frac{\partial H}{\partial \Pi_a} = -\Pi_a \qquad \frac{d\varphi}{d\eta} = \frac{\partial H}{\partial \Pi_\varphi} = \Pi_\varphi$$

$$\frac{d\Pi_a}{d\eta} = -\frac{\partial H}{\partial a} = \epsilon a \qquad \frac{d\Pi_\varphi}{d\eta} = -\frac{\partial H}{\partial \varphi} = -\epsilon\varphi \qquad (2-18)$$

The solution of these equatios are,In the case $\epsilon = 0$:

$$a(\eta) = -p_0\eta + a_0 \qquad \Pi_a(\eta) = p_0$$

$$\varphi(\eta) = q_0\eta + \varphi_0 \qquad \Pi_\varphi(\eta) = q_0 \tag{2-19}$$

This evolution of the function a turns out to be $t^{1/2}$ when expressed in proper time. In the case $\epsilon = 1$

$$a(\eta) = a_0 \cos\eta - p_0 \sin\eta$$

$$\Pi_a(\eta) = p_0 \cos\eta + a_0 \sin\eta$$

$$\varphi(\eta) = \varphi_0 \cos\eta + q_0 \sin\eta$$

$$\Pi_\varphi(\eta) = q_0 \cos\eta - \varphi_0 \sin\eta \tag{2-20}$$

Finally, in the case $\epsilon = -1$,

$$a(\eta) = a_0 \cosh\eta + p_0 \sinh\eta$$

$$\Pi_a(\eta) = p_0 \cosh\eta - a_0 \sinh\eta$$

$$\varphi(\eta) = \varphi_0 \cosh\eta - q_0 \sinh\eta$$

$$\Pi_\varphi(\eta) = q_0 \cosh\eta + \varphi_0 \sinh\eta \tag{2-21}$$

As all these solutions have to satisfy the constraint (2.16), the constants p_0, q_0, a_0 and φ_0 are not independent. In the case $\xi = 0$, we have

$$p_0 = \pm q_0 \tag{2-22}$$

a_0 and φ_0 are independent. Thus we need three independent parameters to fix a classical solution. The relation between $a(\eta)$ and $\varphi(\eta)$ is:

$$\varphi(\eta) = \pm a(\eta) + Const \tag{2-23}$$

In the case $\epsilon = 1$ equation (2.22) reads

$$-a_0^2 - p_0^2 + \varphi_0^2 + q_0^2 = 0 \tag{2-24}$$

so that all the computations can be performed in the same way. This holds also in the case $\epsilon = -1$.

3.1 The "free" $\epsilon = 0$ case

In this case, both constraints (2.15) and (2.16) have the same mathematical structure

$$-\Pi_a^2 + \Pi_\varphi^2 = 0 \qquad (3-1-1)$$

(whereas in the case $\xi = 0$, a is really α). The Wheeler-de Witt equation reads

$$[\frac{\partial^2}{\partial a^2} - \frac{\partial^2}{\partial \varphi^2}]\Phi(a, \varphi) = 0 \qquad (3-1-2)$$

where $\Phi(a, \varphi)$ is the wave function of the universe. Thus, we have obtained what we can consider as the real "free"case of our problem, simply a plane wave equation whose solution is

$$\Phi(a, \varphi) = f(a - \varphi) + g(a + \varphi) \qquad (3-1-3)$$

We have to choose the two functions $f(x)$ and $g(x)$ to obtain a definite wave function of the universe. This problem is well studied in the litterature only in the case of $\epsilon = 1$. Let us nevertheless take the initial conditions inspired by this case to obtain a feeling of what is going one.

a. Hartle-Hawking condition

The Hartle-Hawking initial condition is conceptually meaningless in the $\epsilon = 0$ case. Let us nevertheless explore,in this case, the consequences of the same mathematical requirement, namely

$$a = 0 \rightarrow (\frac{\partial \Phi}{\partial a})_{a=0} = 0$$

$$\varphi = 0 \rightarrow (\frac{\partial \Phi}{\partial \varphi})_{\varphi=0} = 0 \qquad (3-1-4)$$

for the proper time, or

$$\frac{da}{d\eta} = -\frac{\partial S}{\partial a} \qquad (3-1-10)$$

for the conformal time. Using this latter time, the action reads

$$S = -\frac{1}{2}\int d\eta[(a')^2 - (\varphi')^2] \qquad (3-1-11)$$

where the symbol $'$ designates a derivative with respect to the conformal time η. The momenta are:

$$\Pi_a = \frac{\partial \mathcal{L}}{\partial a'} = -\frac{\partial a}{\partial \eta} \qquad \Pi_\varphi = \frac{\partial \mathcal{L}}{\partial \varphi'} = -\frac{\partial \varphi}{\partial \eta} \tag{3-1-12}$$

It results from the Hamilton-Jacobi formalism that

$$\Pi_a = \frac{\partial S}{\partial a} = -\frac{da}{d\eta}$$

$$\Pi_\varphi = \frac{\partial S}{\partial \varphi} = \frac{d\varphi}{d\eta} \tag{3-1-13}$$

where the first equation may be considered as the classical time definition. The solution (3.1.8) leads to:

$$-F'(\varphi - a) = -\frac{\partial a}{\partial \eta}$$

$$F'(\varphi - a) = \frac{\partial \varphi}{\partial \eta} \tag{3-1-14}$$

The classical solution (2.23) then follows from these equations, namely

$$\varphi = a + Const \tag{3-1-15}$$

or using the relation (3.1.8), we obtain

$$G'(\varphi + a) = -\frac{\partial a}{\partial \eta}$$

$$G'(\varphi - a) = \frac{\partial \varphi}{\partial \eta} \tag{3-1-16}$$

namely, the classical solution (eq.(2.23)):

$$\varphi = -a + Const \tag{3-1-17}$$

The classical trajectories are drawn in fig.2. From the classical trajectories, we can obtain the classical time. For instance, let us take a definite classical trajectory, say:

$$\varphi = a + c \tag{3-1-18}$$

It follows from the first equation that:

$$-f'(\varphi) + g'(\varphi) = 0$$

$$f(x) = g(x) + Const$$

The second equation implies

$$f'(-a) + g'(a) = 0$$

The second equation implies namely, $f(x)$ must be an even function.Therefore, the final solution is

$$\Phi(a, \varphi) = f(\varphi - a) + f(\varphi + a) + Const \tag{3-1-5}$$

Although this form does not determine the function completely, it corresponds to two waves moving in opposite directions in the a, φ plane.In consequence, if we consider a wave packet as a classical solution, we are in this case in the presence of at least two wave packets moving in opposite directions in the a, φ plane hence no classical solution.Following the ideas presented in the introduction, this fact would rule out the Hartle-Hawking type condition in this case.

b. Vilenkin's condition.

This condition, as the previous one, is only defined in the case $\epsilon = 1$, but can be translated with some modification to the $\epsilon = 0$ case. For that, we must first solve the Hamilton-Jacobi equation associated with the constraint (3.1.1) :

$$-(\frac{\partial S}{\partial a})^2 + (\frac{\partial S}{\partial \varphi})^2 = 0 \qquad (3-1-6)$$

where S is the Principal Jacobi-Function, i.e.

$$\frac{\partial S}{\partial a} = \pm \frac{\partial S}{\partial \varphi} \qquad (3-1-7)$$

Thus, we have two solutions:

$$S_1(a, \varphi) = F(\varphi - a)$$

$$S_2(a, \varphi) = G(\varphi + a) \qquad (3-1-8)$$

We can now compute the classical time using, as in paper [6],

$$\frac{da}{dt} = -\frac{1}{a}\frac{\partial S}{\partial a} \qquad (3-1-9)$$

where c is given constant. It follows then from equation (3.1.14) that:

$$a = F'(c)\eta + Const \qquad (3-1-19)$$

i.e. equation (2.19) where $F'(c) = -p_0$.

Now, suppose that both functions F and G are growing functions of their arguments.(i.e $F'(c) > 0$ which by eq.3.1.19 implies that a and η have the same direction; also $G'(c) > 0$ which in turn gives by eq.3.1.16 $a' < 0$ hence a and η have opposite directions). Thus, the direction of the gradients of the functions F and G are indicated by the arrows in fig.2. We are now able to use Vilenkin's recipe. The wave function of the universe must be a sum like

$$\Phi = \sum_n c_n(a, \varphi) e^{iS_n(a,\varphi)} \qquad (3-1-20)$$

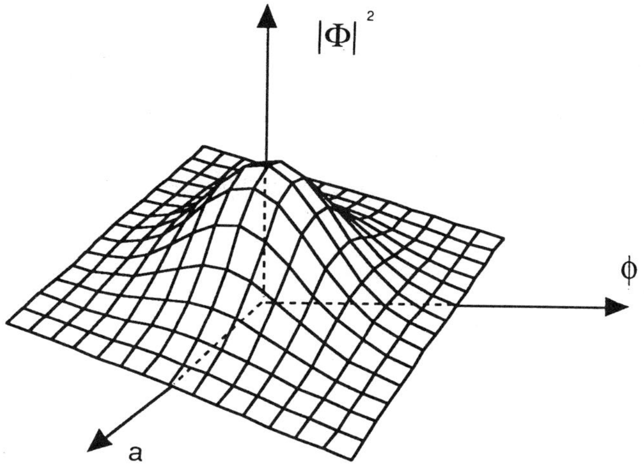

Fig 2

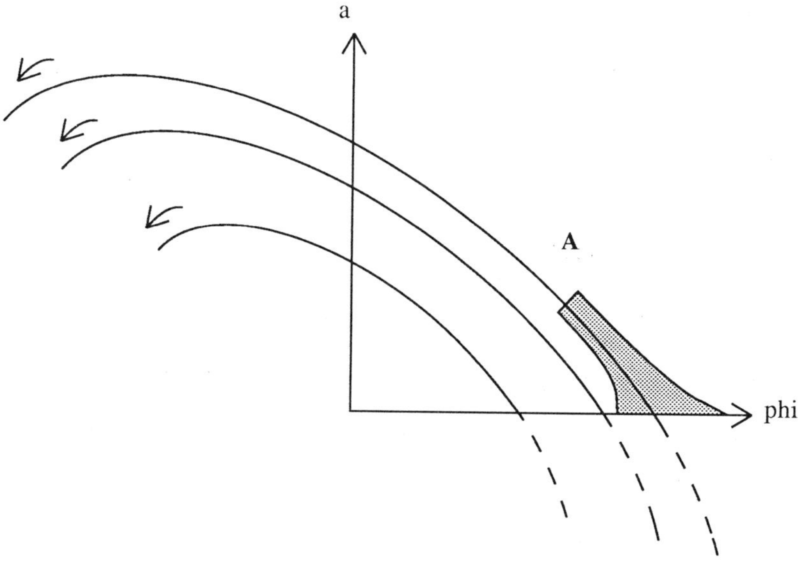

Fig 3

where $S_n(a, \varphi)$ are solutions of the Hamilton-Jacobi equation with gradients such that " at singular boundaries of superspace Φ includes only outgoing modes" (carrying the flux, defined by the gradients, out of superspace). The problem is that we have only singular boundaries.

These are : case $\xi = 1/6$: $a = 0$ or $+\infty$ $\varphi = -\infty$ or $+\infty$
or the case $\xi = 0$ $\alpha = -\infty$ or $+\infty$ $\varphi = -\infty$ or $+\infty$

Therefore, both solutions S_1 and S_1 carry flux into superspace through singular boundaries, thus we cannot choose one solution or the other. In any case, for $\epsilon = 1$, Vilenkin's prescription means that we must only take the expanding universe modes. In the present case, this implies that only the S_1 solution can be used; this can be seen from fig.2 or equally from the fact that for S_1 we have $a' > 0$ (cf.eq.3.1.14 and the fact that F is a growing function of its argument) while for S_2 we have $a' < 0$ (cf.eq.3.1.16,etc..). As Φ must be a solution of the field equation (3.1.2), Vilenkin's wave function reads

$$\Phi(a, \varphi) = \sum_n c_n(\varphi - a) e^{iS_n(\varphi - a)} \qquad (3-1-21)$$

where C_n and S_n are, in this special case restricted to be just functions of one single variable. This prescription gives then many possible functions of the universe. Therefore, even if this proposal cannot be rejected as such, it must be improved in some way.

c. The Wave Packet

Following our proposal, namely that we have to obtain a final classical universe with the classical law of evolution of a and φ, eqs.(3.1.15) or (3.1 17), we are naturally led to introduce the following final condition: when $a \rightarrow \infty$, $\Phi(a, \varphi)$ must be of the following form:

$$\Phi(a, \varphi) = \Delta_\epsilon(\varphi - a - cst) \qquad (3-1-22)$$

(or $\Delta_\epsilon(\varphi + a + cst)$) where Δ_ϵ is a function that peaks strongly at $\varphi = a + const$, in such a way that

$$\lim_{\epsilon \rightarrow 0} |\Delta_\epsilon(\varphi - a - cst)|^2 = \delta(\varphi - a - cst) \qquad (3-1-23)$$

As the solution of the field equation is (3.1.3), equation (3.1.22) gives the wave function of the universe for all a and φ; this would not be the case , of course, for more general field equations. But in our case, the wave function will always peak at $\varphi = a + const$ and we will find always the right classical evolution (3.1.15). At this stage, two remarks have to be made:

- We can choose c_n and S_n in eq.(3.1.21) in such a way that $\Phi(a, \varphi)$ turns out to be a wave packet, like in eq. (3.1.22), namely Vilenkin's condition with some addenda can be the condition we are looking for.
- The wave packet defined by eq.(3.1.22) can be defined in many ways. It can be a "configuration" wave packet as the one defined by eq.(3.1.23) or it can be a "momentum" wave packet that defines only the momentum of

the universe in a Fourier transformation, and not its radius a. In the case of such a "momentum" wave packet, the wave function of the universe must be

$$\Phi(a, \varphi) = e^{-ip_0(\varphi - a)} \qquad (3-1-24)$$

because in this case $S = -p_0(\varphi - a)$ and eq.(3.1.13) delivers a definite momentum for the universe evolution, precisely:

$$\Pi_a = -a' = -\frac{da}{d\eta} = -a^2 H$$

$$= \frac{\partial S}{\partial a} = p_0$$

where H is now the Hubble function, namely

$$p_0 = -a^2 H \qquad (3-1-25)$$

It is important to notice that all the examples wherein Vilenkin's proposal is adopted are constructed with one single term of expansion like (3.1.21), so that the wave function is in fact of the form described by the relation (3.1.24). Perhaps the success of this "single- term Vilenkin's proposal" (it predicts chaotic or new inflation as well as the Harrisson-Zeldovich spectrum) is due to the fact that it represents a particular case of our wave packet prescription.The study of the wave packet determination will be carried out in section 4.2. We have described, up to this point,our general proposal for a new approach to the wave function of the universe. We shall investigate this idea in the framework of more elaborated examples and some more refined techniques,in the rest of this paper.

3.2 The cases $\epsilon = 1$ and $\epsilon = -1$

The Wheeler-deWitt equation takes ,in these cases, the following form:

$$\left[\frac{\partial^2}{\partial a^2} - \frac{\partial^2}{\partial \varphi^2} + \epsilon(\varphi^2 - a^2)\right] \Phi(a, \varphi) = 0 \qquad (3-2-1)$$

It is usefull to introduce the new variables

$$\xi = (\varphi + a)^2 \qquad \eta = (\varphi - a)^2 \qquad (3-2-2)$$

Equation (3.2.1) takes then the form

$$\left[\frac{\partial^2}{\partial \xi \partial \eta} - \frac{\epsilon}{16}\right] \Phi(\xi, \eta) = 0 \qquad (3-2-3)$$

The Fourier transform W_p, defined by

$$\Phi = \int dp e^{ip\xi} W_p(\eta) \qquad (3-2-4)$$

hence satisfies the relation

$$\frac{\partial W_p}{\partial p} = -\frac{i\epsilon}{16p} W_p \qquad (3-2-5)$$

whose general solution is

$$W_p(\eta) = F(p)\exp(-\frac{i\epsilon\eta}{16p}) \qquad (3-2-6)$$

where $F(p)$ is an arbitrary function of p. The wave function of the universe takes accordingly the form

$$\Phi = \int dpF(p)\exp(ip\xi - \frac{i\epsilon}{16p}\eta) \qquad (3-2-7)$$

Reversing the roles of ξ and η in eq.(3.2.4) we are led to the general solution

$$\Phi = \int dpF(p)\exp(ip\xi - \frac{i\epsilon}{16p}\eta) + \int dqG(q)\exp(iq\eta - \frac{i\epsilon}{16q}\xi) \qquad (3.2.8)$$

This reduces, in the case $\epsilon = 0$, to

$$\Phi = \int dpF(p)\exp(ip(\varphi + a)^2) + \int dqG(q)\exp(iq(\varphi - a)^2) =$$

$$= g(\varphi + a) + f(\varphi - a) \qquad (3.2.9)$$

as in equation (3.1.3). In the case $\epsilon = \pm 1$, the solution is redundant because the two arguments of the exponential are equal if

$$pq = -\frac{\epsilon}{16} \qquad (3.2.10)$$

We therefore retain the first term only, i.e.

$$\Phi = \int dpF(p)\exp i[p(a + \varphi)^2 - \frac{\epsilon}{16p}(\varphi - a)^2] \qquad (3.2.11)$$

a. Hartle-Hawking condition

The Hartle-Hawking wave function corresponding to $\xi = 1/6, \epsilon = 1, H = 0$, and $V = 0$ is (see ref.[7]):

$$\Phi(a, \varphi) = \exp(-\frac{1}{2}a^2)\exp(-\frac{1}{2}\varphi^2) \qquad (3.2.12)$$

This follows from (3.2.8) with the choice

$$F(p) = \delta(p - \frac{1}{4}i); \quad G(q) = 0 \qquad (3.2.13)$$

$\Phi(a, \varphi)$ is the product of two gaussian functions,namely the wave function of two harmonic oscillators in their ground state, as depicted in fig.3. From this function, we can only deduce that from a probalistic point of view, the universe will be "located" at $\varphi = 0, a = 0$, which obviously has no link with the classical cosmological motion described by eq.(2.20). Once again we have to reject the Hartle-Hawking wave function in the context of our proposal.

The case $\epsilon = -1$ remains to be investigated, as we do not have the corresponding boundary condition in this case and we will avoid an analogy similar to that of section 3.1a. We have nevertheless some evidence that the conclusion will be similar.

b. Vilenkin's condition

We must reproduce the computations of section 3.1 (from eq. 3.1.6 on) in the framework of this new case. The Hamilton-Jacobi equation reads

$$-(\frac{\partial S}{\partial a})^2 + (\frac{\partial S}{\partial \varphi})^2 + \epsilon(-a^2 + \varphi^2) = 0 \tag{3.2.14}$$

With the change of variables (3.2.2), this leads to

$$\frac{\partial S}{\partial \xi}\frac{\partial S}{\partial \eta} + \frac{\epsilon}{16} = 0 \tag{3.2.15}$$

whose solution is

$$S_p = (p\xi - \frac{\epsilon}{16p}\eta) = p(a+\varphi)^2 - \frac{\epsilon}{16p}(a-\varphi)^2 \tag{3.2.16}$$

In order to obtain the classical trajectories, we must solve equations (3.1.13) which take now the following form

$$\frac{\partial a}{\partial \eta} = -2p(a+\varphi) + \frac{\epsilon}{8p}(a-\varphi)$$

$$\frac{\partial \varphi}{\partial \eta} = 2p(a+\varphi) + \frac{\epsilon}{8p}(a-\varphi) \tag{3.2.17}$$

or defining

$$u = a + \varphi, \quad v = a - \varphi \tag{3.2.18}$$

we obtain

$$\frac{\partial u}{\partial \eta} = \frac{\epsilon}{4p}v$$

$$\frac{\partial v}{\partial \eta} = -4pu \tag{3.2.19}$$

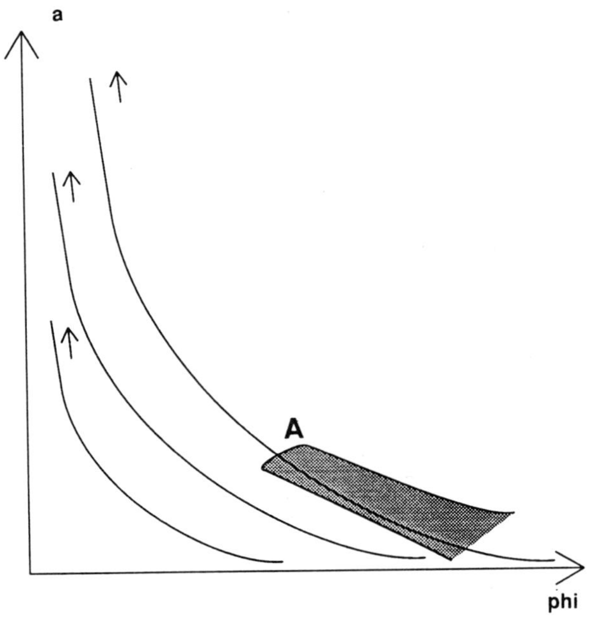

Fig 4

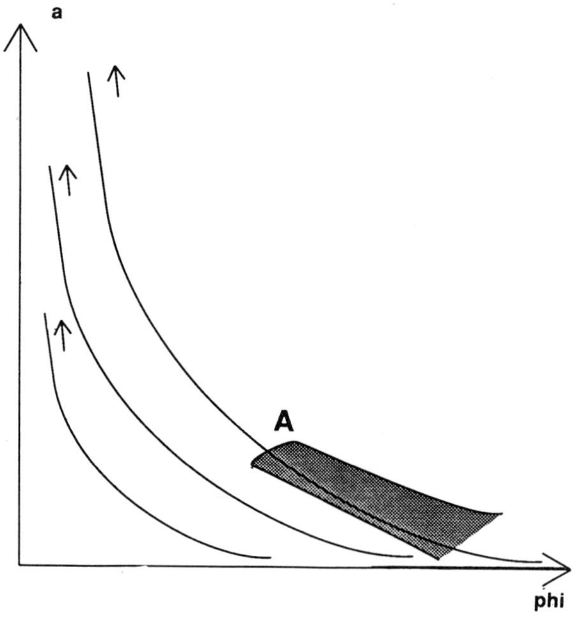

Fig 5

Thus:

$$\frac{\partial^2 u}{\partial \eta^2} = -\epsilon u \tag{3.2.20}$$

and therefore

$$\epsilon = 1$$

$$u = A\cos\eta + B\sin\eta$$

$$v = 4p(-A\sin\eta + B\cos\eta) \tag{3.2.21}$$

$$\epsilon = -1$$

$$u = A\cosh\eta + B\sinh\eta$$

$$v = -4p(A\sinh\eta + B\cosh\eta) \tag{3.2.22}$$

where A and B are arbitrary constants. In fact, p, A and B are the three constants required to define a classical solution. Thus finally, in the case of $\epsilon = 1$ which will be considered first, we obtain

$$a = \frac{1}{2}[(A + 4pB)\cos\eta + (B - 4pA)\sin\eta]$$

$$\varphi = \frac{1}{2}[(A - 4pB)\cos\eta + (B + 4pA)\sin\eta] \tag{3.2.23}$$

i.e. eqs.(2.20) wherein the coefficients p, A, B satisfy the constraint (2.24). We obtain now the classical solutions. If we call

$$A = \rho\cos\theta, \quad B = \rho\sin\theta \tag{3.2.24}$$

the eqs.(3.2.21) take the form

$$u = \rho\cos(\eta - \theta)$$

$$v = -4p\rho\sin(\eta - \theta) \tag{3.2.25}$$

showing that the classical trajectories associated with the function S_p are ellipses, symmetrical in the variables u and v, as shown in fig.4.

The sense in which the trajectories are followed, as classical time η elapses, depends on the sign of p. Obviously, $a = +\infty$ and $\varphi = -\infty$ or $\varphi = +\infty$ are singular boundaries, but $a = 0$ is a non singular boundary (we draw the lower part of fig.4 in dashed lines because we consider only the physical upper part), hence we can try to use Vilenkin's prescription. The problem is that no solution carries flux out of superspace through the singular boundaries. The flux just turns around within superspace and crosses only the non singular boundary. Therefore, Vilenkin's condition cannot be neither used nor improved in this case.

Let us study now the $\epsilon = -1$ case;a and φ are then given by

$$a = \frac{1}{2}[(A - 4pB)\cosh\eta + (B - 4pA)\sinh\eta]$$

$$\varphi = \frac{1}{2}[(A + 4pB)\cosh\eta + (B + 4pA)\sinh\eta] \tag{3.2.26}$$

Defining

$$A = \rho\cosh\theta, \quad B = \rho\sinh\theta \tag{3.2.27}$$

we obtain

$$u = \rho\cosh(\eta + \theta)$$
$$v = -4p\rho\sinh(\eta + \theta) \tag{3.2.28}$$

The corresponding classical trajectories are depicted in fig.5. They are hyperbolae. Once more, the sense followed by the classical motion depends on the sign of p, which is arbitrarily choosen in fig. 5 for the purpose of illustration. The boundaries are the same as in the case of $\epsilon = 1$. Hence, once again Vilenkin's prescription is useless,in its generic version, because there are no modes with outgoing flux only through the singular boundaries of superspace. Nevertheless, we can introduce in this case a modified Vilenkin's condition, as in section 3.1.b.

According to the sign of p we have expanding evolutions (for big a and $p < 0$) or contracting ones ($p > 0$) see fig.5. Thus, using Vilenkin's idea, we can choose only a superposition of expanding evolutions (for big a), as in eq. 3.1.21,

$$\Phi = \sum_{p<0} c_p(\varphi, a)e^{iS_p(\varphi,a)} \tag{3.2.29}$$

where S_p is given by eq. (3.2.16) and c_p must be choosen in such a way as to satisfy the Wheeler-de Witt equation (3.2.1). In this case we know that these functions are just constants (cfr.eq.3.2.11) so that

$$c_p(a, \varphi) = F(p)$$

It follows that the modified Vilenkin's prescription leads to several possible wave functions for the universe. Accordingly, in analogy to the case $\epsilon = 0$, even if this proposal cannot be rejected, it must be improved.

c. The Wave Packet Condition

Let us consider first that our classical universe is closed ($\epsilon = 1$) and radiation dominated.Let our present cosmological localisation be the point A (see fig.4), and the associated physical parameters designated by p, ρ and θ. We may then define around point A a wave packet whose shape fits as adequatly as possible to the values of the above-mentionned parameters, and study its evolution backwards in time , up to the quantum era (represented by the shaded area in fig.4). The optimal way of choosing the wave packet will be analyzed in section 4.2 .The same approach can be applied to the case of $\epsilon = -1$. In this case,let us designate the present universe by the point A in fig.5, and the same procedure consisting of surrounding (in a technical way which will be explained in the next parts of this lecture) the point A by an adequate wave packet will be adopted. Similarily, the evolution of this wave packet backwards in time up to the quantum era (shadowed in fig.5) will then be analyzed. In this case the wave function will be a superposition of expanding solutions (3.2.2) with $C_p(a, \varphi) = F(p)$ choosen in such a way as to insure that $\Phi(a, \varphi)$ be a wave packet. Again probably the wise thing to do is to choose a single-term Vilenkin solution. Suppose that we know that the present universe has the following parameters: a_0, φ_0, H_0, Π_φ are then computed from the constraint (2.16). Then eq.(3.2.17) for $\epsilon = -1$ reads

$$H_0 = -\frac{2p_0}{a_0^2}(a_0 + \varphi_0) + \frac{1}{8p_0 a_0^2}(a_0 - \varphi_0) \tag{3.2.30}$$

We can then compute p_0 from this equation and the single- term Vilenkin wave function of the universe

$$\Phi(a, \varphi) = \exp i[p_0(a + \varphi)^2 + \frac{1}{16p_0}(a - \varphi)^2] \tag{3.2.31}$$

This function could then play the role of a reasonable wave function for the universe, as it corresponds to a well defined H (given by eq.3.2.17) at any time during the classical evolution. It appears therefore that the single-term Vilenkin's solution can be adapted in the $\epsilon = -1$ case.

4. The Time-Coordinate Method

Hartle shows in his paper [5] how to introduce the time coordinate in an approach wherein the geometry as well as the reference frame itself are quantized. The Wheeler-de Witt equation which is in this case,

$$\hat{\mathcal{H}}\Phi(a, \varphi) = 0 \tag{4.1}$$

where $\hat{\mathcal{H}}$ is the operational form of the super -Hamiltonian (2.8) and $\Phi(a, \varphi)$ is the wave function of the universe, must be replaced by the time-dependent equation:

$$i\frac{\partial \Psi(t, a, \varphi)}{\partial t} = N\hat{\mathcal{H}}\Psi(t, a, \varphi) \tag{4.2}$$

where Ψ is now a time-dependent wave function. Moreover, $\Phi(a, \varphi)$ can be obtained as:

$$\Phi(a, \varphi) \simeq \int_{-\infty}^{+\infty} dt \Psi(t, a, \varphi) \tag{4.3}$$

This equation (4.2) is also introduced in papers [8]and[9] by using a different approach. In paper [10] measurement of the t-time coordinate is shown by applying the "probabilistic time method".

Equation (4.2) will be extensively used in this section as it allows computations otherwise impossible outside the scope of the time-independent Wheeler-de Witt equation. As the time-coordinate was absent in all the preceeding reasonings, all observables $\hat{a}, \hat{\varphi}, \hat{\Pi}_a, \hat{\Pi}_\varphi$, were time-independent. In the present context, the wave function $\Psi(t, a, \varphi)$ is time dependent and we are in the Schrodinger picture. As usual, we can take $N = a$, and $t = \eta$ which will be the conformal time. The corresponding Shrodinger picture is:

$$i\frac{\partial \Psi}{\partial \eta} = \hat{H}\Psi \tag{4.4}$$

where $\hat{H}$ is given in eq. (2.17). We are now in the context of usual Quantum Mechanics, with the ordinary Schrodinger product which is invariant by virtue of equation (4.4). Passing to the ket notation, one writes

$$i\frac{\partial}{\partial \eta}|\Psi >= \hat{H}|\Psi > \tag{4.5}$$

The Heisenberg picture is , of course, defined by:

$$\hat{a}(\eta) = e^{i\hat{H}\eta}\hat{a}_0 e^{-i\hat{H}\eta}$$

$$\hat{\varphi}(\eta) = e^{i\hat{H}\eta}\hat{\varphi}_0 e^{-i\hat{H}\eta}$$

$$\hat{\Pi}_a(\eta) = e^{i\hat{H}\eta}\hat{\Pi}^0_a e^{-i\hat{H}\eta}$$

$$\hat{\Pi}_\varphi(\eta) = e^{i\hat{H}\eta}\hat{\Pi}^0_\varphi e^{-i\hat{H}\eta}$$

$$|\Psi_0 >= e^{-i\hat{H}\eta}|Psi > \tag{4.6}$$

and the corresponding mean values are

$$< \hat{a}(\eta) >=< \Psi_0|\hat{a}_0|\Psi_0 >=$$

$$< \Psi|\hat{a}(\eta)|\Psi > \tag{4.7}$$

etc... The operators commutation relations are

$$[\hat{a}_0, \hat{\Pi}^0_a] = [\hat{a}(\eta), \hat{\Pi}_a(\eta)] = i$$

$$[\hat{\varphi}_0, \hat{\Pi}^0_\varphi] = [\hat{\varphi}(\eta), \hat{\Pi}_\varphi(\eta)] = i \tag{4.8}$$

It follows then , that, from these relations and the Hamiltonian (2.17) the evolution equations for the operators are

$$\hat{a}' = i[\hat{H}, \hat{a}] = -\hat{\Pi}_a$$

$$\hat{\Pi}'_a = i[\hat{H}, \hat{\Pi}_a] = \epsilon\hat{a}$$

$$\hat{\varphi}' = i[\hat{H}, \hat{\varphi}] = -\hat{\Pi}_\varphi$$

$$\hat{\Pi}'_\varphi = i[\hat{H}, \hat{\Pi}_\varphi] = -\epsilon\hat{\varphi} \qquad (4.9)$$

This represents the quantum version of the classical equations (2.18). The integration of these equations leads then, in the case $\epsilon = 0$, to

$$\hat{a}(\eta) = -\hat{p}_0\eta + \hat{a}_0, \quad \hat{\Pi}_a(\eta) = \hat{p}_0$$

$$\hat{\varphi}(\eta) = \hat{q}_0\eta + \hat{\varphi}_0, \quad \hat{\Pi}_\varphi(\eta) = \hat{q}_0 \qquad (4.10)$$

namely equation (2.19) where all the classical variables are replaced by the corresponding operators. The solution for the cases $\epsilon = \pm 1$ can be obtained in the same way. If we sandwiche eqs.(4.10) between any quantum state, we obtain

$$< \hat{a}(\eta) >= - < \hat{p}_0 > \eta + < \hat{a}_0 >, \quad < \hat{\Pi}_a(\eta) >=< \hat{p}_0 >$$

$$< \hat{\varphi}(\eta) >=< \hat{q}_0 > \eta + < \hat{\varphi}_0 >, \quad < \hat{\Pi}_\varphi(\eta) >=< \hat{q}_0 > \qquad (4.11)$$

i.e. the mean values obey to the classical equations for all possible wave functions of the universe. Similarily, we can prove the same property for the cases $\epsilon = \pm 1$. Therefore, the requirement that the mean values have to obey to classical solution, does not represent a constraint on the wave function of the universe (at least in the model we are studying).

Let us first review some elementary properties in Quantum mechanics. Consider the operator $\hat{a}_0$ and the basis $|a, 0>$ defined by

$$\hat{a}_0|a, 0 >= a|a, 0 > \qquad (4.12)$$

This equation can be written as

$$e^{iH\eta}\hat{a}_0 e^{-iH\eta}e^{iH\eta}|a, 0 >= ae^{iH\eta}|a, 0 > \qquad (4.13)$$

i.e.

$$\hat{a}(\eta)|a, \eta >= a|a, \eta > \qquad (4.14)$$

We have therefore defined a new basis $|a, \eta >$ owing to operators $\hat{a}(\eta)$. Of course, we have

$$|a, \eta >= e^{i\hat{H}\eta}|a, 0 > \qquad (4.15)$$

The basis $|a, \eta >$ represents a set of solutions of the Schrödinger equation (4.5). If $|A >$ is a ket, constant in conformal time η and

$$-i\frac{\partial}{\partial\eta}|a,\eta>=H|a,\eta> \tag{4.16}$$

Then

$$-i\frac{\partial}{\partial\eta}<A|a,\eta>=<A|H|a,\eta> \tag{4.17}$$

If we call $\Psi_A(a,\eta)=<A|a,\eta>$, then we have

$$i\frac{\partial}{\partial\eta}\Psi_A=H\Psi_A \tag{4.19}$$

and Ψ_A is a solution of the Schrodinger equation (4.5).

4.1 Generalized Coherent States

Consider two conjugate variables like a and Π_a or φ and Π_φ, and define the operator

$$A=\alpha a+\beta\Pi_a \tag{4.1.1}$$

where α and β are arbitrary constants. We shall call a Generalized Coherent State (GCS) an eigenvector of the operator A

$$A|\mu>=\mu|\mu> \tag{4.1.2}$$

This $|\mu>$ is a constant ket that will play the role of $|A>$ in eq.(4.17). α, β and μ are the three constants that define the GCS; they correspond to constant p, A, B in section 3.2. The adjoint of operator $A^\dagger$ is

$$A^\dagger=\alpha^*a+\beta^*\Pi_a \tag{4.1.3}$$

(a and Π_a being observables are self adjoined). Define the determinant

$$D=\alpha\beta^*-\beta\alpha^* \tag{4.1.4}$$

The following commutation relation is easely obtained:

$$[A,A^\dagger]=iD \tag{4.1.5}$$

The inversion of the system (4.1.1) and (4.1.3) leads to

$$a=\frac{1}{D}[\beta^*A-\beta A^\dagger]$$

$$\Pi_a=\frac{1}{D}[-\alpha^*A+\alpha A^\dagger] \tag{4.1.6}$$

We shall call

$$< .. >_\mu = < \mu|..|\mu > \tag{4.1.7}$$

thus

$$< a >_\mu = \frac{1}{D}[\beta^* \mu - \beta \mu^*]$$

$$< \Pi_a >_\mu = \frac{1}{D}[-\alpha^* \mu + \alpha A \mu^*] \tag{4.1.8}$$

These equations show how we can obtain the mean values of a and Π_a from the three constants α, β, μ. It results from (4.1.6) that

$$a^2 = D^{-2}[\beta^{*2} A^2 + \beta^2 A^{\dagger 2} - |\beta|^2 (AA^\dagger + A^\dagger A)] \tag{4.1.9}$$

But from eqs. (4.1.2) and (4.1.5) we know that

$$< A^\dagger A >_\mu = |\mu|^2 \tag{4.1.10}$$

and

$$< AA^\dagger >_\mu = |\mu|^2 + iD \tag{4.1.11}$$

Then

$$< a^2 >_\mu = D^{-2}[\beta^{*2} \mu^2 + \beta^2 \mu^{*2} - |\beta|^2 (2|\mu|^2 + iD)] \tag{4.1.12}$$

Equations (4.1.8) and (4.1.12) lead then to

$$< a^2 >_\mu = < a >_\mu^2 - iD^{-1} |\beta|^2 \tag{4.1.13}$$

We shall define the width of the CGS as the quantity

$$\sigma_a = < a^2 >_\mu - < a >_\mu^2 = -iD^{-1} |\beta|^2 \tag{4.1.14}$$

Define

$$\alpha = \rho_1 e^{i\theta_1}$$

$$\beta = \rho_2 e^{i\theta_2} \tag{4.1.15}$$

σ_a reads

$$\sigma_a = \frac{1}{2} \frac{\rho_2}{\rho_1} \frac{1}{\sin \xi} \tag{4.1.16}$$

where $\xi = \theta_2 - \theta_1$. It turns out that σ_a is a real number (as it should be). In a similar way, we compute

$$\sigma_{\Pi_a} = \frac{1}{2} \frac{\rho_1}{\rho_2} \frac{1}{\sin \xi} \tag{4.1.17}$$

and

$$\sigma_a \sigma_{\Pi_a} = \frac{1}{4} \frac{1}{\sin^2 \xi} \tag{4.1.18}$$

A minimum value for this product corresponds to $\xi = \pi/2$ for all values of ρ_1, ρ_2. The value $\xi = -\pi/2$ must not be considered because $|\mu >$ must be normalized to 1. Therefore,

$$\sigma_a \geq 0 \quad \sigma_{\pi_a} \geq 0 \tag{4.1.19}$$

We would now like to see the shape of this GCS as a wave function at time $\eta = 0$. Thus, we consider the basis $|a, 0>$ already defined in eq.(4.12)

$$a_0|a, 0>= a|a, 0> \tag{4.1.20}$$

From eqs. (4.1.1) and (4.1.2) we deduce

$$(\alpha a_0 + \beta \Pi_a^0)|\mu, 0>= \mu|\mu, 0> \tag{4.1.21}$$

and

$$(\alpha a_0 + \frac{\beta}{i}\frac{\partial}{\partial a_0}) < a, 0|\mu, 0>= \mu < a, 0|\mu, 0> \tag{4.1.22}$$

We shall call

$$\Psi_\mu(a) =< a, 0|\mu, 0> \tag{4.1.23}$$

These functions must satisfy

$$\frac{\beta}{i}\Psi'_\mu = (\mu - \alpha a)\Psi_\mu \tag{4.1.24}$$

Thus, $\Psi'_\mu \sim \Psi_\mu$, which will turn out to be essential. We integrate this last equation, and obtain

$$\Psi_\mu = N \exp[-\frac{i}{2\alpha\beta}(\mu - \alpha a)^2] \tag{4.1.25}$$

We see then that GCS at $\eta = 0$ are pure gaussians (with a phase). We compute their probability, which turns out to be

$$\mathcal{P} = |\psi_\mu(a)|^2 = \tag{4.1.26}$$

$$= |N'|^2 \exp[-\frac{(a- < a >_\mu)^2}{2\sigma_a}]$$

which is a gaussian with its center at $< a_0 >_\mu$ and a width σ_a. From eq. (4.10) in the case $\epsilon = 0$ (there are analogous eqs. for the $\epsilon = \pm 1$ cases), we know that

$$a(\eta) = \xi(\eta)a_0 + \zeta(\eta)\Pi_a^0$$

$$\Pi_a(\eta) = \lambda(\eta)a_0 + \mu(\eta)\Pi_a^0 \tag{4.1.27}$$

where ξ, ζ, λ, μ are known functions of η. We can invert this system (changing in fact η in $-\eta$) and we obtain a_0 and Π_a^0 as linear functions of $a(\eta)$ and $\Pi_a(\eta)$. If we insert these results into eqs.(4.1.1) and (4.1.3), we obtain

$$A_0 = \delta(\eta)a(\eta) + \gamma(\eta)\Pi_a(\eta)$$

$$A^{\dagger} = \delta^* a(\eta) + \gamma^* \Pi_a(\eta) \tag{4.1.28}$$

We must now consider the basis $|a, \eta >$, solutions of the Schrodinger equation (4.5) already introduced in eq. (4.14) and defined by

$$a(\eta)|a, \eta >= a|a, \eta > \tag{4.1.29}$$

and the GCS defined at time $\eta = 0$ (4.1.2)

$$A_0|\mu, 0 >= \mu|\mu, 0 > \tag{4.1.30}$$

Then

$$< a, \eta|A_0|\mu >= \mu < a, \eta|\mu, 0 > \tag{4.1.31}$$

and from (4.1.28)

$$\delta(\eta) a \Psi_\mu(a, \eta) - i\gamma(\eta) \frac{\partial}{\partial a} \Psi_\mu(a, \eta) = \mu \Psi_\mu(a, \eta) \tag{4.1.32}$$

where

$$\Psi_\mu(a, \eta) =< a, \eta|\mu, 0 > \tag{4.1.33}$$

Thus:

$$\frac{\partial}{\partial a} \Psi_\mu(a, \eta) = \frac{i}{\gamma} (\mu - \delta a) \Psi_\mu(a, \eta) \tag{4.1.34}$$

We can now compute the maximum of the density probability $\mathcal{P}$

$$\frac{\partial \mathcal{P}}{\partial a} = |\Psi_\mu(a, \eta)|^2 [\frac{i}{\gamma}(\mu - \delta a) - \frac{i}{\gamma^*}(\mu^* - \delta^* a)] = 0 \tag{4.1.35}$$

namely

$$(\mu\gamma^* - \mu^*\gamma) - (\delta\gamma^* - \delta^*\gamma) a = 0 \tag{4.1.36}$$

In the case $\epsilon = 0$

$$\gamma(\eta) = \beta + \alpha\eta$$

$$\delta(\eta) = \alpha \tag{4.1.37}$$

and eq.(4.1.36) reads

$$(\mu\beta^* - c.c) - (\mu\alpha^* - c.c)\eta - (\alpha\beta^* - c.c)a = 0 \tag{4.1.38}$$

From eqs.(4.1.8) and (4.1,4) it follows that

$$< a_0 >_\mu - < \Pi_a^0 >_\mu \eta - a = 0 \tag{4.1.39}$$

Hence, the maximum of the wave function is located at

$$a =< a_0 >_\mu - < \Pi_a^0 >_\mu \eta \tag{4.1.40}$$

in concordance with the classical motion (2.19). In the case of $\epsilon = 1$, we obtain

$$\gamma(\eta) = -\alpha \sin \eta + \beta \cos \eta$$

$$\delta(\eta) = \alpha \cos \eta + beta \sin \eta \tag{4.1.41}$$

Thus the coefficients in eq. (4.1.36) are

$$(\delta\gamma^* - c.c) = D$$

$$(\mu\gamma^* - c.c) = D < \Pi_a^0 >_\mu \sin \eta + D < a_0 > \mu \cos \eta \tag{4.1.42}$$

It results from eq. (4.1.36) that

$$a = - < \Pi_a^0 >_\mu \sin \eta + < a_0 > \mu \cos \eta \tag{4.1.43}$$

i.e., the classical motion given by eq. (2.20). The same holds true for the case $\epsilon = -1$. In conclusion, in the case of GCS the maximum of the density probability moves exactly according to the classical motion. This was our main objective, and its seems therefore reasonable to use CGS as wave packets.

4.2 The Wave Packet Solution

Our Schrödinger equation reads

$$i\frac{\partial}{\partial\eta}\Psi = [\frac{\partial^2}{\partial a^2} - \frac{\partial^2}{\partial\varphi^2}\epsilon(-a^2 + \varphi^2)]\Psi \tag{4.2.1}$$

We shall solve it by variables separation:

$$\Psi(\eta, a, \varphi) = A(\eta, a)F(\eta, \varphi) \tag{4.2.2}$$

This leads to

$$i[AF' + FA'] = A[-\frac{\partial^2}{\partial\varphi^2} + \epsilon\varphi^2]F + F[\frac{\partial^2}{\partial a^2} - \epsilon a^2]A \tag{4.2.3}$$

Thus

$$\frac{1}{F}[iF' + (\frac{\partial^2}{\partial\varphi^2} - \epsilon\varphi^2)F] =$$

$$\frac{1}{A}[-iA' + (\frac{\partial^2}{\partial a^2} - \epsilon a^2)A] = cst \tag{4.2.4}$$

By adjusting the phase, we can put the constant to zero; F and A satisfy the following equations:

$$-iF' = (-\frac{\partial^2}{\partial\varphi^2} + \epsilon\varphi^2)F$$

$$-iA' = (-\frac{\partial^2}{\partial a^2} + \epsilon a^2)A \tag{4.2.5}$$

F and A can be choosen to be GCS and the solution reads

$$\Psi(\eta, a, \varphi) = A_\mu(\eta, a) F_\mu(\eta, \varphi) \tag{4.2.6}$$

and we know, by the previously obtained results, that the wave packet follows exactly the classical trajectories. We can now eliminate the redundant η coordinate. As $A_\mu(\eta, a)$ is a wave packet sharply peaked at the classical motion, we can measure the time η with the clock A_μ and the measure will be

$$a = a(\eta) \tag{4.2.7}$$

where this function will be one of the classical motions listed in eqs. (2.19), (2.20) and (2.21). According to the cases $\epsilon = 0, 1$ or -1, we can invert this function and obtain

$$\eta = \eta(a) \tag{4.2.8}$$

i.e. the time-coordinate read by the radius of the universe, used a as a clock. We can now introduce $\eta(a)$ in F_μ and obtain

$$\Psi(\eta, a, \varphi) \simeq A_\mu(\eta, a) F_\mu(\eta(a), \varphi) \tag{4.2.9}$$

If we take now into account eq.(4.3) and use a normalized A_μ, we obtain

$$\Phi(a, \varphi) \simeq F_\mu(\eta(a), \varphi) \tag{4.2.10}$$

This would then be the wave function of our universe with the correct classical behaviour;for example, in the case of $\epsilon = 0$, this function will peak at the classical evolution (2.23) and similarily in the two other cases. We still have the problem of choosing the α and β coefficients (eq.4.1.15). We have a reasonable knowledge about the "momentum" of the universe, we know the Hubble function with reasonable error, but really we know very little about the "configuration" of the universe, we do not know which is the real value of a. The wise thing to do, in these circumstances, seems to us to choose a "momentum" wave packet. Thus we must choose for A_μ the ratio ρ_{1a}/ρ_{2a} very small (but not zero) and $\xi = \pi/2$,such as to minimize the product $\sigma_a \sigma_\Pi$. In this way, we will have a precise definition of H and a bad definition of a.

Of course, in the limit $\rho_{1a}/\rho_{2a} \to 0$, we will obtain a function like (3.1.24) or (3.2.31), with constant $|\Phi|^2$ and the classical motion of the maximum will be lost. Thus, our proposal will coincide with the single-term Vilenkin's prescription, in the limit $\rho_{1a}/\rho_{2a} \to 0$, and therefore we believe it would yield the correct results for Inflation and the fluctuation spectrum, in more realistic models. As concerning F_μ , we know more or less φ and Π_φ because φ symbolize matter density. Hence, we can choose the ratio $\rho_{1\varphi}/\rho_{2\varphi}$ according to observed data, and $\xi = \pi/2$ such as to minimize the product $\sigma_\varphi \sigma_\Pi$. In the case of $\epsilon = \pm 1$, the knowledge of Π_a, φ, and Π_φ yield the value of a, according to the constraint (2.16). Therefore, if we choose σ_Π, φ and we can compute σ_a in such a way that eq.(2.16) would be satisfied; but we have not the freedom to choose $\sigma_a \to 0$. This is not the case if $\epsilon = 0$, where only $\sigma_{\Pi a}$ and $\sigma_{\Pi\varphi}$ are related (in fact, in this case a is only a scale factor). The physical consequence of these constraints have to be investigated in more realistic models.

Conclusions

A new prescription for the wave function of the universe was exhibited in this paper, in the framework of a homogeneous pure radiation universe.This function is represented by a specific wave packet: the Generalized Coherent State. The construction of this function is based on the consideration of a boundary final classical condition. It appears that the maximum of this wave packet follows exactly in its evolution backwards in time,including the quantum phase era, the cosmological classical trajectory. The challenge is to extend this technique to more realistic systems wherein the Inflationary phase as well as the Harrisson-Zeldovich fluctuation spectrum will appear as consequences of the new proposed wave function of the universe. The relation between our proposal and the single-term Vilenkin's prescription permits us to be optimistic in this regard.

References

[1] Hartle J.B. and Hawking S.W. **Phys. Rev. D 28** (12), p2960 (1983)
[2] Vilenkin A. **Phys. Rev. D 37** (4) p888 (1988)
[3] Mukhanov V.F. "On many worlds interpretation of quantum theory" Proceeding of the Third Seminar on Quantum Gravity, Moscow 1984 Ed. Markov M.A. et al. World Scientific, Singapore (1985)
[4] Markov M.A., Mukhanov V.F. **Phys. Lett. A 127** (5) p251 (1988)
[5] Hartle J. "Quantum Kinematics of spacetime" III General Relativity Univ. of California, Santa Barbara preprint (1988)
[6] Hartle J.B. "Predictions in Quantum Cosmology". Nato Advance Summer Institute School Cargese (1986)
[7] Hartle J.B. "Quantum Cosmology" Theor. Advance Study Institute in Elem. Part. Phys. TH-71 Yale Univ. preprint (1985)
[8] Henneaux M. and Teitelboom C. **Phys. Lett. B 222** (2) p195 (1988)
[9] Unruh W.G. and Wald R.M. "Time and the Interpretation of Canonical Quantum Gravity" submitted to **Phys. Rev. D**
[10] Castagnino M.A. "The probabilistic time and the semiclassical approximation of Quantum Gravity" this volume
[11] Birell N.D. Davies P.C.W. "Quantum Fields in Curved space" Cambridge Monographs on Mathematical Physics 1982
[12] Castagnino M.A. "Lecture on semiclassical quantum gravity" Proceeding 4 School on Cosmology and Gravitation, Rio de Janeiro (1984) Ed. M. Novello CNPQ Rio de Janeiro (1985)
[13] Brout R. Englert F. Gunzig E. **Ann. Phys. 115** p78 (1978)
[14] Brout R., Englert F., Frere J.M, Gunzig E., Nardone P., Spindel P., Truffin C. **Nucl. Phys. B 170** p228 (1980)
[15] Prigogine I., Geheniau J., Gunzig E., Nardone P. **Proc. Nat. Acad. Sci. USA 85** p7428 (1988)

NONLINEAR SIGMA MODELS IN 4 DIMENSIONS: A LATTICE DEFINITION.*

Bryce S. DeWitt

Center for Relativity and Physics Department
The University of Texas at Austin
Austin, Texas 78712-1081

1. INTRODUCTION

Systems with Curved Configuration Spaces

Consider a nonrelativistic particle of unit mass constrained to move without friction on a smooth curved surface. The surface is the *configuration space* of the particle, and the action that governs the motion of the particle by the principle of least action is the integral

$$S = \frac{1}{2} \int G_{ij}\left(x(t)\right) \dot{x}^i(t)\, \dot{x}^j(t)\, dt \ . \tag{1.1}$$

The $x^i(t)$ are the coordinates at time t of the position of the particle in some appropriate chart, and the G_{ij} are the components of the metric tensor in that chart.

This system can be generalized in an obvious way. The dimensionality of the configuration space need not be restricted to *two;* it may be an arbitrary finite positive integer. Examples of physical systems for which such a generalization is appropriate are complex molecules at temperatures low enough that the internal vibrational degrees of freedom are frozen. In these examples the $x^i(t)$ are generalized (canonical) coordinates, and the dynamical trajectories are geodesics in the configuration space.

---

* Work supported in part by a grant from the U.S. National Science Foundation. Computer services were provided by the Pittsburgh Supercomputing Center, Pittsburgh, Pennsylvania, the National Center for Supercomputing Applications, Champaign, Illinois, the Air Force Weapons Laboratory, Kirtland Air Force Base, New Mexico, and the Center for High Performance Computing, University of Texas at Austin. During part of this research the author was a guest of the Joint Institute for Nuclear Research, Dubna, U.S.S.R., and gratefully acknowledges assistance from D. V. Shirkov, G. V. Efimov, G. A. Vilkovisky, V. G. Kadishevsky, and N. N. Bogolubov.

Another way to generalize system (1.1) is to take the configuration space to be the spatial part of spacetime in general relativity. The demands of relativistic covariance then require that (1.1) be replaced by the arc length (or proper time) in space*time,* and the dynamical trajectories become geodesics in space*time.*

The above systems may be quantized. The quantization of system (1.1) is relatively straightforward and will be briefly discussed in section 3. The quantization of the relativistic particle in curved spacetime is not so simple, for the well-known reason that relativistic quantum theory is not conveniently constructed as a one-particle theory. Rather, it is a theory of one or more quantized fields, describable as systems of *many* particles, the particles appearing as quantized field excitations.

Quantum field theory in curved spacetime is of great interest in its own right, and its study has led to a number of profound insights into quantum field theory as a whole. But our interest here is not in *background* field theory, as this is often called. Rather, we focus on yet another generalization of the action (1.1), suggested by field theory. We replace the coordinates x^i by real scalar fields ϕ^i and the single time variable t by *four* chart coordinates x^μ ($\mu = 0, 1, 2, 3$) in spacetime. The action (1.1) then becomes

$$S = -\tfrac{1}{2}\,\mu^2 \int g^{1/2}(x)\, G_{ij}(\phi(x))\, g^{\mu\nu}(x)\, \phi^i{}_{,\mu}(x)\, \phi^j_{,\nu}(x)\, d^4x \ , \tag{1.2}$$

where $d^4x = dx^0\, dx^1\, dx^2\, dx^3$, $g = -\det(g_{\mu\nu})$, and $g_{\mu\sigma}g^{\sigma\nu} = \delta_\mu{}^\nu$. Here $g^{\mu\nu}$ are the (contravariant) components of the metric of spacetime (in appropriate charts) and commas followed by Greek indices denote differentiation with respect to the chart coordinates. In units with $\hbar = c = 1$, we choose the fields ϕ^i to be dimensionless, which implies that the constant μ has the dimensions of *mass.* The signature of the metric is chosen to be $- + ++$.

The solutions of the classical field equations defined by the action (1.2) are known variously as *dynamical trajectories, on-shell field histories,* or *harmonic maps.* The maps to which the latter terminology refers are from the manifold M of spacetime to the configuration space C of the fields ϕ^i, which carries the metric G_{ij}. Note that the fields described by the action (1.2) are not generally sections of vector bundles, as they so often are in the case of more conventional field theories. The field components ϕ^i must here be regarded as "coordinates" in appropriate charts of configuration space. Only when $C = \mathbf{R}^n$ and the metric G_{ij} is flat do the fields become sections of vector bundles. In that case the configuration space possesses sets of *preferred global coordinates,* related to one another by linear transformations.

Sigma Models

The quantum theory of fields described by actions of the form (1.2) is extremely complicated. On one hand, the fields generally satisfy nonlinear dynamical equations and hence interact with one another. On the other hand their dynamics takes place in a curved background and must be analyzed with the same care and subtlety as that of linear fields in curved spacetime. In order to separate the problems raised by the two aspects of the theory, we shall confine our attention to flat spacetimes. The action then becomes

$$S = -\tfrac{1}{2}\,\mu^2 \int G_{ij}\,(\phi(x))\, \eta^{\mu\nu}\, \phi^i{}_{,\mu}(x)\, \phi^j_{,\nu}(x)\, d^4x \ , \tag{1.3}$$

where $(\eta^{\mu\nu}) = \mathrm{diag}(-1, 1, 1, 1)$. We do not necessarily require that the topology of spatial sections of spacetime be $\mathbf{R}^3$. It could, for example, be T^3. We require only that these sections be coverable by charts in each of which $g^{\mu\nu} = \eta^{\mu\nu}$.

We shall also restrict the metric G_{ij}. Instead of allowing it to be arbitrary, we shall assume that C is a coset space of some finite-dimensional Lie group, that G_{ij} is the natural (group invariant) metric on this coset space, and that G_{ij} is positive definite. The system defined by the action (1.3) is then called a *sigma model.* The configuration space of a sigma model is usually a *symmetric space* with a covariantly constant Riemann tensor R_{ijkl} and a strictly constant curvature scalar $R(= R_{ij}{}^{ij})$. If $R_{ijkl} \neq 0$, then the sigma model is nonlinear. If $R_{ijkl} = 0$ and $C = \mathbf{R}^n$, then one can introduce a global chart in C in which $G_{ij} = \delta_{ij}$, the fields ϕ^i become linear, and the quantum theory is that of a set of free fields, and hence trivial. If $R_{ijkl} = 0$ but $C \neq \mathbf{R}^n$ (for example, if $C = S^1$), the theory is not trivial.

Nonlinear sigma models in two dimensions are currently of considerable interest because of their relevance to string theories. In these models, the base (or background) manifold is not spacetime but a two-dimensional manifold having Lorentzian signature. These models suffer from the usual divergences of quantum field theory, but they are perturbatively renormalizable, the constant μ being dimensionless. In three and four dimensions, nonlinear sigma models are not perturbatively renormalizable and hence cannot be defined by perturbation theory. It is the purpose of the research reported here to explore a nonperturbative definition of them and to determine whether this definition leads to a consistent and nontrivial theory. Interest in the four-dimensional models stems from their similarities to quantum gravity. Like gravity, each has a curved configuration space and a single scale constant μ, and their Feynman graphs have degrees of divergence identical to those of corresponding graphs in gravity theory.

If nonlinear sigma models in four dimensions can be given consistent nontrivial definitions, this will have important implications for quantum gravity. To stress their potential relevance to quantum gravity, we shall adopt the language of quantum gravity from the outset and call the constant μ the *bare Planck mass.* We also describe the sigma models in terms of dimensionless fields ϕ^i, regarding the latter as analogs of the dimensionless fields $g_{\mu\nu}$ of gravity theory. It is sometimes convenient to absorb the constant μ into the fields by redefining

$$\varphi^i = \mu\phi^i \ , \tag{1.4}$$

and to follow the more standard convention of regarding scalar fields in four dimensions as having the dimensions of mass. The redefinition is useful, for example, in a chart in C in which the metric takes the form

$$G_{ij} = \delta_{ij} - \tfrac{1}{3} \overset{0}{R}_{ikjl} \phi^k \phi^l + \cdots \ , \tag{1.5}$$

$\overset{0}{R}_{ikjl}$ being the curvature tensor at $\phi = 0$ in this chart. The action (1.3) then becomes

$$S = -\tfrac{1}{2} \int \eta^{\mu\nu} (\delta_{ij} - \tfrac{1}{3} \mu^{-2} \overset{0}{R}_{ikjl} \varphi^k \varphi^l + \cdots) \varphi^i_{,\mu} \varphi^i_{,\nu} \, d^4x \ , \tag{1.6}$$

and the role of *coupling constant* in the theory is seen to be played by the inverse quantity μ^{-2}.

Conservation Laws

The symmetry of the configuration space of a sigma model gives rise to the existence of conserved quantities. Denote by G the Lie group of which the configuration space is a coset space. (That is, $C = G/H$ for some subgroup H.) Let ξ^α be the coordinates of a chart in G that includes the identity. If the ξ^α are adjusted so that they vanish at the identity, then a group element infinitesimally close to the identity has infinitesimal coordinate values $\delta\xi^\alpha$, and its action on C may be expressed in the form

$$\delta\phi^i = Q^i{}_\alpha(\phi)\delta\xi^\alpha \ , \tag{1.7}$$

where the $Q^i{}_\alpha$ are the components of a complete set of Killing vector fields $\mathbf{Q}_\alpha$ on C. The Killing condition is

$$0 = \mathcal{L}_{\mathbf{Q}_\alpha} G_{ij} = G_{ij,k}\, Q^k{}_\alpha + G_{ik}\, Q^k{}_{\alpha,j} + G_{jk}\, Q^k{}_{\alpha,i} \ , \tag{1.8}$$

where a comma followed by a Latin index denotes differentiation with respect to one of the ϕ^i. Equations (1.7) and (1.8) together assure the group invariance of the integrand of the action (1.3) and, in combination with the dynamical field equations, lead to the conservation laws* (Noether's theorem)

$$j_\alpha{}^\mu{}_{,\mu} = 0 \ , \qquad j_\alpha{}^\mu = -\mu^2 G_{ij} Q^i{}_\alpha \phi^j{}_,{}^\mu \ , \tag{1.9}$$

where $\phi^j{}_,{}^\mu = \eta^{\mu\nu}\phi^j{}_{,\nu}$. The corresponding conserved quantities are the *charges*

$$q_\alpha = \int_\Sigma j_\alpha{}^\mu\, d\Sigma_\mu \ , \tag{1.10}$$

where Σ is an arbitrary complete spacelike hypersurface.

The fact that G is a group implies a closure property for the infinitesimal transformations (1.7), which can be stated as a Lie bracket relation:

$$[\mathbf{Q}_\alpha, \mathbf{Q}_\beta] = -\,\mathbf{Q}_\gamma\, c^\gamma{}_{\alpha\beta} \ . \tag{1.11}$$

Here the $c^\gamma{}_{\alpha\beta}$ are the structure constants of G. In the language of chart coordinates on C, eq. (1.11) takes the form

$$Q^i{}_{\alpha,j}\, Q^j{}_\beta - Q^i{}_{\beta,j}\, Q^j{}_\alpha = Q^i{}_\gamma\, c^\gamma{}_{\alpha\beta} \ . \tag{1.12}$$

Introducing Poisson brackets into the theory, either canonically or covariantly, one can show, using eq. (1.12), that the conserved charges are the generators of group operations,

$$(\phi^i, q_\alpha) = Q^i{}_\alpha \tag{1.13}$$

and have the following Poisson brackets among themselves:

$$(q_\alpha, q_\beta) = q_\gamma\, c^\gamma{}_{\alpha\beta} \tag{1.14}$$

(see DeWitt, 1984).

---

* The same conservation laws hold for the sigma model in curved spacetime (action given by eq. (1.2)) with currents chosen in the density form

$$j_\alpha{}^\mu = -\mu^2 g^{1/2} G_{ij}\, Q^i{}_\alpha\, \phi^j{}_;{}^\mu \ .$$

It should be stressed that the group G acts globally on the ϕ^i, not locally. That is, the $\delta\xi^\alpha$ in eq. (1.7) are independent of position in spacetime. G is not a gauge group and the concept of a connection on a principal fibre bundle has no role to play here as it does in Yang-Mills theory. The field described by the ϕ^i *may* be viewed as a section of a fibre bundle, the fibres being copies of C and G being the structure group, but because G acts globally on the field, the bundle is necessarily untwisted, i.e., trivial. We note also that C *may* be a vector space, but that in interesting cases it is usually not. Finally, we caution that the globality of G constitutes the single greatest difference between nonlinear sigma models and gravity (with its local diffeomorphism group) and may in the end lead to the quantum theories of the two types of fields being quite different.

2. THE $O(1,2)/O(2) \times Z_2$ AND $O(3)/O(2)$ SIGMA MODELS

A sigma model will be called *simple* if its structure group G is both simple and semisimple. All simple sigma models are nonlinear and hence nontrivial. There are no simple sigma models for which $C\,(= G/H)$ is one-dimensional. There are only two for which C has two dimensions: the $O(1,2)/O(2) \times Z_2$ and $O(3)/O(2)$ models. These are the simplest nonlinear sigma models.

The $O(1,2)/O(2) \times Z_2$ Model

The $O(1,2)/O(2) \times Z_2$ model is defined by the action

$$S = -\tfrac{1}{2}\,\mu^2 \int \eta_{ab}\,\phi^a{}_{,\mu}\,\phi^b{}_{,}{}^{\mu}\,d^4x \ , \qquad a,b \in \{0,1,2\} \ , \tag{2.1}$$

with the fields subject to the constraint

$$\eta_{ab}\,\phi^a\phi^b = -1 \ , \qquad \phi^0 \geq 1 \ , \qquad (\eta_{ab}) = \mathrm{diag}\,(-1,1,1) \ . \tag{2.2}$$

This constraint reduces the number of independent field variables from three (ϕ^0, ϕ^1, ϕ^2) to two. These may be chosen to be s and θ, where

$$\left.\begin{aligned} \phi^0 &= \cosh s \ , \\ \phi^1 &= \sinh s \cos\theta \ , \\ \phi^2 &= \sinh s \sin\theta \ , \end{aligned}\right\} \qquad 0 \leq s < \infty \ , \quad 0 \leq \theta < 2\pi \ . \tag{2.3}$$

The configuration space C is a "spacelike" hyperboloid lying in a $(1+2)$-dimensional Minkowski space. The theory is invariant under global boosts and rotations in this space.

The metric line element in C is given by

$$-(d\phi^0)^2 + (d\phi^1)^2 + (d\phi^2)^2 = ds^2 + \sinh^2\! s\, d\theta^2 \ , \tag{2.4}$$

whence it follows that the action may be rewritten in the form

$$S = -\tfrac{1}{2}\,\mu^2 \int (s_{,\mu}\, s_{,}{}^{\mu} + \sinh^2\! s\, \theta_{,\mu}\, \theta_{,}{}^{\mu})\, d^4x \ . \tag{2.5}$$

A useful alternative pair of variables is

$$\left.\begin{aligned} \sigma^1 &= s\cos\theta \ , \\ \sigma^2 &= s\sin\theta \ . \end{aligned}\right\} \tag{2.6}$$

In terms of these, the line element is $G_{ij}\, d\sigma^i\, d\sigma^j$ where

$$\begin{aligned} G_{ij} &= s^{-2}\sinh^2 s\,\delta_{ij} + (1 - s^{-2}\sinh^2 s)s^{-2}\,\sigma^i\sigma^j \ , && (2.7a) \\ &= \delta_{ij} + \tfrac{1}{3}(\delta_{ij}\,\delta_{kl} - \delta_{ik}\,\delta_{jl})\,\sigma^k\sigma^l + \cdots \ , \qquad i,j,k,l \in \{1,2\} \ , && (2.7b) \end{aligned}$$

from which one may infer that the curvature tensor of C is given (see eq. (1.5)) by

$$\left.\begin{aligned} R_{ijkl} &= -(G_{ik}\,G_{jl} - G_{il}\,G_{jk}) \ , \\ R_{ij} &= R_{ikj}{}^{k} = -G_{ij} \ , \\ R &= R_i{}^i = -2 \ . \end{aligned}\right\} \tag{2.8}$$

The σ^i are *Riemann normal coordinates* in C based on the point $\phi^0 = 1$, $\phi^1 = \phi^2 = 0$.

Inserting (2.7) into eq. (1.3), one gets

$$\begin{aligned} S &= -\tfrac{1}{2}\,\mu^2 \int G_{ij}\,\sigma^i{}_{,\mu}\,\sigma^{j}{}_{,}{}^{\mu}\, d^4x \\ &= -\tfrac{1}{2}\,\mu^2 \int \left[\delta_{ij} + \tfrac{1}{3}(\delta_{ij}\,\delta_{kl} - \delta_{ik}\,\delta_{jl})\,\sigma^k\sigma^l + \cdots\right] \sigma^i{}_{,\mu}\,\sigma^{j}{}_{,}{}^{\mu}\, d^4x \ , \end{aligned} \tag{2.9}$$

from which one may readily compute the bare 4-point vertex function of the theory, taken with respect to the Riemann normal coordinates. It is to be noted that the configuration space of the $O(1,2)/O(2) \times Z_2$ model is noncompact and geodesically complete. C is topologically $\mathbf{R}^2$, and the chart with coordinates σ^i is global, covering C totally. Gravity, too, has a noncompact configuration space (topologically $\mathbf{R}^{10}$, although, with the most natural metric, not geodesically complete) and hence the $O(1,2)/O(2) \times Z_2$ model is perhaps the sigma model most relevant to gravity.

Conserved Currents

The action of $O(1,2)$ on C is determined in a neighborhood of the identity by the infinitesimal transformations

$$\delta\phi^a = \delta\xi^a{}_b\,\phi^b \ , \qquad \delta\xi^a{}_b = \delta\xi^{ac}\,\eta_{cb} \ , \qquad \delta\xi^{ab} = -\delta\xi^{ba} \ . \tag{2.10}$$

These can be reexpressed in the form

$$\delta\phi^a = \tfrac{1}{2}Q^a{}_{bc}\,\delta\xi^{bc} \ , \qquad Q^a{}_{bc} = (\delta^a{}_b\,\eta_{cd} - \delta^a{}_c\,\eta_{bd})\,\phi^d \ , \tag{2.11}$$

where we have kept the antisymmetric pair of indices bc explicit rather than trying to map them into a single index α as in eq. (1.7). One can, by a somewhat tedious computation, convert eqs. (2.10) and (2.11) to transformation laws $\delta\sigma^i = \tfrac{1}{2}Q^i{}_{ab}\,\delta\xi^{ab}$

for the σ^i and then substitute (2.7a), together with the forms found for the $Q^i{}_{ab}$, into (1.9) to get explicit forms for the conserved currents $j_{ab}{}^{\mu} = -\mu^2 G_{ij}\, Q^i{}_{ab}\, \sigma^j{}_{,}{}^{\mu}$. But there is an easier way. The constraint (2.2) may be imposed by simply adding to the integrand of (2.1) a term $\lambda(\eta_{ab}\,\phi^a\,\phi^b + 1)$, where λ is a group-invariant supplementary field. The invariance of the integrand under the actions of $O(1,2)$ remains unaffected, and one may apply the general Noether theory directly to it, obtaining the conserved currents in the obvious form

$$j_{ab}{}^{\mu} = -\mu^2\,\eta_{cd}\,Q^c{}_{ab}\,\phi^d{}_{,}{}^{\mu} = \mu^2(\phi_a\,\phi_{b,}{}^{\mu} - \phi_b\,\phi_{a,}{}^{\mu}) \ , \tag{2.12}$$

where $\phi_a = \eta_{ab}\,\phi^b$. One may then reexpress these currents in terms of the σ^i if one wishes.

Geodetic Distance Function

We shall need in the next section an expression for the geodetic distance Δ in C between two points having coordinates (s, θ) and (s', θ'), respectively. The simplest way to compute this is to note first that the coordinate s is the geodetic distance from the origin to the point (s, θ). Therefore, since the geodetic distance between two points remains invariant under the actions of $O(1,2)$, one may apply a boost that brings one of the two points to the origin. This is most easily done in terms of the variables ϕ^a, and one finds

$$\Delta = \cosh^{-1}(-\eta_{ab}\,\phi^a\,\phi'^b) \ , \tag{2.13}$$

the positive branch of the inverse hyperbolic cosine being understood. Substitution of expressions (2.3) into (2.13) yields

$$\Delta = \cosh^{-1}\left[\cosh(s - s')\cos^2\tfrac{1}{2}(\theta - \theta') + \cosh(s + s')\sin^2\tfrac{1}{2}(\theta - \theta')\right] , \tag{2.14}$$

a form that will be useful for computer work because the argument of the $\cosh^{-1}$ function is the sum of two manifestly non-negative terms.

We shall also need the expansion of Δ^2 to fourth order in the σ^i, σ'^i. It is easy to verify that the argument of the $\cosh^{-1}$ function is expressible in the form

$$\begin{aligned} &\cosh s \cosh s' - \sinh s \sinh s'(\cos\theta\cos\theta' + \sin\theta\sin\theta') \\ &\quad = 1 + \tfrac{1}{2}(\sigma^i - \sigma'^i)(\sigma^i - \sigma'^i) + \tfrac{1}{24}\left[(\sigma^i - \sigma'^i)(\sigma^i - \sigma'^i)\right]^2 \\ &\qquad + \tfrac{1}{6}(\varepsilon_{ij}\,\sigma^i\sigma'^j)^2 + \cdots \ , \end{aligned} \tag{2.15}$$

where

$$(\varepsilon_{ij}) = \begin{pmatrix} 0 & 1 \\ -1 & 0 \end{pmatrix} . \tag{2.16}$$

Expression (2.15), together with the expansion

$$(\cosh^{-1} x)^2 = 2(x - 1) - \tfrac{1}{3}(x - 1)^2 + \cdots \ , \tag{2.17}$$

then immediately yields

$$\Delta^2 = (\sigma^i - \sigma'^i)(\sigma^i - \sigma'^i) + \tfrac{1}{3}(\varepsilon_{ij}\,\sigma^i\,\sigma'^j)^2 + \cdots \quad . \tag{2.18}$$

The O(3)/O(2) Model

The $O(3)/O(2)$ sigma model is closely analogous to the $O(1,2)/O(2) \times Z_2$ model. It is defined by the action

$$S = -\tfrac{1}{2}\,\mu^2 \int \phi_{a,\mu}\,\phi_{a,}{}^{\mu}\,d^4x \ , \qquad a \in \{1,2,3\} \ , \tag{2.19}$$

together with the constraint

$$\phi_a\,\phi_a = 1 \ . \tag{2.20}$$

Here the indices on the fields ϕ_a are written in the lower position because they take their values in $\mathbf{R}^3$, which has δ_{ab} as its $O(3)$-invariant metric. The configuration space C is the unit 2-sphere, which is compact. The compactness of C gives rise to major differences in the physical behavior of the $O(3)/O(2)$ model from that of the $O(1,2)/O(2) \times Z_2$ model.

Again one may introduce variables s, θ and Riemann normal coordinates σ^1, σ^2. In the present case they are given by

$$\left.\begin{aligned} \phi_1 &= \sin s \cos\theta \ , \\ \phi_2 &= \sin s \sin\theta \ , \\ \phi_3 &= \cos s \ , \end{aligned}\right\} \tag{2.21}$$

$$\left.\begin{aligned} \sigma^1 &= s\cos\theta \ , \\ \sigma^2 &= s\sin\theta \ . \end{aligned}\right\} \tag{2.22}$$

In terms of these variables the action takes the form

$$S = -\tfrac{1}{2}\,\mu^2 \int (s_{,\mu}\,s_{,}{}^{\mu} + \sin^2 s\,\theta_{,\mu}\,\theta_{,}{}^{\mu})\,d^4x \tag{2.23}$$

$$= -\tfrac{1}{2}\,\mu^2 \int G_{ij}\,\sigma^i{}_{,\mu}\,\sigma^j{}_{,}{}^{\mu}\,d^4x \ , \qquad i,j \in \{1,2\} \ , \tag{2.24}$$

where

$$G_{ij} = s^{-2}\sin^2 s\,\delta_{ij} + (1 - s^{-2}\sin^2 s)s^{-2}\sigma^i\,\sigma^j \tag{2.25a}$$

$$= \delta_{ij} - \tfrac{1}{3}(\delta_{ij}\,\delta_{kl} - \delta_{ik}\,\delta_{jl})\sigma^k\,\sigma^l + \cdots \ . \tag{2.25b}$$

The curvature tensor of C is

$$\left.\begin{aligned} R_{ijkl} &= G_{ik}\,G_{jl} - G_{il}\,G_{jk} \ , \\ R_{ij} &= R_{ikj}{}^{k} = G_{ij} \ , \\ R &= R_i{}^i = 2 \ . \end{aligned}\right\} \tag{2.26}$$

Since C is compact and boundaryless, it cannot be covered by a single chart. In particular, the Riemann normal coordinates σ^i cannot be used globally. They are restricted to the region $s < \pi$.

The action of $O(3)$ on C is expressed in a neighborhood of the identity by infinitesimal transformations of the form

$$\delta\phi_a = \delta\xi_{ab}\,\phi_b = \tfrac{1}{2} Q_{abc}\,\delta\xi_{bc} \;, \tag{2.27}$$

where

$$Q_{abc} = \delta_{ab}\,\phi_c - \delta_{ac}\,\phi_b \;, \qquad \delta\xi_{ab} = -\delta\xi_{ba} \;, \tag{2.28}$$

leading to the conserved currents

$$j_{ab}{}^{\mu} = -\mu^2\,\delta_{cd}\,Q_{cab}\,\phi_{d,}{}^{\mu} = \mu^2(\phi_a\,\phi_{b,}{}^{\mu} - \phi_b\,\phi_{a,}{}^{\mu}) \;. \tag{2.29}$$

The geodetic distance between the points (s, θ) and (s', θ') is given by

$$\Delta = \cos^{-1}\phi_a\,\phi'_a \tag{2.30}$$

$$= \cos^{-1}\left[\cos(s-s')\cos^2\tfrac{1}{2}(\theta-\theta') + \cos(s+s')\sin^2\tfrac{1}{2}(\theta-\theta')\right] \;, \tag{2.31}$$

the positive branch of the arc cosine being understood. Expanding the argument of the arc cosine and using

$$(\cos^{-1}x)^2 = 2(1-x) + \tfrac{1}{3}(1-x)^2 + \cdots \;, \tag{2.33}$$

one also finds

$$\Delta^2 = (\sigma^i - \sigma'^i)(\sigma^i - \sigma'^i) - \tfrac{1}{3}(\varepsilon_{ij}\,\sigma^i\,\sigma'^j)^2 + \cdots \;. \tag{2.34}$$

3. LATTICE QUANTIZATION

The Feynman Functional Integral

Since the nonlinear sigma models in four dimensions are not perturbatively renormalizable, their quantum theories must be defined by nonperturbative means. The only way presently known to do this is via the Feynman functional integral or sum over histories. When using the Feynman integral one does not need to manipulate field operators directly or to construct a Hilbert space or Fock space explicitly. Amplitudes for all physical processes can, in principle, be expressed in terms of in-out matrix elements of chronological averages of appropriate functionals $A[\boldsymbol{\phi}]$ of the field operators $\boldsymbol{\phi}^i$ (here written in boldface), and these are expressible as Feynman integrals:

$$\langle A[\boldsymbol{\phi}]\,\rangle \equiv \frac{\langle \text{out}|\,T(A[\boldsymbol{\phi}])\,|\text{in}\rangle}{\langle \text{out}|\text{in}\rangle} \tag{3.1a}$$

$$= \frac{\int A[\phi]\,e^{iS[\phi]}\,\mu[\phi]\,d\phi}{\int e^{iS[\phi]}\,\mu[\phi]\,d\phi} \;. \tag{3.1b}$$

The boundary conditions imposed by the state vectors $|\text{in}\rangle$ and $\langle\text{out}|$ are reflected in (a) the inclusion of appropriate boundary terms in the action $S[\phi]$ (if necessary) and (b) the imposition of appropriate boundary behavior on the "classical" fields ϕ_i over which the functional integrals are taken.

The volume element $d\phi$ and *measure functional* $\mu[\phi]$ in (3.1b) are formal continuous infinite products over spacetime:

$$d\phi = \prod_x \prod_i d\phi^i(x) \ , \tag{3.2}$$

$$\mu[\phi] = \prod_x G^{1/2}(\phi(x)) \ , \qquad G \equiv \det(G_{ij}) \ , \tag{3.3}$$

where G_{ij} is the metric of configuration space. If perturbation theory cannot be used, the only way the functional integrals can be defined is by replacing spacetime with a finite discrete lattice and by passing to the limits as the size of the lattice becomes infinite and the lattice spacing goes to zero. This is the procedure that has been adopted by a small group at the University of Texas: *Jorge de Lyra, See Kit Foong, Timothy Gallivan,* and the author.

Euclidean Quantization

It is easy to write down the lattice versions of the infinite products (3.2) and (3.3). It is not so easy to evaluate the functional integrals themselves, for two reasons. First, to attain any sort of useful approximation to a real theory, one must work with at least 10,000 lattice sites and hence with several tens of thousands of integration variables. Second, the integrands are oscillatory, and no computer is accurate enough to prevent round-off errors from completely dominating the sums by which the integrals are approximated.

The first difficulty is overcome with the use of Monte Carlo techniques to be described in the next section. The second difficulty is also overcome by a well-known method: analytic continuation from Minkowski to Euclidean spacetime. This method is so well known that no attempt will be made here to describe it in detail or to justify it. We shall simply state a few facts: The time variable x^0 is replaced by x^4 where

$$x^0 = -i\,x^4 \ , \tag{3.4}$$

and the Fourier transform, or momentum, variable p_0, which appears in the product $p \cdot x = p_\mu\, x^\mu$ in the exponential $e^{ip\cdot x}$, is replaced by p_4 where

$$p_0 = ip_4 \ . \tag{3.5}$$

If either of these variables appears explicitly in the functional $A[\phi]$, it will appear analytically, and the Minkowski average (3.1) can be obtained from the corresponding Euclidean average by analytic continuation, with proper account taken of branch points. As far as the state vectors $|\text{in}\rangle$ and $\langle\text{out}|$ are concerned, it is possible to enforce appropriate boundary conditions on the functional integration variables ϕ^i even in the Euclidean regime. In practice $|\text{in}\rangle$ and $\langle\text{out}|$ are the in and out ground or vacuum state vectors, and the ϕ^i may be either set equal to classical ground-state values at the boundary of the lattice or, if periodic boundary conditions are imposed, allowed to drift, with the action seeking its own minimum.

The relevant action in the Euclidean regime is the so-called *Euclidean action* $S_E[\phi]$, and the Minkowski average (3.1) is replaced by the Euclidean average

$$\langle A[\phi] \rangle = \frac{\int A[\phi] e^{-S_E[\phi]} \mu[\phi]\, d\phi}{\int e^{-S_E[\phi]} \mu[\phi]\, d\phi} \ . \tag{3.6}$$

Since we shall from now on be dealing exclusively with the Euclidean action, the subscript E will be omitted. The Euclidean action for sigma models has the form

$$S = \tfrac{1}{2} \mu^2 \int G_{ij}(\phi)\, \phi^i{}_{,\mu}\, \phi^j{}_{,\mu}\, d^4x \ , \qquad d^4x = dx^1\, dx^2\, dx^3\, dx^4 \ . \tag{3.7}$$

It is seen to be nonnegative, bounded by zero from below. It attains its lower bound whenever ϕ^i is a constant field. As is well known, the Euclidean action for quantum gravity is *not* bounded from below, and this constitutes a major difference between gravity and sigma models. It is a source of many difficulties for the computer analysis of quantum gravity, but these will not be discussed here.

The Massless Scalar Field. Lattice Action and Lattice Fourier Transforms

Before tackling the full nonlinear problem on the lattice, let us look first at a completely soluble system: the free massless scalar field. This system is described by a single real scalar field ϕ, and its Euclidean action is

$$S = \tfrac{1}{2} \mu^2 \int \phi_{,\mu}\, \phi_{,\mu}\, d^4x \ . \tag{3.8}$$

Here ϕ is chosen to be dimensionless and the scale factor μ is made explicit in order to maintain as close a similarity as possible to the sigma models. The lattice is most simply chosen to be a cubical one having N^4 sites with periodic boundary conditions (4-torus). On such a lattice the action (3.8) takes the differenced form

$$S = \tfrac{1}{2} \mu^2 a^2 \sum_{\text{links}} \left[\phi(\text{one end of link}) - \phi(\text{other end of link})\right]^2 \tag{3.9}$$

where a is the lattice spacing and the links are between adjacent sites in the four principal lattice directions, with periodicity conditions taken into account.

The size, or periodicity distance, of the lattice is given by

$$L = Na \ , \tag{3.10}$$

but to the computer this length is irrelevant. The only adjustable constants in the lattice version of the theory are the dimensionless quantities N and β, where β is the coefficient in (3.9):

$$\beta \equiv \mu^2 a^2 \ . \tag{3.11}$$

The system (3.9) is most easily studied by introducing the *lattice Fourier transform.* Denote by $\phi_{\alpha\beta\gamma\delta}$ the value of the field ϕ at the lattice site having coordinates $\alpha a, \beta a, \gamma a, \delta a$, with the origin of coordinates being taken at a corner of the periodic cube, the coordinate axes being oriented along the principal lattice directions, and $\alpha, \beta, \gamma, \delta$ being integers ranging from 0 to $N-1$. The lattice Fourier transform is defined by

$$\tilde{\phi}_{klmn} = N^{-2} \sum_{\alpha,\beta,\gamma,\delta} \phi_{\alpha\beta\gamma\delta} \exp\left[\frac{2\pi i}{N}(k\alpha + l\beta + m\gamma + n\delta)\right] , \tag{3.12}$$

where k, l, m, n are integers ranging from $-N/2+1$ to $N/2$ (N being here assumed to be even). The original periodic field may be regained by taking the inverse Fourier transform

$$\phi_{\alpha\beta\gamma\delta} = N^{-2} \sum_{k,l,m,n} \tilde{\phi}_{klmn} \exp\left[-\frac{2\pi i}{N}(k\alpha + l\beta + m\gamma + n\delta)\right] . \tag{3.13}$$

It is well known that imposition of periodicity in (Euclidean) time corresponds to studying the field ϕ in a thermal state at temperature $T = 1/kL$, where k is Boltzmann's constant. However, since our focus will be mainly on the behavior of our models at *cutoff energies* corresponding to temperatures of order $1/ka = NT$, T may effectively be regarded as vanishing, the thermal state being indistinguishable from the ground state for large N. In the following sections we *shall* encounter a nonvanishing *dimensionless* "temperature," namely, the quantity $1/\beta$, but this will be a formal, *discretization* temperature, not a physical temperature.

In order to reexpress the lattice action (3.9) in terms of the Fourier transforms $\tilde{\phi}_{klmn}$, it will be convenient to introduce some abbreviations. We shall denote by $\boldsymbol{\alpha}$ the ordered set of integers $(\alpha, \beta, \gamma, \delta)$ and by $\mathbf{k}$ the ordered set (k, l, m, n), and we shall write

$$\mathbf{k} \cdot \boldsymbol{\alpha} \equiv k\alpha + l\beta + m\gamma + n\delta , \tag{3.14}$$

$$\boldsymbol{\alpha} + \boldsymbol{\alpha}' \equiv (\alpha + \alpha', \beta + \beta', \gamma + \gamma', \delta + \delta') , \tag{3.15}$$

$$-\mathbf{k} \equiv (-k, -l, -m, -n) , \text{ etc.} \tag{3.16}$$

If we now let the symbol $\boldsymbol{\lambda}$ (for *link*) range over the eight values $(\pm1, 0, 0, 0)$, $(0, \pm1, 0, 0)$, $(0, 0, \pm1, 0)$, $(0, 0, 0, \pm1)$, regard all integers as defined *modulo* N, and remember that there are four times as many links as lattice sites, we may rewrite expression (3.9) in the form

$$\begin{aligned} S &= \tfrac{1}{4}\beta \sum_{\boldsymbol{\alpha}} \sum_{\boldsymbol{\lambda}} (\phi_{\boldsymbol{\alpha}} - \phi_{\boldsymbol{\alpha}+\boldsymbol{\lambda}})^2 \\ &= \tfrac{1}{4}\beta N^{-4} \sum_{\boldsymbol{\alpha}} \sum_{\boldsymbol{\lambda}} \sum_{\mathbf{k}} \sum_{\mathbf{k}'} \tilde{\phi}_{\mathbf{k}} \tilde{\phi}_{\mathbf{k}'} [e^{-(2\pi i/N)\mathbf{k}\cdot\boldsymbol{\alpha}} - e^{-(2\pi i/N)\mathbf{k}\cdot(\boldsymbol{\alpha}+\boldsymbol{\lambda})}] \\ &\qquad \times [e^{-(2\pi i/N)\mathbf{k}'\cdot\boldsymbol{\alpha}} - e^{-(2\pi i/N)\mathbf{k}'\cdot(\boldsymbol{\alpha}+\boldsymbol{\lambda})}] \\ &= \tfrac{1}{4}\beta \sum_{\boldsymbol{\lambda}} \sum_{\mathbf{k}} \tilde{\phi}_{\mathbf{k}} \tilde{\phi}^*_{\mathbf{k}} [2 - e^{(2\pi i/N)\mathbf{k}\cdot\boldsymbol{\lambda}} - e^{-(2\pi i/N)\mathbf{k}\cdot\boldsymbol{\lambda}}] \\ &= \tfrac{1}{2}\beta \sum_{\mathbf{k}} \overline{K}^2(\mathbf{k}) |\tilde{\phi}_{\mathbf{k}}|^2 , \end{aligned} \tag{3.17}$$

where

$$\overline{K}(\mathbf{k}) \equiv \left(2\sin\frac{\pi k}{N},\ 2\sin\frac{\pi l}{N},\ 2\sin\frac{\pi m}{N},\ 2\sin\frac{\pi n}{N}\right) , \tag{3.18}$$

$$\overline{K}^2(\mathbf{k}) \equiv \overline{K}(\mathbf{k})\cdot\overline{K}(\mathbf{k}) . \tag{3.19}$$

In passing to the last line of (3.17), we have used the fact that $\tilde{\phi}_{-\mathbf{k}} = \tilde{\phi}^*_{\mathbf{k}}$ together with the easily verified identity

$$\sum_{\boldsymbol{\lambda}}[1 - e^{(2\pi i/N)\mathbf{k}\cdot\boldsymbol{\lambda}}] = \sum_{\boldsymbol{\lambda}}[1 - e^{-(2\pi i/N)\mathbf{k}\cdot\boldsymbol{\lambda}}] = \overline{K}^2(\mathbf{k}) . \tag{3.20}$$

Note that although the physical continuum momentum corresponding to the integers k, l, m, n is $p = a^{-1}K(\mathbf{k})$ where

$$K(\mathbf{k}) = \frac{2\pi}{N}(k, l, m, n) = \frac{2\pi}{N}\mathbf{k} , \tag{3.21}$$

it gets replaced on the lattice by

$$\bar{p} = a^{-1}\overline{K}(\mathbf{k}) , \tag{3.22}$$

which equals p only in the limit $N \to \infty$. The sine functions appearing in equation (3.18) are lattice artifacts stemming from the replacement of derivatives by differences.

The 2-point Function

It is not difficult to verify that the transformation from the variables $\phi_{\alpha\beta\gamma\delta}$ to the variables $\tilde{\phi}_{klmn}$ is a unitary transformation of unit Jacobian. Since, for the free field, the integrals in equation (3.6) are Gaussian and since the transformation to the variables $\tilde{\phi}_{klmn}$ effectively diagonalizes the Gaussian matrix (equation (3.17)), the integrals are readily evaluated. One may write down the so-called *2-point function* by inspection:

$$\langle\tilde{\phi}_{klmn}\,\tilde{\phi}_{k'l'm'n'}\rangle = \frac{\delta_{k,-k'}\,\delta_{l,-l'}\,\delta_{m,-m'}\,\delta_{n,-n'}}{\beta\overline{K}(\mathbf{k})^2} . \tag{3.23}$$

The general *n-point functions* are equally easy to obtain. They vanish when n is odd and are expressible as sums of permuted products of 2-point functions when n is even, which is another way of saying that all higher correlation functions vanish for Gaussian probability distributions.

Actual machine computations of the 2-point function are found to agree very accurately with equation (3.23) for all values of $\mathbf{k}$ and $\mathbf{k}'$ except $(0,0,0,0)$. The reason $(0,0,0,0)$ is special (apart from the fact that expression (3.23) becomes singular at this value) is that $\tilde{\phi}_{0000}$ represents the *constant component* of ϕ, being N^2 times the average of ϕ over the lattice. We have already remarked that the Euclidean action attains its lower bound, namely, zero, whenever ϕ is a constant. Moreover, the addition of a constant to ϕ leaves the action unchanged, and hence the integral over the variable $\tilde{\phi}_{0000}$ diverges. On the computer this divergent degree of integration freedom is eliminated by repeated addition of constants to ϕ so that its lattice average remains zero. This has the consequence that the computer average (*i.e.*, the Monte Carlo average) $\langle \tilde{\phi}_{0000}{}^n \rangle$ vanishes for all positive integers n. $\tilde{\phi}_{0000}$ could, of course, be held fixed at any other value. Zero is chosen merely for simplicity.

A Fundamental Difficulty

The success achieved in simulating the massless free field on a computer prompts one to attempt an analogous simulation of the $O(1,2)/O(2) \times Z_2$ model. The Euclidean action of the latter is (see equation (2.1))

$$S = \tfrac{1}{2}\,\mu^2 \int \eta_{ab}\, \phi^a{}_{,\mu}\, \phi^b{}_{,\mu}\, d^4x \ , \tag{3.24}$$

with the ϕ^a subject to the constraint (2.2). Comparison with (3.8) suggests that one choose for the differenced form of this action the following analog of (3.9):

$$\begin{aligned} S = \tfrac{1}{2}\,\beta \sum_{\text{links}} \eta_{ab}\, &[\phi^a(\text{one end of link}) - \phi^a(\text{other end of link})] \\ &\times\ [\phi^b(\text{one end of link}) - \phi^b(\text{other end of link})] \ , \end{aligned} \tag{3.25}$$

with the constraint (2.2) again being imposed. This was, in fact, the choice initially made and extensively studied by the Texas group. But *it is wrong.* All results obtained with its use are worthless!

The reason it is wrong can be understood by considering the fields that make the dominant contributions to the lattice versions of the integrals in equation (3.6) Let ϕ be such a field, and let $\Delta\phi$ be the difference between the values of ϕ at opposite ends of one of the links in the sum (3.25). The contribution that this link makes to the sum is of order $\beta(\Delta\phi)^2$. As long as this contribution is of order 1 or less for all links, the field will contribute significantly to (3.6). The dominant fields are therefore characterized by the condition $\beta(\Delta\phi)^2 \sim 1$ or

$$\Delta\phi \sim \frac{1}{\sqrt{\beta}} = \frac{1}{\mu a} \ . \tag{3.26}$$

In the continuum limit $a \to 0$ these fields are seen to be wildly discontinuous. Their derivatives are even more singular, diverging nearly everywhere like $1/a^2$ as $a \to 0$.

It is easy to see that in d dimensions the derivatives of the dominant fields diverge nearly everywhere like $a^{-d/2}$ as $a \to 0$. Thus even when $d = 1$, *i.e.*, in the case of the system (1.1), the dominant trajectories are nondifferentiable. This, of course, is well known, but it has a consequence that is sometimes forgotten. Classically, a system may be described by any convenient variables, and in a transformation from one set of variables to another, derivatives of one set are obtained from those of the other by application of the chain rule, which leads to the appearance of Jacobian matrices. The chain rule is valid, however, only when the derivatives actually exist.

In the case of one-dimensional systems like (1.1), it has been known for years that the transformation laws for lattice derivatives do not involve merely simple Jacobians. Extra terms often have to be added to the naive finite differences suggested by the continuum classical action.

In the case of nonlinear sigma models in four dimensions, we do not have any already-worked-out theorems to help us in choosing an appropriate lattice action for approximating a given model. That is because the quantum theory of these models does not yet exist as a coherent discipline. We are presently in the position of having to *define* the quantum theory by choosing a specific lattice action *a priori.*

The $\mathbf{O(1,1)/Z_2 \times Z_2}$ Model

In order to orient ourselves on this problem, it is helpful to look first at the $O(1,1)/Z_2 \times Z_2$ sigma model. This model is obtained from the $O(1,2)/O(2) \times Z_2$ model by simply omitting the field ϕ^2. Equations (2.2) and (3.24) continue to hold, but η_{ab} is now given by $(\eta_{ab}) = \operatorname{diag}(-1,1)$. The configuration space of the $O(1,1)/Z_2 \times Z_2$ model is simply $\mathbf{R}$, and it can be parameterized by a single variable s, where

$$\left.\begin{aligned} \phi^0 &= \cosh s \ , \\ \phi^1 &= \sinh s \ . \end{aligned}\right\} \tag{3.27}$$

Inserting this into (3.24) we get

$$S = \tfrac{1}{2}\,\mu^2 \int s_{,\mu}\, s_{,\mu}\, d^4x \ , \tag{3.28}$$

and we see that the $O(1,1)/Z_2 \times Z_2$ model is just the massless free field, which we have already studied.

The use of two variables, ϕ^0 and ϕ^1, to describe the massless free field allows us to ask questions that would not have occurred to us if we had confined our atention to only one variable, questions that are relevant also for the $O(1,2)/O(2) \times Z_2$ model. Note that for both models ϕ^0 is always greater than or equal to zero. On the lattice the dominant fields in the integrals (3.6) are those for which ϕ^0 oscillates between larger and larger values as β tends to zero (see equation (3.26)). One therefore expects that the average $\langle \phi^0 \rangle$ will diverge in the limit $\beta \to 0$. Let us try to compute this average on the assumption, based on equation (3.28), that the correct lattice action for the $O(1,1)/Z_2 \times Z_2$ model is expression (3.9) with ϕ replaced by s.

We first compute $\langle s^2 \rangle$. Because of the displacement invariance of the actions (3.9) and (3.28), we may expect this average to be independent of position. Therefore, using

$$s_{\alpha\beta\gamma\delta} = N^{-2} \sum_{\mathbf{k}}{}' \, \tilde{s}_{\mathbf{k}}\, e^{-(2\pi i/N)\mathbf{k}\cdot\boldsymbol{\alpha}} \ , \tag{3.29}$$

where the prime on the summation sign indicates that the term with $\mathbf{k} = (0,0,0,0)$ is to be omitted (on the assumption that $\tilde{s}_{0000}$ is held equal to zero), we get

$$\begin{aligned}
\langle s^2 \rangle &= \langle s_{\alpha\beta\gamma\delta}{}^2 \rangle = N^{-4} \sum_{\alpha,\beta,\gamma,\delta} \langle s_{\alpha\beta\gamma\delta}{}^2 \rangle \\
&= N^{-4} \sum_{\mathbf{k},\mathbf{k}',\boldsymbol{\alpha}}{}' \langle \tilde{s}_{\mathbf{k}} \tilde{s}_{\mathbf{k}'} \rangle e^{-(2\pi i/N)(\mathbf{k}+\mathbf{k}')\cdot\boldsymbol{\alpha}} \\
&= N^{-4} \sum_{\mathbf{k}}{}' \langle \tilde{s}_{\mathbf{k}} \tilde{s}_{\mathbf{k}}^* \rangle = N^{-4}\beta^{-1} \sum_{\mathbf{k}}{}' \overline{K}^{-2}(\mathbf{k}) \quad \text{(see eq. (3.23))} \\
&= F(N)/\beta
\end{aligned} \tag{3.30}$$

where

$$F(N) = \tfrac{1}{4} N^{-4} \sum_{k,l,m,n}{}' \left(\sin^2 \frac{\pi k}{N} + \sin^2 \frac{\pi l}{N} + \sin^2 \frac{\pi m}{N} + \sin^2 \frac{\pi n}{N} \right)^{-1}. \tag{3.31}$$

For a free field we also have

$$\langle s^{2n} \rangle = \frac{(2n)!}{2^n n!} \langle s^2 \rangle^n \tag{3.32}$$

and hence

$$\begin{aligned}
\langle \phi^0 \rangle &= \langle \cosh s \rangle = \sum_{n=0}^{\infty} \frac{1}{(2n)!} \langle s^{2n} \rangle = \sum_{n=0}^{\infty} \frac{1}{2^n n!} \langle s^2 \rangle^n \\
&= e^{1/2\langle s^2 \rangle} = e^{F(N)/2\beta} ,
\end{aligned} \tag{3.33}$$

which shows that $\langle \phi^0 \rangle$ does indeed diverge (exponentially!) in the limit $\beta \to 0$.

Comparison with Computer Results

For comparison with computer results it is easier to plot $\langle \phi^0 \rangle^{-1} = e^{-F(N)/2\beta}$. Figure 1 displays this function for $N = \infty$. One cannot, of course, work with an infinite lattice on the computer, but it happens that $F(N)$ converges very rapidly to its limiting value

$$F(\infty) = .15 \ldots \tag{3.34}$$

for N greater than 4. This is an example of a remarkable stability property that recurs repeatedly in the study of sigma models: Many of the important quantities that one computes become quickly independent of N as N is increased. In the present case this has the consequence that one can compare the machine-computed curves for $\langle \phi^0 \rangle^{-1}$ directly with the figure. For $N > 8$ the machine-computed curves cannot be distinguished from the curve shown in the figure.

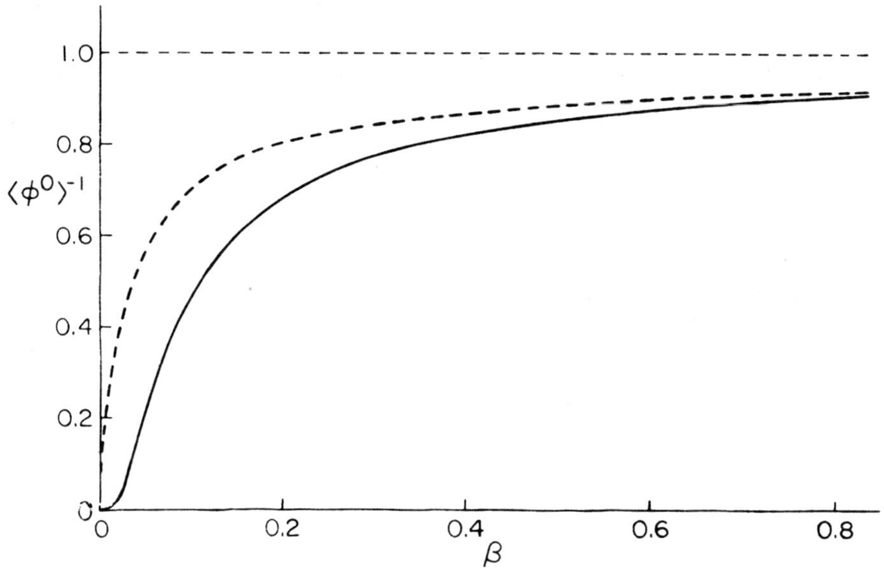

Figure 1. The function $\langle\phi^0\rangle^{-1}$ when N > 8. The solid curve shows the function $\exp\big[-F(\infty)/2\beta\big]$. It is indistinguishable from the machine-computed curves for $\langle\phi^0\rangle^{-1}$ when $N > 8$. The dotted curve is the curve for $\langle\phi^0\rangle^{-1}$ (with $N > 8$) that one gets when using the incorrect lattice action (3.25).

The dotted curve in Figure 1 is the curve for $\langle\phi^0\rangle^{-1}$ when it is computed with the lattice action (3.25) and $N > 8$. It differs dramatically from the solid curve as $\beta \to 0$. The solid curve displays the behavior appropriate to the existence of an essential singularity at $\beta = 0$ while the dotted curve follows a fractional power law. Since our ultimate interest is in precisely the continuum limit $\beta \to 0$ (or $a \to 0$), it is crucial for us to use the "correct" lattice action.

In the present case we know that the correct lattice action is (3.9) (with ϕ replaced by s) because we know that the continuum theory of the $O(1,1)/Z_2 \times Z_2$ model is really that of a massless free field, and we have seen that (3.9) describes correctly this theory, yielding the well-known $1/p^2$ behavior of the 2-point function in the continuum limit $N \to \infty$, $a \to 0$. But what should we choose for the lattice action when the theory is intrinsically nonlinear?

A Proposal

We here make a proposal, which is somewhat arbitrary in that one can think of other possibilities, but which has the virtue of being simple (other proposals are more complicated), of being independent of the choice of basic field variables, and of being "free-field consistent," *i.e.*, of yielding (3.9) and its generalization to $C = \mathbf{R}^n$ when the theory is linear. We propose to replace the lattice action (3.25) by

$$S = \tfrac{1}{2}\,\beta \sum_{\text{links}} \Delta^2\left(\phi(\text{one end of link}),\ \phi(\text{other end of link})\right) \tag{3.35}$$

where $\Delta(\phi, \phi')$ is the geodetic distance in configuration space between the point with coordinates ϕ^i and the point with coordinates ϕ'^i. The value of Δ is obviously independent of the choice of basic field variables (*i.e.*, of charts in C). Equations (2.14), (2.18), (2.32), and (2.35), for example, express Δ for the $O(1,2)/O(2) \times Z_2$ and $O(3)/O(2)$ sigma models alternately in terms of the variables s, θ, s', θ' or the variables σ^i, σ'^i.

From now on the quantum theory of nonlinear sigma models will be understood as *defined* by taking the continuum limit of the functional integrals (3.6), with Euclidean action chosen in the form (3.35) and the measure functional and volume element given respectively by

$$\mu[\phi] = \prod_{\text{sites}} G^{1/2}(\phi(\text{site})) \ , \tag{3.36}$$

$$d\phi = \prod_{\text{sites}} \prod_i d\phi^i(\text{site}) \ . \tag{3.37}$$

It will be noted that with β appropriately defined, this definition can serve in any number of spacetime dimensions. In fact, it is one of the standard definitions used for quantizing the system (1.1) and is worth discussing, for a moment, in connection with this system.

Years ago the author showed that if the above definition is used for computing transition amplitudes for system (1.1) in the continuum limit, the resulting quantum theory is that of a system described by a wave function ψ satisfying the Schrödinger equation

$$i\partial\psi/\partial t = (-\tfrac{1}{2}\Box + \tfrac{1}{6}R)\psi \ , \tag{3.38}$$

where $\Box$ is the covariant Laplacian operator on the configuration space and R is its curvature scalar (DeWitt 1957). Equation (3.38) yields a perfectly acceptable quantum theory of system (1.1) in that it conserves probability and yields the correct classical theory (geodetic motion) in the extreme short wavelength limit, but it is not the only equation that does. The quantum theory of system (1.1) leads to no divergences in perturbation theory, and one may, for example, compute transition amplitudes by using the loop expansion. If the quantum theory is defined by the loop expansion, one finds that a computation to no higher than 2-loop order suffices to show that the Schrödinger equation for system (1.1) takes the form

$$i\partial\psi/\partial t = (-\tfrac{1}{2}\Box + \tfrac{1}{8}R)\psi \ , \tag{3.39}$$

which differs from (3.38) in the coefficient of the term in R (DeWitt 1984, problem 35).

No known experiment can determine which Schrödinger equation is right. One cannot, for example, appeal to the properties of complex molecules at low temperatures, for in the last analysis the particles of which laboratory molecules are composed exist in a flat space (*i.e.*, flat compared to R) and are bound together by potentials in this space. However, there is an important case in which it does not matter which equation is right, namely, when R is everywhere constant, for then the only difference between (3.38) and (3.39) is the choice of energy zero point.

Constant curvature of C is a generic property of sigma models. In one-dimensional "spacetime" different definitions of the theory lead to differences in the way R enters. In higher dimensions one may expect other scalar quantities, such as $R_{ijkl}R^{ijkl}$, R^3, and their covariant derivatives, to appear in the effective Hamiltonian, with only their coefficients differing from one definition to another. But all these quantities are constants, and hence it is plausible to suppose *(although not proved)* that the definition that uses the lattice action (3.35) will be as good as any other free-field-consistent definition for determining whether a consistent nontrivial quantum theory of nonlinear sigma models in four dimensions exists in the continuum limit.

4. MONTE CARLO INTEGRATION AND THE METROPOLIS ALGORITHM

The General Problem

Suppose we are working with a lattice having $N = 10$. If the configuration space C of our model is 2-dimensional, then the integrations in expression (3.6) are with respect to 20,000 variables. In this section we shall describe how to perform such integrations.

We begin by developing a general procedure applicable to a wide variety of examples. The task before us is the evaluation of averages of the form

$$\langle A[\phi]\rangle = \frac{\int A[\phi]\,F[\phi]\,d\phi}{\int F[\phi]\,d\phi} \ . \tag{4.1}$$

In the specific case of lattice quantum field theory we have

$$F[\phi] = e^{-S[\phi]}\,\mu[\phi] \ , \tag{4.2}$$

where the subscripts E appearing in (3.6) have been dropped. More generally, $F[\phi]$ may be any function satisfying

$$F[\phi] > 0 \quad \text{for all } \phi \text{ except, possibly, at isolated zeros}\,, \tag{4.3}$$

$$0 < \int F[\phi]\,d\phi < \infty \ , \tag{4.4}$$

all integrations being understood to be over the space Φ of all points ϕ. Φ will always be a differentiable manifold. In field theory it is called the *space of field histories* and is given by

$$\Phi = C^{N^4} \ , \tag{4.5}$$

which is compact or noncompact according as C is compact or noncompact. In the noncompact case $e^{-S[\phi]}$ generally vanishes sufficiently rapidly at "infinity" that one can map Φ into a compact manifold by introducing appropriate new variables ϕ^i and a corresponding modification in the measure function $\mu[\phi]$. It will be convenient in what follows to assume that Φ is compact.

The symbol ϕ in the above equations may be understood as standing for a set of generic variables. In the case of the $O(1,2)/O(2) \times Z_2$ and $O(3)/O(2)$ models, it has been found convenient, on the computer, to work with the variables s and θ. Using the lattice action (3.35) together with equation (2.14) and the notation of equations (3.14) to (3.23), we have, for the $O(1,2)/O(2) \times Z_2$ model,

$$\begin{aligned} F[s,\theta] & \\ = \exp\Big(- \tfrac{1}{4}\beta \sum_{\boldsymbol{\alpha}} \sum_{\boldsymbol{\lambda}} \Big\{ & \cosh^{-1} \big[\cosh(s_{\boldsymbol{\alpha}} - s_{\boldsymbol{\alpha}+\boldsymbol{\lambda}}) \cos^2 \tfrac{1}{2}(\theta_{\boldsymbol{\alpha}} - \theta_{\boldsymbol{\alpha}+\boldsymbol{\lambda}}) \\ & + \cosh(s_{\boldsymbol{\alpha}} + s_{\boldsymbol{\alpha}+\boldsymbol{\lambda}}) \sin^2 \tfrac{1}{2}(\theta_{\boldsymbol{\alpha}} - \theta_{\boldsymbol{\alpha}+\boldsymbol{\lambda}})\big] \Big\}^2 \Big) \\ \times \prod_{\boldsymbol{\alpha}} \sinh s_{\boldsymbol{\alpha}} \ . & \end{aligned} \tag{4.6}$$

A similar expression holds for the $O(3)/O(2)$ model.

The most efficient method so far discovered for evaluating averages like (4.1) is to scatter points randomly in Φ with a probability distribution proportional to $F[\phi]$, and to sum the values that $A[\phi]$ takes at these points. Because of the aleatory nature of the scatter process, the method is known as *Monte Carlo integration.* We now describe how to set up the probability distribution.

The Metropolis Algorithm

We begin by choosing a point ϕ_0 in Φ. At this *zeroth* stage we know exactly where ϕ_0 is, so our probability distribution is a delta function

$$P_0[\phi] = \delta(\phi, \phi_0) \ , \tag{4.7}$$

satisfying

$$\int P_0[\phi]\, d\phi = 1 \ . \tag{4.8}$$

We next introduce a "convenient" *joint probability distribution function* $p[\phi, \phi']$ satisfying

$$p[\phi, \phi'] = p[\phi', \phi] > 0 \ , \tag{4.9}$$

$$\int p[\phi, \phi']\, d\phi = 1 \ , \tag{4.10}$$

for all ϕ, ϕ' in Φ. We then choose a random sequence of points in Φ, starting with ϕ_0, by the following prescription known as the *Metropolis algorithm.* Suppose we have reached the $(n-1)$st point in the sequence. Call it ϕ_{n-1}. To get the nth point ϕ_n we first choose a point ϕ by a random process with probability distribution $p[\phi, \phi_{n-1}]$, and then choose a random number ξ with uniform weight in the interval 0 to 1. If $F[\phi] \geq \xi F[\phi_{n-1}]$, we set $\phi_n = \phi$. If $F[\phi] < \xi F[\phi_{n-1}]$, we set $\phi_n = \phi_{n-1}$.

We know that ϕ_{n-1} lies *somewhere* in Φ with a probability distribution $P_{n-1}[\phi]$ that depends on the functions $p[\phi, \phi']$, $F[\phi]$, and $P_0[\phi]$. It is not difficult to see, from the Metropolis algorithm, that the probability distribution for the point ϕ_n is given by

$$P_n[\phi] = \int d\phi' \int_0^1 d\xi \Big\{ p[\phi,\phi']\, \theta\big(F[\phi] - \xi F[\phi']\big) + \delta(\phi,\phi') \int d\phi''\, p[\phi'',\phi']\, \theta\big(\xi F[\phi'] - F[\phi'']\big)\Big\} P_{n-1}[\phi'] . \tag{4.11}$$

Carrying out the ξ integration, we can reexpress this in the form

$$P_n[\phi] = \int M[\phi,\phi']\, P_{n-1}[\phi']\, d\phi' \tag{4.12}$$

where

$$M[\phi,\phi'] = p[\phi,\phi'] \Big\{ \frac{F[\phi]}{F[\phi']} \theta\big(F[\phi'] - F[\phi]\big) + \theta\big(F[\phi] - F[\phi']\big)\Big\} + \delta(\phi,\phi') \int p[\phi'',\phi'] \Big(1 - \frac{F[\phi'']}{F[\phi']}\Big) \theta\big(F[\phi'] - F[\phi'']\big)\, d\phi'' . \tag{4.13}$$

The step functions θ in expression (4.13) are to be understood as vanishing whenever their arguments are less than or *equal* to zero, and this prevents ambiguities from arising when the denominators $F[\phi']$ vanish.

Using equation (4.10) one may easily show that

$$\int M[\phi,\phi']\, d\phi = 1 \qquad \text{for all } \phi' \text{ in } \Phi \tag{4.14}$$

and hence that

$$\int P_n[\phi]\, d\phi = \int P_{n-1}[\phi]\, d\phi = \cdots = \int P_0[\phi]\, d\phi = 1 . \tag{4.15}$$

Using the symmetry condition (4.9) as well as equation (4.10), one also verifies easily the relation

$$\int M[\phi,\phi']\, F[\phi']\, d\phi' = F[\phi] . \tag{4.16}$$

$M[\phi,\phi']$ may be regarded as an element of a continuous matrix M on Φ. Equation (4.16) says that $F[\phi]$ is a right eigenfunction of this matrix, corresponding to the eigenvalue unity. We shall show presently that unity is a nondegenerate eigenvalue of M and that all other eigenvalues λ satisfy $|\lambda| < 1$. From the iterated corollary of equation (4.12), namely,

$$P_n[\phi] = \int M^n[\phi,\phi']\, P_0[\phi']\, d\phi' , \tag{4.17}$$

it then follows that

$$\lim_{n\to\infty} P_n[\phi] = \frac{F[\phi]}{\int F[\phi']\,d\phi'} \ . \tag{4.18}$$

This means that if one eliminates a sufficient number of initial points in the *Metropolis sequence* $\phi_0, \phi_1, \phi_2, \ldots$, the remaining points will be scattered over Φ with a probability distribution that is arbitrarily close to what is needed for computing the average (4.1). The discarded points will in practice be determined by inspection of computer subaverages. They are called *transient* field configurations, and their number may be minimized by astute choice of the function $p[\phi, \phi']$.

Proof of the Metropolis Algorithm

The matrix M, although real, is not symmetric, and hence some of its eigenvalues may be complex. The complex eigenvalues come in conjugate imaginary pairs. Since the manifold Φ is assumed to be compact, the eigenvalues are necessarily discrete, although infinite in number (counting all multiplicities). Let $G[\phi]$ be any right eigenfunction of M that is not a multiple of $F[\phi]$, and let λ be the corresponding eigenvalue. If λ is complex, then $G[\phi]$ is necessarily complex, and its complex phase cannot be constant over Φ. If $\lambda \neq 0$, then $G[\phi]$ must vanish as fast as $F[\phi]$ wherever $F[\phi]$ vanishes. This follows from the equation

$$\lambda G[\phi] = \int M[\phi, \phi']\, G[\phi']\, d\phi' \tag{4.19}$$

and the easily verified fact that $M[\phi, \phi']$ vanishes as fast as $F[\phi]$ whenever ϕ approaches a zero of $F[\phi]$.

Suppose $\lambda \neq 0$. Since Φ is compact, there exists at least one point ϕ_* where $|G[\phi]|/F[\phi]$ (defined, at any zeros of $F[\phi]$, by taking limits) assumes its maximum value in Φ. We have

$$\infty > \frac{|G[\phi_*]|}{F[\phi_*]} \geq \frac{|G[\phi]|}{F[\phi]} \quad \text{for all } \phi \text{ in } \Phi \tag{4.20}$$

and, if λ is real,

$$\infty > \frac{|G[\phi_*]|}{F[\phi_*]} > \frac{|G[\phi]|}{F[\phi]} \quad \text{for all } \phi \text{ in some open subset of } \Phi \ . \tag{4.21}$$

If ϕ_* is a zero of $F[\phi]$, then it follows from the easily verified limit

$$\lim_{\phi\to\phi_*} \frac{M[\phi, \phi']}{F[\phi]} = \frac{p[\phi_*, \phi']}{F[\phi']} \ , \tag{4.22}$$

that

$$
\begin{aligned}
|\lambda| \, \frac{|G[\phi_*]|}{F[\phi_*]} &= \lim_{\phi \to \phi_*} \left| \int \frac{M[\phi, \phi'] \, G[\phi']}{F[\phi]} \, d\phi' \right| \\
&\leq \int p[\phi_*, \phi'] \, \frac{|G[\phi']|}{F[\phi']} \, d\phi' \leq \frac{|G[\phi_*]|}{F[\phi_*]} \ ,
\end{aligned}
\tag{4.23}
$$

the first $\leq$ sign being replaceable by $<$ when λ is complex and the second by $<$ when λ is real, by virtue of (4.21) and the fact that $p[\phi, \phi']$ is strictly positive (see (4.9)). Evidently, $|\lambda| < 1$. If ϕ_* is not a zero of $F[\phi]$, then neither is it a zero of $G[\phi]$, and one may write

$$
\begin{aligned}
&|\lambda| \ |G[\phi_*]| \\
&\quad = \left| \int M[\phi_*, \phi] \, G[\phi] \, d\phi \right| \\
&\quad \leq \int M[\phi_*, \phi] \, |G[\phi]| \, d\phi \\
&\quad = \int p[\phi_*, \phi] \, |G[\phi]| \left\{ \frac{F[\phi_*]}{F[\phi]} \, \theta(F[\phi] - F[\phi_*]) + \theta(F[\phi_*] - F[\phi]) \right\} d\phi \\
&\qquad + |G[\phi_*]| \int p[\phi, \phi_*] \left(1 - \frac{F[\phi]}{F[\phi_*]} \right) \theta(F[\phi_*] - F[\phi]) \, d\phi \\
&\quad \leq \int p[\phi_*, \phi] \, \{ |G[\phi_*]| \, \theta(F[\phi] - F[\phi_*]) + |G[\phi]| \, \theta(F[\phi_*] - F[\phi]) \\
&\qquad + |G[\phi_*]| \, \theta(F[\phi_*] - F[\phi]) - |G[\phi]| \, \theta(F[\phi_*] - F[\phi]) \} \, d\phi \\
&\quad = |G[\phi_*]| \int p[\phi_*, \phi] \, d\phi = |G[\phi_*]| \ ,
\end{aligned}
\tag{4.24}
$$

the first $\leq$ sign again being replaceable by $<$ when λ is complex and the second by $<$ when λ is real. Again, we have $|\lambda| < 1$.

We note in passing that this result, in combination with the corollary

$$
\lambda \int G[\phi] \, d\phi = \int G[\phi] \, d\phi \tag{4.25}
$$

of equations (4.14) and (4.19), implies

$$
\int G[\phi] \, d\phi = 0 \ . \tag{4.26}
$$

Since M is not a symmetric matrix, its right eigenfunctions do not generally constitute a complete basis on Φ. However, the theory of the canonical form of a matrix tells us that there does exist a complete set of independent basis functions $G_{\alpha\beta}[\phi]$, with indices ranging over values

$$
\left.
\begin{aligned}
\beta &= 0, \ldots, n_\alpha \qquad (n_\alpha \geq 0 \text{ for each } \alpha) \\
\alpha &= 0, 1, 2, \ldots \ ,
\end{aligned}
\right\}
\tag{4.27}
$$

which satisfy, for all α, the more general equations

$$\left.\begin{aligned} \int M[\phi,\phi']G_{\alpha 0}[\phi']\,d\phi' &= \lambda_\alpha G_{\alpha 0}[\phi] \ , \\ \int M[\phi,\phi']G_{\alpha\beta}[\phi']\,d\phi' &= G_{\alpha\beta-1}[\phi] + \lambda_\alpha G_{\alpha\beta}[\phi] \ , \quad \beta = 1,\ldots,n_\alpha \ . \end{aligned}\right\} \tag{4.28}$$

If $n_\alpha \geq 1$ for some α, or if $\lambda_\alpha = \lambda_\beta$ for two distinct indices α and β, then the eigenvalue λ_α is degenerate.

We may assume that the λ_α are ordered according to

$$1 = \lambda_0 > |\lambda_1| \geq |\lambda_2| \geq |\lambda_3| \geq \cdots \tag{4.29}$$

and that

$$G_{00}[\phi] = F[\phi] \ . \tag{4.30}$$

We have already seen that λ_0 cannot be equal to any of the other λ_α. Hence, the only way in which the eigenvalue unity could be degenerate would be if $n_0 \geq 1$. But in this case the second of equations (4.28), together with equation (4.14), implies

$$\int G_{01}[\phi]\,d\phi = \int F[\phi]\,d\phi + \int G_{01}[\phi]\,d\phi \ , \tag{4.31}$$

which contradicts (4.4). It follows that the eigenvalue unity is isolated and nondegenerate. Equations (4.14) and (4.28) consequently imply the following generalization of (4.26):

$$\int G_{\alpha\beta}[\phi]\,d\phi = 0 \qquad \text{for all } \alpha \neq 0 \ . \tag{4.32}$$

Since the functions $G_{\alpha\beta}[\phi]$ form a complete basis, the initial δ-function probability distribution function $P_0[\phi]$ can be expanded in terms of them:

$$P_0[\phi] = c_{00}F[\phi] + \sum_{\alpha=1}^{\infty}\sum_{\beta=0}^{n_\alpha} c_{\alpha\beta}\,G_{\alpha\beta}[\phi] \ , \tag{4.33}$$

where the $c_{\alpha\beta}$ are certain coefficients. Equations (4.8) and (4.32) together imply

$$c_{00} = 1\Big/\int F[\phi]\,d\phi \ . \tag{4.34}$$

Moreover, equations (4.28) imply

$$\int M^2[\phi,\phi']\,G_{\alpha\beta}[\phi']\,d\phi' = G_{\alpha\beta-2}[\phi] + 2\lambda_\alpha\,G_{\alpha\beta-1}[\phi] + {\lambda_\alpha}^2\,G_{\alpha\beta}[\phi] \ , \tag{4.35}$$

$$\int M^n[\phi,\phi']\,G_{\alpha\beta}[\phi']\,d\phi' = \sum_{r=0}^{n} \frac{n!}{r!(n-r)!}\,{\lambda_\alpha}^{n-r}\,G_{\alpha\beta-r}[\phi] \ , \tag{4.36}$$

where it is understood that $G_{\alpha\beta-r}[\phi] = 0$ if $r > \beta$. Since $|\lambda_\alpha| < 1$ for all $\alpha \neq 0$, it follows that

$$\lim_{n\to\infty} \int M^n[\phi,\phi']\, G_{\alpha\beta}[\phi']\, d\phi' = 0 \qquad \text{when } \alpha \neq 0 \tag{4.37}$$

and hence

$$\lim_{n\to\infty} \int M^n[\phi,\phi']\, P_0[\phi']\, d\phi' = c_{00}\, F[\phi] \ , \tag{4.38}$$

which is just equation (4.18).

Modified Metropolis Algorithm

The above proof that the Metropolis algorithm produces the correct probability distribution in the limit $n \to \infty$ makes use of the assumed strict positivity (4.9) of the joint probability distribution function $p[\phi, \phi']$. Equation (4.38) in fact holds under a weaker condition. When Φ is compact it holds if merely

$$\left.\begin{array}{r} p[\phi,\phi'] \geq 0 \\ p^n[\phi,\phi'] > 0 \end{array}\right\} \quad \text{for all } \phi, \phi' \text{ in } \Phi \text{ for some positive integer } n, \tag{4.39}$$

where p^n denotes the nth power of the continuous matrix p having elements $p[\phi, \phi']$. That is, only some finite power of p needs to be strictly positive. When Φ is noncompact, the appropriate condition is

$$\left.\begin{array}{r} p[\phi,\phi'] \geq 0 \\ \lim\limits_{n\to\infty} n^\alpha\, p^n[\phi,\phi'] > 0 \end{array}\right\} \quad \begin{array}{l}\text{for all } \phi, \phi' \text{ in } \Phi \text{ and all } \alpha \text{ greater than} \\ \text{or equal to some positive real number.}\end{array} \tag{4.40}$$

The proof of (4.38) under these weaker conditions is relatively straightforward but tedious in detail. To gain an intuitive understanding of what goes on during the iterative process that generates the probability distributions $P_n[\phi]$, it is helpful to think of it as a kind of random walk guided by the functions $\varphi[\phi, \phi']$ and $F[\phi]$.

For most computer work, even these weaker conditions are too strong, or at least inconvenient. Instead of working with only a single function $p[\phi, \phi']$, what one does in practice is use a sequence $\{p_{\boldsymbol{\alpha}}[\phi, \phi']\}$ of joint probability distribution functions, one for each lattice site $\boldsymbol{\alpha}$. Let $M_{\boldsymbol{\alpha}}$ be the matrix defined by eq. (4.13) with $p[\phi, \phi']$ replaced by $p_{\boldsymbol{\alpha}}[\phi, \phi']$.

Redefine the matrix M by

$$M \equiv \prod_{\boldsymbol{\alpha}} M_{\boldsymbol{\alpha}} \ , \tag{4.41}$$

where the factors in the product are taken in any convenient order. This new matrix will satisfy eq. (4.38) provided the compound matrix

$$p \equiv \prod_{\boldsymbol{\alpha}} p_{\boldsymbol{\alpha}} \qquad \text{(factors in same order)} \tag{4.42}$$

satisfies condition (4.39) or (4.40). In the case of the $O(1,2)/O(2) \times Z_2$ model, a possible choice for the $p_{\boldsymbol{\alpha}}$ is

$$p_{\boldsymbol{\alpha}}[s,\theta;s',\theta'] = \left[\theta(\zeta - s_{\boldsymbol{\alpha}} - s'_{\boldsymbol{\alpha}}) + \theta(\zeta - |s_{\boldsymbol{\alpha}} - s'_{\boldsymbol{\alpha}}|)\right] \prod_{\boldsymbol{\beta} \neq \boldsymbol{\alpha}} \delta(s_{\boldsymbol{\beta}} - s'_{\boldsymbol{\beta}})\delta(\theta_{\boldsymbol{\beta}} - \theta'_{\boldsymbol{\beta}}) \,. \quad (4.43)$$

This corresponds, in the application of the Metropolis algorithm, to choosing the tentative new field s, θ in the following way: Leave the field values unchanged (i.e., equal to the values s_{n-1}, θ_{n-1}) at all lattice sites other than $\boldsymbol{\alpha}$. At the site $\boldsymbol{\alpha}$, choose $\theta_{\boldsymbol{\alpha}}$ randomly with uniform weight in the interval $0 \leq \theta_{\boldsymbol{\alpha}} < 2\pi$ and set $s_{\boldsymbol{\alpha}} = |s_{n-1\boldsymbol{\alpha}} + \Delta s|$ where Δs is chosen randomly with unit weight in the interval $-\frac{1}{2}\zeta < \Delta s < \frac{1}{2}\zeta$. The width ζ of the latter interval is adjusted by the computer (for given β and N) in such a way as to yield approximately a 50% chance that $(s_n, \theta_n) = (s, \theta)$ and a 50% chance that $(s_n, \theta_n) = (s_{n-1}, \theta_{n-1})$ when the random number ζ is chosen (see eq. (4.11)).

Expression (4.43) is just one of several choices being tested by the Texas group, which have the following feature in common. Each generates a Metropolis sequence in which each field differs from its predecessor at no more than one lattice site. This feature, which is almost standard in lattice field theory calculations, allows for very efficient computer programming. By introducing a so-called "checkerboard sequence" on a supercomputer, one can "vectorize" the calculation so that many different arithmetical operations are carried out simultaneously. The reader is referred to the technical computer literature for details.

To implement the consequences of eq. (4.38), with M given by (4.41), it is necessary that the computer *sweep* repeatedly over the lattice, using each of the functions $p_{\boldsymbol{\alpha}}[\phi, \phi']$ exactly once in each sweep. At the end of the nth sweep, the computer arrives at the point ϕ_{nN^4} of the Metropolis sequence in Φ. The (approximate) value that the computer obtains for the average (4.1) is

$$\langle A[\phi] \rangle \approx (n_1 - n_0)^{-1} \sum_{n=n_0+1}^{n_1} A[\phi_{nN^4}] \,, \qquad (4.44)$$

where n_1 is the total number of sweeps in the computer run (typically 100,000 or more) and n_0 is the number of initial sweeps deleted as transients (typically several thousand). The likely error of the result (4.44) can be estimated by dividing the sweeps into groups of a few thousand each, calculating a subaverage for each group, and computing the standard deviation of the subaverages.

Critical Slowing Down—Autocorrelation Intervals

The number n_0 of sweeps that need to be deleted as transients depends primarily on β. Actual computer runs show that a kind of critical behavior often sets in at certain values of β. When the configuration space is compact, the critical behavior is found to occur at a positive value of β, denoted by β_c, and a phase transition occurs as β passes through β_c. When the configuration space is noncompact, the critical behavior is found to occur at $\beta = 0$.

When β is just above its critical value, the quantized field is found to have long-range order, as revealed, for example, by the behavior of the 2-point function (see §5). The modified Metropolis algorithm, which generates a series of fields each differing from its predecessor at not more than a single lattice site, is poorly suited to the computation of averages $\langle A[\phi] \rangle$ when β is near β_c. The computer is allowed to probe the space Φ only in a very "timid," essentially *local* way, lattice site by lattice site, whereas the dominant fields in the integral (4.1) have correlations reaching across many lattice sites. The consequence is that the matrix M of eq. (4.41) acquires many eigenvalues near unity, and the limit in eq. (4.38) is reached only very slowly.

There are various ways of overcoming this problem, at least partially. One way is to shift back and forth between coarser and finer lattices, use of the coarser lattices allowing the long-range order to appear more easily and use of the finer lattices permitting precision to be maintained. Other methods are too technical to be described here.

No matter what method is used, this problem of *critical slowing down,* as it is known, is always present to some extent. It is important to know to what extent. One way to determine this is to compute *autocorrelation intervals.* These are defined as follows. Let $\langle A[\phi] \rangle$ be the Monte Carlo average (over all sweeps) of some functional $A[\phi]$ of the field ϕ. Define

$$I_{A[\phi]}(m) \equiv N_R^{-1}(n_1 - m)^{-1} \sum_{\text{runs}} \sum_{n=1}^{n_1-m} \{A[\phi_{nN^4} - \langle A[\phi] \rangle\}$$
$$\times \{A[\phi_{(n+m)N^4}] - \langle A[\phi] \rangle\} \;, \qquad (4.45)$$

where the sum on the right is over N_R computer runs of n_1 sweeps each. The function $I_{A[\phi]}(m)$ is an *autocorrelation function.* It typically assumes its largest value at $m = 0$ and decreases as m increases. If $N_R = 1$, it will often pass through zero and then oscillate about zero at moderate amplitudes and irregular periods. If N_R is sufficiently large, these oscillations, which are a kind of random walk effect, disappear, and $I_{A[\phi]}(m)$ is essentially a monotonically decreasing function.

The value of m at which $I_{A[\phi]}(m)$ drops to zero (to some prescribed degree of accuracy) is called the *autocorrelation interval associated with* $A[\phi]$ and is a measure of how severe the critical slowing down problem is. Let $\bar{m}$ be the maximum of the autocorrelation intervals associated with all the functionals whose averages one wishes to compute. The number of sweeps in the subaverages used to estimate errors, as well as the number n_0 of initial sweeps deleted as transients, should all be at least an order of magnitude greater than $\bar{m}$.

5. DATA ANALYSIS AND THE EFFECTIVE ACTION

"Magnetization"

The Euclidean action (3.7) for the general sigma model, like the Euclidean acton (3.8) for the massless free field, vanishes for constant fields and remains unchanged in value under the actions of the global structure group. When the group is noncompact, this gives rise to a divergent degree of integration freedom in the integrals (3.6) even when the model is simulated on a lattice. In the case of the massless free field, we eliminate this degree of freedom by adding a positive or negative constant to the field at the end of each sweep so that its lattice average remains zero. When the massless free field is viewed as an $O(1,1)/Z_2 \times Z_2$ model, this corresponds to applying an $O(1,1)$ operation at the end of each sweep.

A similar operation is performed on every sigma model that we study. An arbitrary point ϕ_* of the configuration space C is chosen, and Riemann normal coordinates σ^i based on this point are introduced. (Since C is a symmetric space, it does not matter what point is chosen.) At the end of each sweep, a global group operation is applied, which brings the lattice averages of the fields σ^i back to zero.

If this operation were not performed, the lattice averages would tend to drift and, when β is large, the fields at individual lattice sites would tend to cluster around these averages, drifting with them. On a very large lattice the clustering would tend to occur around different points in C in different parts of the lattice. These lattice regions would be analogous to magnetic domains, the "magnetization" becoming more pronounced as one increases the constant β, which here plays the role of an inverse lattice-computation, or *discretization*, temperature. The "magnetization" is an expression of spontaneous symmetry breaking. When β is above β_c (low discretization temperature), the model possesses a degenerate ground state at $N \to \infty$ (vanishing physical temperature). When working with a finite N, as we must on the computer, instead of allowing the lattice averages to drift, we essentially force the field to choose one of its ground states by repeatedly resetting the average of the σ^i to zero.

When C is compact, the "magnetization" disappears for $\beta < \beta_c$ (high discretization temperature) and the ground state becomes nondegenerate. When C is noncompact, some residual "magnetization" remains at all discretization temperatures. In the case of both the $O(1,1)/Z_2 \times Z_2$ and $O(1,2)/O(2) \times Z_2$ models, a good measure of the "magnetization" is given by the inverse of the average

$$\langle \phi^0 \rangle = \langle \cosh s \rangle = \langle \cosh s_{\alpha\beta\gamma\delta} \rangle = N^{-4} \Big\langle \sum_{\alpha,\beta,\gamma,\delta} \cosh s_{\alpha\beta\gamma\delta} \Big\rangle \ . \tag{5.1}$$

In §3 we have computed and plotted $\langle \phi^0 \rangle^{-1}$ for the $O(1,1)/Z_2 \times Z_2$ model (Fig. 1). Figure 2 displays the functions $\langle \phi^0 \rangle^{-1}$ for both the $O(1,1)/Z_2 \times Z_2$ and $O(1,2)/O(2) \times Z_2$ models. The two curves are seen to be quite similar in shape. The fact that the curve for the $O(1,2)/O(2) \times Z_2$ model lies below that for the $O(1,1)/Z_2 \times Z_2$ model is easily understood from the fact that volume of configuration space in which the field has room to fluctuate, as $\beta \to 0$, increases linearly with s in the $O(1,1)/Z_2 \times Z_2$ case and exponentially with s in the $O(1,2)/O(2) \times Z_2$ case, and hence ϕ^0 is typically larger in the latter case.

The 2-point Function

The $O(1,1)/Z_2 \times Z_2$ model is linear and trivial. The $O(1,2)/O(2) \times Z_2$ model is nonlinear and nonrenormalizable by perturbative means. Information on whether it might nevertheless be renormalizable nonperturbatively can be obtained by studying the 2-point function $\langle \tilde{\sigma}^i{}_{klmn} \tilde{\sigma}^j{}_{k'l'm'n'}{}^* \rangle$ in the continuum limit. The Texas group is presently attempting to fit the computer results for this function to the following Ansatz, suggested by perturbation theory:

$$\langle \tilde{\sigma}^i{}_{klmn} \tilde{\sigma}^j{}_{k'l'm'n'}{}^* \rangle = \frac{\delta_{ij}\, \delta_{kk'}\, \delta_{ll'}\, \delta_{mm'}\, \delta_{nn'}}{\beta_R \overline{K}^2(\mathbf{k}) + \alpha \overline{K}^4(\mathbf{k})\, \ell n\, [\overline{K}^2(\mathbf{k})/\lambda \beta_R]} \ , \tag{5.2}$$

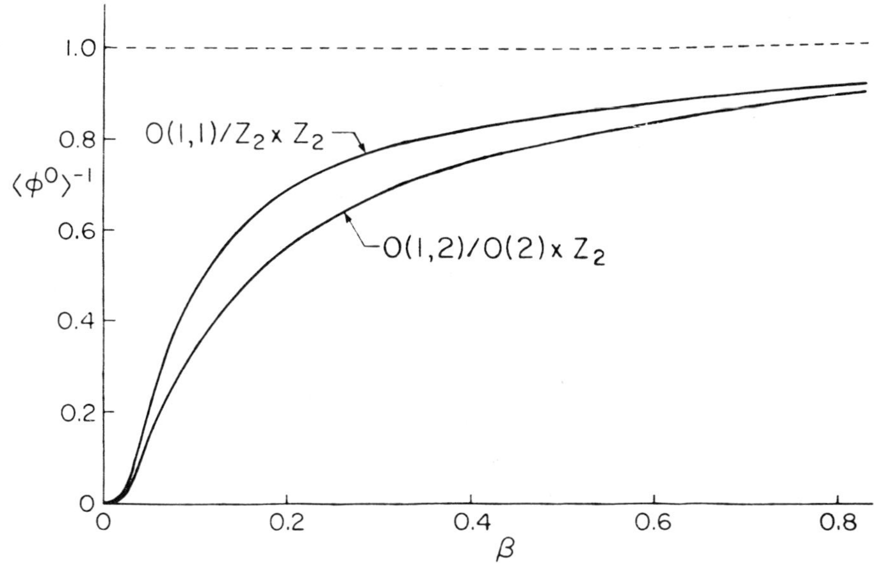

Figure 2. The function $\langle\phi^0\rangle^{-1}$ for the $O(1,1)/Z_2 \times Z_2$ and $O(1,2)/O(2) \times Z_2$ models The upper curve is the theoretical curve for $\langle\phi^0\rangle^{-1}$ for the $O(1,1)/Z_2 \times Z_2$ model. It is identical with the solid curve shown in Fig. 1. The lower curve is the curve for $\langle\phi^0\rangle^{-1}$ for the $O(1,2)/O(2) \times Z_2$ model, obtained from the computer when $N > 8$.

where $\overline{K}(\mathbf{k})$ is defined by eq. (3.18). Least-squares best fits to the computer-generated data are being used to determine the coefficients β_R, α, and λ as functions of β and N. In the continuum limit ($a \to 0, N \to \infty$), the Ansatz (5.2) corresponds to a momentum behavior for the 2-point function of the form

$$\frac{1}{\mu_R^2\, p^2 + \alpha\, p^4\, \ell n\, (p^2/\lambda\, \mu_R^2)} \tag{5.3}$$

where

$$\mu_R^2 = \beta_R/a^2 \; . \tag{5.4}$$

Preliminary computer runs have yielded three significant results.

(1) The curves for the 2-point function depend smoothly on $\overline{K}^2(\mathbf{k})$, even when the orientation of the vector $\mathbf{k}$ ranges over a wide variety of lattice direction. This implies that the basic $O(4)$ symmetry of the Euclidean sigma model (and hence the Lorentz invariance of the Minkowski model) is not destroyed by the lattice and is regained in the continuum limit.

(2) The curves for β_R and α as functions of β tend to asymptotes rapidly as N increases. This is an example of the stability property of lattice simulations that was mentioned in §3. It allows us, knowing the behavior of β_R and α for modest values of N, to infer their behavior for an infinite lattice and to investigate the continuum limit $a \to 0$ without worrying that we are simultaneously shrinking the size of our 4-torus "universe."

(3) β_R appears to tend to zero as $\beta \to 0$. This is extremely important, for it means that the $O(1,2)/O(2) \times Z_2$ model may well be nonperturbatively renormalizable and make consistent sense as a continuum theory. (More accurately, one should say that the $O(1,2)/O(2) \times Z_2$ model may be definable, and make consistent sense, as a continuum limit of a lattice theory.) The reason for this is that if β_R tends to zero as $\beta \to 0$, then the quantity μ_R of eq. (5.4), which we shall call *the renormalized Planck mass,* may be held fixed at any finite ("experimental") value we like. However, this free parameter must be the only free parameter of the quantum theory. (Since there is only one free parameter, namely, μ, in the classical action, there can be no more than one in the quantum theory.) If the theory is to be consistent, all n-point functions must be completely determined as soon as μ_R is fixed, and all form factors must satisfy simple universe scaling laws.

The Effective Action

Renormalization and consistency questions are best dealt with in terms of the so-called *effective action.* The theory of the effective action has been set forth by many authors, but usually in the Minkowskian context. For convenience, we give here a resumé of the Euclidean version of the theory.

The starting point in the classical Euclidean action $S[\phi]$ modified by the addition of a term $J_i\phi^i$ where J_i is known as an *external source.* Here the symbol x, which labels the points of spacetime, is to be understood as lumped with the generic index i, and the latter is to be understood as doing double duty as a discrete label for the field components and as a continuous label for the points of spacetime. The summation convention for repeated Latin indices therefore now includes an integration over spacetime, and a comma followed by a Latin index will denote *functional* differentiation with respect to a field component at a point of spacetime.

When the source term is added, eq. (3.6) gets modified to

$$\langle A[\phi] \rangle_J = \frac{\int A[\phi]\, e^{-S[\phi]-J_i\phi^i}\, \mu[\phi]\, d\phi}{\int e^{-S[\phi]-J_j\phi^j}\, \mu[\phi]\, d\phi} \quad . \tag{5.5}$$

If $A[\phi]$ is a polynomial in the ϕ^i, its average, with the source present, is expressible in terms of a generating functional $W[J]$ defined by

$$e^{-W[J]} \equiv Z \int e^{-S[\phi]-J_i\phi^i}\, \mu[\phi]\, d\phi \ , \tag{5.6}$$

the coefficient Z being an arbitrary constant that serves to set the zero point of $W[J]$. The relevant equations follow from the identity

$$\sum_{n=0}^{\infty} \frac{(-1)^n}{n!} \langle \phi^{i_1} \dots \phi^{i_n} \rangle_J \Delta J_{i_1} \dots \Delta J_{i_n}$$

$$= \langle e^{-\phi^i \Delta J_i} \rangle_J$$

$$= e^{-(W[J+\Delta J]-W[J])}$$

$$= \exp\Big\{ -\bar{\phi}^i \Delta J_i + \sum_{n=2}^{\infty} \frac{(-1)^n}{n!} \Gamma^{i_1 \dots i_n}[\bar{\phi}] \Delta J_{i_1} \dots \Delta J_{i_n} \Big\} \tag{5.7}$$

where

$$\bar{\phi}^i \equiv \frac{\delta}{\delta J_i} W[J] , \tag{5.8}$$

$$\Gamma^{i_1 \dots i_n}[\bar{\phi}] \equiv -(-1)^n \frac{\delta}{\delta J_{i_1}} \cdots \frac{\delta}{\delta J_{i_n}} W[J] . \tag{5.9}$$

Picking out terms of equal degree in the ΔJ_i on both sides of eq. (5.7), one finds

$$\langle \phi^i \rangle_J = \bar{\phi}^i , \tag{5.10}$$

$$\langle \phi^i \phi^j \rangle_J = \bar{\phi}^i \bar{\phi}^j + \Gamma^{ij} , \tag{5.11}$$

$$\langle \phi^i \phi^j \phi^k \rangle_J = \bar{\phi}^i \bar{\phi}^j \bar{\phi}^k + P_3 \, \bar{\phi}^i \Gamma^{jk} + \Gamma^{ijk} , \tag{5.12}$$

$$\langle \phi^i \phi^j \phi^k \phi^l \rangle_J = \bar{\phi}^i \bar{\phi}^j \bar{\phi}^k \bar{\phi}^l + P_6 \, \bar{\phi}^i \bar{\phi}^j \Gamma^{kl} + P_4 \, \bar{\phi}^i \Gamma^{jkl} + P_3 \, \Gamma^{ij} \Gamma^{kl} + \Gamma^{ijkl} , \tag{5.13}$$

$$\dots$$

where "P" indicates that a summation is to be performed over all distinct permutations of the indices appearing in the term to which it is affixed, and the subscript on P indicates the number of permutations required in each case.

It is to be noted that the $\Gamma^{i_1 \dots i_n}$ are regarded as functionals of the $\bar{\phi}^i$. This is possible only if eqs. (5.8) can be solved for the J_i as functionals of the $\bar{\phi}^i$, and this in turn depends on Γ^{ij} $(= -\delta\bar{\phi}^i/\delta J_j)$ being a nonsingular continuous matrix. Assuming Γ^{ij} to be nonsingular, one defines the effective action by

$$\Gamma[\bar{\phi}] \equiv W[J] - J_i \bar{\phi}^i . \tag{5.14}$$

With the aid of eq. (5.8), it is easily verified that

$$\Gamma_{,i}[\bar{\phi}] = \frac{\delta J_i}{\delta \bar{\phi}^i} \frac{\delta}{\delta J_j} W[J] - \frac{\delta J_j}{\delta} \bar{\phi}^i \bar{\phi}^j - J_i = -J_i , \tag{5.15}$$

$$\Gamma_{,ij}[\bar{\phi}] = -\delta J_i / \delta \bar{\phi}^j , \tag{5.16}$$

which, since $\Gamma^{ij} = -\delta\bar{\phi}^i/\delta J_j$, implies that $\Gamma_{,ij}$ is the matrix inverse to Γ^{ij}:

$$\Gamma_{,ik}\,\Gamma^{kj} = \delta_i^{\ j} \; . \tag{5.17}$$

Here $\delta_i^{\ j}$ denotes a combined δ-function and Kronecker delta.

Functional differentiation of (5.17) yields

$$\Gamma^{ij}{}_{,k} = -\Gamma^{il}\,\Gamma^{jm}\,\Gamma_{,lmk} \; . \tag{5.18}$$

Combining this result with

$$\Gamma^{i_1 \ldots i_n} = -\frac{\delta}{\delta J_{i_n}} \Gamma^{i_1 \ldots i_{n-1}} = \Gamma^{i_n j} \frac{\delta}{\delta \phi^j} \, \Gamma^{i_1 \ldots i_{n-1}} \; , \tag{5.19}$$

one finds

$$\Gamma^{ijk} = -\Gamma^{il}\,\Gamma^{jm}\,\Gamma^{kn}\,\Gamma_{,lmn} \; , \tag{5.20}$$

$$\Gamma^{ijkl} = \Gamma^{im}\,\Gamma^{jn}\,\Gamma^{kr}\,\Gamma^{ls}(P_3\,\Gamma_{,mnp}\,\Gamma^{pq}\,\Gamma_{,prs} - \Gamma_{,mnrs}) \; , \tag{5.21}$$

$$\ldots$$

If $S[\phi]$ is an even function of the ϕ^i, then when $J_i = 0$, the averages of products of odd numbers of ϕ^i vanish. Together with eqs. (5.10) to (5.13) and eqs. (5.20) and (5.21), this implies that $\Gamma[\bar{\phi}]$ is an even function of the $\bar{\phi}^i$. In this case we have

$$\langle \phi^i \phi^j \rangle = \Gamma^{ij}[0] \; , \tag{5.22}$$

$$\langle \phi^i \phi^j \phi^k \phi^l \rangle = P_3\,\Gamma^{ij}[0]\,\Gamma^{kl}[0] + \Gamma^{ijkl}[0] \; , \tag{5.23}$$

whence

$$\begin{aligned} \Gamma_{,ijkl}[0] &= -\Gamma_{,im}[0]\,\Gamma_{,jn}[0]\,\Gamma_{,kr}[0]\,\Gamma_{,ls}[0]\,\langle \phi^m \phi^n \phi^r \phi^s \rangle \\ &\quad + P_3\,\Gamma_{,ij}[0]\,\Gamma_{,kl}[0] \; . \end{aligned} \tag{5.24}$$

In Minkowskian spacetime it is well known that S-matrix elements are obtainable by performing the Lehmann-Symanzik-Zimmermann operation on averages of products of the ϕ^i. As we have just seen, the latter are completely expressible in terms of the functional derivatives of $\Gamma[\bar{\phi}]$. The effective action, in fact, yields a complete description of the quantum system. In particular, it contains within itself all experimentally determinable parameters—masses, coupling constants, etc.—which, themselves, are often found by scattering experiments. In the case of sigma models there is only one experimental parameter, the renormalized Planck mass. It, together with the geometry of the configuration space, completely determines the quantum theory. We must now examine how it appears in the effective action, but for this we need first to develop the formalism a bit further.

Equations (5.5), (5.6), (5.14), and (5.15) together imply

$$e^{-\Gamma[\bar\phi]} = Z\int e^{-S[\phi]+\Gamma_{,i}[\bar\phi](\phi^i-\bar\phi^i)}\,\mu[\phi]\,d\phi\ , \tag{5.25}$$

$$\langle A[\phi]\rangle_J = Ze^{\Gamma[\bar\phi]}\int A[\phi]\,e^{-S[\phi]+\Gamma_{,i}[\bar\phi](\phi^i-\bar\phi^i)}\,\mu[\phi]\,d\phi\ . \tag{5.26}$$

The theory of the effective action may be based on this pair of equations alone, for there is nothing in the theory that cannot be derived from them. For example, functionally differentiating eq. (5.25), one finds

$$\begin{aligned}\Gamma_{,i}[\bar\phi] &= -e^{\Gamma[\bar\phi]}\frac{\delta}{\delta\bar\phi^i}\,e^{-\Gamma[\bar\phi^i]}\\ &= -Ze^{\Gamma[\bar\phi]}\int\{\Gamma_{,ij}[\bar\phi](\phi^j-\bar\phi^j)-\Gamma_{,i}[\bar\phi]\}\,e^{-S[\phi]+\Gamma_{,k}[\bar\phi](\phi^k-\bar\phi^k)}\,\mu[\phi]\,d\phi\\ &= -\Gamma_{,ij}\big(\langle\phi^j\rangle_J-\bar\phi^j\big)+\Gamma_{,i}[\bar\phi]\ ,\end{aligned} \tag{5.27}$$

which, in view of the assumed nonsingularity of $\Gamma_{,ij}[\bar\phi]$, leads to eq. (5.10). Repeated functional differentiation of eq. (5.25) yields eqs. (5.11), (5.12), etc.

An advantage of basing the theory on eqs. (5.25) and (5.26) is that it permits $\Gamma[\bar\phi]$ to be defined without the introduction of external sources and the functional $W[J]$ as auxiliary quantities. Equation (5.25) leads immediately to an iterative scheme for the computation of $\Gamma[\bar\phi]$ and to a diagrammatic representation of $\Gamma[\bar\phi]$ that is useful in perturbation theory, in which only one-particle-irreducible graphs appear.

Another advantage of using eqs. (5.25) and (5.26) is that they display, in a particularly visible form, their own defects. It has been pointed out by Vilkovisky (1984) that these equations do not make good geometrical sense. In the classical theory one may regard the action $S[\phi]$ as a scalar field on the space Φ of field histories and its functional derivatives $S_{,i}[\phi]$ as the components of the gradient or differential of S at the point ϕ of Φ, in the "coordinate system" defined by the particular choice of variables ϕ^i used to describe the field. If S is a scalar field on Φ, then its value $S[\phi]$ at the point ϕ must be independent of the choice of coordinates in Φ; not so, however, the value of the effective action $\Gamma[\bar\phi]$ at the point $\bar\phi$. This is because the difference $\phi^i-\bar\phi^i$, which appears in the exponents of the integrands in both eqs. (5.25) and (5.26), is a difference of *coordinates* and not a component of a contravariant vector.

This difficulty can be surmounted by choosing the ϕ^i to be a set of *preferred* coordinates in Φ and referring all other coordinates back to these. When the field is a section of a vector bundle, the preferred coordinates can be the natural vector-space coordinates in each fibre. In the case of nonlinear sigma models, Vilkovisky proposes to use as preferred coordinates the Riemann normal coordinates based on a fixed point ϕ_*.

In §2 we introduced Riemann normal coordinates in the configuration space C. Because the bundle of which the points of Φ are sections is a trivial bundle, the concept of Riemann normal coordinates is easily extended to Φ itself. First one defines the distance ds between two infinitesimally close fields, ϕ and $\phi + d\phi$, by

$$ds^2 = \int G_{ij}(\phi(x))\, d\phi^i(x)\, d\phi^j(x)\, d^4x \;, \tag{5.28}$$

which defines a metric tensor on Φ. Let $\sigma[\phi, \phi']$ be one-half the square of the geodetic distance, determined by this metric, between fields ϕ and ϕ' in Φ. It is easy to see that this is given by

$$\sigma[\phi', \phi] = \tfrac{1}{2} \int \Delta^2(\phi'(x), \phi(x))\, d^4x \;, \tag{5.29}$$

where Δ is the geodetic distance function introduced in §2. If one chooses for the fixed point ϕ_* a constant field in Φ,

$$\phi^i_*(x) = \phi^i_* \qquad \text{for all } x \text{ in } \Phi,$$

then the Riemann normal coordinates σ^i of §2 are given by

$$\sigma^i(x) = -\Big\{G^{ij}(\phi'(x)) \frac{\delta}{\delta\phi'^j(x)} \sigma[\phi', \phi]\Big\}_{\phi'=\phi_*} = G^{ij}(\phi_*)\sigma_j(x) \;, \tag{5.30}$$

$$\sigma^i(x) = -\Delta(\phi_*, \phi(x)) \frac{\partial}{\partial\phi^i_*} \Delta(\phi_*, \phi(x)) \;. \tag{5.31}$$

Return now to the notation that suppresses the spacetime point labels $x, x', \ldots$ and absorbs them into the Latin indices. Introduce the continuous matrix

$$D_{ij}[\phi_*, \phi] \equiv -\Big\{\frac{\delta}{\delta\phi'^i} \frac{\delta}{\delta\phi^j} \sigma[\phi', \phi]\Big\}_{\phi'=\phi_*} = \frac{\delta\sigma^i}{\delta\phi^j} \tag{5.32}$$

and its inverse

$$D^{-1\,ij}[\phi, \phi_*] = \frac{\delta\phi^i}{\delta\sigma_j} \;. \tag{5.33}$$

Vilkovisky's effective action is obtained by replacing ϕ^i and $\bar{\phi}^i$ everywhere in eqs. (5.25) and (5.26) by σ^i and $\bar{\sigma}^i$. In terms of the arbitrary field variables ϕ^i and $\bar{\phi}^i$, this is equivalent to writing

$$\Gamma[\bar{\sigma}] = \hat{\Gamma}[\phi_*, \bar{\phi}] \tag{5.34}$$

where

$$e^{-\hat{\Gamma}[\phi_*, \bar{\phi}]} = Z \int \exp\Big\{ - S[\phi] - \hat{\Gamma}[\phi_*, \bar{\phi}] \frac{\overleftarrow{\delta}}{\delta\bar{\phi}^i} D^{-1\,ij}[\bar{\phi}, \phi_*] \times \frac{\overrightarrow{\delta}}{\delta\phi^j_*} \big(\sigma[\phi_*, \phi] - \sigma[\phi_*, \bar{\phi}]\big)\Big\} \mu[\phi]\, d\phi \;, \tag{5.35}$$

where $\bar{\phi}$ is defined by

$$\frac{\delta}{\delta\phi^i_*} \sigma[\phi_*, \bar{\phi}] \equiv \Big\langle \frac{\delta}{\delta\phi^i_*} \sigma[\phi_*, \phi] \Big\rangle_J \;. \tag{5.36}$$

In Appendix A it is shown that $\hat{\Gamma}[\phi_*, \bar{\phi}]$, in addition to being manifestly a scalar function of $\bar{\phi}$ on Φ, is also invariant under global actions of the group G, more precisely under simultaneous group-generated displacements of ϕ_* and $\bar{\phi}$. Now the functional form of the classical action S, when expressed in terms of the Riemann normal fields σ^i, is independent of where the fixed point ϕ_* is chosen. This is because the configuration space C is a symmetric space, and all its points are equivalent. It follows that $\Gamma[\bar{\sigma}]$, too, is independent of ϕ_* and is invariant under group-generated displacements of $\bar{\sigma}$ alone. We now return to the problem of data analysis and make use of this invariance.

Classical and Nonclassical Parts of $\Gamma[\bar{\sigma}]$

Because of our ultimate concern with quantum gravity, the nonlinear sigma models are of interest to us only if they, like gravity, behave classically in an appropriate domain. The classical domain, for any field theory that has one, is the domain of low energies and long wavelengths. In this domain, the effective action $\Gamma[\bar{\sigma}]$ is replaceable by the classical action, i.e., by the very functional that *defines* the quantum theory (through its appearance in the exponents of the Feynman functional integrals) but with renormalized (i.e., experimentally determined) parameters. If a nonlinear sigma model is to have a classical domain, the effective action must take the form

$$\Gamma[\bar{\sigma}] = \tfrac{1}{2}\,\mu_R^2 \int G_{ij}(\bar{\sigma})\,\bar{\sigma}^i{}_{,\mu}\,\bar{\sigma}^j{}_{,\mu}\,d^4x + \Delta\Gamma[\bar{\sigma}] \ . \tag{5.37}$$

The first term on the right is the classical part of $\Gamma[\bar{\sigma}]$. It makes a local contribution to the so-called *effective field equation* (5.15) and is expected to dominate at low energies. The second term is the nonclassical part of $\Gamma[\bar{\sigma}]$. It generally makes a nonlocal contribution to the effective field equation and becomes important at intermediate and high energies.

In some field theories the classical part of the effective action contains terms of a form that may not have been included in the original classical action. This is not true for sigma models. From eqs. (5.17) and (5.22), one easily sees that the first term of expression (5.37), if it is to dominate at low energies (small p^2), must account completely for the term in p^2 in the denominator of expression (5.3). Moreover, it is the only term that can do so. Its form is unique, being entirely determined by the group invariance of the effective action.

The nonclassical part $\Delta\Gamma[\bar{\phi}]$ of the effective action must account for the term in α in the denominator of (5.3), a term that becomes negligible when p^2 is small. Preliminary data from the computer seem to indicate that α may vanish for the $O(1,2)/O(2) \times Z_2$ model, although further computer runs will be necessary to determine this for sure. If α *is* equal to zero, this does not mean that the quantized field is a free field and hence trivial; indeed, the classical term in (5.37) already prevents this. What it does mean is that $\Delta\Gamma_{,ij}[0] = 0$, i.e., that $\Delta\Gamma[\bar{\sigma}]$ is of at least fourth order in the $\bar{\sigma}^i$ and that one must study the 4-point function in order to get any information about the nonclassical part.

The 4-point Function

Since the classical part of (5.73) corresponds to a nonlinear theory, it will contribute to the 4-point function through the 4-point vertex function $\Gamma_{,ijkl}[0]$ (see eq. (5.24)). It will, in fact, determine the quadratic-in-momentum (low-energy) part of the Fourier transform of $\Gamma_{,ijkl}[0]$. The high-energy part of the Fourier transform should depend solely on $\Delta\Gamma$. As a check on the consistency of the lattice-defined theory in the continuum limit, one must obtain the Fourier transform of the right side of eq. (5.24) from the computer and extract its quadratic-in-momentum part. For consistency this part must agree with the quadratic-in-momentum part of the Fourier transform of the left side of eq. (5.24) as determined by the classical part of expression (5.37).

The classical part of $\Gamma_{,ijkl}[0]$ is readily obtained upon inserting the metric tensor (2.7b) into the classical part of expression (5.37). Defining

$$\Gamma^{\rm cl}_{,ij'k''l'''}[\bar{\sigma}] \equiv \frac{\delta^4}{\delta\bar{\sigma}^i(x)\,\delta\bar{\sigma}^j(x')\,\delta\bar{\sigma}^k(x'')\,\delta\bar{\sigma}^l(x''')}\,\tfrac{1}{2}\,\mu_R^2 \int G_{ij}(\bar{\sigma})\,\bar{\sigma}^i{}_{,\mu}\,\bar{\sigma}^j{}_{,\mu}\,d^4x\,, \quad (5.38)$$

one finds, after a straightforward computation,

$$\begin{aligned}\Gamma^{\rm cl}_{,ij'k''l'''}[0] = \tfrac{1}{3}\,\mu_R^2\,\delta_{ij}\,\delta_{kl}\Big\{ &-2\delta(x,x')\big[\delta(x,x'')\,\delta(x,x''')\big]_{,\mu\mu} \\ &+ \delta(x,x'')\big[\delta(x,x')\,\delta(x,x''')\big]_{,\mu\mu} \\ &+ \delta(x,x''')\big[\delta(x,x')\,\delta(x,x'')\big]_{,\mu\mu}\Big\} \\ &+ \text{cyclic permutation of } jx', kx'', lx'''. \end{aligned} \quad (5.39)$$

The 4-torus Fourier transform of this follows immediately:

$$\begin{aligned}\tilde{\Gamma}^{\rm cl}_{ij'k''l'''} &\equiv L^{-8}\int d^4x \int d^4x' \int d^4x'' \int d^4x'''\,\Gamma^{\rm cl}_{,ij'k''l'''}[0]\,e^{i(p\cdot x + p'\cdot x' + p''\cdot x'' + p'''\cdot x''')} \\ &= \tfrac{1}{3}\mu_R^2\,L^{-4}\,\delta_{0\,p+p'+p''+p'''}\big\{\delta_{ij}\delta_{kl}\big[2(p''+p''')^2 \\ &\qquad - (p'+p''')^2 - (p'+p'')^2\big] \\ &\qquad + \text{cyclic permutation of } jp', kp'', lp'''\big\}\;. \end{aligned} \quad (5.40)$$

Here the momenta p, p', p'', p''' are restricted to the discrete values $(2\pi/L)(k,l,m,n)$, where k, l, m, n are integers, positive, negative, or zero.

On the computer one must deal with the lattice versions of these expressions. It is easy to see that the theory of the effective action can be worked out for the lattice simulation in complete analogy with the continuum theory. The lattice effective action, like the continuum effective action, should have its classical part, which in the present case should be given by

$$\Gamma^{\rm cl}[\bar{\sigma}] = \tfrac{1}{4}\beta_R \sum_{\boldsymbol{\alpha}} \sum_{\boldsymbol{\lambda}} \Delta^2(\bar{\sigma}_{\boldsymbol{\alpha}}, \bar{\sigma}_{\boldsymbol{\alpha}+\boldsymbol{\lambda}}) \quad (5.41)$$

(see eq. (3.35)). Making use of expression (2.18) and defining

$$\Gamma^{\rm cl}_{,i'j'k''l'''}[\bar{\sigma}] = \frac{\partial^4\Gamma^{\rm cl}[\bar{\sigma}]}{\partial\bar{\sigma}^i_{\boldsymbol{\alpha}}\,\partial\bar{\sigma}^j_{\boldsymbol{\alpha}'}\,\partial\bar{\sigma}^k_{\boldsymbol{\alpha}''}\,\partial\bar{\sigma}^l_{\boldsymbol{\alpha}'''}}\,, \quad (5.42)$$

one finds, after a straightforward computation,

$$
\begin{aligned}
\Gamma^{\rm cl}_{,i'j'k''l'''}[0] &= \tfrac{1}{3}\beta_R \sum_{\boldsymbol{\lambda}} \big[\delta_{ij}\delta_{kl}(2\delta_{\boldsymbol{\alpha}\boldsymbol{\alpha}'}\,\delta_{\boldsymbol{\alpha}+\boldsymbol{\lambda},\boldsymbol{\alpha}''}\,\delta_{\boldsymbol{\alpha}+\boldsymbol{\lambda},\boldsymbol{\alpha}'''} \\
&\qquad - \delta_{\boldsymbol{\alpha}\boldsymbol{\alpha}''}\,\delta_{\boldsymbol{\alpha}+\boldsymbol{\lambda},\boldsymbol{\alpha}'}\,\delta_{\boldsymbol{\alpha}+\boldsymbol{\lambda},\boldsymbol{\alpha}'''} - \delta_{\boldsymbol{\alpha}\boldsymbol{\alpha}'''}\,\delta_{\boldsymbol{\alpha}+\boldsymbol{\lambda},\boldsymbol{\alpha}'}\,\delta_{\boldsymbol{\alpha}+\boldsymbol{\lambda},\boldsymbol{\alpha}''}) \\
&\qquad + \text{cyclic permutation of } j\boldsymbol{\alpha}', k\boldsymbol{\alpha}'', l\boldsymbol{\alpha}'''\big] \; . \qquad (5.43)
\end{aligned}
$$

With use of (3.20) the lattice Fourier transform of this is readily found to be

$$
\begin{aligned}
\tilde{\Gamma}^{\rm cl}_{ij'k''l'''} &\equiv N^{-8} \sum_{\boldsymbol{\alpha},\boldsymbol{\alpha}',\boldsymbol{\alpha}'',\boldsymbol{\alpha}'''} \Gamma^{\rm cl}_{,ij'k''l'''}[0]\, e^{(2\pi i/N)(\mathbf{k}\cdot\boldsymbol{\alpha}+\mathbf{k}'\cdot\boldsymbol{\alpha}'+\mathbf{k}''\cdot\boldsymbol{\alpha}''+\mathbf{k}'''\cdot\boldsymbol{\alpha}''')} \\
&= \tfrac{1}{3}\beta_R N^{-4}\delta_{0\,\mathbf{k}+\mathbf{k}'+\mathbf{k}''+\mathbf{k}'''}\big\{\delta_{ij}\delta_{kl}\big[2\overline{K}^2(\mathbf{k}''+\mathbf{k}''') \\
&\qquad - \overline{K}^2(\mathbf{k}'+\mathbf{k}''') - \overline{K}^2(\mathbf{k}'+\mathbf{k}'')\big] \\
&\qquad + \text{cyclic permutation of } j\mathbf{k}', k\mathbf{k}'', l\mathbf{k}'''\big\} , \qquad (5.44)
\end{aligned}
$$

which is an obvious lattice analog of (5.40). In using this expression, as well as expression (5.45) below, one must remember that the integral components of lattice vectors such as $\mathbf{k}''+\mathbf{k}'''$ or $\mathbf{k}+\mathbf{k}'+\mathbf{k}''+\mathbf{k}'''$ are defined *modulo* N.

Consistency

Consistency of the lattice definition of the theory will be strongly supported if one can verify the relation (see eq. (5.24))

$$
\begin{aligned}
\tilde{\Gamma}^{\rm cl}_{ij'k''l'''} = \big\{ &-\tilde{\Gamma}_2(\mathbf{k})\tilde{\Gamma}_2(\mathbf{k}')\tilde{\Gamma}_2(\mathbf{k}'')\tilde{\Gamma}_2(\mathbf{k}''')\,\langle\tilde{\sigma}^i_{\mathbf{k}}\,\tilde{\sigma}^j_{\mathbf{k}'}\,\tilde{\sigma}^k_{\mathbf{k}''}\,\tilde{\sigma}^l_{\mathbf{k}'''}\rangle \\
&+ \delta_{ij}\,\delta_{kl}\,\delta_{0\,\mathbf{k}+\mathbf{k}'}\,\delta_{0\,\mathbf{k}''+\mathbf{k}'''}\,\tilde{\Gamma}_2(\mathbf{k})\tilde{\Gamma}_2(\mathbf{k}'') \\
&+ \delta_{ik}\,\delta_{jl}\,\delta_{0\,\mathbf{k}+\mathbf{k}''}\,\delta_{0\,\mathbf{k}'+\mathbf{k}'''}\,\tilde{\Gamma}_2(\mathbf{k})\tilde{\Gamma}_2(\mathbf{k}') \\
&+ \delta_{il}\,\delta_{jk}\,\delta_{0\,\mathbf{k}+\mathbf{k}'''}\,\delta_{0\,\mathbf{k}+\mathbf{k}''}\,\tilde{\Gamma}_2(\mathbf{k})\tilde{\Gamma}(\mathbf{k}')\big\}_{\rm quad} \; , \qquad (5.45)
\end{aligned}
$$

where the subscript "quad" denotes the part quadratic in $\overline{K}(\mathbf{k}''+\mathbf{k}''')$, $\overline{K}(\mathbf{k}'+\mathbf{k}''')$, etc., $\tilde{\sigma}^i_{\mathbf{k}}$ is an abbreviation for $\tilde{\sigma}^i_{klmn}$, etc., and $\tilde{\Gamma}_2(\mathbf{k})$ denotes the denominator of expression (5.2):

$$
\tilde{\Gamma}_2(\mathbf{k}) \equiv \beta_R\overline{K}^2(\mathbf{k}) + \alpha\overline{K}^4(\mathbf{k})\,\ell n[\overline{K}^2(\mathbf{k})/\lambda\beta_R] \; . \qquad (5.46)
$$

In the comparison of the left and right sides of eq. (5.45), one must obtain $\langle\tilde{\sigma}^i_{\mathbf{k}}\,\tilde{\sigma}^j_{\mathbf{k}'}\,\tilde{\sigma}^k_{\mathbf{k}''}\,\tilde{\sigma}^l_{\mathbf{k}'''}\rangle$ directly from the computer and, for the factors $\tilde{\Gamma}_2(\mathbf{k})$, use the results of the least-squares best fits to the computer-generated 2-point function. For the left side one must use expression (5.44) with only the constant β_R taken from the computer.

It is not practicable on the computer to test eq. (5.45) for all possible combinations of indices and momenta. Instead, the Texas group proposes to check the following special cases:

$$\left\{ \left[\tilde{\Gamma}_2(\mathbf{k})\right]^4 \langle |\tilde{\sigma}^i_{\mathbf{k}}|^4 \rangle - 2\left[\tilde{\Gamma}_2(\mathbf{k})\right]^2 \right\}_{\text{quad}} = 0 \qquad (i = 1, 2) \ , \tag{5.47}$$

$$\left\{ \left[\tilde{\Gamma}_2(\mathbf{k})\right]^4 \langle |\tilde{\sigma}^1_{\mathbf{k}}|^2 \, |\tilde{\sigma}^2_{\mathbf{k}}|^2 \rangle - \left[\tilde{\Gamma}_2(\mathbf{k})\right]^2 \right\}_{\text{quad}} = \tfrac{1}{3} \beta_R N^{-4} \overline{K}^2(2\mathbf{k}) \ , \tag{5.48}$$

$$\left\{ \left[\tilde{\Gamma}_2(\mathbf{k})\right]^4 \langle (\tilde{\sigma}^1_{\mathbf{k}})^2 \, (\tilde{\sigma}^2_{-\mathbf{k}})^2 \rangle \right\}_{\text{quad}} = -\tfrac{2}{3} \beta_R N^{-4} \overline{K}^2(2\mathbf{k}) \ , \tag{5.49}$$

$$\left\{ \left[\tilde{\Gamma}_2(\mathbf{k})\right]^4 \langle |\tilde{\sigma}^i_{\mathbf{k}}|^2 \, |\tilde{\sigma}^i_{\mathbf{k}_\perp}|^2 \rangle - \left[\tilde{\Gamma}_2(\mathbf{k})\right]^2 \right\}_{\text{quad}} = 0 \qquad (i = 1, 2) \ , \tag{5.50}$$

$$\left\{ \left[\tilde{\Gamma}_2(\mathbf{k})\right]^4 \langle |\tilde{\sigma}^1_{\mathbf{k}}|^2 \, |\tilde{\sigma}^2_{\mathbf{k}_\perp}|^2 \rangle - \left[\tilde{\Gamma}_2(\mathbf{k})\right]^2 \right\}_{\text{quad}} = \tfrac{2}{3} \beta_R N^{-4} \overline{K}^2(\mathbf{k} + \mathbf{k}_\perp) \ , \tag{5.51}$$

$$\left\{ \left[\tilde{\Gamma}_2(\mathbf{k})\right]^4 \langle \tilde{\sigma}^1_{\mathbf{k}} \, \tilde{\sigma}^2_{-\mathbf{k}} \, \tilde{\sigma}^1_{\mathbf{k}_\perp} \, \tilde{\sigma}^2_{-\mathbf{k}_\perp} \rangle \right\}_{\text{quad}} = \tfrac{1}{3} \beta_R N^{-4} \overline{K}^2(\mathbf{k} + \mathbf{k}_\perp) \ . \tag{5.52}$$

In these equations $\mathbf{k} = (k, l, m, n)$ and $\mathbf{k}_\perp = (m, n, -k, -l)$ (so that $\mathbf{k}^2 = \mathbf{k}_\perp^2$ and $\mathbf{k} \cdot \mathbf{k}_\perp = 0$) where k, l, m, n are arbitrary integers not all simultaneously equal to $\frac{1}{2} N \,(\text{mod}\, N)$. Verification of these equations is presently under way.

APPENDIX

A. Group Invariance of the Vilkovisky Effective Action

Since the metric of configuration space is group invariant, the geodetic distance function $\Delta(\phi_*(x), \phi(x))$, and hence the functional $\sigma[\phi_*, \phi]$, is invariant under simultaneous group displacements of the points ϕ_* and ϕ. The formal statement of this fact is

$$\sigma[\phi_*, \phi] \frac{\overleftarrow{\delta}}{\delta \phi^i_*} \, Q^i{}_\alpha[\phi_*] + \sigma[\phi_*, \phi] \frac{\overleftarrow{\delta}}{\delta \phi^i} \, Q^i{}_\alpha[\phi] = 0 \ , \tag{A.1}$$

where ϕ_* and ϕ are arbitrary points of Φ, and the functional $Q^i{}_\alpha[\phi]$ is an obvious extension (to Φ) of the components $Q^i{}_\alpha(\phi)$ of the killing vector fields on C, introduced in eq. (1.7). That is, the $Q^i{}_\alpha[\phi]$ are components of the analogous Killing vector fields on Φ. Functional differentiation of eq. (A.1) with respect to ϕ^j_* yields

$$\begin{aligned} &\sigma[\phi_*, \phi] \frac{\overleftarrow{\delta}}{\delta \phi^j_*} \frac{\overleftarrow{\delta}}{\delta \phi^i_*} \, Q^i{}_\alpha[\phi_*] \\ &\quad = -\sigma[\phi_*, \phi] \frac{\overleftarrow{\delta}}{\delta \phi^i_*} \, Q^i{}_{\alpha,j}[\phi_*] - \sigma[\phi_*, \phi] \frac{\overleftarrow{\delta}}{\delta \phi^j_*} \frac{\overleftarrow{\delta}}{\delta \phi^i} \, Q^i{}_\alpha[\phi] \ , \end{aligned} \tag{A.2}$$

for all ϕ_*, ϕ in Φ.

Next note that the Killing condition (1.8) on the metric of C implies an analogous condition on the metric of Φ, which, when applied to the measure (3.3) yields

$$\frac{\delta}{\delta\phi^i}\left(\mu[\phi]\, Q^i{}_\alpha[\phi]\right) = 0 \ . \tag{A.3}$$

This has the consequence that the volume element $\mu[\phi]\, d\phi$ in the functional integral (5.35) is group invariant. We may use this to obtain an important identity. Let the dummy variable of integration in (5.35) be changed from ϕ to $\phi + Q^i{}_\alpha[\phi]\delta\xi^\alpha$, where the $\delta\xi^\alpha$ are arbitrary constant infinitesimals. Since both the volume element and the classical action are group invariant, the only change induced in the quantity following the integral sign in (5.35) is in the term in $\sigma[\phi_*, \phi]$. But since the variable is a dummy, this must leave the value of the integral itself unchanged. It is easy to see that this leads to

$$\hat{\Gamma}[\phi_*, \bar{\phi}]\frac{\overleftarrow{\delta}}{\delta\bar{\phi}^i} D^{-1\,ij}[\bar{\phi}, \phi_*]\left\langle \frac{\overrightarrow{\delta}}{\delta\phi_*^j}\sigma[\phi_*, \phi]\frac{\overleftarrow{\delta}}{\delta\phi^k}\, Q^k{}_\alpha[\phi]\right\rangle = 0 \ . \tag{A.4}$$

Finally note that, in view of eq. (5.36), if we functionally differentiate eq. (5.35) with respect to ϕ_*^i, the only contribution we get from the right hand side is from terms in which the functionals $\sigma[\phi_*, \phi]$ and $\sigma[\phi_*, \bar{\phi}]$ are differentiated twice with respect to the ϕ_*'s. Thus, making use of eqs. (5.32), (5.36), (A.2), and (A.4), we obtain

$$\begin{aligned}
\hat{\Gamma}[\phi_*, \bar{\phi}]\frac{\overleftarrow{\delta}}{\delta\phi_*^k}\, Q^k{}_\alpha[\phi_*] &= -e^{-\hat{\Gamma}[\phi_*,\bar{\phi}]}\frac{\overleftarrow{\delta}}{\delta\phi_*^k}\, Q^k{}_\alpha[\phi_*]\, e^{\hat{\Gamma}[\phi_*,\bar{\phi}]} \\
&= \hat{\Gamma}[\phi_*, \bar{\phi}]\frac{\overleftarrow{\delta}}{\delta\bar{\phi}^i} D^{-1\,ij}[\bar{\phi}, \phi_*] \\
&\quad \times \left\langle \frac{\overrightarrow{\delta}}{\delta\phi_*^j}\left(\sigma[\phi_*, \phi] - \sigma[\phi_*, \bar{\phi}]\right)\frac{\overleftarrow{\delta}}{\delta\phi_*^k}\, Q^k{}_\alpha[\phi_*]\right\rangle \\
&= -\hat{\Gamma}[\phi_*, \bar{\phi}]\frac{\overleftarrow{\delta}}{\delta\bar{\phi}^i} D^{-1\,ij}[\bar{\phi}, \phi_*] \\
&\quad \times \left\{\left\langle\left(\sigma[\phi_*, \phi] - \sigma[\phi_*, \bar{\phi}]\right)\frac{\overleftarrow{\delta}}{\delta\phi_*^k}\, Q^k{}_{\alpha,j}[\phi_*]\right\rangle\right. \\
&\qquad \left. + \left\langle \frac{\overrightarrow{\delta}}{\delta\phi_*^j}\sigma[\phi_*, \phi]\frac{\overleftarrow{\delta}}{\delta\phi^k}\, Q^k{}_\alpha[\phi]\right\rangle - \frac{\overrightarrow{\delta}}{\delta\phi_*^j}\sigma[\phi_*, \bar{\phi}]\frac{\overleftarrow{\delta}}{\delta\bar{\phi}^k}\, Q^k{}_\alpha[\bar{\phi}]\right\} \\
&= -\hat{\Gamma}[\phi_*, \bar{\phi}]\frac{\overleftarrow{\delta}}{\delta\bar{\phi}^i}\, D^{-1\,ij}[\bar{\phi}, \phi_*]\, D_{jk}[\bar{\phi}, \bar{\phi}]\, Q^k{}_\alpha[\bar{\phi}] \\
&= -\hat{\Gamma}[\phi_*, \bar{\phi}]\frac{\overleftarrow{\delta}}{\delta\bar{\phi}^k}\, Q^k{}_\alpha[\bar{\phi}] \ .
\end{aligned} \tag{A.5}$$

When $\hat{\Gamma}[\phi_*, \bar{\phi}]$ is independent of the ϕ_*^i, as it is when the variables ϕ^i are chosen to be the Riemann normal coordinates σ^i, this equation reduces to

$$\Gamma_{,i}[\bar{\sigma}]\, Q^i{}_\alpha[\bar{\sigma}] = 0 \ , \tag{A.6}$$

which is the statement of the group invariance of the Vilkovisky effective action used in the text.

B. Phase Transition in Compact Sigma Models

We remarked in §4, in the discussion of critical slowing down, that a lattice-simulated sigma model with a compact configuration space undergoes a phase transition at a finite value of β denoted by β_c. The phase transition occurs at the value of β at which the fields $\phi_n, \phi_{n+1}, \ldots$ in the Metropolis sequence cease to have internal correlations. In each member of the sequence (after discard of the transients), the field typically varies in a completely random manner from one site to the other, ranging freely over the whole configuration space.*

It is not difficult to give a rough estimate of β_c for the cases in which the configuration spaces are unit spheres, the so-called $O(n)$ models. For a completely random field, the average of the square of the geodetic distance function between the field at one site and the field at a neighboring site is roughly

$$\bar{\Delta}^2 \sim \left(\frac{\pi}{2}\right)^2 . \tag{B.1}$$

As β approaches β_c from above, such fields begin to contribute significantly to the Feynman functional integral when $\exp(-\frac{1}{2}\beta\bar{\Delta}^2) \sim \frac{1}{2}$, and this determines β_c:

$$e^{-\frac{1}{2}\beta_c\bar{\Delta}^2} \sim \frac{1}{2} . \tag{B.2}$$

Equations (B.1) and (B.2) together imply

$$\beta_c \sim \frac{8}{\pi^2} \ell n\, 2 = .562\ldots . \tag{B.3}$$

Actual computer runs yield values of β_c between .3 and .6, depending on the model. In particular, the $O(2)/Z_2$ model, for which the configuration space is the circle, displays a phase transition despite the fact that it is locally (in configuration space) linear.

The way the critical value of β_c is determined for the $O(n)$ models is as follows. The action for these models has the form (2.19), where the index a ranges from 1 to n, and the field satisfies the constraint (2.20). As usual, on the computer we force the system into one of its degenerate ground states by holding the averages $\langle\phi_1\rangle, \ldots, \langle\phi_{n-1}\rangle$, for example, equal to zero. We then compute $\langle\phi_n\rangle$, choosing the positive branch. For $\beta < \beta_c$, we find that $\langle\phi_n\rangle$ drops rapidly to zero, with increasing abruptness as N increases. For $\beta > \beta_c$, we find that it mounts steadily to unity, the field becoming essentially frozen at $\phi_n = 1$ at large values of β (low discretization temperatures). The picture that emerges in the limit $N \to \infty$ is as follows. $\langle\phi_n\rangle$ for an infinite lattice rigorously vanishes for all $\beta < \beta_c$. For values of β modestly greater than β_c, it obeys a law of the form

$$\langle\phi_n\rangle = A(\beta - \beta_c)^\nu \qquad \beta_c < \beta \leq 1 \tag{B.4}$$

where the exponent ν is close to $\frac{1}{2}$.

---

* When the configuration space is noncompact, the field can never range over the whole of it. This is the reason why the $O(1,2)/O(2) \times Z_2$ model has no phase transition at a finite value of β.

On an infinite lattice all n-point functions vanish (i.e., all correlations vanish) for $\beta < \beta_c$. For $\beta > \beta_c$ the Texas group plans to compute the 2-point function for the $O(3)/O(2)$ model and attempt to fit it to the function (5.2), but no results are yet available. Since the Riemann normal coordinates σ^i for the $O(3)/O(2)$ model are limited to the range $-\pi \leq \sigma^i \leq \pi$, it is difficult to see how the coefficient β_R could tend to zero near *any* value of β, and it therefore seems likely that the $O(3)/O(2)$ model has no consistent continuum limit. But this remains to be seen.

REFERENCES

DeWitt, B. S., 1984, The Spacetime Approach to Quantum Field Theory, *in:* "Relativity, Groups and Topology II," B. DeWitt and R. Stora, eds., Elsevier Science Publishers, North Holland.

———————, 1957, *Rev. Mod. Phys.* **29**, 377.

Vilkovisky, G. A., 1984, *in:* "Quantum Theory of Gravity," S. M. Christenson, ed., Adam Hilger, Bristol.

BERRY'S PHASE AND PARTICLE INTERFEROMETRY IN WEAK GRAVITATIONAL FIELDS

Giorgio Papini

Department of Physics
University of Regina
Regina, Sask., S4S 0A2, Canada

INTRODUCTION

Particle interferometry has long been recognized as an invaluable tool to carry out delicate measurements of physical quantities that can be related to the difference in phase of the interfering beams. This is the case, for instance, in photon interferometry where displacements a fraction of the photon wavelength determine the magnitude of the phase difference. The latter is here classical.

The observation of quantum mechanical phases provides, on the other hand, accurate information about quantities such as the flux of a magnetic field, or allows the testing of the quantum behaviour of a system. For these reasons the traditional photon beam interferometers have recently been joined by a host of new devices. These include superconducting interferometers[1] and others using neutrons[2,3], neutral superfluids[4] and molecular beams[5].

Despite the obvious differences among the devices, a comprehensive theory of interferometry is possible, insofar as the phase is concerned, thanks to the work of Berry[6].

Two of Berry's fundamental results are here recalled.

i) A quantum system whose Hamiltonian $H(\lambda)$ depends continuously on the slowly varying parameters λ_a and has a set of n levels $|\phi>$ for each value of the λ's, acquires a geometrical phase

$$\gamma(C) = i \oint_C < \phi | \frac{\partial}{\partial \lambda_a} | \phi > d\lambda_a, \qquad a = 1, 2, ...k \tag{1}$$

for every closed path C in parameter space. In (1) summation over the repeated index a is understood.

ii) The evolution of the system is accompanied by the presence of Abelian gauge potentials (or non-Abelian ones for degenerate levels[7]) represented by

$$A^a = < \phi | \frac{\partial}{\partial \lambda_a} | \phi > . \tag{2}$$

These results can not be blindly applied even to the simpler but physically important case of particle interferometry in weak gravitational fields, the only ones considered in this work. It is in fact important to observe the following. First, Eqs. (1) and (2) are derived from non-relativistic quantum mechanics, while the problem at hand requires in general the use of covariant wave equations. Second, parameter space does not necessarily coincide with spacetime, which is necessary to transform the potentials (2) into ordinary gauge potentials. Finally, a line of action aimed at resolving these difficulties does not, per se, guarantee the

Quantum Mechanics in Curved Space-Time
Edited by J. Audretsch and V. de Sabbata
Plenum Press, New York, 1990

existence of a solution to the problem. For one thing, the gravitational potential would appear in (1) as a vector potential, contrary to general relativity. Heuristically, this approach may nonetheless be followed. It is known, in fact, that weak gravitational fields do behave, in some instances, as four-vectors. This is shown in the next section. A covariant generalization of Berry's results is then given, followed by the explicit calculation of the gravitational Berry phase and some applications.

PARTICLE WAVE FUNCTIONS IN WEAK GRAVITATIONAL FIELDS

An example that is particularily appropriate for this section is represented by the quantum behaviour of a particle of mass m in a weak, stationary gravitational field. By starting from the usual definition of the Hamiltonian $H = p_i\dot{x}^i - L$ where $p_i = \frac{\partial L}{\partial \dot{x}^i}$ and $L = -mc\int ds$, one arrives at the Schrödinger equation[8]

$$i\hbar\frac{\partial\psi}{\partial t} = [\frac{1}{2m}(p_i - mc\gamma_{0i})^2 - \frac{1}{2}mc^2\gamma_{00}]\psi, \tag{3}$$

where

$$\gamma_{\mu\nu} = g_{\mu\nu} - \eta_{\mu\nu}, \qquad |\gamma_{\mu\nu}| << 1 \tag{4}$$

is the metric deviation and $\eta_{\mu\nu}$ has signature +2. Since under the coordinate transformations

$$x'^{\mu} = x^{\mu} + \xi^{\mu}, \quad \xi_{\mu} = \eta_{\mu\nu}\xi^{\nu}, \tag{5}$$

where ξ^{μ} is smooth enough to keep $\gamma_{\mu\nu}$ small, the quantities $\gamma_{\mu\nu}$ transform as

$$\gamma'_{\mu\nu} = \gamma_{\mu\nu} - \xi_{\mu,\nu} - \xi_{\nu,\mu}, \tag{6}$$

it immediately follows that Eq.(3) is covariant under (5) to first order in ξ_{μ} and also gauge invariant if $\gamma_{\mu\nu}$ is stationary. Effectively $\gamma_{0\mu}$ does therefore satisfy the same gauge transformations of a vector potential. For closed spacetime paths one finds

$$\psi = exp(i\frac{mc}{\hbar}\oint \gamma_{0\mu}dx^{\mu})\psi_0 \tag{7}$$

where ψ_0 is the field-free solution of (3). The effect of an electromagnetic field can be easily incorporated by adding to the exponential in (7) the term $-\frac{ie}{\hbar c}\oint A_{\mu}dx^{\mu}$. The further addition to the r.h.s. of (3) of a term $V_I\psi$ to account for electron pairing makes the equation apt to describe superconducting interferometers[9]. The generalization of this result to time-dependent fields and relativistic particles has been attempted by several authors [10,11,12]. These approaches contain in general additional assumptions which appear too restrictive. The method presented here requires only knowledge of the solutions in the field free case. Formally, as shown below, one does not even need the explicit field-free solutions in order to write an expression for the phase.

In view of the wide variety of applications, in particular to interferometers, it is convenient to study systems whose wave functions satisfy the equation [13]

$$\left[(\nabla_{\mu} - i\frac{e}{c}A_{\mu})^2 - \frac{m^2c^2}{\hbar^2}\right]\phi(x) = \beta\,|\,\phi(x)\,|^2\,\phi(x), \tag{8}$$

where β is a constant and $A_{\mu}(x)$ represents the total electromagnetic potential of all external and gravity induced fields present. The quantity $\phi(x)$ here represents a scalar. Eq.(8) is the fully covariant version of the Landau-Ginzburg equation, it reduces to the Gross-Pitaevskii equation when A_{μ} vanishes and to the Klein-Gordon equation for $\beta = 0$. It is therefore well suited to discuss a number of systems, from superfluids to scalar particles. Generalizations to include particles with spin different from zero can also be given [14], as shown below.

A solution of Eq.(8) with $e = 0$ can be obtained by introducing a new field $\Phi(x)$ related to $\phi(x)$ above by

$$\Phi(x) = [e^{i\chi}, \phi(x)] = e^{i\chi}\phi(x), \tag{9}$$

where the phase operator χ is defined by

$$\chi \equiv -\frac{1}{4}\int_P^x dz^\lambda(\Gamma_{\alpha,\lambda\beta}(z) - \Gamma_{\beta,\lambda\alpha}(z))J^{\alpha\beta}(z) + \frac{1}{2}\int_P^x dz^\lambda \gamma_{\lambda\beta}P^\beta \tag{10}$$

with

$$[J^{\alpha\beta}(z), F(x)] \equiv i((x^\alpha - z^\alpha)\partial^\beta F(x) - (x^\beta - z^\beta)\partial^\alpha F(x)), \tag{11}$$

$$[P^\alpha, F(x)] \equiv i\partial^\alpha F(x). \tag{12}$$

The quantities

$$\Gamma_{\alpha,\lambda\beta} = \frac{1}{2}(\gamma_{\alpha\lambda,\beta} + \gamma_{\alpha\beta,\lambda} - \gamma_{\lambda\beta,\alpha}) \tag{13}$$

represent the Christoffel symbols of the first kind constructed in terms of $\gamma_{\mu\nu}$. Rewriting Eq.(9) in the form

$$\Phi(x) \equiv \phi(x) + \int_P^x dz^\lambda G_\lambda(z, \phi(x)) \tag{14}$$

and using the expansion

$$\gamma_{\mu\nu} \sim \gamma_{\mu\nu}^{(1)} + \gamma_{\mu\nu}^{(2)} + \cdots \tag{15}$$

where $1 >\mid \gamma_{\mu\nu}^{(1)} \mid>\mid \gamma_{\mu\nu}^{(2)} \mid \cdots$, one can show that

$$\begin{aligned} \eta^{\mu\nu}\Phi,_{\mu\nu} - \frac{m^2c^2}{\hbar^2}\Phi - \beta \mid \Phi \mid^2 \Phi &= \eta^{\mu\nu}\phi,_{\mu\nu} - \gamma_{\mu\beta}^{(1)}\phi,^{\beta\mu} - \Gamma_\beta^{(1)}\phi,^\beta - \frac{m^2c^2}{\hbar^2}\phi - \beta \mid \phi \mid^2 \phi + \\ &+ (-\gamma_{\mu\beta}^{(1)}\phi,^{\beta\mu} - \Gamma_\beta^{(1)}\phi,^\beta + \Sigma) + \\ &+ \gamma_{\mu\beta}^{(1)}\phi,^{\beta\mu} + \Gamma_\beta^{(1)}\phi,^\beta = 0 \end{aligned} \tag{16}$$

where Σ collects all higher order terms. Eq.(16) yields to first order

$$\eta^{\mu\nu}\Phi,_{\mu\nu} - (\frac{m^2c^2}{\hbar^2} + \beta \mid \Phi \mid^2)\Phi = \phi;_\mu{}^\mu - (\frac{m^2c^2}{\hbar^2} + \beta \mid \phi \mid^2)\phi = 0. \tag{17}$$

Higher order corrections can also be calculated as shown in Ref.[15].

In turn, one can prove that if $\Phi(x)$ satisfies the equation

$$\eta^{\mu\nu}\Phi,_{\mu\nu} - \frac{m^2c^2}{\hbar^2}\Phi = \beta \mid \Phi \mid^2 \Phi, \tag{18}$$

then

$$\phi(x) = [e^{-i\chi}, \Phi(x)] = e^{-i\chi}\Phi(x) \tag{19}$$

satisfies Eq.(8) with $e = 0$. Thus the effect of weak gravitational fields can be described by a phase factor once the solution of the gravity-free equation is known, and Φ need not be a constant as in Ref.[12]. The effect of the electromagnetic field can again be incorporated by adding to Eq.(10) the term $-\frac{e}{\hbar c}\int_P^x dz^\lambda A_{\lambda(z)}$.

It is easy to see that the first order solution (19) is gauge-invariant and invariant under the transformations (5) and (6) only for closed spacetime paths. In this case, Stokes' theorem gives

$$\chi = -\frac{1}{4}\int_{\Sigma_P} R_{\mu\nu\alpha\beta}J^{\alpha\beta}d\tau^{\mu\nu} - \frac{e}{\hbar c}\int_{\Sigma_P} F_{\mu\nu}d\tau^{\mu\nu}, \tag{20}$$

where Σ_P is the surface bound by the closed path, $F_{\mu\nu} = -A_{\mu,\nu} + A_{\nu,\mu}$ and $R_{\mu\nu\alpha\beta}$ represents the linearized Riemann curvature tensor

$$R_{\mu\nu\alpha\beta} = -\frac{1}{2}(\gamma^{(1)}_{\mu\alpha,\nu\beta} - \gamma^{(1)}_{\nu\alpha,\mu\beta} - \gamma^{(1)}_{\mu\beta,\nu\alpha} + \gamma^{(1)}_{\nu\beta,\mu\alpha}). \tag{21}$$

In the absence of electromagnetic fields, the result is that of Ref.[16]. $G_\lambda(z, \Phi(x))$ does in fact behave as a four-vector only for weak fields and gauge-invariance is assured only for closed paths.

In applications, the phase operator (10) can be rewritten to first order in $\gamma^{(1)}_{\mu\nu}$ as

$$\chi = -\frac{1}{4}\int_P^x dz^\lambda(\gamma^{(1)}_{\alpha\lambda,\beta}(z) - \gamma^{(1)}_{\beta\lambda,\alpha}(z))J^{\alpha\beta}(z) + \frac{1}{2}\int_P^x dz^\lambda \gamma^{(1)}_{\lambda\beta}P^\beta \tag{22}$$

A COVARIANT GENERALIZATION OF BERRY'S PHASE

In the fully covariant generalization given below use is made of the formalism originally introduced by Fock[17], and sometimes referred to, inappropriately, as 'proper time' formalism. The essential features of the formalism are here repeated for convenience. The Klein-Gordon equation for spin-0 particles in an external electromagnetic field

$$(\partial_\mu - \frac{ie}{c}A_\mu)^2\phi(x) = m^2\phi(x) \tag{23}$$

can be re-cast in the form of a Schrödinger equation

$$i\frac{\partial\Phi}{\partial\tau} = H\Phi \tag{24}$$

where τ is a relativistic invariant evolution parameter and

$$H \equiv -(\partial_\mu - \frac{ie}{c}A_\mu)^2 \tag{25}$$

plays the role of a relativistic Hamiltonian. Since the potentials $A_\mu(x)$ are functions of x^μ only and do not depend explicitly on τ, a particular solution of Eq.(19) is

$$\Phi(x,\tau) = exp(im^2\tau)\phi(x), \tag{26}$$

where $\phi(x)$ satisfies Eq.(23). This shows that the Hilbert space of the system described by Eq.(24) can be decomposed into a sum of Hilbert sub-spaces corresponding to different values of m^2. When condition (26) is imposed, a particular Hilbert sub-space is chosen to represent particles of mass m. Thus Eq.(24) is equivalent to Eq.(23) if the solutions of Eq.(23) have the form

$$\phi(x) = \int_{-\infty}^{\infty} exp(-im^2\tau)\Phi(x,\tau)d\tau. \tag{27}$$

The procedure can even be extended to spin-0 particles in an arbitrary potential $U(x)$ by using the method of Cheng, Coon and Zhu[18]. In fact Eq.(23) can be rewritten as

$$(\Box - m^2)\phi(x) = U(x)\phi(x) \tag{28}$$

with $\Box \equiv \partial_\mu\partial^\mu$ and

$$U(x) \equiv -\frac{ie}{c}(\partial_\mu A^\mu + A_\mu\partial^\mu) + \frac{e^2}{c^2}A_\mu A^\mu. \tag{29}$$

If now $U(x)$ takes an arbitrary form to include interactions other than the electromagnetic one, the Green's function $G(x,x')$ for Eq.(28) can be found:

$$G(x,x') = \int \frac{d^4q}{(2\pi)^4}\frac{exp[-iq\cdot(x-x')]}{q^2+m^2+i\epsilon}. \tag{30}$$

Thus Eq.(28) with Feynman's boundary conditions[18] can be transformed into the integral equation

$$\phi(x) = exp(-i\kappa \cdot x) - \int d^4x' G(x,x')U(x')\phi(x'). \tag{31}$$

If one now considers Eq.(24) with

$$H \equiv -(\Box - U(x)), \tag{32}$$

its solution is

$$\Phi(x,\tau) = e^{i(m^2\tau - \kappa\cdot x)} - \int d^4x' d\tau' G(x,\tau;x',\tau')U(x')\Phi(x',\tau') \tag{33}$$

with $G(x,\tau;x',\tau')$ given by

$$G(x,\tau;x',\tau') = \int \frac{d^4q}{(2\pi)^4}\frac{d\omega}{2\pi}\frac{exp[i\omega(\tau-\tau') - iq\cdot(x-x')]}{\omega + q^2 + i\epsilon}. \tag{34}$$

Since $U(x)$ is independent of τ, one finds that

$$\Phi(x,\tau) = e^{im^2\tau}\phi(x) \tag{35}$$

is a solution of Eq.(33). In particular the Dirac equation

$$(-i\gamma^\mu\partial_\mu - \frac{e}{c}\gamma^\mu A_\mu + m)\psi = 0 \tag{36}$$

can be rewritten in the form of Eq.(19) by introducing a new spinor function ψ' and defining

$$\psi \equiv (-i\gamma^\mu\partial_\mu - \frac{e}{c}\gamma^\mu A_\mu - m)\psi'. \tag{37}$$

Schwinger[19] has also shown how to derive the relativistic Schrödinger equation from a quantized Dirac field in interaction with an external potential A_μ. Vector fields, satisfying the Proca equation[20], can also be rewritten in the form of Eq.(28). In this case ϕ is a column matrix of ten components.

It is obvious that τ here plays a role analogous to that of time in non-relativistic quantum mechanics and a corresponding formalism can be constructed[21,22]. Although mathematically τ is determined unambiguously by its commutation relation with the conjugate variable H, its physical interpretation is not unique[23]. One can however treat τ as a pure parameter and circumvent the problem by imposing the constraint (26) on the solution of Eq.(24). In this way τ can be eliminated.

If one now considers the relativistic invariant Schrödinger equation

$$i\frac{\partial\Psi}{\partial\tau} = H(\lambda_a(\tau))\Psi \qquad a = 1,2,\cdots k \tag{38}$$

where $\lambda_a(\tau)$ are k functions of τ and one sets, in particular, $k = 4$ and $\lambda_a(\tau) = x^\mu(\tau)$ with a = 1,2,3,4 corresponding to μ = 0,1,2,3 respectively, the constraint Eq.(26) for a particle of mass m becomes

$$H(\lambda)\psi(\lambda) = -m^2\psi(\lambda). \tag{39}$$

The adiabatic condition normally used in the derivation of Berry's phase for non-relativistic systems is here replaced by the inequality

$$\frac{< m'' \mid \frac{\partial H}{\partial\tau} \mid m' >}{< m' \mid \frac{\partial H}{\partial\tau} \mid m' >} \ll 1 \qquad m'' \neq m', \tag{40}$$

where $\frac{\partial H}{\partial\tau} \equiv \frac{\partial H}{\partial\lambda_a}\frac{d\lambda_a}{d\tau}$ and $\mid m' >$ denotes an eigenstate of H with eigenvalue m'^2. Condition (40), which is independent of τ, amounts to requiring that the system, originally in a pure mass square eigenstate, stays in that state during the evolution. Following Berry[6], one then finds that the total phase change of Ψ around the closed path C in the parameter space of $\lambda_a(\tau)$ is

$$\Psi(\tau) = exp(i\gamma(C))exp\{i\int_o^T m^2 d\tau\}\Psi(0), \tag{41}$$

where

$$\gamma(C) = i\oint_C < \psi|\frac{\partial}{\partial\lambda_a}|\psi > d\lambda_a \tag{42}$$

is the covariant generalization of Berry's phase[24] that corresponds to (1). If in particular the system is represented by charged particles in an external electromagnetic potential A_μ and one chooses $\lambda_a \equiv x'^\mu$, the eigenstate wave function ψ acquires a Dirac phase factor

$$\psi \rightarrow exp\{\frac{ie}{c}\int_{x'}^{x} A_\mu dx^\mu\}\psi. \tag{43}$$

Thus

$$\gamma(C) = \frac{e}{c}\oint_C A_\mu(x')dx'^\mu \int d^4x\psi^*(x)\psi(x) = \frac{e}{c}\oint_C A_\mu dx^\mu, \tag{44}$$

or

$$\gamma(C) = \frac{e}{c}\int_{\Sigma_C} F_{\mu\nu}dS^{\mu\nu}, \tag{45}$$

which gives, in appropriate circumstances, the Bohm-Aharonov effect[25] in covariant form.

In summary, the Klein-Gordon, Proca and Dirac equations can be re-cast in the form of the relativistic Schrödinger equation. The corresponding Berry phase can be calculated. The Hamiltonian operator of the ordinary Schrödinger equation is generalized to the rest mass square operator H of Eq.(38). H has a set of eigenvalues, the square of the rest masses of the possible particles in the system considered. Then the constraint (26) chooses the system to contain only particles with mass m. As long as the mass of the particles remain constant, condition (40) is trivially satisfied. These results can be further extended to non-linear field theories[24].

BERRY'S PHASE FOR WEAK GRAVITATIONAL FIELDS

The main point of the covariant generalization given in the previous section is represented by the fact that *it is now possible to identify parameter space with ordinary spacetime by setting* $k = 4$. The parameters are then the collective coordinates of the system considered and A^a becomes a gauge field in the usual sense.

Now consider a system of scalar particles interacting with a gravitational field and satisfying Eq.(8). In the gravity-free case, Eq.(8) becomes

$$\Box\phi_o - m^2\phi_o - \beta \mid \phi_o \mid^2 \phi_o = 0. \tag{46}$$

In the weak field approximation (4) one can rewrite Eq.(8) with $\beta = 0$ in the form of Eq.(28) and discard terms of second order and higher in $\gamma_{\mu\nu}$:

$$(\Box - m^2)\phi(x) = [\gamma_{\mu\nu}\partial^\mu\partial^\nu - (\frac{1}{2}\gamma_\mu{}^\mu{}_{,\nu} - \gamma_\nu{}^\mu{}_{,\mu})\partial^\nu]\phi(x), \tag{47}$$

By using Eq.(30) for $G(x, x')$, one obtains

$$\phi(x) = \phi_o(x) + \int d^4x' G(x, x')U(x')\phi(x'), \tag{48}$$

where ϕ_o satisfies Eq.(46) with $\beta = 0$. In the first order approximation, $\phi(x')$ in (48) can be replaced by $\phi_o(x')$. By substituting the expression of $U(x)$ into (48) and integrating by parts, one gets

$$\begin{aligned}
\phi(x) &= \phi_o + \int d^4x' \partial^{\mu}_{x'} G(x,x') \partial_{x'\mu} [\frac{1}{4} \int_X^{x'} dz^{\lambda} (\gamma_{\alpha\lambda,\beta}(z) - \gamma_{\beta\lambda,\alpha}(z))((x'^{\alpha} - z^{\alpha}) \partial^{\beta} \phi_o(x') - \\
&- (x'^{\beta} - z^{\beta}) \partial^{\alpha} \phi_o(x')) - \frac{1}{2} \int_X^{x'} dz^{\lambda} \gamma_{\alpha\lambda}(z) \partial^{\alpha} \phi_o(x')] + \\
&+ \int d^4x' m^2 G(x,x') [\frac{1}{4} \int_X^{x'} dz^{\lambda} (\gamma_{\alpha\lambda,\beta}(z) - \gamma_{\beta\lambda,\alpha}(z))((x'^{\alpha} - z^{\alpha}) \partial^{\beta} \phi_o(x') - \\
&- (x'^{\beta} - z^{\beta}) \partial^{\alpha} \phi_o(x')) - \frac{1}{2} \int_X^{x'} dz^{\lambda} \gamma_{\alpha\lambda}(z) \partial^{\alpha} \phi_o(x')] \\
&= \phi_o - \int d^4x' (\Box_x - m^2) G(x,x') [\frac{1}{4} \int_X^{x'} dz^{\lambda} (\gamma_{\alpha\lambda,\beta}(z) - \gamma_{\beta\lambda,\alpha}(z))((x'^{\alpha} - z^{\alpha}) \partial^{\beta} \phi_o(x') - \\
&- (x'^{\beta} - z^{\beta}) \partial^{\alpha} \phi_o(x')) - \frac{1}{2} \int_X^{x'} dz^{\lambda} \gamma_{\alpha\lambda}(z) \partial^{\alpha} \phi_o(x')]. \qquad (49)
\end{aligned}$$

One finally arrives at the first order solution

$$\phi(x) = e^{-i\chi} \phi_o(x) \qquad (50)$$

with

$$\chi \equiv -\frac{1}{4} \int_X^x dz^{\lambda} (\gamma_{\alpha\lambda,\beta}(z) - \gamma_{\beta\lambda,\alpha}(z)) J^{\alpha\beta}(z) + \frac{1}{2} \int_X^x dz^{\lambda} \gamma_{\alpha\lambda}(z) P^{\alpha}. \qquad (51)$$

In (51) the integration is taken along a path from X to the field point x and the operators $J^{\alpha\beta}(z)$ and P^{α} are defined as in (11) and (12) with $F(x)$ replaced by $\phi_o(x)$. It is not difficult to verify that this solution is also valid when $\beta \neq 0$. Eq.(51) coincides with (22). Similarily, when the integration path is closed, as in the important case of interferometry, one obtains Eq.(20). The effect of a minimally coupled electromagnetic field may be included in (51) in the usual way.

SOME APPLICATIONS

The phase shifts calculated below are those induced by the gravitational field of the earth, rotation and gravitational waves. In all cases it is assumed for simplicity that ϕ_o in (50) may be represented by plane-wave solutions.

Field of the Earth

If earth is assumed spherical, then its gravitational field may be described by the Schwarzschild metric [26,15]

$$ds^2 = -(1 - \frac{2MG}{c^2 r}) dt^2 + \frac{1}{c^2}(1 - \frac{2MG}{c^2 r})^{-1} dr^2 + \frac{r^2}{c^2} d\theta^2 + \frac{r^2}{c^2} sin^2\theta d\phi^2, \qquad (52)$$

where M represents the earth mass. Eq.(51) is now applied to a square interferometer of side ℓ and vertices A, B, C, D, in which a beam of particles is split at A and the resulting beams interfer at A again after travelling along the opposite paths ABCDA and ADCBA. Vertex A coincides with the origin of a tern of orthogonal axes x, y, z. The unit vector $\hat{z}$ is directed as the outward normal to the earth surface. Side $\bar{AB}$ lies along the z-axis while side $\bar{AD}$ lies along the y-axis. If the particles are neutral, the contribution to the change in phase may be divided into a part $\Delta\chi_1$ containing $J_{\mu\nu}$ and the remainder $\Delta\chi_2$. Since the particles are assumed to move with constant speed v, the integration over the time portion of the spacetime loop may be reduced to space integrations by choosing the limits of integration appropriately. A detailed calculation shows that $\Delta\chi_1$ vanishes for both paths ABCDA and ADCBA. Assuming that the linear dimension of the interferometer ℓ is such that $\frac{\ell}{R} << 1$, where R is the earth radius, expanding $\frac{1}{r}$ in the neighborhood of $\frac{1}{R}$ up to the third order term and assuming that the particles are non-relativistic so that

$$\kappa^{\circ} \approx \frac{mc}{\hbar} + \frac{\hbar\kappa^2}{2mc} \qquad \text{with } \kappa = \frac{mv}{\hbar} \tag{53}$$

one finds

$$\Delta\chi = \{\frac{m^2}{\hbar^2\kappa} + \frac{\kappa}{2c^2} - \frac{3}{2}\frac{m^2\ell}{\hbar^2\kappa R} - \frac{\kappa\ell}{c^2 R}\frac{3}{4}\} \mid \vec{A} \times \vec{g} \mid, \tag{54}$$

where

$$\vec{A} = \ell^2 \hat{N}, \quad \vec{g} = -\frac{MG}{R^2}\hat{z} \tag{55}$$

and $\hat{N}$ is the normal to the interferometer plane. The first term coincides with the corresponding term of Anandan's result [12]. When the interferometer is rotated by 180°, the contribution from this term gives a phase shift

$$\Delta\chi = \frac{2m^2}{\hbar^2\kappa} \mid \vec{A} \times \vec{g} \mid = \frac{4\pi m^2 Ag\lambda}{h^2} \tag{56}$$

which is just what has been observed in the COW experiment [3]. The ratios of the various terms in (54) are $1 : \frac{v^2}{c^2} : \frac{\ell}{R} : \frac{v^2\ell}{c^2R}$. When $v = 10^{-5}c$, $\ell = 1cm$, $R = 7 \times 10^8$ cm, one obtains $1 : 10^{-10} : 10^{-6} : 10^{-16}$. The second term represents a special relativistic correction, which is smaller than the general relativistic effect represented by the third term.

Rotation

The non-inertial effects due to the rotation of a frame of reference can be described from the point of view of a co-rotating observer by the metric [27,15]

$$ds^2 = -(1 - \frac{\omega^2 r^2}{c^2})dt^2 + 2\frac{\omega}{c^2}(-ydx + xdy)dt + \frac{1}{c^2}(dx^2 + dy^2 + dz^2) \tag{57}$$

with $r^2 \equiv x^2 + y^2$ and where ω is the angular velocity about the z-axis. The interferometer considered above now lies in the (xy)-plane and interference occurs at vertex C opposite to A. Eliminating again the time intergration and choosing appropriate integration limits along the space part of the loop, one arrives at the result $\Delta\chi_1 = 0$ and

$$\Delta\chi = \Delta\chi_2 = -\frac{\omega\ell^2}{c}(\kappa^{\circ} + \kappa\frac{c}{v}). \tag{58}$$

By using (53) and (55), Eq.(58) becomes

$$\Delta\chi = -\frac{2m\vec{\omega} \cdot \vec{A}}{\hbar} - \frac{\hbar\kappa^2\vec{\omega} \cdot \vec{A}}{2mc^2}, \qquad \vec{\omega} \equiv \omega\hat{z} \tag{59}$$

where the first term agrees with the result of other non-relativistic and relativistic approaches[15] and the second term represents a relativistic correction. When ω is the angular velocity of rotation of the earth, the first term agrees with the experimental result of Ref.[12]. For the particular case of the Sagnac effect, beam splitting and interference take place at A. Then

$$\Delta\chi = \Delta\chi_2 = -\frac{2\omega\ell^2}{c}(\kappa^{\circ} + \kappa\frac{c}{v}) \tag{60}$$

which, with the usual approximations, becomes

$$\Delta\chi = -\frac{4m\vec{\omega} \cdot \vec{A}}{\hbar} - \frac{\hbar\kappa^2\vec{\omega} \cdot \vec{A}}{mc^2}. \tag{61}$$

Assume for simplicity that the interferometer is a square of side ℓ in the (xy)-plane and that the wave travels in the z-direction. In the TT-gauge, one has

$$\gamma^{(1)}_{xx} = -\gamma^{(1)}_{yy} = f_{+}cos\omega(t - z/c); \quad \gamma^{(1)}_{xy} = \gamma^{(1)}_{yx} = f_{\times}cos\omega(t - z/c). \tag{62}$$

These expressions must now be substituted in (51) and the integrals evaluated, taking into account that the beams are split at A at time t and interfere at C. For small velocities, setting $f_{+} = f_{\times}$ and using (53) one obtains

$$\begin{aligned}\Delta\chi = \Delta\chi_2 + \Delta\chi_1 &= \frac{4fv\kappa}{\omega}\{(1 + \frac{\hbar v\kappa}{2mc^2})sin(\omega t + \frac{\ell\omega}{v})sin\frac{\ell\omega}{2v} - \\ &- \frac{\ell\hbar\kappa\omega}{4mc^2}sin(\omega t + \frac{\ell\omega}{2v})\}sin\frac{\ell\omega}{2v},\end{aligned} \tag{63}$$

where of the three terms in curly brackets the first one only refers to $\Delta\chi_2$. In the case of photons of frequency $\frac{\tilde{\omega}}{2\pi}$, one gets

$$\Delta\chi = \Delta\chi_2 = \frac{4f\tilde{\omega}}{\omega}sin^2\frac{\omega\ell}{2c}sin(\omega t + \frac{\omega\ell}{c}). \tag{64}$$

The $\Delta\chi_2$ term in (63) above agrees with Stodolsky's result [11] when a misprint in the latter is corrected by inserting an overall factor $\frac{1}{2}$. The second term is a correction of order $(\frac{v}{c})^2$ and can be safely neglected. The third term is new and plays a role at high frequencies. Eq.(63) gives an entirely different phase shift from that calculated by Anandan and Chiao [28] for a superfluid loop in the (xz)-plane. This case can be easily treated by applying (20) to a spacelike square path of length $2a$ in the x-direction and length $2b$ in the z-direction. A simple calculation performed in the TT-gauge yields

$$\Delta\chi = -\frac{2m\omega^2 a^2 b}{\hbar c}f_{+}cos\omega t \tag{65}$$

in the approximation $a, b << \frac{c}{\omega}$. When the latter restriction is removed, the result is

$$\Delta\chi = -\frac{m\omega a^2}{\hbar}f_{+}sin\omega(t - \frac{2b}{c}) \tag{66}$$

and is more suitable for the high frequency range of the spectrum. Both (65) and (66) must be multiplied by a factor N if the superfluid path consists of N loops. Formula (65) agrees completely with the result of [28] obtained however using Fermi normal coordinates which entail the condition $f(\frac{a}{\lambda})^2 < 1$, where λ is the wavelength of the incoming radiation. This restriction is here removed.

CONCLUSIONS

As shown in the previous sections, particle interferometry in weak gravitational fields can be entirely discussed in terms of Berry's phase. Whenever a solution of the free wave equation is known, the contribution of the external gravitational fields to first order in $\gamma_{\mu\nu}$ can be calculated explicitly by means of path integrals. This contribution is obviously amplified by the number of complete paths the particles travel before interference. According to (10), Berry's phase may be also considered as arising from a continuous sequence of boosts and rotations applied to the particles in displacing them along their path. In the approximation used and in the particular problem considered, the gravitational field behaves effectively as a vector potential $G_\lambda(z, \phi_o(x))$[14,15].

The present results also apply to inertial fields, at least to the extent that these are adequately treated in general relativity. This requires that, unlike the Newtonian theory, the inertial potentials transform as the components of a true gravitational field and undergo the gauge changes (6) under the transformations (5). Though the phase change can be written in the form (20), the existence of a linearized Riemann curvature tensor is no warranty for the existence of true curvature. For rotation, e.g., terms quadratic in the Christoffel symbols can be of the same order as the second order derivatives of $g_{\mu\nu}$. Thus interferometry, despite its manifest non-locality, does not provide a good way to distinguish between true and fictitious gravitational fields, unless terms of second order at least can be measured. The successfull treatment of the rotational case for which experimental data exist[29] indicates that the gravitational Berry phase introduced *provides the appropriate theoretical basis for both optical and non-optical gyroscopy.* This has been suggested as a possibility[30] but never substantiated theoretically.

The phase shift induced by an incoming gravitational wave offers an alternative detection scheme which is independent of any resonance condition and thus not necessarily limited to the *KHz* region of the spectrum. This scheme looks ideally suited for the detection of high frequency radiation, possibly produced by bremsstrahlung or photoproduction in stars [31]. In this case, however, the response time τ of the antenna becomes critical. A calculation based on the sudden aproximation [32] indicates that a change in the state of the system cannot take place before a time $\tau \sim \frac{\hbar}{\Delta E}$ has elapsed, where ΔE is the change in the Hamiltonian induced, in this case by the gravitational perturbation. Since

$$\Delta E \approx \frac{\gamma^{(1)}_{\mu\nu}\phi,^{\mu\nu}\hbar^2}{m} \sim fmv^2 cos\omega t, \tag{67}$$

one finds a lower limit

$$\tau \sim \frac{\hbar}{fmv^2} \sim \frac{10^{-13}}{f}. \tag{68}$$

A more complete understanding of the difficulties involved in the detection of high frequency gravitational radiation can be gained from the quantum theory of the antenna given by Weber [33].

The ratio of the third to the first term in (63) give $\sim \frac{\ell v\omega}{4c^2} > 1$ for $\omega > \frac{4c^2}{\ell v} \sim 4 \times 10^{14} s^{-1}$ for $v \sim 10^5$cm/s and $\ell \sim 10^2$ cm. Thus the third term prevails in magnitude at very high frequencies. A comparison with photon interferometers can now be made. At frequencies $\omega < 4 \times 10^{14} s^{-1}$ the ratio of the dominant terms in (63) and (64) is $\sim \frac{mv^2}{\hbar\tilde{\omega}} << 1$, so that optical or laser interferometers appear advantageous, as detectors of gravitational radiation, over usual particle interferometers where m is at most of the order of a nucleon mass and $v \sim 10^5 cm/s$. An x-ray microscope could, in principle, constitute a rather sensitive detector because of the high photon frequency involved. When $\omega > 4\times10^{14}s^{-1}$ the ratio of the dominant terms is $\sim \frac{m\ell v^3\omega}{4\hbar c^2\tilde{\omega}} \sim \frac{1}{4} \times 10^{-4}\frac{\omega}{\tilde{\omega}}\gtrless 1$ for $\omega \gtrless 4 \times 10^4\tilde{\omega}$. Therefore photon interferometers are more advantageous even in the range $4 \times 10^{14}s^{-1} < \omega < 4 \times 10^4\tilde{\omega}$. Only for graviton frequencies $\omega > 4 \times 10^4\tilde{\omega} > 4 \times 10^{14}s^{-1}$ particle interferometers become in principle more sensitive.

Acknowledgement

The author wishes to thank Mr. W.R. Wood for his invaluable help in the preparation of the manuscript. This work was supported in part by Natural Sciences and Engineering Research Council of Canada.

References

[1] J.E. Zimmerman and J.E. Mercereau, Phys. Rev. Lett. 14:887 (1965).

[2] S.A. Werner, J.-L. Staudenman and R. Colella, Phys. Rev. Lett. 42:1103 (1979).

[3] R. Colella, A.W. Overhauser and S.A. Werner, Phys. Rev. Lett. 34:1472 (1975); J.-L. Staudenman, S.A. Werner, R. Colella and A.W. Overhauser, Phys. Rev. A21:1419 (1980).

[4] M. Cerdonio and S. Vitale, Phys. Rev. B29:481 (1984) and in: "Josephson Effect - Achievements and Trends," A. Barone ed., World Scientific, Singapore (1986); M. Cerdonio, A. Goller and S. Vitale, in: "SQUID '85," Walter de Gruyter & Co., Berlin (1985); R. Tommasini, L. Vanzo, S. Vitale and S. Zerbini, in: "Proc. 4th M. Grossmann Meeting on General Relativity," R. Ruffini ed., North-Holland, Armsterdam (1986).

[5] G.I. Opat, in: "Proc. 3rd M. Grossman Meeting on General Relativity," H. Ning ed., Science Press, North-Holland (1983).

[6] M.V. Berry, Proc. Roy. Soc. London A392:45 (1984).

[7] F. Wilczek and A. Zee, Phys. Rev. Lett. 52:2111 (1984).

[8] G. Papini, Nuovo Cim. 52B:136 (1967).

[9] B.S. Dewitt, Phys. Rev. Lett. 16:1092 (1966); G. Papini, Nuovo Cim. 45B:66 (1966). Though these works consider specifically charged superfluids, their application to neutral ones is immediate.

[10] B. Linet and P. Tourrenc, Can. J. Phys. 54:1129 (1976).

[11] L. Stodolsky, Gen. Rel. Grav. 11:391 (1979).

[12] J. Anandan, Phys. Rev. D15:1448 (1977).

[13] V. Meyer and N. Salier, Theoret. Math. Phys. 38:270 (1979).

[14] Y.Q. Cai and G. Papini, "Berry's Phase for Weak Gravitational Fields," in: "Proc. 3rd Canadian Conf. General Relativity and Relativistic Astrophysics," World Scientific, Singapore (in press).

[15] Y.Q. Cai and G. Papini, Class Quant. Grav. 6:407 (1989).

[16] G. Papini, Nuovo Cim. 68B:1 (1970).

[17] V. Fock, Physik Zeits. Sowjetunion 12:404 (1937); Y. Nambu, Prog. Theor. Phys. 5:82 (1950); R.P. Feynman, Phys. Rev. 80:440 (1950); J. Schwinger, Phys. Rev. 82:664 (1951); See also C. Itzykson and J.-B. Zuber, "Quantum Field Theory," McGraw-Hill, New York (1980).

[18] H. Cheng, D.D. Coon and X.Q. Zhu, Phys. Rev. D26:896 (1982).

[19] J. Schwinger, Phys. Rev. 82:664 (1951).

[20] P. Roman, "Theory of Elementary Particles," North-Holland, Amsterdam (1964).

[21] J.H. Cooke, Phys. Rev. 166:1293 (1968).

[22] A. Kyprianidis, Phys. Rep. 155:1 (1987).

[23] J.R. Fanchi, Phys. Rev. A34:1677 (1986).

[24] Y.Q. Cai and G. Papini, Mod. Phys. Lett. A (in press).

[25] Y. Aharonov and D. Bohm, Phys. Rev. 115:485 (1959).

[26] L.Z. Fang and R. Ruffini, "Basic Concepts in Relativistic Astrophysics," World Scientific, Singapore (1983).

[27] C. Möller, "The Theory of Relativity," Oxford Univ. Press, London (1952).

[28] J. Anandan and R.Y. Chiao, Gen. Rel. Grav. 14:515 (1982); J. Anandan, Phys. Rev. D24:338 (1984).

[29] S.A. Werner, J.-L Staudenman and R. Colella, Phys. Rev. Lett. 42:1103 (1979).

[30] A.L. Robinson, Science 234:424 (1986).

[31] G. Papini and S.R. Valluri, Phys. Rep. 33:51 (1977); Astron. Astroph. 208:345 (1989); R.J. Gould, Astrophys. J. 288:789 (1985).

[32] A. Messiah, "Quantum Mechanics," John Wiley & Sons, New York (1968), Vol. II, p.740.

[33] J. Weber, Found. Phys. 14:1185 (1984); in: "Proc. Sir A. Eddington Symposium," Vol.3, T.M. Karade and J. Weber eds., World Scientific, Singapore (1984).

THE FINAL STATE OF AN EVAPORATING BLACK HOLE

AND THE DIMENSIONALITY OF THE SPACE-TIME

Venzo de Sabbata and C. Sivaram

World Lab, Lausanne - Switzerland
Dept.of Physics, Ferrara University - Italy
Institute of Astrophysics, Bangalore - India

1. Introduction

One of the main paradoxes of the contemporary physics of elementary particles is the apparent incompatibility of two main theoretical foundations: on the one hand the theory of general relativity which connects the force of gravity to the structure of space-time, on the other hand the theory of quantum mechanics. General relativity has been developed mainly to understand phenomena on a cosmic scale and the evolution of universe, while quantum mechanics regards mainly the atomic, subatomic and subnuclear world; this latter theory has been formulated for three of the four forces of nature, namely the strong, weak and electromagnetic interactions. In recent times unification of electromagnetic and weak interactions has been reached through gauge theories and one can try to include also strong interactions. Gravitation appears to be the more elusive to include in a true unification with all the other interactions and this may be due to the fact the energy at which gravity and quantum effects become of comparable strength, that is the energy at which one may hope to have unification of gravitational interaction with the other interactions is given by the so-called Planck energy

$$E_{Pl} = \left(\frac{\hbar c^5}{G}\right)^{1/2} \approx 10^{19}\text{Gev}$$

But perhaps the main difficulty for unifying gravitation with other interactions lies in more deep property of this phenomenon known as 'force of gravity'. In fact, according to Einstein, gravity is not a force at all but is an intrinsic property of the space and time: this follows from the chief, very peculiar fact, that is the famous "equivalence principle" for which inertial and gravitational masses are equivalent.

So we are faced with the situation for which while 3/4 of modern physics (the physics of strong, weak and electromagnetic interactions acting at a microscopical

Quantum Mechanics in Curved Space-Time
Edited by J. Audretsch and V. de Sabbata
Plenum Press, New York, 1990

level) are successfully described at present in the framework of a flat and rigid space-time structure, the remaining 1/4 (the macroscopic physics of gravity) needs the introduction of a curved, dynamic geometrical background.

In order to overcome this dichotomy, first of all we will try to extend the geometrical principles of general relativity also to microphysics, with the aim to establish a direct comparison, and possibly a connection, between gravity and the other interactions. Now we know that in general relativity matter is represented by the energy-momentum tensor which provides a description of the mass density distribution in space time, so that the mass-energy concept is sufficient to define the properties of the classical, macroscopic bodies; but if we go down to a microscopical level, we find that matter is formed by elementary particles which are characterized not only by mass but also by a spin. In that case, therefore, the energy-momentum tensor alone is no longer sufficient to characterize dynamically the matter sources, but also the spin density tensor is needed and the simplest and more natural way to take account of the spin in the Einstein theory, is the introduction of torsion, that is the antisymmetric part of the affine connection.

Of course this does not exhaust the problem to reconcile quantum mechanics with gravity and the aim of this course is to understand better the influence of quantum effects on gravity, or, viceversa the influence of gravitational fields on the quantum mechanics. This route seems to constitute a further step toward unification of all forces, including gravity and we believe also that the argument of quantum mechanics in curved space constitutes a necessary step to go toward the more difficult subject of quantum gravity. We like to understand better the influence of external gravitational fields on quantum matter and for that reason we have chosen to inquire what happens in the final state of an evaporating black hole. It is a problem that is strictly connected with the study of quantum fields in the presence of strong gravitational fields. We can say that the massive modes of closed superstring theories may play a crucial role in the last stages of black hole evaporation, and if the Bekenstein-Hawking entropy describes the true statistical entropy (the true degeneracy) of an evaporating black hole, it becomes favorable (entropically) that the black hole makes a transition to an excited state of massive string which, in turn, can decay to massless radiation, avoiding the naked singularity and also preserving quantum coherence as we will see.

2. Black hole evaporation

Are well known the consideration of Hawking [1] about the evolution of quantum field in background metric of classical black hole. He find that particle creation takes place and his semiclassical treatment of evaporation is valid if Schwarzschild radius >> Compton wave length i.e. $GM/c^2 \gg \hbar/Mc$, or $GM^2/(\hbar c) \gg 1$.

This quantity is also proportional to the entropy or area of the black hole. Emitted particles build a thermal spectrum and are uncorrelated among themselves. As black hole loses masses, its temperature rises, and evaporation accelerates (the rate is proportional to M^{-2}) as mass approaches the Planck mass = $(\hbar c^5/G)^{1/2}$ ≈ 10^{19}Gev). However at this stage the Compton length is comparable to the Schwarschild radius, i.e. $GM/c^2 \approx \hbar/Mc$ for $M = M_{Pl}$, i.e. $GM_{Pl}^2/\hbar c \approx 1$, so that a quantum treatment of the gravitational field is required. The semiclassical approach would suggest explosive decay of black hole of mass M_{Pl} (temperature ≈ $(1/k)(\hbar c^5/G)^{1/2}$ ≈ 10^{32}) in a burst of duration $(\hbar G/c^5)^{1/2}$ ≈ 10^{-43}sec. In the absence of any conservation law prohibiting the decay one might expect decay into various elementary field quanta.

This explosive disappearance when M reaches M_{Pl} would leave behind residue of thermal radiation which is a mixed state,in quantum mechanics term, i.e. the initial pure state has evolved into a mixed state of thermal radiation as seen by an observer at infinity. Such conversion would violate the fundamental tenets of quantum mechanics in flat space such as loss of quantum coherence; it appears also inconsistent with quantum unitary postulate: "time evolution is governed by unitary operator in Hilbert space of states" which will be maintained in a pratically stable remnant. The emitted quanta are an uncorrelated ensemble and cannot be described by a single wave function; some information about emitted radiation is contained inside event horizon in the form of correlations between photons absorbed by black holes and those emitted to infinity; i.e. unitary postulate is maintained by correlating each state ψ_n of radiation with a corresponding state Φ_n of black hole so that the joint system has a well defined overall wave function $|\psi_{tot}\rangle = \Sigma A_n |\psi_n^{rad}\rangle \times |\phi_n^{bh}\rangle$ that is we have a quantum state inside a black hole combined with thermal state of radiation to construct a pure state $|\psi_{tot}\rangle$ so that the unitarity is maintained.

Hawking process consists of pair creation in strong gravitational field, that is one member is emitted to infinity and the other is staying near horizon. When the degrees of freedom of the black hole are integrated out, external radiation is described by a density matrix $\bar{\rho}_{rad} = \Sigma |A_n|^2 |\psi_n^{rad}\rangle \langle\psi_n^{rad}|$ (only stationary states found for a black hole are consistent with the statistical nature of thermal radiation suggesting continous random emission of thermal radiation). If the residual mini black hole disappears completely by decaying into ordinary particles we will be deprived of the large reservoir of states in the black hole which is required for the construction of the overall wave function; the whole system will be described by the density matrix and unitary postulated is violated. Again as the initial and final states do not contain black holes we have to abandon even the weak assumption that an S-matrix exists between arbitrary initial and final asymptotic states, that is as the black hole evaporates, the entropy $\propto (GM^2/\hbar c)\cdot k$, decreases continously.

In order to preserve unitarity and coherence, residual mini black hole should not decay completely. Stable remnant should survive. Final state is not known but it is reasonable to assume from time reversal invariance that final state consists of particles not more exotic than the ones that made the black hole in the beginning.

There are essentially three possibilities:

a) the final state of evaporation may leave behind a naked singularity. This entails violation of cosmic censorship at quantum level [2]

b) black hole may evaporate completely leaving no residue giving rise to serious problem with quantum consistency described above; moreover by CPT theorem if there is no singularity initially, system must return to state with no singularity

c) stable remnant or residue of around Planck mass might remain; semiclassical back reaction and surface correction terms suggest that emission process might stop, but still better, and surely more physical, one may consider the possibility of a M_{Pl} black hole with spin ($\hbar$).

For a black hole with mass and spin of a = s/Mc, s being total spin, the black hole temperature is given by

$$T_{bh} = \frac{\hbar^2 (M^2 - a^2)^{1/2}}{32\,\pi\, K_B M\, [M + (M^2 - a^2)^{1/2}]} \qquad (1)$$

and can be zero for M = a.

For $M = M_{Pl}$, indeed, we have $GM_{Pl}/c^2 \simeq \hbar/M_{Pl}c$, i.e. M = a that is $T_{bh} = 0$ for Planck mass black holes provided they have a spin of $\sim \hbar$. So in that case black holes may indeed be stable! With the attainment of zero temperature such black holes would stop radiating and may be stable.

A natural way of understanding spin effects in gravitation is through torsion. The modification of the metric by inclusion of torsion can be expressed as:

$$g_{oo} = 1 - \frac{2GM}{Rc^2} + \frac{3G^2s^2}{2R^4c^4} \qquad (2)$$

The surface gravity of a black hole with torsion can be written as [3]:

$$\chi = \left(-\frac{GM}{R^2} + \frac{3G^2s^2}{2c^4R^5}\right)\cdot \text{const.} \qquad (3)$$

For a mass $M = (\hbar c/G)^{1/2}$ and radius $R = (\hbar G/c^3)^{1/2}$, the Planck length, zero surface gravity would correspond to

$$s^2 = (2/3)(c^4/G)(\hbar c/G)^{1/2}(\hbar G/c^3)^{3/2} = (2/3)\hbar^2 \qquad (4)$$

implying that for a spin $s \simeq \hbar$, for such a black hole, the surface gravity and hence the temperature vanishes. Thus in this case the torsion effects which enter with opposite sign, cancel those of gravity. This gives a minimum radius of $\sim (3/4)^{1/3} R_{Pl}$. Writing $s = \sigma R^3$, σ being the spin density, the torsion term in eq.(2) would correspond to that of an effective cosmological term of type $\Lambda_{eff} R^2$, the corresponding temperature being of the form $\left(\frac{\hbar c^3}{8\pi G K_B M} - \frac{\Lambda_{eff} M G \hbar}{K_B c}\right)$, vanishing for particular Λ_{eff}, i.e.for Schwarzschild-de Sitter metric. The entropy of this residual black hole of Planck mass and spin $\hbar$, would be $\sim K_B$, so that the entropy of these black holes is quantized in units of K_B, just as their spin is quantized in units of $\hbar$.

It is to be noted that only black holes with spin can transform into a massive string (also an object of negative specific heat) as massive strings inevitably have angular momentum proportional to M^2, the same as for black holes with maximum spins. Otherwise we would have violation of angular momentum conservation. In all such transition, entropy would change in discrete units of K_B. Now, if this were the case then all primordial black holes formed with mass $< (\hbar c^4/G^2 H_o)^{1/3}$ ($\sim 5\cdot 10^{14}$g., H_o being the Hubble's constant, quite insensitive to H_o going as $H_o^{1/3}$) would now be Planck mass remnants! For the scale invariant perturbations in the early universe the number density of black holes in the initial mass range (M, M + dM) (with $P = \gamma\rho$ equation of state) is $dn = (\beta - 2)\, \Omega_{bh} \rho_c M_{H_o}^{\beta-2} M^{-\beta} dM$ where ρ_c is the critical density and $\beta = (3\gamma-1)/(2+\gamma)$, $M_{H_o} = c^3 t$ the horizon mass, $\gamma = 1/3$ $\beta = 5/2$ in early universe. That means that the number density of Planck mass relicts left by evaporation of primordial holes is [4]:

$$n_{Pl} = (1/3)\Omega_{bh}\rho_c M_{H_o}^{1/2} M_{in}^{-3/2} \qquad (5)$$

where M_{in} is the mass of initially formed black holes.

$$n_P = \Omega_{Pl}\rho_c/M_{Pl} \simeq 1\cdot 10^{-25}\Omega_{Pl}(H_o/10^2)^2 \text{ cm}^{-3}. \qquad (6)$$

3. Black holes and strings

So we have seen that Planck mass black hole's may undergo a transition to a massive superstring. As string theory yields general relativity in low energy limit, this suggests that the Riemann geometry is embedded in a more general geometric structure. As a black hole is an excited state of gravitational field in general relativity, it can make a quantum transition to a new geometry. Again string theories unify all interactions supposedly at about Planck energies and that too in ten dimensions. So our understanding of the final state must incorporate these aspects. Fundamental object in string theories is one-dimension extended structure characterized by a string tension $\alpha = 1/M_{st}^2$ and interaction strength g. Their excitation spectrum includes an infinite tower of massive states with exponentially rising level starting at M_{Pl}!

Now as regards the string tension α we have some interesting implications in the context of strong gravity theory when considering also the dual models of hadrons [5]. In fact in the zero slope limit dual resonance models reduce to lagrangian field theories of the Yang-Mills or gravitational type.

In particular, the quantum theory of gravity (for e.g. the Gupta-Feynman theory) can be obtained as a zero-slope limit of the generalized Virasaro-shapiro model where the Regge intercept is fixed at two. One way to demonstrate such a correspondence involves computing the amplitude for graviton Compton scattering both in the Virasaro model as well as in linerized general relativity through Gupta-Feynman quantization [6]. The two expressions agree provided the identification

$$g^2 \alpha_R \equiv G \tag{7}$$

is made. Here g^2 is the strong interaction coupling (given as $g^2/\hbar c \simeq 1$), α_R is the universal Regge slope (related to the supestring tension μ) and G is the gravitational constant. So in general the corrispondence between dual resonance model and Yang-Mills-Einstein theory leads to such a relation between the gauge (strong) coupling constant and the gravitational constant through the Regge slope α_R. As $g^2 \approx \hbar c$, we see that this gives:

$$\alpha_R = G/\hbar c \tag{8}$$

and in order to agree with the observed slope of Regge trajectory (i.e. $\alpha_R = 1\ (\text{Gev})^{-2}$), it is necessary that G be the strong gravity coupling constant $G = G_f \simeq 6.7\cdot 10^{30}$ c.g.s.units.

The Regge trajectories of hadrons given by a relation of type $J(M) = \alpha_R M^2$, is consistent with the observed mass spectroscopy of a large number of hadronic resonance states lying on these trajectories. The above spectrum can also be interpreted as the spectrum of the rotational states of a string with the string tension μ related to the strong gravity constant by [7],[8]:

$$\alpha_R = c(2\pi\mu)^{-1}, \qquad \alpha_R \simeq G_f/\hbar c \tag{9}$$

We can also observe that dual models give an upper limiting temperature as related to α_R as: $T \propto (1/\alpha_R)^{1/2}$, the constant of proportionality being the Boltzmann constant K_B. The degeneracy of the Nth energy level of the string is given for large N by asymptotic formula P(N) = constantA(N) $\exp(M_N/T_H)$ where T_H is the Hagedorn limiting temperature; if we use for α_R the expression given by strong gravity we get [9]

$$T \simeq (1/K_B)(\hbar c^5/G)^{1/2} \simeq 10^{13}\ {}^{\circ}K \qquad (10)$$

which agrees with the Hagedorn temperature given by the dual model.

Entropy of a black hole of energy E is $S_{bh} \approx 4\pi E^2$ and entropy of a massive string mode of energy E is: $S_{string} = -a \ln E + bE$ with $a \simeq 10$ and $b \simeq \pi(2 + \sqrt{2})\alpha^{1/2}$ for heterotic string. All units are in Planck units.

In order that there is no information loss, true degeneracy of a black hole which has evaporated down to Planck mass is $\sim \exp(4\pi M^2) >>$ degeneracy of massive string modes of energy E given by $E^{-10}\exp(bE)$ where $E \simeq E_{Pl}$ so it is unlikely for a black hole to transform into massive string modewithout also radiation, but when its mass is less than γM_p, $S_b(E) < S_s(E)$, then the transition becomes highly probable.

Specific heat of massive string $C = -(1/T^2)(E^2/a)$ is negative! Specific heat of a black hole $dM/dT \propto 1/T^2$ is also negative. Gas consisting of massive superstring excitations behaves in many respects like a black hole, i.e. black hole and massive string excitations can never be in thermal equilibrium with an infinite heat reservoir. A body with negative specific heat can be in equilibrium with a finite heat bath. For a black hole the condition for a stable equilibrium is that energy of heat bath (i.e.radiation) be less than 1/4 mass of the black hole. The corresponding condition for equilibrium of massive excitations of heterotic string with massless modes of a radiation can be obtained.

In ten dimension space-time energy of massless gas of bosons and fermions is

$E = \sigma V T^{10}$; $\sigma = (8\pi^5/3465)\left[n_b + (1 - 1/2^9)n_f\right]$ where n_b is the number of bosons and n_f is the number of fermions. The most probable values of E_{st}, E_γ maximize $S_{st} + S_{rad}$ and this happens when the second derivative of total S is negative.

(for $n_b = n_f = 4032$, $\sigma = 6 \cdot 10^4$ and S is very high)

We can study the various phases of a three component system (blach holes, string and radiation) [10]. We have three distinct phases: (1) black hole and radiation, (2) string and radiation (3) pure radiation.

Total energy E in a volume V is:

Phase (1) $E_{bh+rad} = \sigma V T^4 + 1/(8\pi T)$

$$S_{bh+rad} = 4/3(\sigma V T^3) + 1/(16\pi T^2) \qquad (11)$$

Phase (2) $E_{st+rad} = \sigma V T^4 + aT/(bT-1)$

$$S_{st+rad} = 4/3(\sigma V T^3) + baT/(bT-1) - (1/T)(bT-1) \qquad (12)$$

Phase (3) $E_{rad} = \sigma V T^4$

$$S_{rad} = 4/3(\sigma V T^3) \qquad (13)$$

We have S_{bh+rad}, S_{st+rad} and S_{rad} as function of E and V. Given E and V the system with higher S would be preferred. There is a critical volume above which only the radiation can be in thermal equilibrium for both systems: black hole + radiation and massive string + radiation. For black holes + radiation, as is known, $V_c = 9 \cdot 2^{20} \sigma E^5/125$ and for strings + radiation $\sigma V_c = (E + 3a/2b - D)b^4(D - 5a/2b)^4/(D - 3a/2b)^4$ with $D = [4Ea/b + (3a/2b)^2]^{1/2}$. The two critical volume curves intersect at $E \approx 7\ E_{Pl}$ and $V \simeq 10^5$ in Planck units. Below this value of E, V_c for string plus radiation is higher than that for black hole + radiation. Thus a black hole bathed in radiation with total $E \approx 5\ M_{Pl}$ can, if the volume increases slowly, undergo a phase transition to enter a string and radiation phase. For further increase in volume, the massive strings will evaporate to pure radiation.

Thus in this thermodynamic process black hole has evaporated through its transition to a string phase and if the string theory is free of singularity (but this is not yet known also if conceivable) no singularity will be left. If superstring unification of gravity with other interactions takes place at $\approx M_{Pl}$, then it is necessary to include other degrees of freedom associated with massive string excitation to understand final stages of black hole evaporation.

4. Model for string gas

The above energy levels and corresponding degeneracies are confined to a space volume taken as 9 dimensions. Energy and entropy are correlated to volume and temperature like photon gas except that the space is 9-dimensional and the 2 polarization states of photons replaced by 8064 fold degeneracy of massless string (1/2 of the states are fermionic).

$$E_{st(9)} = 9Q\ V\ T^{10}_{st(9)} \qquad (14)$$

$$S_{st(9)} = 10Q\ V\ T^{9}_{st(9)}, \qquad Q = 8\ \pi^6/15$$

Gas temperature $\neq$ String temperature

For subsystem consisting of all massless strings and heaviest strings the energy is
Energy is

$E = M_N + E_g$ $\qquad M_N$ = mass of heaviest string. The dependence of entropy is:

$S(N,V,M) = S_N + S_g$ at fixed (V,M) or N or M_N/M

we have the cases:

a) $M > M_N$: system dominated by gas of masless strings

b) for increasing N, entropy increases: system dominated by one string

c) Entropy is not a monotonic function of N: in general both massive string and gas of massless strings in equilibrium are important. For $T_g < T_H$, massless strings dominate.

Even at $T_g = T_H$ massive strings contribute less than 0.3% to entropy. Gas must be in state for which

$M_{tot} < 9\,Q\,V\,T_H^{10}$. The result depends on

$\gamma = M_{Pl}\sqrt{\alpha}$. We choose $\gamma = 10$

Using mean square radius for bosonic string in 9 dimension for high N, is natural to assume the volume (V) larger than volume of 9-dimensional sphere of radius R(N), otherwise the string would not fit the radiation box.

This gives

$$M_{st} < 0.2\ V^{2/9}$$

$$M_{st}/M_{tot} < 7 \cdot 10^5\ V^{-7/9}$$

which is a negligible fraction of total energy of string gas. Possibility of the black hole going over directly into heavy string without massless strings has problems with conservation laws like angular momentum. Black hole nearing the end of is life usually is assumed to have zero angular momentum and heavy strings with $J = 0$ don't exist.

The most probable state of system of non-interacting heterotic strings in a box of volume V, may be a single string, or a heavy string in equilibrium with a gas of massless strings, or a gas of massless strings.

Each of these at low mass has an entropy which is higher than the entropy of a black hole having the same mass.

The d-dimensional Schwarzschild-de Sitter metrics:

$$ds^2 = -\left(1 - \frac{2\,G\,M\,F(d)}{\gamma^{d-3}} - \frac{\Lambda\,r^2}{d-1}\right) dt^2 + \left(1 - \frac{2\,G\,M\,F(d)}{\gamma^{d-3}} - \frac{\Lambda\,r^2}{d-1}\right)^{-1} dr^2 + r^2 d\Omega^2_{d-1} \qquad (15)$$

generalize to higher dimensions the standard asymptotic analysis of solutions to Einstein equations in 4 dimensions. Higher dimensional de Sitter space behaves as though it has intrinsic temperature

$$T = (2\pi)^{-1}\,[\,(d-1)/\Lambda\,]^{-1/2}. \qquad (16)$$

For given values of Λ, the number of space dimensions tends to decrease the de Sitter temperature.

For d-dimensional black holes:
entropy scales as $\propto M^{(d-2)/(d-3)}$ (that is as M^2 in 4-dimensions)
Schwarzschild radius scales as $\propto (2GM)^{1/(d-3)}$
and temperature scales as $\propto 1/M^{d-3}$.
(2GM and 1/M respectively for d = 4).

Effect of shadow state particles in $E_8 \times E_8'$ superstrings is to accelerate the decay of black holes by a factor of about 2 [12]. Compactified manifold with radius of curvature $\sim R_{Pl}/\alpha^{1/2} > R_{Pl}$ imply black holes with $R > R_{Pl}$ or $M > M_{Pl}$.

Considering higher dimensional black holes, the Schwarzschild solution for d-dimensions is given by (15). The temperature of higher dimensional black holes scales as

$T \simeq \dfrac{1}{M^{d-3}}$ that for d = 4 gives $T \simeq 1/M$.

Entropy scales as $S \propto M^{(d-2)/(d-3)}$ that for d = 4 gives $S \propto M^2$. Evaporation rate is $\propto M^{(4d-d^2-2)}$ that for d = 4 gives M^{-2}.

Lifetime is $\propto \int M^{-(4d-d^2-1)} dM \propto M^{d^2-4d+3}$. For d = 4, i.e. the usual four-dimensional case, the lifetime is given by:

$$t = \frac{G^2 M^3}{\hbar c^4}. \tag{17}$$

For the d-dimensional case, this generalizes to:

$$t_d = \frac{\hbar M^{(d^2-4d+3)} G^{(d^2-4d+4)/2}}{c^2 \hbar^{(d^2-4d+4)/2} c^{(d^2-4d+4)/2}} \tag{18}$$

$$= \frac{\hbar M^{(d^2-4d+3)}}{c^2} \left(\frac{G}{\hbar c}\right)^{(d^2-4d+4)/2}.$$

For d = 4, we set the usual formula. As $G/\hbar c \approx M_{Pl}^{-2}$, this can be written

$$t_d = \frac{\hbar M^{(d^2-4d+3)}}{c^2 M_{Pl}^{(d^2-4d+4)}} = \frac{\hbar}{M_{Pl} c^2}\left(M/M_{Pl}\right)^{(d^2-4d+3)} \tag{19}$$

Then for $M \approx 10^2 M_{Pl}$ black hole, the lifetime in d = 10 dimension, for instance, is >> age of the universe, so it is practically stable unlike in 4-dimensions. This raises the possibility of a stable remnant of the order of a few Planck masses of the evaporating black hole in higher dimensions, solving problems of unitarity and coherence.

The temperature of a 4-dimensional black hole is:

$$T \approx \frac{\hbar c^3}{8\pi G K_B M} . \qquad (20)$$

Correspondingly for a d-dimensional black hole it is:

$$T_d \approx \frac{\hbar^{d-3} c^{d-1}}{8\pi G^{d-3} K_B M^{d-3}}, \qquad (21)$$

thus much reducing the temperature for a black hole with $M >> M_{Pl}$ as compared to four dimensional case.

For $d = 10$ entropy $\propto M^{8/7}$, lifetime $\propto M^{63}$ and temperature $T \propto M^{-7}$. Thus temperatures are much lowered for higher dimensional black holes and life times much enlarged. Similarly for d-dimensional de Sitter space

$T \approx (2\pi)^{-1} [(d - 1)/\Lambda]^{-1/2}$ i.e. temperature is much lower for increase in d.

For d-dimensional black hole solutions of these models, we have that metric tensor has the form

$g_{oo} = 1 - 2GM/R^{(d-3)}$ and the Schwarzschild radius is $(2GM)^{1/(d-3)}$. Entropy of higher d black holes scales respect to M as $S \propto M^{(d-2)/(d-3)}$ that for $d = 4$ gives the usual $S \propto M^2$. Each Planck mass black hole has one unit of entropy. If we assume that in the earliest epoch the universe began with $\sim 10^{60}$ Planck mass black holes in d-dimensions which evaporated to give the observed 4-dimensional entropy, then the total entropy S of the universe of $\sim 10^{60}$ M_{Pl} black holes would scale in d-dimensions as $(M/M_{Pl})^{(d-2)/(d-3)} \sim (10^{60})^{(d-2)/(d-3)}$. This entropy when released by black hole evaporation cannot exceed $S_{microwave}$ in microwave background, i.e. $S_m \sim 10^{88}$. For $d = 6$ this gives $S \sim 10^{60 \cdot 4/3} \sim 10^{80}$; for $d = 5$ we have $\sim 10^{90}$ which is a few orders larger than that of the observed entropy S_m of the microwave background. For $d = 2$ no entropy release (an impossible situation) and for $d = 3$ $S \Rightarrow$ infinite!; $d = 7$ gives too small an entropy. So as far as the entropy release in the d-dimensional black holes arising from the compactified dimensional space in the early universe is concerned, $d = 6$ seems to be the optimum dimension of the compactified space.

The scenario we have here is as follows: we already have an expanding four-dimensional space-time with several compactified objects of mass $\sim M_{Pl}$ (strings, membranes etc.), in d compact dimensions which collapse to form black holes;

as when excited, these objects produce an energy spectrum of states with energies several times Planck mass. So the compactified objects collapse to form higher dimensional (d-dimensional) black holes which then evaporate to produce entropy, particles etc. This total entropy released on their evaporation must be comparable to that in the microwave background observed now. This would require d = 6 for the dimension of the compact space as other values of d , as seen above, are inconsistent with the observed entropy. Again as entropy of black hole is connected with area, this might have to do with geometrical property of area in d-dimensions which as we shall see below is maximized for d = 6. So the two pictures can be connected.

5. Ten dimension ?

We have in fact another argument [13] that also leads to d = 6: it is an 'a priori' argument that can privilege 10 space-time dimensions. 10 dimensions in all means 9 + 1 that is 9 spatial dimensions. As we live in a three dimensional space, six dimensions must be hidden from view, thereby leaving only the four familiar dimensions of space-time to be observed. The six extra spatial dimensions must be curled up to form a structure so small that it cannot directly be seen.

Now why six compact extra dimensions appear to be preferred? We can observe that when we calculate the area of an hypersurface, we find that (referring to a unitary radius) it is maximum when it is an esasurface [14]. We know that there is a general principle in physics for which every physical system tends to put itself in the state of minimum energy. More generally one uses the superlative in order to express in concise form a general principle which covers a great variety of phenomena. In this sense we say that a straight line is the "shortest" distance between to points (or in a non-euclidean, curved space-time we speake of the "geodesic"), on the sphere such a path is the great-circle route between two points: in this sense the statement that a physical system so acts that some function of its behaviour is least (or greatest) is the starting point for theoretical investigations.

The mathematical formulation of the superlatives is usually that the integral of some function has a smaller (or larger) value for the actual performance of the system. This is the case of the action integral and we are led to the variational method: certain integral has to be minimized or maximized; in other words we search for an 'extreme' value of this integral so that it has either a minimum or a maximum (or a point of inflexion). Usually we can tell from the physical situation which of these cases are true.

Now if we investigate the area of an hypersurface of a unitary radius we find out that it has a maximum value when it has six dimensions. To be more precise: let E_n be an n-dimensional space and S_{n-1} an hypersurface whose volume is V_n and surface area A_n. If we indicate V_2 as the area of the circle, V_3 the volume of the sphere and V_n the n-hypervolume we find:

TABLE 1

n	A_n	$A_n(1)$
2	$2\pi r$	6.28
3	$4\pi r^2$	12.56
4	$2\pi^2 r^3$	19.73
5	$8\pi^2 r^4/3$	26.31
6	$\pi^3 r^5$	31.00
7	$16\pi^3 r^6/15$	33.07
8	$2\pi^4 r^7/6$	32.46
9	$32\pi^4 r^8/105$	29.68
10	$2\pi^5 r^9/24$	25.50
⋮	⋮	
40	$2\pi^{20} r^{39}/19$	$1.44 \cdot 10^{-7}$

$$V_{2k} = \frac{\pi^k r^{2k}}{k!} \; ; \qquad V_{2k+1} = \frac{(2\pi)^{k+1} r^{2k+1}}{\pi(2k+1)!!} \qquad (22)$$

($n = 2k$ or $n = 2k + 1$)

where $k! = 1 \cdot 2 \cdot 3 \ldots\ldots k$ and $(2k+1)!! = 1 \cdot 3 \cdot 5 \cdot 7 \ldots\ldots (2k + 1)$, while if we indicate A_2 as the circumference of a circle, A_3 the area of a sphere surface and A_n the area of the S_{n-1} hypersurface in a n-space we find:

$$A_{2k} = \frac{2\pi^k r^{2k-1}}{(k-1)!} \; ; \qquad A_{2k+1} = \frac{(2\pi)^{2k+1} r^{2k}}{\pi\,(2k-1)!!} \qquad (23)$$

($n = 2k$ or $n = 2k + 1$).
For unitary radius, (see Tab.1), the maximum area is for hypersurface S_6 (embedded in a 7-dimensional space).
In this context the natural unit of measure is the Planck length.

6. Impossibility of deflation

As regards the vacuum dominance in a collapsing universe, for a Lorentz invariant and general covariant vacuum $T_{\mu\nu}$ we require at all temperatures $(\varepsilon + p)_{vac} = 0$, so vacuum domination is characterized by $p < 0$, that is by negative pressure. In early universe the vacuum energy dominates radiation energy being $\varepsilon + 3p < 0$, whenever temperature T_U falls below T_V, that is $\varepsilon_U(T_U) \leq \varepsilon_V(T_V)$. The period of the exponential expansion is driven by negative pressure, negative energy becoming larger with increasing volume, resulting in creation of positive energy paticles. Entropy multiplication took place (to present $\sim 10^{88}$) at reheating epoch when latent heat was released.
Vacuum dominated de Sitter phase is only a supercooled state of metastable equilibrium. Supercooled state becomes vacuum dominated as soon as radiation temperature falls below T during the expansion of early universe.

For contracting universe we would have the corresponding metastable superheated phase, which always has $p > 0$, which must be always radiation dominated as only for $p < 0$, vacuum can dominate. Thus in any contraction of the universe there can no deflation that is vacuum domination superheating leads to instability aiding collapse.

The mass of black holes accreting black body radiation in a collapsing universe, would diverge in a finite time; the rate of accretion is [15]:

$$dM(t)/dt \simeq 4\pi R_S^2(t)\rho(t)/c \; ; \quad R_S(t) = 2GM(t)/c^2 \; ; \qquad (24)$$
$$\rho(t) = aT^4(t) \; ; \quad R(t) \propto 1/T(t) \propto (t - t)^{1/2}.$$

Substituting, we see that M(t) diverges so does entropy;

black hole accretion causes instability in a collapsing universe.

Now superstring theories are valid at very early expansion; in late stage of collapse as the universe contracts there would be phase transition to a system of massive string excitations and we will have a corresponding rise in entropy. The gas consisting of massive superstring excitations behaves like a black hole with negative specific heat proportional to $1/T^2$. In ten dimension space-time of superstrings entropy density of a gas of bosons and fermions is

$$S = 10\,\sigma T^9 \; ; \; \sigma = (8\pi^5/3465)[n_b + (1 - 1/2^9)n_f]$$

(that is $(1 - 1/8)n_f = (7/8)n_f$ for 4-dimension space-time)

$n_b \sim n_f = 4032$ massless modes of heterotic strings

$\sigma = 6\cdot10^4$; $T \approx M_{Pl} \approx 10^{32}K$ and S is very high.

Also in Klein Kaluza theories, the (4 + d) dimensional scalar curvature R is the Lagrangian and the (4 + d) $g_{\mu\nu}(x)$ describes general relativity as well as gauge field in a compactified manifold with radius of curvature given by $R_c^2 = \alpha_g R^2_{Pl}$ where α_g is the gauge coupling constant. The entropy for n-dimension is $S \sim const/\hbar^n$.

In the d-dimensional space, Planck spectral density is of the form:

$$E=\sigma_d\, A\, T^{d+1} \qquad (25)$$

where
$$\sigma_d = \frac{2\pi^{(3d+1)/2}}{\Gamma[(d+1)/2]}\,\Gamma(d+1)\zeta(d+1)\,\frac{K_B^{d+1}}{h^d\, c^{d-1}}$$

is the d-dimensional Stefan-Boltzmann constant and $\zeta(x)$ the Riemann zeta function.

σ_3 is usual Stefan-Boltzmann constant for three space dimensions that is:

$$\sigma_3 = 2\pi^5 K_B^4 / 15\, h^3 c^2 = 5.67\cdot10^{-5} \text{ erg cm}^{-2} \text{ deg}^{-4}$$

$$\sigma_1 = \frac{\pi^2 K_B^2}{3h}, \quad \left(E_1 = \frac{\pi^2 K_B^{\,2} T^2}{6h}\right.$$ is the well-known case of one-dimensional thermal radiation i.e. Johnson noise or Nyquist noise in electrical networks!)

$$\sigma_9 = \frac{32\,\pi^{14}\,K_B^{10}}{99\, h^9\, c^8},$$

$$\sigma_{25} = \frac{10779541504\, \pi^{38}\, K_B^{26}}{1403325\, h^{25}\, c^{24}},$$ appropriate for 26-dimensional bosonic strings! etc.

Again the Schwarzschild metric at $\approx R_{Pl}$ may be modified by quantum corrections. So horizon is different. For m-loop terms, corrections are:

$$ds^2 = -(1 - \sum_n \alpha_n R_{Sch.}^n r^{-n} + \sum_n \sum_m \beta_{nm} R_{Sch.}^n R_{Pl}^{2m} r^{-(n+m)})dt^2 - \text{etc.}$$

For $r >> R_{Pl}$ one recovers the usual solution. These corrections would make the horizon fluctuate [12].

Again low energy limit of string theories give gravity with higher order curvature terms with dilatons. Here the corresponding Schwarzschild solution has smaller horizon. G.'t Hooft [16] has pointed out that black hole theory is related to string theory provided the string constant equal $T = 1/(8\pi G)$, with negative sign however! We point out the correspondence between action leading to equation for oscillations of black hole horizon and the string action used in describing Veneziano amplitude, provided T is identified as $1/(8\pi G)$. Thus end point of a black hole evaporating may be tied up with unificationo of interactions!

It turns out that string corrections reduce the black hole temperature [17] with

$T = T_{bh}(1 - F_D\, \alpha'^3/M^6]$, with F_D constant and α' the string slope expansion parameter. One sets a vanishing value for temperature at a particular value of mass of black hole. Gauss-Bonnett corrections (R^2 corrections) also have similar effects. Again back reaction effect of Hawking radiation on the Schwarzschild geometry in the presence of massless graviton, lowers the black hole temperature.

For a number N of weak boson fields, with a vacuum contribution due to polarization of $\rho = 41 \cdot N/7680\ \pi^2 M^4$, the black hole temperature approaches zero as mass $M \Rightarrow (41 \cdot N/240\ \pi)^{1/2}$ which implies that $T \Rightarrow 0$ for $M \simeq 5M_{Pl}$ in the case of superstring theory with N = 496. In short there are several ways by which the terminal stage of black hole evaporation may be indefinitely prolonged!

A mechanism for considerably prolonging the lifetime of the remnant state of mass $\sim M_{Pl}$ was suggested for instance by Carlitz [18] where the period of Hawking radiation is followed by much larger period during which remnant is radiated away, producing a pure state with unusual long range correlations.

There are also arguments against the decay: in fact if the remnant evaporates into N quanta, the average wave length of final N quanta is $\approx \lambda \approx (M_{Pl}/N)^{-1} \approx N\, R_{Pl}$ that is larger by a factor of N than the size of decaying system which is $\sim R_{Pl}$. The wave function overlap factor for each of the final quanta with initial size $\sim R^3_{Pl}$ is $\sim R^3_{Pl}/\lambda^3 \approx N^{-3}$. The rate for simultaneous emission of N quanta is suppressed by N^{-3N} and resulting system may hence be practically stable. So the number of primordial black holes may be adjusted so that remnants do not conflict with cosmological or astrophysical constraints.

For super-membranes or super-d-branes, the spin-mass relation is of the form

$$J = A\, T_d^{-1/d} (Mc)^{(d+1)/d} + B\hbar \qquad (26)$$

which reduces for d = 1 (the superstring) to the familiar $J = \alpha M^2 + \alpha_o$ formula. As eq. (26) is very different from that for a spinning black hole, (which has the same relation as for a superstring) the possibility of black holes going over to supermembranes is ruled out [19].

REFERENCES

[1] W.Hawking - Nature (London) 246, 30 (1974)

[2] N.Birrel and P.Davis - "Quantum fields in curved space" CUP, 1982

[3] V.de Sabbata, C.Sivaram and D.Wang - to appear in Annalen der Physik, 1989

[4] B.Carr - Ap.J.201, 1 (1975) and 206, 8 (1986)

[5] C.Sivaram and K.Sinha - Physics Reports 51,111 (1979)

[6] T.Yoneya - Progr.Theor.Phys.51, 1907 (1974)

[7] C.Sivaram - Nature 327, 108 (1987)

[8] V.de Sabbata and C.Sivaram - Nuovo Cimento 100A, 919 (1989)

[9] C.Sivaram - Am.J.Phys.51, 277 (1983)

[10] V.de Sabbata and P.Rizzati - Lett. Nuovo Cimento 20, 525 (1977)

[11] M. Bowick et al. - GRG 19, 113 (1987)

[12] C.Sivaram - paper presented at Goa ICGRC, 1987

[13] V.de Sabbata - Proc.Einstein Found.Int.4, 31 (1987)

[14] G.Arcidiacono - "Oltre la quarta dimensione" ed.il Fuoco, Roma, 1984

[15] S.Shapiro and S.Teukolsky - "Black holes, white dwarfs and neutron stars" Freeman Comp. 1962

[16] G.t'Hooft - Utrecht preprint, 1987

[17] C.O.Lousto and N.Sanchez - Phys.Letters B212, 411 (1988)

[18] R.D.Carlitz and R.S.Willey - Phys.Rev.D36, 2336 (1987)

[19] C.Sivaram - Gravity Research Found.Essay, 1988

INFLATION WITH MASSIVE SPIN-2 FIELD IN CURVED SPACE-TIME

C. Sivaram and Venzo de Sabbata

World Laboratory, Lausanne - Switzerland
Institute of Astrophysics, Bangalore - India
Dipartimento di Fisica, Università di Ferrara-Italy

1. Introduction

The hot big bang model of Cosmology is generally accepted as providing a correct description of the evolution of the universe. It naturally accounts for the isotropic cosmic microwave black-body background $3^{\circ}K$ radiation. Moreover it gives a good quantitative estimate of the amounts of light elements such as helium and deuterium synthesized several seconds after the expansion started. These estimates involve very little input of physics with no additional assumptions and agree quantitatively with the actually observed abundances of these light elements in stars and in interstellar matter. However when one extrapolate the model to early epochs one encounters rather puzzling aspects regarding the initial conditions. For instance we have to do with the expansion proceeding at a critical rate ($H^2_{crit} = 8\pi G\rho/3$) to a very high degree of precision at the early epochs implying that the universe was close to critical density (ρ_c) to very high degree of precision (i.e. to within one part in 10^{16} at epoch of nucleosynthesis and to one part in 10^{60} at the Planck epoch of $t \approx 10^{-43}$s! To make this more precise, we note that the observations indicate that the present value of $\Omega \equiv \rho/\rho_c$ (which measures ratio of energy density of universe to the critical energy density) though not known with great precision lies in the range $0.01 \leq \Omega \leq$ few units The luminous matter in the universe would indicate $\Omega \approx 0.1 \div 0.3$ Again from the uncertainties in the deceleration parameter defined as $q_o \equiv -(R/R)H^2 = \Omega/2$, one could restrict Ω to at most a few times unity. From the Robertson-Walker equation:

$$K/R^2H^2 = \rho/(3\pi^2/8\pi G) - 1 \qquad (1)$$

Quantum Mechanics in Curved Space-Time
Edited by J. Audretsch and V. de Sabbata
Plenum Press, New York, 1990

one can write Ω in a time dependent form:

$$\Omega = 1/(1 - y(t)) \qquad (2)$$

where $y(t) = (K/R^2)/(8\pi G\rho/3)$.
Ω is not constant but varies with time since $y(t) \propto R(t)^n$ ($n = 1$ for matter dominated universe and $n = 2$ for radiation domination). Equation (2) implies that at epoch of nucleosynthesis, value of y ($=y_N$) was $y_N \leq 10^{-16}$ which means $\Omega_N \simeq 1 + 0(\leq 10^{-16})$ and at Planck epoch $y_{Pl} \leq 10^{-60}$ so that consequently $\Omega_{Pl} = 1 + 0(10^{-60})$. If this ratio was not infinitesimally small at early epochs, the universe would have recollapsed long ago (for $K > 0$) or began a coasting phase ($K < 0$) with $R \propto t$. This extreme smallness of the ratio y if required as an initial condition is very strange, as in other words it would mean that the kinetic term $(\dot{R}/R)^2$ and the potential term $(8\pi G\rho/3)$ in the R-W equation balanced each other to arbitrarily high degree of precision (one part in 10^{60} at Planck epoch!) at early epochs. It is as if from very early epochs on, the ratio of curvature term to density term was extremely small (see eq.(2)), that is the universe began as extremely flat (with Ω arbitrarily close to one) which is a very special initial condition.

Another problem is the horizon problem. As is evident from the microwave background the universe on the largest scales is extremely homogeneous and isotropic (to better than one part in 10^4). However, as is known, standard cosmology has particle horizons. When matter and radiation last interacted vigourously (at $t \simeq 10^{13}$s, and Temperature $\simeq 1/3$ ev), what was to become the presently observable universe was comprised of $\simeq 10^6$ causally distinct regions. The particle horizon at decoupling only subtends an angle of about $(1/2)^\circ$ on the sky today; then how is that the microwave background temperature is so uniform on angular scales $\gg (1/2)^\circ$? At early epochs the number of causally distinct regions keeps increasing. For instance one second after the big bang the size of the universe currently observable was $\approx 10^{19}$cm. So there were about $(10^{14}/3\cdot 10^{10})^3 \sim 10^{27}$ causally distinct regions not communicating with each other. As the universe expanded at earliest epochs as $t^{1/2}$ whereas the horizon expands with light velocity as ct, the number of incommunicable regions $\sim t^{1/2}/ct \Rightarrow \infty$ as $t \Rightarrow 0$. With so many causally distinct regions in the early universe why is the present universe so homogeneous and isotropic all over?

Then we have the magnetic monopole problem. There should have been a glut of monopoles produced with densities several orders larger than the critical density at the Guts spontaneous symmetry breaking (GSSB) phase transition. Then why don't we see any monopoles?

The so called inflationary universe paradigm [1],[2] was invented to take care of the above problems confronting big bang cosmology at its earliest epochs. This invokes a vacuum dominated exponential expansion rate for the universe at an early phase with $H_{Infl} \simeq (8\pi V(0)/3M^2_{Pl})^{1/2} \approx M^2_G/M_{Pl}$, $V(0)$ assumed as $\approx M^4_G$, where M_G is the mass scale of scalar field which drives the expansion. While H_I is constant, R grows as exp(Ht). So a typical homogeneous region can expand physically by a factor of e^{100}, to encompass the whole of the observed universe, therefore taking care of the homogeneous or horizon problem, i.e. a single causally connected region can expand exponentially to give rise to the observed universe. Such an expansion also accounts for the curvature term becoming vanishingly small after inflation, i.e. the y term as defined in eq.(2) tends to zero after inflation and the inflationary scenario predicts a $\Omega \approx 1.0 \mp O(10^{-BIG})$. So that $\Omega \approx 1$ to a very high degree of precision. Again the inflationary expansion would have exponentially diluted away any large relic monopole density thus removing the monopole problem. The additional bonus is that quantum fluctuations of the scalar field [2] would give rise to scale-invariant perturbations which seem to be required to account for the formation of large scale hierarchy of structure in the universe. However the amplitude $(\delta\rho/\rho)_I$ of the fluctuations seems too large $\approx 10^2$ in most scenarios.

Of course one could have alternatives to the conventional inflationary scenarios requiring massive scalar fields with very 'flat' potential wells. One such alternative could be modification of general relativity at the Planck scale. The monopole and flatness problems can be solved by producing large amounts of entropy. Again if during an early epoch ($t \approx 10^{-43}$s), R, the scale factor, increased as rapidly as or more rapidly than t (for eg. $t^{1.2}$ or more) then d_H (horizon distance) $\Rightarrow \infty$, eliminating horizon problem. One such possibility of modification is to consider the Weyl type lagrangian for high energy gravity at the Planck scale [3]. This would be of type $L_W \sim \alpha C^2 + \beta R^2$, i.e. quadratic in the curvatures C and R with dimensionless constants α and β (appropriate for a renormalizable theory of gravity in contrast to the dimensional Newtonian constant for the Einstein non-renormalizable gravity). The field equations would be of fourth order i.e. of form $\alpha \ \nabla^4\Phi = km\delta^3(r)$ with a solution for the potential rising with r as Φ = const r = ar. The corresponding solution for the scale factor would be of type $R = at^2$ rather than the usual R = const $t^{1/2}$ type of solution. As R now increases faster than t, the horizon problem is eliminated. Again the flatness problem is also solved in this theory as quadratic curvature lagrangians of above type are known to have classical solutions with zero total energy, which means a K = 0 cosmological model, i.e. complete equality of kinetic and potential energy terms in the R-W expansion [3]. Similar situation holds for lagrangians with quadratic torsion terms, so it is possible to solve the flatness and horizon problems in the framework of such models [4]. Moreover if we consider lagrangians of the type

$$L \approx \gamma_o k^{-2} R + \alpha R_{\mu\nu} R^{\mu\nu} - \beta R^2 ,$$

their solutions are of type [5]

$$V = -\frac{k^2 M}{8\pi\gamma r} + \frac{k^2 M}{6\pi\gamma}\frac{e^{-m_2 r}}{r} - \frac{k^2 M}{42\pi\gamma}\frac{e^{-m_o r}}{r} ;$$

with $m_2 = \gamma^{1/2}(\alpha k^2)^{-1/2}$; $m_o = \gamma^{1/2}\left[2\ (3\beta - \alpha)k^2\right]^{-1/2}$,

i.e. their particle spectrum also contains massive tensor particles with mass m_2 and massive scalar particles with mass m_o. Massive scalar particles are contained in theories with lagrangians of type $L = k^{-2}R - \beta R^2$. These are precisely of the type used by Starobinsky [6] for inflationary models. In general these models are equivalent to those using massive scalar fields, there being a general transformation due to Whit [7] linking the two types of theories. Also the massive spin-2 field in the above relation for V, enters with an opposite sign (for energy) to that for the massive scalar field. This raises the possibility of having a zero energy momentum tensor for appropriate choice of constants with such a tensor naturally giving rise to a de Sitter solution. This will be elaborated in the next section. Again a lagrangian with non-minimal coupling $e^{\phi}R^2$ can be transformed to Einstein's theory with two scalar fields. We can also consider a general lagrangian L(R) with an arbitrary function of R [8]. Then scale invariant solutions g_{ij} give rise to a one-parameter family ($e^{2\alpha}g_{ij}$, α = const) of homothetically equivalent solutions. For lagrangian of type $L = R^m$, $m \neq 0$, the expanding solution $R(t) = t^2$ is an attractor solution for $L = R^{3/2}$ in the set of spatially flat Friedmann models. For arbitrary m we analogously have an attractor $R(t) \simeq t^n$, with $n = -(m-1)(2m-1)/(m-2)$; when $n \Rightarrow \infty$ and $m \Rightarrow 2$, this gives the usual attractor property of de Sitter space-time.

2. Massless and massive spin-2 fields in curved space-time

The massless spin-2 field can be described by a rank-4 tensor $\Phi_{\mu\nu\rho\sigma}$ which being a Weyl tensor [9], satisfies

$$\partial_\mu \Phi^{\mu\nu\rho\sigma} = 0 \qquad (1)$$

and

$$\partial_\lambda \Phi_{\mu\nu\rho\sigma} + \partial_\mu \Phi_{\nu\lambda\rho\sigma}\ \partial_\nu \Phi_{\lambda\mu\rho\sigma} = 0 \qquad (2)$$

If the potential $\psi_{\mu\nu} = \psi_{\nu\mu}$ is defined as

$$\Phi_{\rho\mu\nu} = \partial_\mu \psi_{\nu\rho} - \partial_\nu \psi_{\mu\rho} \qquad (3)$$

we have

$$\Box\psi_{\mu\nu} + \partial_{\mu\nu}\psi - \partial_{\nu\rho}\psi^{\rho}_{\mu} - \partial_{\mu\rho}\psi^{\rho}_{\nu} = 0 \qquad (4)$$

$(\psi = \psi^{\mu}_{\mu})$.

The gauge transformations

$$\psi_{\mu\nu} \Rightarrow \psi_{\mu\nu} - \partial_{\mu}\vartheta_{\nu} - \partial_{\nu}\vartheta_{\mu}$$

leaves $\Phi_{\mu\nu\rho\sigma}$ unchanged. The Lorentz gauge $\Box\vartheta_{\mu} = 0$, reduces to

$$\Box\psi_{\mu\nu} = 0 \ ; \qquad \partial_{\mu}\psi^{\mu\nu} - (1/2)\partial^{\nu}\psi = 0 \qquad (5)$$

The gauge conditions eliminate the spin-one and spin-zero components.

The appropriate lagrangianfor a massive spin-2 field in flat space-time is (mass m_2):

$$L(f_{\mu\nu}) = (1/2)f_{\mu\nu}\left[P^{\mu\nu\alpha\beta} + m_2^2(\eta^{\mu\nu}\eta^{\alpha\beta} \quad \eta^{\mu(\alpha}\eta^{\beta)\nu}\right]f_{\alpha\beta} \qquad (6)$$

where $P^{\mu\nu\alpha\beta}$ is the massless spin-2 inverse propagator [10]

$$P^{\mu\nu\alpha\beta} = (g^{\mu\nu}g^{\alpha\beta} - g^{\mu(\alpha}g^{\beta)\nu})\Box - g^{\mu\nu}\partial^{(\alpha}\partial^{\beta)}$$
$$- g^{\alpha\beta}\partial^{(\mu}\partial^{\nu)} + g^{\alpha(\mu}\partial^{\nu)}\partial^{\beta} + g^{\beta(\mu}\partial^{\nu)}\partial^{\alpha} . \qquad (7)$$

The field equation for $f_{\mu\nu}$ is

$$\left[P^{\mu\nu\alpha\beta} + m_2^2\ (\eta^{\mu\nu}\eta^{\alpha\beta} - \eta^{\mu(\alpha}\eta^{\beta)\nu})\right]f_{\alpha\beta} = 0 \qquad (8)$$

The trace gives

$$\left[2(\eta^{\alpha\beta}\Box - \partial^{\alpha}\partial^{\beta}) + 3\ m_2^2\ \eta^{\alpha\beta})\right]f_{\alpha\beta} = 0$$

The divergence is
$$m_2^2\ (\eta^{\alpha\beta}\partial^{\mu} - \eta^{\mu(\alpha}\partial^{\beta)})f_{\alpha\beta} = 0 \qquad (9)$$

An alternative form is (massive spin-2 field) [9]:

$$\Box f_{\mu\nu} - \partial_{\mu}\partial_{\gamma}f^{\gamma}_{\nu} - \partial_{\nu}\partial_{\gamma}f^{\gamma}_{\mu} + \partial_{\mu}\partial_{\nu}f + g_{\mu\nu}(\partial_{\gamma}\partial_{\rho}f^{\gamma\rho} - \Box f)$$
$$- m_2^2\ (f_{\mu\nu} - g_{\mu\nu}f) = 0 \qquad (10)$$

the divergence condition is: $\partial^{\gamma}f_{\gamma\mu} = \partial_{\mu}f$

and the trace conditions on f: $f = 0 \ ; \quad \partial^{\gamma}f_{\gamma\mu} = 0$

then (10) becomes

$$\Box f_{\mu\nu} - m_2^2\ f_{\mu\nu} = 0 \qquad (12)$$

These constraints can be generalized to curved space time as:

$$f_{\gamma\mu;\gamma} = 0 \, , \qquad f \quad 0 \qquad (13)$$

which reduce to $\partial^{\gamma} f_{\gamma\mu} = 0$ and $f = 0$ in flat space-time.

The most general covariant field equation for $f_{\mu\nu}$ in curved space-time can be written:

$$\Box f_{\mu\nu} - R_{\mu\gamma} f^{\gamma}_{\nu} - R_{\nu\gamma} f^{\gamma}_{\mu} + 2R_{\mu\gamma\nu\rho} f^{\gamma\rho} + H_{\mu\nu\gamma\rho} f^{\gamma\rho} = 0 \qquad (14)$$

We can try to write in the form

$$\Box f_{\mu\nu} + J_{\mu\nu\gamma\rho} f^{\gamma\rho} = 0 \qquad (15)$$

(as an appropriate generalization of eq.(12)), where we define (analogous to Weyl tensor for massless case above):

$$J_{\mu\nu\gamma\rho} = H_{\mu\nu\gamma\rho} - (1/2)(g_{\mu\gamma} R_{\nu\rho} + g_{\mu\rho} R_{\nu\gamma} + g_{\nu\gamma} R_{\mu\rho} + g_{\nu\rho} R_{\mu\gamma})$$
$$+ 2R_{\mu\nu\gamma\rho} \qquad (16)$$

Tensors J and H have the same symmetries. In fact:

$$\nabla_{\gamma} \nabla_{\mu} f^{\gamma}_{\nu} + \nabla_{\gamma} \nabla_{\nu} f^{\gamma}_{\mu} = R_{\mu\gamma} f^{\gamma}_{\nu} + R_{\nu\gamma} f^{\gamma}_{\mu} - 2R_{\mu\gamma\nu\rho} f^{\gamma\rho} \qquad (17)$$

imposed with corresponding constraints (in curved space):

$$\nabla^{\gamma} f_{\gamma\mu} = 0 \quad \text{or} \quad f_{\gamma\mu;\gamma} = 0 \qquad (18)$$

The divergence equation now read (in curved space, analogue to eq.(9)):

$$\Box \nabla^{\gamma} f_{\gamma\mu} - R_{\gamma\mu} \nabla^{\rho} f^{\gamma}_{\rho} + (\nabla_{\mu} R_{\gamma\rho} - \nabla_{\gamma} R_{\rho\mu} - \nabla_{\rho} R_{\gamma\mu} + \nabla^{\alpha} H_{\alpha\mu\gamma\rho}) f^{\gamma\rho}$$
$$+ H_{\mu\nu\gamma\rho} \nabla^{\nu} f^{\gamma\rho} \qquad (19)$$

Again we have the following postulated relations:

$$(\nabla_{\mu} R_{\gamma\rho} - \nabla_{\gamma} R_{\rho\mu} - \nabla_{\rho} R_{\gamma\mu} + \nabla^{\alpha} H_{\alpha\mu\gamma\rho}) f^{\gamma\rho} = G_{\mu} f \qquad (20)$$

$$H_{\mu\nu\gamma\rho} \nabla^{\nu} f^{\gamma\rho} = K_{\mu\varepsilon} \partial^{\varepsilon} f + 4I_{\mu\varepsilon} \nabla_{\gamma} f^{\gamma\varepsilon} \qquad (21)$$

where G_{μ} , $K_{\mu\varepsilon}$ and $I_{\mu\varepsilon}$ are functions of the background metric and derivatives of the metric. The symmetry of $H_{\mu\nu\gamma\rho}$ (i.e. $\gamma \Leftrightarrow \rho$ and also $\mu \Leftrightarrow \nu$) enables us to write (with (21)):

$$H_{\mu\nu\gamma\rho} = 2(I_{\mu\gamma} g_{\nu\rho} + I_{\mu\rho} g_{\nu\gamma}) + K_{\mu\nu} g_{\gamma\rho} \qquad (22)$$

$$I_{\mu\gamma} = (1/4) g_{\mu\gamma} I \quad \text{with} \quad I = I^{\alpha}_{\alpha} \qquad (23)$$

The trace should vaanish with respect to γ, ρ of bracketed expressions. Combinations of the above relations leads to:

$$\nabla_\mu R_{\gamma\rho} - \nabla_\gamma R_{\rho\mu} - \nabla_\rho R_{\gamma\mu} + \nabla^\alpha H_{\alpha\mu\gamma\rho} - (1/4) g_{\gamma\rho} g^{\beta\varepsilon} \nabla^\alpha H_{\alpha\mu\beta\varepsilon} = 0$$

and (24)

$$\nabla_\mu R_{\gamma\rho} - \nabla_\gamma R_{\rho\mu} - \nabla_\rho R_{\gamma\mu} + (1/2)[g_{\mu\gamma}\partial_\rho + g_{\mu\rho}\partial_\gamma - (1/2) g_{\gamma\rho}\partial_\mu] I = 0$$

Taking trace with respect to ρ and μ:

$$\partial_\gamma R = (9/4)\ \partial_\gamma I \qquad (25)$$

where

$$I \quad (4/9)\ R - A^2 \ , \ A^2 = \text{constant} \qquad (26)$$

and solving for $\nabla_\mu R_{\gamma\rho}$, from eq. (24), using eqs.(25),(26) we finally get:

$$\nabla_\mu R_{\gamma\rho} = (1/9)\left[(1/2)(g_{\mu\rho}\partial_\gamma + g_{\mu\gamma}\partial_\rho) + 2 g_{\gamma\rho}\partial_\mu\right] R \qquad (27)$$

The equation (14) for $f_{\mu\nu}$ is now in any arbitrary curved space:

$$\Box f_{\mu\nu} - R_{\mu\gamma} f^\gamma_\nu - R_{\nu\gamma} f^\gamma_\mu + 2 R_{\mu\nu\gamma\rho} f^{\gamma\rho} + \left[(4/9)R - A^2\right] f_{\mu\nu} = 0 \qquad (28)$$

with the trace equation

$$\Box f + \left[(4/9)R - A^2\right] f = 0 \qquad (29)$$

For a space of constant curvature (i.e. a de Sitter background) with:

$$R_{\mu\gamma\nu\rho} = \Lambda/3 (g_{\mu\nu} g_{\gamma\rho} - g_{\mu\rho} g_{\nu\gamma}) \qquad (30)$$

equation (28) becomes:

$$\Box f_{\mu\nu} - (2/3)\Lambda f_{\mu\nu} = 0 \quad \text{with} \quad A^2 = -2\Lambda/9 \qquad (31)$$

with trace equation $\qquad (\Box - 2\Lambda) f = 0 \qquad (32)$

The above is a generalization of the procedure in ref.[9] which was specialized for space of constant curvature, particularly for case of de Sitter space. In fact the equations for massive spin-2 field (with coupling K_f) when linearized on de Sitter background become the de Sitter covariant theory of massive spin-2 and spin-0 particles with

$(\Box - 2\Lambda/3) f_{\mu\nu} \sim -2K_f (T_{\mu\nu} - \eta_{\mu\nu} T/4)$ and $(\Box - 2\Lambda f) \sim 2K_f T$.
The opposite signs of the source terms in case of massive spin-2 and spin-0 [5],[9], enables the possibility of having a net $T_{\mu\nu}$ of zero, i.e. a vacuum dominated phase.

3. Inflationary solutions of above equations

We can generalize tha above Klein-Gordon equations with source terms by considering a perturbation $f'_{\mu\nu}$ of $f_{\mu\nu}$ and writing the equation of motion for $f'_{\mu\nu}$ as [10],[11]:

$$G^{\mu\nu\alpha\beta} f'_{\alpha\beta} = 0 \tag{33}$$

Following ref.[10],[11] we can write:

$$G^{\mu\nu\alpha\beta} = P^{\mu\nu\alpha\beta} + M^{\mu\nu\alpha\beta} \tag{34}$$

where $P^{\mu\nu\alpha\beta}$ is the massless spin-2 propagator (eq.7) and $M^{\mu\nu\alpha\beta}$ is an effective mass tensor defined as [10]:

$$M^{\mu\nu\alpha\beta} = 2K_f \Big[(g^{\mu\nu}T^{\alpha\beta} + g^{\alpha\beta}T^{\mu\nu})/4 - g^{\mu(\alpha}T^{\beta)\nu} - g^{\nu(\alpha}T^{\beta)\mu}$$

$$+ T^{\lambda}_{\lambda}\, g^{\mu(\alpha}g^{\beta)\nu}/2 - T^{\lambda}_{\lambda}\, g^{\mu\nu}g^{\alpha\beta}/4$$

$$- (1/2)\left[\delta T^{\mu\nu}/\delta g_{\alpha\beta} + \delta T^{\alpha\beta}/\delta g_{\mu\nu}\right]\Big]$$

so that the Jacobi equation for $f'_{\alpha\beta}$ can be written:

$$\int d^4x\sqrt{-g}\; f'_{\mu\nu}(P^{\mu\nu\alpha\beta} + M^{\mu\nu\alpha\beta})f'_{\alpha\beta} = 0 \tag{35}$$

with the usual conditions [11]:

$$\left.\begin{aligned} f_{\mu\nu}G^{\mu\nu\alpha\beta}f'_{\alpha\beta} &= 0 \\ \nabla_{\nu}G^{\mu\nu\alpha\beta}f'_{\alpha\beta} &= 0 \end{aligned}\right\} \tag{36}$$

and

$$u_{\nu}G^{\mu\nu\alpha\beta}f'_{\alpha\beta} = 0$$

where u_ν is an arbitrary four vector;
with the harmonic coordinate condition [9],[11]:

$$\nabla_{\mu}f'^{\mu\nu} = (1/2)\nabla^{\nu}f'^{\mu} \tag{37}$$

the constraints becoming

$$f^{\alpha\beta}f'_{\alpha\beta} = 0, \quad \nabla^{\alpha}f'_{\alpha\beta} = 0\;; \quad u^{\alpha}f'_{\alpha\beta} = 0 \tag{38}$$

we have

$$P^{\mu\nu\alpha\beta}f'_{\alpha\beta} = -\,g^{\mu(\alpha}g^{\beta)\nu}(\Box - 2\Lambda)f'_{\alpha\beta} \tag{39}$$

$$M^{\mu\nu\alpha\beta}f'_{\alpha\beta} = -\,g^{\mu(\alpha}g^{\beta)\nu}K_f(\rho + p)f'_{\alpha\beta} \tag{40}$$

giving the effective Klein-Gordon equation for $f'_{\alpha\beta}$ as:

$$\left[\Box - 2\Lambda + K_f(\rho + p) + \ldots\right]f'_{\alpha\beta} = 0 \tag{41}$$

with an effective mass $M_f^2 = -K_f(\rho + p) - 2\Lambda$.
For a vanishing mass tensor (possible for appropriate combination of spin-2 and spin-0 massive fields), the equation (40) implies a vanishing $(\rho + p)$ or a negative

pressure corresponding to a de Sitter type situation with exponential expansion.

An explicit exponentially expanding solution of the massive spin-2 field equations in a R-W background can also be obtained, Consider the metric:

$$ds^2 = -dt^2 + R^2(t)(dx^2 + dy^2 + dz^2) \qquad (42)$$

For this metric we have:

$$R^{o}_{\alpha o \beta} = R\ddot{R}\,\delta_{\alpha\beta} \qquad (43)$$

$$R^{\alpha}_{\beta\gamma\delta} = \dot{R}^2 (\delta^{\alpha}_{\gamma}\,\delta_{\beta\delta} - \delta^{\alpha}_{\beta}\,\delta_{\delta\gamma}) \qquad (44)$$

$$R_{oo} = -3\,\ddot{R}/R \qquad (45)$$

$$R_{\alpha\beta} = (R\ddot{R} + 2\dot{R}^2)\delta_{\alpha\beta} \qquad (46)$$

$$R = 6\left[(\dot{R}/R)^2 + \ddot{R}/R\right]$$

Substituting above relations in eq.(27) we have:

$$\frac{d}{dt}\left[2(\dot{R}/R)^2 - \ddot{R}/R\right] = 0 \qquad (47)$$

using D = lnR

$$\frac{d}{dt}(\ddot{D} - \dot{D}^2) = 0 \qquad (48)$$

$$\ddot{D} - \dot{D}^2 = -m_f^2\,a^2 \qquad (49)$$

$$D = \pm m_f \;; \qquad m_f > 0 \ \text{(positive mass)}$$

or

$$R = R_o \exp(a\,m_f t) \qquad (50)$$

a = const. and $a\,m_f$ has dimension of inverse time.

The duration of the inflation corresponds to the decay of the massive spin-2 particles, or the oscillation time to massless spin-2 particle, i.e., as was pointed out in ref.[12], every spin-2 particle produced in interaction is to be regarded as a combination of massive and massless states corresponding to the eigenvalues of the mass matrix. The mass-mixing term in curved space can be written [13]:

$$L_m = -g^{1/2}(m_2^2/4k_1^2)(f^{\mu\nu} - g^{\mu\nu})(f^{\alpha\beta} - g^{\alpha\beta})(g_{\mu\alpha}g_{\nu\beta} - g_{\mu\nu}\,g_{\alpha\beta}) \qquad (51)$$

where as before $f^{\mu\nu} = \eta^{\mu\nu} + k_1 f'^{\mu\nu}$ (k_1 being the coupling constant for massive field) and $g_{\mu\nu} = k_2 h_{\mu\nu}$.

Up to quadratic terms, L_m becomes:

$$L_m = (m_2^2/4k_1^2)(k_1 f'^{\mu\nu} - k_2 h^{\mu\nu})(k_1 f'^{\alpha\beta} - k_2 h^{\alpha\beta}) \cdot \cdot(\eta_{\mu\alpha}\eta_{\nu\beta} - \eta_{\mu\nu}\eta_{\alpha\beta}) \qquad (52)$$

Introducing 2-component vectors [12]:

$$\psi_{\mu\nu} = \begin{bmatrix} f'_{\mu\nu} \\ h_{\mu\nu} \end{bmatrix}, \quad \psi^T_{\mu\nu} = \begin{bmatrix} f'_{\mu\nu}, & h_{\mu\nu} \end{bmatrix},$$

the mixing term may be written as:

$$L_m = \psi^T_{\mu\nu} H \psi^{\mu\nu} - \psi^T H \psi, \quad \psi = \psi^{\mu}_{\mu}$$

$$H = H^T \quad \text{is} \quad H = \frac{m_2^2}{4k_1^2} \begin{bmatrix} k_1^2 & -k_1 k_2 \\ -k_1 k_2 & k_2^2 \end{bmatrix} \qquad (53)$$

which is not diagonal with respect to $f'_{\mu\nu}$ and$h_{\mu\nu}$ i.e. eigenvalue of H are $m_2^2/4\ (1 + k_2^2/k_1^2)$ and 0 with eigenstates $\bar{f}'_{\mu\nu}$, $\bar{h}_{\mu\nu}$ related to $f'_{\mu\nu}$, $h_{\mu\nu}$ by a rotation angle θ (mixing angle)

$$\bar{f}_{\mu\nu} = \cos\theta\, f'_{\mu\nu} - \sin\theta\, h_{\mu\nu}$$

$$h'_{\mu\nu} = \sin\theta\, f_{\mu\nu} + \cos\theta\, h_{\mu\nu}$$

$$\cos\theta = k_1/\sqrt{k_1^2+k_2^2}, \qquad \sin\theta = k_2/\sqrt{k_1^2+k_2^2}$$

Thus the time evolution of massive spin-2 particles in the early universe is described by:

$$|f'_{\mu\nu}(t)\rangle = \cos\theta \exp(-iE_1 t)\exp[-(\Gamma/2)t](\cos\theta|f'_{\mu\nu}\rangle - \sin\theta|h_{\mu\nu}\rangle$$

$$+ \sin\theta \exp(-iE_2 t)(\sin\theta|f'_{\mu\nu}\rangle + \cos\theta|h_{\mu\nu}\rangle \qquad (54)$$

$$E_1 = (|p_1|^2 + m_2^2)^{1/2}, \qquad E_2 = |p_2|$$

p_1, p_2 are momenta of the massive and massless spin-2 particles respectively. The exponential damping of the state $|f'_{\mu\nu}\rangle$ in time is due to decay of massive tensor field with mean life $\tau = \Gamma^{-1}$. Massless gravitons have infinite lifetime and its amplitude does not decay. If at t = 0 we have purely massive spin-2 particles at any later time the number of massless spin-2 particles would be:

$$|\langle h_{\mu\nu}|f'_{\mu\nu}(t)\rangle|^2 = \frac{k_1^2 k_2^2}{(k_1^2+k_2^2)^2}[1 + \exp(-\Gamma t) - 2\exp(-\Gamma t/2)\cos\Delta E t \qquad (55)$$

$\Delta E = E_1 - E_2$. For $t \gg \tau$, we have only massless spin-2 particles with a total intensity reduced by factor $k_1^2 k_2^2/(k_1^2+k_2^2)^2$ with respect to initial intensity of massive spin-2 particles.

We did not consider the possibility of decay to a massive scalar, i.e. we assumed the spin-2 massive field to have only five degrees of freedom. For the above type of mass lagrangian we can introduce Steuckelberg [14] fields to constrain the spin-2 field to have only five degrees of freedom. If we make the substitution:

$$f_{\mu\nu} \Rightarrow f_{\mu\nu} - \partial_\mu A_\nu - \partial_\nu A_\mu \equiv K_{\mu\nu}$$

the mass term is replaced by

$$f_{\mu\nu} \Rightarrow -(1/2)m_2^2[(f_{\mu\nu} - \partial_\mu A_\nu - \partial_\nu A_\mu)(h^{\mu\nu} + \partial^\mu A^\nu + \partial^\nu A^\mu)$$

$$- (h^\mu_\mu - 2\partial_\mu A^\mu)^2]$$

This is now gauge invariant under $f_{\mu\nu} \Rightarrow f_{\mu\nu} + \partial_\mu \zeta_\nu + \partial_\nu \zeta_\mu$ provided A_μ transforms as : $A_\mu \Rightarrow A_\mu - \zeta_\mu$ giving the field equations $\partial^\mu K_{\mu\nu} - \partial_\nu K^\mu_\mu = 0$, which gives four of the required conditions on $f_{\mu\nu}$ when going to the gauge $A_\mu = 0$. By requiring a new invariance under which only A_μ transforms as $A_\mu \Rightarrow A_\mu - \partial_\mu \zeta$, the variation of the mass term above is then $\partial L_m = 2m_2^2(\partial^\mu\partial^\nu f_{\mu\nu} - \partial^\mu\partial_\mu f^\nu_\nu)\zeta$ + surface terms. Terms that are quadratic in A_μ (in L_m) take the form $-(m_2^2/2)F^{\mu\nu}F_{\mu\nu}$, $F_{\mu\nu}$ is the usual spin-1 field strength. The right side of δL_m equals $2m_2^2\ R^{(\ell)}\zeta$, where $R^{(\ell)}$ is the linearized Ricci scalar. The variation δL_m may therefore be cancelled by adding the term $m^2\phi\ R^{(\ell)}$, where ϕ transforms as $\delta\phi = -2\zeta$. Field equation for ϕ is $R^{(\ell)} = 0$ and this together with trace of the $f_{\mu\nu}$ equation gives the required constraint $f^\mu_\mu = 0$ which eliminates the massive scalar.

4. Conclusions

What is the initial source of the massive spin-2 particles in the early universe? To answer this, it is to be noted that in most models of space compactification in higher dimensional unified theories such as Klein-Kaluza theories, one obtains consistent lower-dimensional theories with infinite towers of massive spin-2 particles interacting with gravity. For instance the five-dimensional metric $g_{AB}(x.y)$ may be expanded in Fourier modes [15]:

$$g_{AB}(x,y) = \sum_n g_{(n)AB}(x)\ e^{iny} ,$$

x denotes the co-ordinates of four-dimensional space-time and y is a co-ordinate on the circle with period 2π. The five dimensional general co-ordinate transformation parameters $\zeta^A(x,y)$ may similarly be expanded in Fourier series giving rise to an infinite number of four dimensional gauge symmetries $\zeta^A(x,y) = \sum_n \zeta^A_{(n)}(x)\ e^{iny}$. The n = 0 term in above sum, describes gravity and a massless spin-one field. Each term with $n \neq 0$, describes a massive spin-2 field with mass ~ n/R_c, R_c = compactification radius. The vector and scalar parts of $g_{(n)AB}(x)$, when $n \neq 0$ will be absorbed by Higgs mechanism, as all but the $n = 0$ symmetries are spontaneously broken. These are the A and ϕ fields discussed before. Again in superstrings theories one also obtains infinite towers of massive spin-2 fields and also higher spins and masses that increase indefinitely. So in the early universe when all the fundamental forces were unified one had a description in terms of superstring or Klein-Kaluza type of framework. As the universe expanded, the internal space became compactified generating the infinite towers of massive spin-2 particles. So if one had inflation induced by the massive spin-2 fields in the curved R-W space as described above, one gets a natural way of diluting away all the indefinitely large spectrum of higher spins and masses to very low values, so that they do not contribute much to the present background density. Without such a mechanism, all these indefinitely large remnants (of masses and spins) of compactification would have created very serious problems for cosmological observations! Many of the massive spin-2 particles could have larger masses than M_{Pl}, so that they would have formed miniblackholes of spin 2. The evaporation time of a 10 M_{Pl} miniblackhole would be ~ $10^3\ t_{Pl}$ so that their decay over this time scale would give an expansion factor of ~ e^{1000} (cf.eq.50), which is more than sufficient inflation.

Again the evaporation of the several miniblackholes, would generate sufficient amount of entropy during the inflationary phase. As shown in the other paper of these proceedings [16] the evaporation of ~ 10^{60} blackholes of masses ~ M_{Pl} would generate an entropy comparable to that seen in the microwave background. The generation of such a large amount of entropy in a time scale of a few times t_{Pl} would naturally resolve the flatness and horizon problems.

Moreover the evaporation of these blackholes would be most likely to violate CP invariance and also baryon number, as we know in any case that baryon number is not conserved in black hole decay or collapse. A small violation of CP of ~ 10^{-9} in such decays is sufficient to produce the observed baryon asymmetry. In other words the evaporation of these miniblackholes in the early universe is capable not only of producing the observed entropy but also the observed net baryon number.

The present zero value of the cosmological constant can also be understood in the above picture. The effective cosmological constant driving the inflation was in this model related to the mass of the massive spin-2 particles generated in the early universe, i.e. $\Lambda \approx m_2^2 c^2/\hbar^2$ (see eqs.31,32,41 and 50). So when the massive particles decay to massless spin-2 particles (eqs.51-55), Λ tends to drop to zero ($m_2 \Rightarrow 0$) and the inflation stops. So the end of the inflationary phase and the vanishing of the effective cosmological constant are both smoothly connected in this picture. In the more conventional models these are difficult questions to resolve. In short the inflationary phase that can be induced by the coupling of massive spin-2 particles to curved space-time may resolve several difficulties associated with early universe cosmology and particle physics.

REFERENCES

[1] A.Guth - Phys.Rev.D23, 347 (1981)
[2] A.Linde - Phys.Lett. 116B, 335 (1982)
[3] C.Sivaram - Bull.Astr.Soc.Ind. 13, 339(1985)
Gravity Found.Essay (1985)
Astrophys.Spc.Sci. 125, 111 (1986)
[4] V.de Sabbata and C.Sivaram - Int.Journ.Theor Phys. (in preparation) (1989)
[5] K.S.Stelle - Gen Rel.Grav. 9, 353 (1978)
[6] A.Starobinsky - Phys.Lett. 117B, 175 (1982)
[7] B.Whit - Phys.Lett.145B, 176 (1984)
[8] H.J.Schmidt - Cl.Q.G.6, 557 (1989)
[9] E.A.Lord, K.P.Sinha and C.Sivaram - Prog.Theor.Phys. 52, 161 (1974)
[10] D.Gross, M.Perry and L.Yaffe - Phys.Rev. D25, 330 (1982)
[11] B.Holstein et al.- Ann.Phys.N.Y. 190, 149 (1989)
[12] V.de Sabbata and M.Gasperini - Gen.Rel.Grav. 18, 669 (1986)
[13] C.Isham, A.Salam and J.Strathdee - Phys.Rev.D3, 867 (1971)
[14] C.Aragone and S.Deser - Nuovo Cimento 57B, 33 (1980) and 3A, 709 (1971)
[15] A.Salam and J.Strathdee - Ann.Phys.N.Y. 141, 316 (1982)
[16] V.de Sabbata and C.Sivaram - these proceedings.

RENORMALIZATION OF FIELD THEORIES IN RIEMANN-CARTAN SPACE-TIME

P.I. Pronin

Moscow State University

Moscow, USSR

INTRODUCTION

Riemann-Cartan geometry with curvature and torsion arises naturally within the framework of Poincaré gauge theory of gravity. The simplest example is given by the Einstein-Cartan theory (Kibble, 1961; Sciama, 1962; Trautman, 1973; Hehl et al., 1976) in which the coupling of spin and torsion is realised in a degenerate algebraic manner. More general models are based on the Yang-Mills type Lagrangians, quadratic both in torsion and curvature. Classical dynamics of the Poincaré gravitational fields is at present intensively studied (some bibliography can be found in books by Ivanenko et al.,(1985) and Ponomariev et al.(1985)).

However, one can say very little about quantum dynamics of the Poincaré gauge theory of gravity. In fact there only few works in which the quantization of the gravitational gauge field is considered. Renormalization problems are discussed in (Tseytlin, 1982; Yan, 1983; Obukhov and Nazarovsky 1984) while in (Gvozdev and Pronin, 1985) an attempt is made to understand the space-time torsion as a quantum collective phenomenon, arising from the interactions of quantised spinor matter. Stochastic quantization method has been applied to the Einstein-Cartan theory and some restrictions on the gauge transformations were discussed (Baulieu, 1986; Ivanenko et al., 1988). The problem of unitarity for gravity theories with dynamical torsion is treated on the tree level by

Quantum Mechanics in Curved Space-Time
Edited by J. Audretsch and V. de Sabbata
Plenum Press, New York, 1990

Neville(I98I), Sezgin and Nieuwenhuizen (I980), Sezgin (I98I) Kunfuss and Nitsch (I986). Partially the lack of progress here is connected both with enormous technical difficulties, arising in quantum models with torsion and with the problems of the gravitational Lagrangian choice. There are more than I50 functionally independent terms constructed from the curvature and torsion tensors, which have the canonical dimension sm^{-4} and can be included in the action functional. Clearly, that investigation of the physical consequences of such a general model is a difficult problem. Hence it is reasonable to study at first non-selfconsistent problems, in which only matter fields are quantised while the geometrical gravitational background is treated clasically. Previously this class of problems has been throughly discussed for the case of purely Riemannian geometry without torsion. The obtained results (for review see the books by Birrell and Davis (I982) Grib et al.(I980)) provided important progress in the theory of black holes, astrophysics and cosmology, as well as in the quantum theory of fields.

The case of the Riemann-Cartan geometry is currently under investigation. The problem of particle creation in spacetimes with torsion has been considered by Rumpf (I979) for fermions and by Ponomariev and Pronin (I979) for bosons. In the latter case the anisotropisation effect in cosmological models with torsion and scalar particles is predicted. Vacuum polarisation effects and renormalisation problems are discussed in (Pronin, I984; Denardo and Spallucci, I987). Recently some general calculational methods for covariantly quantised fields in the Riemann-Cartan spacetime have been developed (Goldthrope, I980; Kimura, I98I; Nieh and Yan, I982; Obukhov, I983; Barth, I987). These techniques enable one to compute gravitational anomalies and counterterms for arbitrary material field in space with torsion.

Present lectures are devoted to the discussion of renormalization of fields theories in the Riemann-Cartan spacetime.

According to the common opinion the investigation of renormalization properties of field models on an arbitrary geometrical background is the first and essential step on the way of constructing a quantum gravitational theory. It is very important to know if such theories are renormalizable in curved spacetime as well as in flat one. Some time ago the theorem on the renormalizability of models in curved spacetime has been formulated (Freedan et al., 1974). However, its general proof is absent and so far the attempts to find the examples of theories in which this theorem does not hold are made (Gass, 1986). Therefore, we belive that the proposal and investigation of new models will be usefull and will help us to solve not only the renormalization tasks but also the setting of problems concerned with the origin of quantum theory in general.

In the past decade numerous papers were devoted to the study of renormalization problems of interacting fields in curved spacetime. The issue of renormalizability has been addressed for several situations, including: massless self-interacting scalar fields on an arbitrary Riemannian manifold (Bunch and Parker, 1979; Bunch, 1981; Birrell, 1980; Toms, 1982), conformal invariant quantum electrodynamics in de Sitter spacetime (Drummond and Shore, 1979), QED in an arbitrary curved spacetime with the use of local momentum space representation (Panangaden, 1981), quantum theory of non-abelian gauge fields in the Riemannian manifold (Toms, 1983). The method of calculation of gauge invariant effective action in one- and two-loop approximation was proposed and applied by several authors (Omote and Ichinose, 1983; Jack, 1984; Jack and Osborn, 1985) to the interacting scalar, spinor and vector fields. Some progress was achieved in the use of renormalization group technique to solve the set of problems (Nelson and Panangaden, 1982; Toms, 1983b) and to investigate scaling behaviour of non-Abelian models on non-Euclidean manifolds (Leen, 1983).

The situation becomes more complicated in the theories with non-minimal interaction terms of material and gravitational fields (Aragone and Deser, 1980; Benn et al., 1980; Bukhdal, 1962; 1982). The well known example is the selfinteracting scalar field. We will see that it is not unique example in non-Einsteinian gravitational theories, because the minimal interaction between gravity and matter fields has been found not sufficient to garantee the renormalizability of field theories in the Riemann-Cartan spacetime (Obukhov and Pronin, 1987; 1988; Buchbinder et al., 1985). Besides, it is very essential that non-minimal terms appear not only in scalar sector but also in fermion sector due to quantum corrections. Moreover some difficulties in renormalization of Abelian models off-shell were revealed (Obukhov and Pronin, 1987).

The least investigated aspect is the interaction of electromagnetic and Yang-Mills fields with torsion. Today there is no unanimous ansver to the question whether the fields interact with torsion or not. The direct application of the minimal coupling principle, i.e. the replacement of partial derivatives by covariant ones with respect to the Riemann-Cartan connection in the gauge field strength tensor, gives such interaction. These terms destroy the gauge invariance in the usual sence. But previous works indicated the way to overcome these difficulties and to keep the generalised gauge invariance of the action functionals of gauge fields (Hojman et al., 1978; Mukky and Sayed, 1979). The method is not simple mathematical tool but predicts new physical effects, particularly, the short range modification of Coulomb potential and the dependence of the electromagnetic charge value on the geometrical structure of spacetime (De Sabbata and Gasperini 1980; 1981). These results hint at intriguing linkages of external and internal symmetries of fundamental physical interactions.

Some other authors consider the interaction of gauge fields with torsion as a defect of kinematical origin of the minimal coupling principle since in their opinion these terms call the renormalizability in question (Trautman, 1973; Hehl 1974; Novello, 1976).

The aim of my lectures is to provide a brief summary of the main results concerning the problems of renormalization of interacting fields in the Riemann-Cartan spacetime and examine the one-loop renormalization of Abelian and non-Abelian models in the details.

The lectures are organized as follows. We briefly discuss the geometry of the Riemannian-Cartan manifold and classical material fields on its (Section I).

In Section 2 the relevant expression for the one-loop effective action obtained via the background field method is given.

The problems of renormalizability of the Abelian gauge

model with non-minimal torsion coupling will be investigated in the third section.

The structure of divergeces of the one-loop effective action of general non-Abelian theory is obtained and two particular cases are discussed in Section 4.

We propose a new treatment of modified gauge transformations in the Yang-Mills fields interacting with torsion and discussed both the meaning of the link equation of torsion and "effective constant" and problem of renormalization in the fifth section.

The last section contains the summary of results concerning the interaction quantum fields with the Riemann-Cartan geometry.

I. CLASSICAL FIELDS IN RIEMANN-CARTAN SPACE-TIME

In this section we give our notations and conventions and briefly discusse the interaction of material fields with the Riemann-Cartan geometry.

I.I The Riemann-Cartan space-time

The Riemann-Cartan spacetime U_4 is assumed to be the four-dimensional smooth compact manifold without the boundary, which is supplied with a pseudo-Riemannian metric $g_{\mu\nu}(x)$ with the signature (+I, -I, -I, -I) and with a (world) affine connection $\tilde{\Gamma}^{\lambda}{}_{\mu\nu}$. The latter is compartible with the metric, i.e. $\tilde{\nabla}_{\mu} g_{\alpha\beta} = 0$, but is in general non-symmetric. The skew-symmetric part is the torsion

$$Q^{\alpha}{}_{\mu\nu} = \frac{1}{2}\left(\tilde{\Gamma}^{\alpha}_{\mu\nu} - \tilde{\Gamma}^{\alpha}{}_{\nu\mu}\right). \qquad (I.I)$$

It is always possible to split the Riemann-Cartan connection into (torsion independent) Riemannian connection and the contorsion tensor

$$\tilde{\Gamma}^{\alpha}_{\mu\nu} = \{^{\alpha}{}_{\mu\nu}\} + K^{\alpha}{}_{\mu\nu}, \quad K^{\alpha}_{\mu\nu} = Q^{\alpha}_{\mu\nu} + Q_{\mu\nu}{}^{\alpha} + Q_{\nu\mu}{}^{\alpha}. \qquad (I.2)$$

Here the first term denotes the standard Christoffel symbol

$$\{^{\alpha}{}_{\mu\nu}\} = \frac{1}{2} g^{\alpha\lambda}\left(g_{\lambda\mu,\nu} + g_{\lambda\nu,\mu} - g_{\mu\nu,\lambda}\right). \qquad (I.3)$$

As we will discuss the models of interacting fields including scalar, spinor, vector etc., we have to assume a spinor structure on U_4 so as to be able to define spinor fields globally on it. To define spinor fields on U_4 let us introduce a tangent space at any point of U_4 and let $\vec{e}=\{\vec{e}_k\}$, (k=0,I,2,3) be a Lorentz basis of the tangent space and $\vec{E}=\{\vec{E}_\mu\}=\{\frac{\partial}{\partial x^\mu}\}$($\mu$ =0,I,2,3) be a co-ordinate basis on the tangent space. Then the metric assigns at any point an inner production for any pairs of vectors belonging to the tangent space.

The orthonormal vierbeins consists of those I6 functions $h^k{}_\mu(x)$, which relate the co-ordinate basis to the Lorentz basis and

$$\vec{e}_k = h_k{}^\mu(x)\,\vec{E}_\mu\,, \tag{I.4a}$$

and

$$\vec{E}_\mu = h^k{}_\mu(x)\,\vec{e}_k\,. \tag{I.4b}$$

Combaining (I.4a) and (I.4b) we find the next relations

$$h^a{}_\mu h_b{}^\mu = \delta^a_b\,, \quad h^a{}_\nu h_a{}^\alpha = \delta^\alpha{}_\nu\,,$$
$$g_{\mu\nu}(x) = h^a{}_\mu h^b{}_\nu \eta_{ab}\,, \quad \eta_{ab} = h_a{}^\mu h_b{}^\nu g_{\mu\nu}\,. \tag{I.5}$$

To define the covariant derivative of the material fields of any tensorial dimension we have to introduce the local Lorentz connection $\omega^a{}_{b\mu}$. The explicit form of the last can be obtained from the following expression

$$\tilde{\nabla}_\mu h^a{}_\alpha \equiv \partial_\mu h^a{}_\alpha + \omega^a{}_{b\mu} h^b{}_\alpha - \tilde{\Gamma}^\lambda{}_{\alpha\mu} h^a{}_\lambda = 0\,. \tag{I.6}$$

Inserting the metric-affine connection (I.2) into (I.6) and taking into account the (I.5) we find

$$\omega^a{}_{b\mu} = \Delta^a{}_{b\mu} + h^a{}_\alpha h_b{}^\beta \left(Q^\alpha{}_{\beta\mu} + Q_{\beta\mu}{}^\alpha + Q_{\mu\beta}{}^\alpha\right), \tag{I.7}$$

where Δ_{abc} is the Ricci rotation coefficients defined by

$$\Delta_{abc} = \tfrac{1}{2}\left(\lambda_{abc} - \lambda_{bac} - \lambda_{cab}\right), \quad \lambda_{abc} = 2\, h_a{}^{[\alpha} h_b{}^{\beta]} h_{c\beta,\alpha} \tag{I.8}$$

If one displace a vector parallelly along the closed circle we can obtained the expression for the curvature tensor

$$\tilde{R}^\alpha{}_{\mu\beta\nu} = \tilde{\Gamma}^\alpha{}_{\mu\nu,\beta} - \tilde{\Gamma}^\alpha{}_{\mu\beta,\nu} + \tilde{\Gamma}^\lambda{}_{\mu\nu}\tilde{\Gamma}^\alpha{}_{\lambda\beta} - \tilde{\Gamma}^\lambda{}_{\mu\beta}\tilde{\Gamma}^\alpha{}_{\lambda\nu}\,. \tag{I.9}$$

This tensor can be split into the Riemannian curvature and torsion contributions

$$\tilde{R}^\alpha{}_{\mu\beta\nu} = R^\alpha{}_{\mu\beta\nu} + \nabla_\beta K^\alpha{}_{\mu\nu} - \nabla_\nu K^\alpha{}_{\mu\beta} + K^\lambda{}_{\mu\nu}K^\alpha{}_{\lambda\beta} - K^\lambda{}_{\mu\beta}K^\alpha{}_{\lambda\nu} \tag{I.IO}$$

Hereinafter

$$\nabla_\nu V^\alpha \equiv V^\alpha{}_{;\nu} = \partial_\nu V^\alpha + \left\{{}^{\alpha}_{\lambda\nu}\right\} V^\lambda\,. \tag{I.II}$$

It is well known fact that curvature and torsion can be decomposed into the irreducible parts. For example, decomposition of torsion into irreducible parts reads

$$Q^{\alpha}{}_{\mu\nu} = \bar{Q}^{\alpha}{}_{\mu\nu} + \frac{1}{3}(\delta^{\alpha}_{\nu} Q_{\mu} - \delta^{\alpha}_{\mu} Q_{\nu}) + \epsilon^{\alpha}{}_{\mu\nu\beta} \check{Q}^{\beta} \quad (I.I2)$$

where $Q_{\mu} = Q^{\alpha}_{\mu\alpha}$, is the torsion trace, $\check{Q}_{\mu} = \frac{1}{6} \epsilon_{\mu\nu\alpha\lambda} Q^{\nu\alpha\lambda}$ is the so-called pseudotrace, and $\bar{Q}^{\alpha}{}_{\mu\nu}$ is the traceless and pseudotraceless reducible tensor $\bar{Q}^{\alpha}{}_{\nu\alpha} = 0$, $\epsilon^{\alpha\mu\nu\lambda}\bar{Q}_{\mu\nu\lambda} = 0$ which can be decomposed into self-dual (antiself-dual) parts.

I.2 The minimal coupling.principle

To introduce the minimal interaction of material fields with the Riemann-Cartan geometry it is necessary to replace the partial derivatives in the special relativistic material action functional with covariant ones, i.e. $\partial_{\mu} \to \tilde{\nabla}_{\mu}$ and to use the covariant volume $d\tau(x) = \sqrt{-g}\, d^4x$.

For any field $\varphi^A(x)$ which is transformed according to the irreducible (A,B) representation of Lorentz group

$$\varphi^A(x) = S^A{}_B\, \varphi^B(x)\ ,\ S = exp(\varepsilon^{ab} \Sigma_{ab}),$$

where $(\Sigma_{ab})^i_j$ $(a, b = 0,1,2,3)$ are Lorentz generators in the (A,B) representation and i,j,.. ranging over I to D=(2A + I)(2B + I), the covariant derivative take the following form

$$\tilde{\nabla}_{\mu} \varphi^A = (\partial_{\mu} + \frac{i}{2} \omega^{ab}{}_{\mu} \Sigma_{ab})^A{}_B\, \varphi^B. \quad (I.I3)$$

The generators obey the equations

$$[\Sigma_{ab}, \Sigma_{cd}] = i(\eta_{ac}\Sigma_{bd} + \eta_{bd}\Sigma_{ac} - \eta_{ad}\Sigma_{bc} - \eta_{bc}\Sigma_{ad}) \quad (I.I4)$$

There are some relevant examples. For a scalar field Σ is vanishing and there is no minimal interaction.

For the case of a Dirac spinor field ψ , Σ_{ab} is given in the terms of the Dirac gamma matrices as

$$\Sigma_{ab} = \frac{i}{4}[\gamma_a, \gamma_b]\ ,\quad \{\gamma_a, \gamma_b\} = 2\eta_{ab}. \quad (I.I5)$$

The covariant derivative for an adjont Dirac spinor field $\bar{\psi} = \psi^{+}\gamma_0$ is given by

$$\tilde{\nabla}_{\mu} \bar{\psi} = \bar{\psi}(\partial_{\mu} - \frac{i}{4} \omega^{ab}{}_{\mu} \Sigma_{ab}). \quad (I.I6)$$

For a vector field V^{μ} , Σ_{ab} is a 4×4 matrix whose element are given by

$$(\Sigma^{ab})_{ij} = i(\delta^a_i \delta^b_j - \delta^b_i \delta^a_j). \qquad \text{(I.17)}$$

According to this prescription the vector fields (Maxwell, Proca, Yang-Mills) are interacted with torsion and it can lead to various inconsistentcies, which will be considered in the next sections.

I.3 Interacting fields in U_4

The physical systems we are interested in are the systems of scalar, spinor and vector fields. First we will start with the Lagrangians of these free fields in the U_4.

Applying the minimal principle to the Lagrangian of spinor field in the Minkowski space

$$L_{(1/2)} = \frac{i}{2}(\bar{\psi}\gamma^k \partial_k \psi - \partial_k \bar{\psi}\gamma^k \psi) - m\bar{\psi}\psi, \qquad \text{(I.18)}$$

we obtain the following Lagrangian

$$L_{(1/2)} = \frac{i}{2} h_a{}^\mu (\bar{\psi}\gamma^a \tilde{\nabla}_\mu \psi - \hat{\nabla}_\mu \bar{\psi}\gamma^a \psi) - m\bar{\psi}\psi. \qquad \text{(I.19)}$$

We can reduce the $L_{1/2}$ after some algebra to the next form

$$L_{(1/2)} = \bar{\psi}\{h_a{}^\mu \gamma^a (\nabla_\mu + \gamma_5 \check{Q}_\mu) - m\}\psi. \qquad \text{(I.20)}$$

According to the minimal principle the Maxwell tensor $F_{\mu\nu} = \partial_\mu A_\nu - \partial_\nu A_\mu$, defined by the four-potential A_ν, must be replaced by

$$\tilde{F}_{\mu\nu} = \hat{\nabla}_\mu A_\nu - \hat{\nabla}_\nu A_\mu = F_{\mu\nu} + 2Q^\alpha{}_{\mu\nu} A_\alpha. \qquad \text{(I.21)}$$

For the massless vector field the last term in (I.21) evidently breaks down the gauge invariance of the theory, and hence A_μ can no longer be identified with the electromagnetic potential. The classical dynamics of this field has been considered by Ponomariev and Smetanin (1978) within the framework of the Einstein-Cartan theory. Thus obtained model possesses some features of nonlinear electrodynamics, for example it predicts approximately Maxwell-like behavior of $\tilde{F}_{\mu\nu}$ at large distaces, but modifys the vector field near the source in such a way that the classical field energy of a point source becomes finite. However the quantum version of this theory faces many serious difficulties.

For the case of massive vector field the last term in (I.21) seems harmless, since there is no gauge invariance. However one can show that tha wave propagation of the vector field in U_4 is in general acausal. Acausal anomalies disappear only when the torsion is represented by the trace or pseudotrace. On the other hand, in the first order description of higher spin fields in U_4 the propagation of waves is causal, but instead one encounters (for spin I and higher) algebraic inconsistencies, which eliminate torsion (Obukhov, 1983b).

We are intersted in the quantum properties of vector fields interacting with torsion. The vacuum polarization of the Maxwell field has been investigated recently (Obukhov and Pronin, 1988) in the proposal of special case of torsion, namely $Q^{\alpha}{}_{\mu\nu} = \frac{1}{3}(\delta^{\alpha}{}_{\nu} Q_{\mu} - \delta^{\alpha}{}_{\mu} Q_{\nu})$. The one-loop divergences were calculated and the effective gravitational action was constructed. But I consider that this picture can be generalized and in the fifth section I will consider the problem of renormalization of the Yang-Mills fields interacting with torsion.

Due to the minimal principle a scalar field does not interact with torsion and curvature. But it was log ago recognised, that the minimal coupling recipe is not sufficient to preserve renormalizability. This is clearly seen for those models, which include interacting scalar fields φ : it is well known an additional non-minimal coupling term of the form $\xi R \varphi^2$ is necessarily required.

The same is expected for scalars in the Riemannian-Cartan spacetime with torsion. For example, the simple analysis for the quantum theory of interacting scalar φ and spinor fields ψ in U_4 with the Lagrangian

$$L = \frac{1}{2} g^{\mu\nu} \partial_{\mu}\varphi \partial_{\nu}\varphi + \frac{1}{2}\xi R \varphi^2 + \frac{\lambda}{4!}\varphi^4 + h\varphi\bar{\psi}\psi + i\bar{\psi}\gamma^{\mu}(\nabla_{\mu} + \alpha\gamma_5 \check{Q}_{\mu})\psi, \quad \lambda, h, \xi, \alpha = const, \tag{I.22}$$

shows that the effective Lagrangian, determined by loop contributions, containes divergences of type $\check{Q}_{\mu}\check{Q}^{\mu}\varphi^2$. Hence the latter non-minimal torsion coupling term should be included in the bare action, in order to achieve renormalizability in the Riemann-Cartan space-time.

As compared to the purely Riemannian case, the number of different non-minimal torsion interaction terms (which could naturally arise in various models) is rather great. Therefore it would be usefull to study the general problem: which types of non-minimal torsion coupling terms do not disturb renormalizability of given flat space theory, when non-minimal coupling includes for scalar and spinors respectively

$$\frac{1}{2}\Lambda(R, Q^2)\varphi^2,$$

$$\Lambda = \xi_0 R + \xi_1 Q_\mu Q^\mu + \xi_2 \check{Q}_\mu \check{Q}^\mu + \xi_3 \bar{Q}^\alpha{}_{\mu\nu} \bar{Q}_\alpha{}^{\mu\nu}, \qquad (I.23)$$

$$\alpha Q_\mu \bar{\psi}\gamma^\mu\psi + \beta \check{Q}_\mu \bar{\psi}\gamma^\mu\gamma_5\psi, \qquad (I.24)$$

where $\xi_0, \ldots, \xi_3$, α, β are the coupling constants.

II. EVALUATION OF ONE-LOOP FUNCTIONAL INTEGRAL

In covariantly quantised theory the fundamental vacuum-to-vacuum amplitude is given by the functional integral

$$Z = \int [dq]\, e^{iS[g_{\mu\nu}, Q^\alpha{}_{\mu\nu}, \hat{\varphi}+q]}, \qquad (2.I)$$

where the classical action S is assumed to be a functional of the background gravitational fields $g_{\mu\nu}$ and $Q^\alpha{}_{\mu\nu}$, and material fields $\varphi = \hat{\varphi} + q$ (we suppress all tensor and spinor indices). The latter, in accordance with the background field method (DeWitt, 1965) is the sum of classical $\hat{\varphi}$ and quantum q parts, and S is expanded in powers of q ; If S is invariant under the action of a local group, the integral (2.I) becomes undetermined. Then adds to S an appropriate gauge-breaking term, and takes into account the Faddeev-Popov ghost contribution. Then the one-loop effective action

$$\Gamma_{(1)} = -i \ln Z_{(1)}, \qquad (2.2)$$

is determined by the gaussian integral

$$Z_{(1)} = \int [dq]\, e^{i\frac{1}{2} q \Delta q} (\det \Delta_{gh}), \qquad (2.3)$$

where $\Delta = \delta^2(S + S_{gh})/\delta q^2$ is the operator for small disturbances (DeWitt, 1965) and Δ_{gh} is the usual Faddeev-Popov ghost operator.

For the real bosonic fields q the integral (2.3) gives

$$\Gamma_{(1)} = \frac{i}{2}\{\ln\det\Delta - 2\ln\det\Delta_{gh}\}. \qquad (2.4)$$

The purely fermionic case is also well-known.

However, as concerns the general boson-fermion systems, only recently the correct method of computation has been proposed (Lee and Rim, 1985). In brief the idea is very simple. Let the matter fields q be arranged into a column, the upper part of which are boson variables, while the rest - fermion ones. Then the operator Δ takes the form of supermatrix

$$\Delta = \left(\begin{array}{c|c} \Delta_{(B)} & 2N \\ \hline 2L & 2\Delta_{(F)} \end{array}\right), \qquad (2.5)$$

diagonal elements of which are respectively the boson-boson operator $\Delta_{(B)}$ and the fermion-fermion operator $\Delta_{(F)}$. Usually the former is the second-order differential operator (Laplace or D'Alembert like one) while the latter is the first-order (Dirac) operator. The integral (2.3) for (2.5) can be easily calculated with the help of usual Beresin rules, to give

$$Z_{(1)} = \{\det(\Delta_{(B)} - 2N\Delta_{(F)}^{-1}L\}^{-1/2} (\det\Delta_{(F)})(\det\Delta_{gh}). \qquad (2.6)$$

Since it is rather difficult (though in general not impossible) to deal with the first determinant, which depends on the fermion propagator $\Delta_{(F)}^{-1}$, one proceeds as follows (Lee and Rim, 1985). With the help of conjugate operator one defines the fermionic determinant to be

$$\det\Delta_{(F)} = (\det\Delta_{(F)}\Delta_{(F)}^{*})^{1/2},$$

and then, combining the first two factors in (2.6), one recognises in them the superdeterminant of the supermatrix

$$\Delta_{(S)} = \left(\begin{array}{c|c} \Delta_{(B)} & 2N\Delta^{*}_{(F)} \\ \hline L & \Delta_{(F)}\Delta^{*}_{(F)} \end{array} \right), \qquad (2.7)$$

which is now the second-order differential operator both in boson and fermion sectors. Hence

$$Z_{(1)} = (S\det\Delta_{(S)})^{-1/2}(\det\Delta_{gh}), \qquad (2.8)$$

and the structure of divergences of the effective action (2.2) is determined by infinities of determinant (2.7)

$$\Gamma_{(1)} = \frac{i}{2}\{\ln S\det\Delta_{(S)} - 2\ln\det\Delta_{gh}\}. \qquad (2.9)$$

It can be shown (Lee and Rim, 1985) that divergences of (2.9) may be obtained by mens of the heat kernel method (Gilkey, 1975), if one succeeds in finding supermatrices $\mathcal{D}_\mu$ and X, such that the former is linear in first (covariant) derivatives, the latter does not contain derivatives, and

$$\Delta_{(S)} = -(\mathcal{D}_\mu\mathcal{D}^\mu + X). \qquad (2.10)$$

Then divergences of $\Gamma_{(1)}$ are determined by the Minakshisundaram (Seeley - DeWitt) coefficients of asymptotic expansion of the heat kernel for the operator (2.10), and in dimensional regularization

$$\Gamma^{\infty}_{(1)} = \frac{1}{16\pi^2(n-4)}\{B_4(\Delta_{(S)}) - 2B_4(\Delta_{gh})\}, \qquad (2.11)$$

where the B_4-coefficient for an operator of the form (2.10) is given by

$$B_4(\Delta_{(S)}) = \int d^4x\sqrt{-g}\Big\{\frac{1}{180}(R_{\alpha\beta\mu\nu}R^{\alpha\beta\mu\nu} - R_{\mu\nu}R^{\mu\nu})\,\mathrm{Str}\,\mathbb{1} +$$

$$+ \frac{1}{2}\mathrm{Str}\Big(X + \frac{R}{6}\Big)^2 + \frac{1}{12}\mathrm{Str}([\mathcal{D}_\mu,\mathcal{D}_\nu][\mathcal{D}^\mu,\mathcal{D}^\nu])\Big\}. \qquad (2.12)$$

It is easy to show that in the Riemann-Cartan spacetime U_4 the same technique is applicable to an arbitrary field theory: one should only split the original Riemann-Cartan differential operators into purely Riemannian part and the torsion dependent part. This always can be done with the help of (I.2). Now let us apply the developed technique to the some models.

III. RENORMALIZATION OF ABELIAN MODEL

In this section we investigate the one-loop renormalizability of an abelian gauge model with spinors and scalars, and non-minimal interactions (I.23-24). This is the spinor electrodynamics coupled through a Yukawa type interaction with a real nonlinear scalar field. Its Lagrangian in U_4 reads

$$L = -\frac{1}{4}F_{\mu\nu}F^{\mu\nu} + i\bar{\psi}\gamma^{\mu}\nabla_{\mu}\psi + eA_{\mu}\bar{\psi}\gamma^{\mu}\psi - m\bar{\psi}\psi + \\ + \frac{1}{2}\partial_{\mu}\varphi\partial^{\mu}\varphi - \frac{1}{2}M^2\varphi^2 - \frac{f}{4!}\varphi^4 + h\varphi\bar{\psi}\psi + \\ + \alpha Q_{\mu}\bar{\psi}\gamma^{\mu}\psi + \beta\bar{\psi}\gamma^{\mu}\gamma_5\psi\check{Q}_{\mu} + \frac{1}{2}\Lambda(R, Q^2)\varphi^2. \qquad (3.I)$$

Here m and M are respectively masses of spinor ψ and scalar fields φ , e , h and f are the coupling constants. Electromagnetic field A_{μ} do not interact with torsion , hence as usually $F_{\mu\nu} = \partial_{\mu}A_{\nu} - \partial_{\nu}A_{\mu}$.

In the framework of the background field method we deccompose all fields into classical and quantum parts

$$A_{\mu} = \hat{A}_{\mu} + a_{\mu}, \quad \varphi = \hat{\varphi} + \phi, \quad \psi = \hat{\psi} + \eta, \quad \bar{\psi} = \hat{\bar{\psi}} + \bar{\eta}. \qquad (3.2)$$

The over cups denote classical fields, however in what follows we will drop them, since only background parts enter the formulas, given below. The one-loop effective action is determined by the quadratic term in the expansion of (3.I) in powers of quantum parts. We compute $\Gamma_{(1)}^{\infty}$, using the algorithm by Lee and Rim (I985), which we outlined in the previous section. Detailes of calculation are given in our work (Obukhov and Pronin, I988), and here we present only the form of supermatrices $\mathcal{D}_{\mu}$ and X , which determine superoperator (2.IO) for the Lagrangian (3.I). They are as follows, with the upper left, middle and lower right bloks refering respectively to electromagnetic field, scalars and spinors

$$\mathcal{D}_\mu = \left(\begin{array}{c|c|c} g^{\alpha\beta}\nabla_\mu & 0 & ie\bar{\psi}\gamma^\alpha\gamma_\mu \\ \hline 0 & \overset{\circ}{\nabla}_\mu & ih\bar{\psi}\gamma_\mu \\ \hline 0 & 0 & D_\mu \end{array}\right) \tag{3.3}$$

$$X = \left(\begin{array}{c|c|c} -R^{\alpha\beta} & 0 & -e(i\overset{1/2}{\nabla}_\mu\bar{\psi}\gamma^\alpha\gamma^\mu+\bar{\psi}\gamma^\alpha N) \\ \hline 0 & M^2-\Lambda+\frac{f}{2}\varphi^2 & h(i\overset{1/2}{\nabla}_\mu\bar{\psi}\gamma^\mu+\bar{\psi}N) \\ \hline -e\gamma^\beta\psi & -h\psi & Z \end{array}\right) \tag{3.4}$$

where

$$N=-eA_\mu\gamma^\mu-\alpha Q_\mu\gamma^\mu+\beta\check{Q}_\mu\gamma^\mu\gamma_5+2h\varphi-2m, \tag{3.4}$$

$$Z = m^2-\frac{R}{4}-i\sigma^{\mu\nu}\left(\frac{e}{2}F_{\mu\nu}+\alpha\nabla_\mu Q_\nu\right)+3h^2\varphi^2-$$

$$-i\beta\gamma_5\nabla_\mu\check{Q}^\mu+4\beta h\varphi\check{Q}_\mu\gamma^\mu\gamma_5-2\beta^2\check{Q}_\alpha\check{Q}^\alpha, \tag{3.6}$$

$$D_\mu = \overset{1/2}{\nabla}_\mu-ieA_\mu+i\alpha Q_\mu-i\beta\sigma_{\mu\nu}\gamma_5\check{Q}^\nu-ih\varphi\gamma_\mu, \tag{3.7}$$

and $\overset{1/2}{\nabla}_\mu$, $\overset{\circ}{\nabla}_\mu$ and ∇_μ denote covariant derivatives respectively for spinors, scalars and vectors.

The ghost operator is much simpler: $\Delta_{gh} = -\overset{\circ}{\nabla}_\mu\overset{\circ}{\nabla}^\mu$.

After calculation of all necessary traces of X , X^2 and $[\mathcal{D}_\mu,\mathcal{D}_\nu]$, $\mathbb{1}$, and inserting them into (2.I2) we finally obtain

$$\Gamma^{\infty}_{(1)} = \frac{1}{16\pi^2(n-4)}\int d^4x\sqrt{-g}\Big\{-\frac{17}{360}R_{\mu\nu\alpha\beta}R^{\mu\nu\alpha\beta}+\frac{91}{180}R_{\mu\nu}R^{\mu\nu}-$$

$$-\frac{5}{36}R^2+\frac{1}{2}\Lambda^2-\frac{1}{6}R\Lambda-\Lambda M^2+R\left(\frac{M^2}{6}+\frac{m^2}{3}\right)+\frac{M^4}{2}-2m^4+$$

$$+8m^2\beta^2\check{Q}_\mu\check{Q}^\mu-\frac{2}{3}\beta^2(\nabla_\mu\check{Q}_\nu-\nabla_\nu\check{Q}_\mu)(\nabla^\mu\check{Q}^\nu-\nabla^\nu\check{Q}^\mu)-$$

$$-\frac{2}{3}\alpha^2(\nabla_\mu Q_\nu-\nabla_\nu Q_\mu)(\nabla^\mu Q^\nu-\nabla^\nu Q^\mu)-\frac{8}{3}\alpha e F^{\mu\nu}\nabla_\mu Q_\nu-$$
$$-\frac{2}{3}e^2F_{\mu\nu}F^{\mu\nu}+2h^2\partial_\mu\varphi\partial^\mu\varphi+\left(\frac{fM^2}{2}-12m^2h^2\right)\varphi^2+$$
$$+R\varphi^2\left(\frac{f}{12}+\frac{h^2}{3}\right)+8\beta^2h^2\check{Q}_\mu\check{Q}^\mu\varphi^2-\frac{1}{2}\Lambda f\varphi^2+$$
$$+\left(\frac{f^2}{8}-2h^4\right)\varphi^4+(2e^2+h^2)\left(i\bar{\psi}\gamma^\mu\overset{1/2}{\nabla}_\mu\psi+eA_\mu\bar{\psi}\gamma^\mu\psi+\right.$$
$$\left.+\alpha Q_\mu\bar{\psi}\gamma^\mu\psi\right)+(2e^2-h^2)\beta\check{Q}_\mu\bar{\psi}\gamma^\mu\gamma_5\psi+$$
$$+(8e^2-2h^2)(h\varphi\bar{\psi}\psi-m\bar{\psi}\psi)\}. \tag{3.8}$$

Let us briefly analysis this result. The purely gravitational part is very complicated and today it seems impossible to give some reasonable explanation (in the framework of the Poincaré gauge theory) to this sum of curvature and torsion quadratic terms. As concerns material interacting fields contribution, one notice an unpleasant term $F^{\mu\nu}\nabla_\mu Q_\nu$. In its absence the theory would be indeed renormalizable, similarly to the case of flat space. This term disappears only on the mass shell, when the classical Maxwell equations for the electromagnetic field $\nabla_\mu F^{\mu\nu}=-e\bar{\psi}\gamma^\nu\psi$ are used. Elimination of divergences, which vanish on the mass shell, can be considered as a shift of a quantised field (t'Hooft, 1975)

$$A_\mu \rightarrow A_\mu + \varepsilon Q_\mu$$

by divergent contribution of the background geometry. In two and higher loops this leads to new complications. Hence we conclude, that the model (3.1) cannot be considered as multiplicatively renormalizable in presence of non-minimal term $\alpha Q_\mu\bar{\psi}\gamma^\mu\psi$. Such a torsion coupling should be excluded in abelian gauge models.

In the absence of the non-minimal interaction torsion with spinor field $\alpha=0$ the one-loop divergences (3.8) can be eliminated by renormalization of fields and coupling constants:

$$e_B=\mu^{2-\frac{n}{2}}Z_e e,\ (A_\mu)_B=\mu^{\frac{n}{2}-2}Z_A^{1/2}A_\mu,\ Z_e Z_A^{1/2}=1, \tag{3.9a}$$

$$\psi_B = \mu^{\frac{n}{2}-2} Z_\psi \psi, \quad \bar\psi_B = \mu^{\frac{n}{2}-2} Z_\psi \bar\psi, \quad m_B = Z_m \cdot m, \tag{3.9b}$$

$$\varphi_B = \mu^{\frac{n}{2}-2} Z_\varphi^{1/2} \varphi, \quad M_B^2 = Z_M M^2, \quad f_B = \mu^{4-n} Z_f f, \tag{3.9c}$$

$$h_B = \mu^{2-\frac{n}{2}} Z_h h, \quad (\xi_i)_B = Z_{\xi_i} \xi_i, \ i=0,..,3, \quad \beta_B = Z_\beta \cdot \beta. \tag{3.9d}$$

Here the subscript B denotes bare quantities (renormalized ones are without additional marks), μ is an arbitrary mass scale parameter which makes the action dimensionless in dimension n $\neq$ 4. Formulas (3.9) give renormalization of the material gauge theory (3.I), as for renormalization of gravity action, it is trivial and we will not write out it explicitly.

From (3.8) one easily obtaines renormalization constants $\delta Z = Z - 1$. Denoting $\varepsilon = 16\pi^2(n-4)$ we get

$$\delta Z_e = -\frac{1}{\varepsilon}\frac{4}{3} e^2, \tag{3.IOa}$$

$$\delta Z_\varphi = \frac{1}{\varepsilon} 4 h^2, \tag{3.IOb}$$

$$f \delta Z_f = \frac{1}{\varepsilon}(-3f^2 - 8h^2 f + 48 h^2), \tag{3.IOc}$$

$$M^2 \delta Z_M = \frac{1}{\varepsilon}[24 m^2 h^2 - M^2(f + 4h^2)], \tag{3.IOd}$$

$$\xi_0 \delta Z_{\xi_0} = -\frac{1}{\varepsilon}(\xi_0 - \frac{1}{6})(f + 4h^2), \tag{3.IOe}$$

$$\delta Z_{\xi_1} = \delta Z_{\xi_3} = -\frac{1}{\varepsilon}(f + 4h^2), \tag{3.IOf}$$

$$\xi_2 \delta Z_{\xi_2} = \frac{1}{\varepsilon}[16\beta^2 h^2 - \xi_2(f + 4h^2)], \tag{3.IOg}$$

$$\delta Z_\psi = \frac{1}{\varepsilon}(2e^2 + h^2), \quad \delta Z_m = \frac{1}{\varepsilon}(6e^2 - 3h^2), \tag{3.IOh}$$

$$\delta Z_h = \frac{1}{\varepsilon}(6e^2 - 5h^2), \tag{3.IOi}$$

$$\delta Z_\beta = -\frac{1}{\varepsilon} 2h^2 \tag{3.IOk}$$

IV. GENERAL RENORMALIZABLE FIELD THEORY IN RIEMANN-CARTAN SPACE-TIME

We consider here the renormalization properties of the general non-Abelian gauge models , taking into account the non-minimal coupling terms (I.23) and (I.24). At first, we will obtain the general expression for evaluation of one-loop divergences and then we will regard two SU(2)-symmetrical models.

4.I Master formula for one-loop counterterms

We write the most general Lagrangian involving spin 0, I/2 and I fields in U_4 as

$$L_{gen.} = -\frac{1}{4} F^{a}_{\mu\nu} F_{a}^{\mu\nu} + \frac{1}{2} (\overset{0}{D}_{\mu}\varphi)_i (\overset{0}{D}{}^{\mu}\varphi)^i + V(\varphi, Q, R) +$$
$$+ i \bar{\psi}_{\ell} \gamma^{\mu} (\overset{1/2}{D}_{\mu}\psi)^{\ell} - \bar{\psi}_{\ell} M^{\ell}{}_{k} \psi^{k} - \bar{\psi}_{\ell} W^{\ell}_{k}(\varphi) \psi^{k} + \tag{4.I}$$
$$+ \alpha Q_{\mu} \bar{\psi}_{\ell} \gamma^{\mu} \psi^{\ell} + \beta^{\ell}{}_{k} \check{Q}_{\mu} \bar{\psi}_{\ell} \gamma^{\mu} \gamma_5 \psi^{k} ,$$

where

$$F^{a}_{\mu\nu} \equiv \partial_{\mu} A^{a}_{\nu} - \partial_{\nu} A^{a}_{\mu} + g f^{a}{}_{bc} A^{b}_{\mu} A^{c}_{\nu} , \tag{4.2a}$$

$$(\overset{0}{D}_{\mu}\varphi)^{i} \equiv \partial_{\mu} \varphi^{i} + g A^{a}_{\mu} (T_a)^{i}_{j} \varphi^{i} , \tag{4.2b}$$

$$(\overset{1/2}{D}_{\mu}\psi)^{\ell} \equiv \overset{1/2}{\nabla}_{\mu} \psi^{\ell} + g A^{a}_{\mu} (U_a)^{\ell}_{k} \psi^{k} . \tag{4.2c}$$

We take φ_i to be real scalar fields and ψ^{ℓ} Dirac spinors (Here the subscripts i,j,.... represent internal degrees of freedom and when no confusion arises they will be suppressed, e.g. $\varphi_i (T^a)^{ij} \varphi_j \to \varphi^{+} T^a \varphi$). The function $V(\varphi, Q, R)$ is polynomial of order four in φ_i and includes the discussed types of the non-minimal torsion and curvature couplings and $W(\varphi)$ is at most linear in φ_i . We propose that the generators of an internal symmetry group satisfy to the algebra

$$[T^{a}, T^{b}] = f^{ab}{}_{c} T^{c}, \quad [U^{a}, U^{b}] = f^{ab}{}_{c} U^{c} .$$

By our choice of scalar field variables the matrices T^{a} are real and antisymmetric. Also note the relation

$$(U_a)^{\ell}{}_{\ell'} (W_i)^{\ell'}_{k} - (W_i)^{\ell}{}_{\ell'} (U_a)^{\ell'}_{k} = (W_j)^{\ell}_{k} (T_a)^{j}{}_{i} ,$$

due to the gauge invariance of $\bar{\psi} W(\varphi) \psi$.

In the framework of the background field method all fields should be decomposed into a classical and quantum part

$$A^a_\mu = \hat{A}^a_\mu + a^a_\mu\,,\ \varphi_i = \hat{\varphi}_i + \phi_i\,,\ \psi_l = \hat{\psi}_l + \eta_l\,,\ \bar{\psi}_l = \bar{\psi}_l + \bar{\eta}_l\,. \quad (4.3)$$

For one-loop effective action, we need to know explicitly the terms bilinear in quantum variables from L_{gen}. After some manipulations the bilinear term from (4.I) can be written as

$$L_{bil} = \bar{\eta}'\Delta_F\eta' + \frac{1}{2}(\overset{1}{D}_\mu a^{a\mu} + g\varphi^+ T^a\varphi)^2 +$$

$$+ \frac{1}{2}[a^a{}_\alpha, \varphi^i]\left(\begin{array}{c|c} N_1 & N_2 \\ \hline N_3 & N_4 \end{array}\right)\left(\begin{array}{c} a^b{}_\nu \\ \varphi^j \end{array}\right) \quad (4.4)$$

where

$$N_1 = g^{\mu\nu}(\overset{1}{D}_\alpha \overset{1}{D}{}^\alpha)_{ab} - R^{\mu\nu}\delta_{ab} - 2gf^c{}_{ab}F_c{}^{\mu\nu} -$$

$$- g^2 g^{\mu\nu}(\varphi^+ T_a T_b\varphi) + 2g^2\bar{\psi}\gamma^\mu U_a\Delta_F^{-1}\gamma^\nu U_b\psi, \quad (4.5)$$

$$N_2 = 2g((\overset{0}{D}{}^\mu\varphi)^+ T_a)_j + 2ig\bar{\psi}\gamma^\mu U_a\Delta_F^{-1}W_j\psi, \quad (4.6)$$

$$N_3 = 2g((\overset{0}{D}{}^\nu\varphi)^+ T_b)_i + 2ig\bar{\psi}W_i\Delta_F^{-1}\gamma^\nu U_b\psi, \quad (4.7)$$

$$N_4 = -(\overset{0}{D}_\alpha\overset{0}{D}{}^\alpha)_{ij} - V_{ij} - g^2(\varphi^+ T_a)_i(\varphi^+ T^a)_j -$$

$$- 2\bar{\psi}W_i\Delta_F^{-1}W_j\psi. \quad (4.8)$$

Here we denote

$$(\Delta_F)^l{}_k = (i\gamma^\mu d_\mu - B)^l{}_k\,,\quad B^l{}_k = M^l{}_k + W^l{}_k(\varphi), \quad (4.9a)$$

$$(d_\mu)^l{}_k = \delta^l_k\overset{1/2}{\nabla}_\mu + gA^a_\mu(U_a)^l{}_k - i\alpha Q_\mu\delta^l_k - i\beta^l{}_k\gamma_5\check{Q}_\mu. \quad (4.9b)$$

We have used the fermionic shift in the form

$$\eta \to \eta' = \eta + ig\Delta_F^{-1}\gamma^\mu U^a\psi a^a_\mu + \Delta_F^{-1}W_j\psi\phi^j, \quad (4.10a)$$

$$\bar{\eta} \to \bar{\eta}' = \bar{\eta} - ig a^a_\mu\bar{\psi}\gamma^\mu U_a\Delta_F^{-1} + \phi^i\bar{\psi}W_i\Delta_F^{-1}. \quad (4.10b)$$

If we chose $V(\varphi)$ in the form

$$V(\varphi, Q, R) = \frac{1}{2}m^2\varphi_i\varphi^i + \Lambda(R, Q^2)\varphi_i\varphi^i + \frac{\lambda}{4}(\varphi_i\varphi^i)^2, \quad (4.11)$$

then

$$V_{ij} = \delta_{ij}(m^2 + \Lambda(R, Q^2)) + \lambda(\delta_{ij}\varphi^2 + 2\varphi_i\varphi_j). \qquad (4.12)$$

To secure a manifestly gauge-invariant effective action we may adopt the background gauge-fixing term

$$L_{gf} = -\frac{1}{2}\left(\overset{1}{\mathcal{D}}_\mu a_a{}^\mu + g\varphi^+ T_a \varphi\right)^2, \qquad (4.13)$$

and then the corresponding Faddeev-Popov ghost Lagrangian will be

$$L_{FP} = -\bar{C}^a\{(\overset{1}{\mathcal{D}}_\mu \overset{1}{\mathcal{D}}_\mu)_{ab} - g^2(\varphi^+ T_a T_b \varphi)\}C^b. \qquad (4.14)$$

We omit here the details of calculation and present only the final form of supermatrices $\Delta_{(S)}$, $\mathcal{D}_\mu$ and X to develop the algorithm, which has been discussed in the second section. They are as follows

$$\Delta_{(S)} = \begin{pmatrix} \begin{array}{c} g^{\mu\nu}(\overset{1}{\mathcal{D}}_\alpha \overset{1}{\mathcal{D}}{}^\alpha)_{ab} - R^{\mu\nu}\delta_{ab} \\ -2gf^c{}_{ab}F_c{}^{\mu\nu} - \\ -g^2 g^{\mu\nu}(\varphi^+ T_a T_b \varphi) \end{array} & 2g((\overset{0}{\mathcal{D}}{}^\mu\varphi)^+ T_a)_j & 2ig\bar\psi\gamma^\mu U_a \Delta_F^* \\ 2g((\overset{0}{\mathcal{D}}{}^\nu\varphi)^+ T_b)_i & \begin{array}{c} -(\overset{0}{\mathcal{D}}_\alpha \overset{0}{\mathcal{D}}{}^\alpha)_{ij} - \\ -V_{ij} - \\ -g^2(\varphi^+ T_a)_i(\varphi^+ T^a)_j \end{array} & -2\bar\psi W_i \Delta_F^* \\ -ig\gamma^\nu U_b \psi & -iW_i\psi & -(\overset{1/2}{\mathcal{D}}_\mu \overset{1/2}{\mathcal{D}}{}^\mu + Z) \end{pmatrix}$$

here

$$(\overset{1/2}{\mathcal{D}}_\mu)^\ell{}_k = \delta^\ell_k(\overset{1/2}{\nabla}_\mu - i\alpha Q_\mu) - i\beta^\ell{}_k\gamma_5\sigma_{\mu\nu}\check{Q}^\nu + gA^a_\mu(U_a)^\ell{}_k \qquad (4.16)$$

and

$$\begin{aligned}(Z)^\ell{}_k = {} & (B^2)^\ell_k - i\gamma^\mu(W_i)^\ell_k(\overset{0}{\mathcal{D}}_\mu\varphi)^i + \\ & + \check{Q}_\mu\gamma^\mu\gamma_5[W,\beta]^\ell{}_k - 2\check{Q}_\mu\check{Q}^\mu(\beta^2)^\ell{}_k - \\ & - i\gamma_5\nabla_\mu\check{Q}^\mu\beta^\ell_k - \frac{i}{2}\alpha\sigma^{\mu\nu}(\nabla_\mu Q_\nu - \nabla_\nu Q_\mu)\delta^\ell_k - \\ & - \frac{R}{4}\delta^\ell_k + \frac{g}{2}\sigma^{\mu\nu}(U_a)^\ell{}_k F^a{}_{\mu\nu}, \end{aligned} \qquad (4.17)$$

$$
\mathcal{D}_\mu = \begin{pmatrix} g^{\alpha\beta} \overset{1}{D}_\mu & 0 & -g\bar{\psi} U_a \gamma^\alpha \gamma_\mu \\ 0 & \overset{0}{D}_\mu & i\bar{\psi} W_i \gamma_\mu \\ 0 & 0 & \overset{1/2}{D}_\mu \end{pmatrix} \qquad (4.17)
$$

Here $\overset{1/2}{D}_\mu \bar{\psi} = \overset{1/2}{\nabla}_\mu \bar{\psi} - g A^a_\mu \bar{\psi} U_a$.

$$
X = \begin{pmatrix} \begin{matrix} -R^{\mu\nu}\delta_{ab} - \\ -2gf^c{}_{ab}F_c^{\mu\nu} - \\ -g^2 g^{\mu\nu}(\varphi^+ T_a T_b \varphi) \end{matrix} & 2g((\overset{0}{D}{}^\mu \varphi)^+ T_a)_j & \begin{matrix} g(\overset{1/2}{D}_\alpha \bar{\psi}) U_a \gamma^\mu \gamma^\alpha + \\ +2ig\bar{\psi}\gamma^\mu U_a B + \\ +i\alpha g Q_\lambda \bar{\psi} U_a \gamma^\mu \gamma^\lambda + \\ -ig\bar{\psi}\beta U_a \gamma^\mu \gamma^\lambda \gamma_5 \check{Q}_\lambda \end{matrix} \\ -2g((\overset{0}{D}{}^\nu \varphi)^+ T_b)_i & \begin{matrix} V_{ij} + \\ +g^2(\varphi^+ T_a)_i (\varphi^+ T_b)_j \end{matrix} & \begin{matrix} -i(\overset{1/2}{D}_\alpha \bar{\psi})\gamma^\alpha W_i + \\ +2\bar{\psi} W_i B + \\ +\alpha Q_\lambda \bar{\psi} W_i \gamma^\lambda - \\ -\bar{\psi} W_i \beta \gamma^\lambda \gamma_5 \check{Q}_\lambda \end{matrix} \\ -ig\gamma^\nu U_b \psi & W_j \psi & Z \end{pmatrix} \qquad (4.19)
$$

After calculation of traces of X, X^2 and $([\mathcal{D}_\mu, \mathcal{D}_\nu])^2$ both material and ghost operators and inserting the results into (2.12) we obtain the master formula for the one-loop contributions in the effective action

$$
\begin{aligned}
\Gamma_{(1)}^{\infty} = \frac{1}{\varepsilon}\int d^4x\sqrt{-g}\Big\{\Big[& R_{\alpha\beta\mu\nu}R^{\alpha\beta\mu\nu}\Big(-\frac{13n_A}{180} + \frac{n_S}{180} + \frac{7n_F}{360}\Big) + \\
& + R_{\mu\nu}R^{\mu\nu}\Big(\frac{22n_A}{45} - \frac{n_S}{180} + \frac{n_F}{45}\Big) + R^2\Big(-\frac{5n_A}{36} + \frac{n_S}{72} - \frac{n_F}{72}\Big) + \\
& + \frac{n_S}{2}(m^2+\Lambda)^2 + n_S\frac{R}{6}(m^2+\Lambda) - n_F\frac{2\alpha^2}{3}Q_{\mu\nu}Q^{\mu\nu} - \\
& - \frac{2}{3}tr(\beta^2)\check{Q}_{\mu\nu}\check{Q}^{\mu\nu} + \frac{1}{3}R\,tr(M^2) + 8\check{Q}_\alpha\check{Q}^\alpha tr(\beta^2 M^2) + \\
& + 2R\,tr(W\cdot M) + 16\check{Q}_\alpha\check{Q}^\alpha tr(\beta^2 W\cdot M)\Big] + \\
& + g^2F^a_{\mu\nu}F_b^{\mu\nu}\Big(\frac{11}{6}f_{acd}f^{bcd} + \frac{2}{3}tr(U_aU^b) + \frac{1}{12}(T_aT^b)\Big) +
\end{aligned}
$$

$$+4g^2(\overset{\circ}{D}_\mu\varphi^+T_aT^a\overset{\circ}{D}\varphi)+2\,tr(W_jW_i)(\overset{\circ}{D}_\mu\varphi)^j(\overset{\circ}{D}^\mu\varphi)^i+$$

$$+m^2\lambda(n_s+2)\varphi^2-m^2g^2(\varphi^+T_aT^a\varphi)-2\,tr(B^4)+$$

$$+\frac{\lambda^2}{2}(n_s+8)\varphi^4+\frac{3}{2}g^4(\varphi^+T_aT_b\varphi)^2-\lambda\varphi^2(\varphi^+T_aT^a\varphi)+$$

$$+\frac{1}{3}R\,tr(W^2)+4\check{Q}_\mu\check{Q}^\mu(tr(\beta^2W^2)+tr(\beta^2W^2))+$$

$$+\varphi^2\Lambda\cdot\lambda(n_s+2)+g^2\Lambda(\varphi^+T_aT^a\varphi)+\frac{1}{2}g^2R(\varphi^+T_aT^a\varphi)+$$

$$+\frac{1}{6}\lambda(n_s+2)R\varphi^2-2g^2i\bar{\psi}U_aU^a\gamma^\mu\overset{1/2}{D}_\mu\psi+i\bar{\psi}W_iW^i\gamma^\lambda\overset{1/2}{D}_\lambda\psi-$$

$$-2g^2\alpha Q_\mu\bar{\psi}U_aU^a\gamma^\mu\psi+\alpha Q_\mu\bar{\psi}W_iW^i\gamma^\mu\psi-2g^2\check{Q}_\nu\bar{\psi}\gamma^\nu\gamma_5U_aU^a\beta\psi-$$

$$-\check{Q}_\mu\bar{\psi}\gamma^\mu\gamma_5W_j\beta W^j\psi+8g^2\bar{\psi}U_aWU^a\psi+2\bar{\psi}W_iWW^i\psi+$$

$$+8g^2\bar{\psi}U_aU^aM\psi+2\bar{\psi}W_iMW^i\psi\Big\}, \qquad (4.20)$$

where

$$Q_{\mu\nu}=\partial_\mu Q_\nu-\partial_\nu Q_\mu,\quad \check{Q}_{\mu\nu}=\partial_\mu\check{Q}_\nu-\partial_\nu\check{Q}_\mu, \qquad (4.2I)$$

and

$$n_A=tr(f_{abc}f^{dbc}),\quad n_F=tr(U_aU^a),\quad n_s=tr(T_aT^a). \qquad (4.22)$$

The obtained master formula could be usefull to evaluate the one-loop divergences in concrete non-Abelian models.

4.2 Two examples of SU(2)-symmetrical models in U_4

Let us consider two particular models. The Lagrangian of the first model includes the scalar triplet $\{\varphi^i\}$, i = I,2,3 and the spinor triplets $\{\psi_I^l\}$, I = I, 2, ...P , interacting with the Yang-Mills fields. This Lagrangian take the form

$$L_{(1)}=-\frac{1}{4}F^a_{\mu\nu}F_a^{\mu\nu}+\frac{1}{2}(\overset{\circ}{D}_\mu\varphi)_i(\overset{\circ}{D}^\mu\varphi)^i+\frac{1}{2}m^2\varphi^2-\frac{1}{2}\Lambda(R,Q^2)\varphi^2$$
$$-\frac{\lambda}{4}(\varphi_i\varphi^i)^2+\sum_{I=1}^{P}\Big\{i\bar{\psi}^l_I\gamma^\mu\overset{1/2}{D}_\mu\psi^l_I-\bar{\psi}^l_I M^{IJ}_{lk}\psi^k_J+$$
$$+\alpha Q_\mu\bar{\psi}^l_I\gamma^\mu\psi^l_I+\beta^{IJ}_{lk}\check{Q}_\mu\bar{\psi}^l_I\gamma^\mu\gamma_5\psi^k_J-ih\epsilon_{lik}\bar{\psi}^l\varphi^i\psi^k. \qquad (4.23)$$

As the all spinor triplets are included in the Lagrangian

in the symmetrical manner, we may consider that

$$\beta_{lk}^{IJ} = \beta \delta^{IJ} \delta_{ek} , \quad M_{lk}^{IJ} = M \delta^{IJ} \delta_{ek}, \tag{4.24}$$

and

$$(Ta)_{ij} = \epsilon_{aij} , \quad (Ua)_{ek}^{IJ} = \epsilon_{alk} \delta^{IJ},$$
$$(Wa)_{ek}{}^{IJ} = ih \epsilon_{alk} \delta^{IJ}. \tag{4.25}$$

Using the master formula and takig into consideration the propositions (4.24) and (4.25) we obtain the effective action in one-loop aproximation for this model

$$\Gamma_{(1)}^{\infty} = \frac{1}{\varepsilon} \int d^4x \sqrt{-g} \Big\{ \text{grav. terms} - g^2 \Big(\frac{21-8P}{6} \Big) F_a^{\mu\nu} F_{\mu\nu}^a +$$
$$+ (8g^2 - 4Ph^2)(D_\mu^c \varphi)^2 + \varphi^2 (24Ph^2M^2 - 5\lambda m^2 - 2g^2 m^2) +$$
$$+ \varphi^4 \Big(4Ph^4 - \frac{11}{12}\lambda^2) 3g^4 - 2\lambda g^2\Big) + R\varphi^2 \Big(-\frac{2}{3}Ph^2 - 5\lambda\xi_0 - 2g^2\xi_0 +$$
$$+ g^2 - \frac{5\lambda}{6}\Big) - (5\lambda + 2g^2)(\xi_1 Q_\alpha Q^\alpha + \xi_3 \bar{Q}_{\alpha\mu\nu} \bar{Q}^{\alpha\mu\nu}) \varphi^2 -$$
$$- \check{Q}_\alpha \check{Q}^\alpha \varphi^2 (16P^2h^2\beta^2 + 5\lambda\xi_2 + 2g^2\xi_2) - i\bar{\psi}_I^l \gamma^\mu \overset{1/2}{D}_\mu \psi_I^l (4g^2 + 2h^2) -$$
$$- \alpha Q_\mu \bar{\psi}_I^l \gamma^\mu \psi_I^l (4g^2 + 2h^2) + \beta \check{Q}_\nu \psi_I^l \gamma^\nu \gamma_5 \psi_I^l (-4g^2 + 2h^2) +$$
$$+ (8g^2 - 2h^2) \bar{\psi}_I^l W_{lk}^{IJ} \psi_J^k + (16g^2 - 4h^2) M \bar{\psi}_I^l \psi_I^l \Big\},$$
$$\varphi^2 = \varphi_i \varphi^i , \quad \varphi^4 = (\varphi_i \varphi^i)^2. \tag{4.26}$$

The sum over I, J, ... is supposed.

One-loop divergences can be eliminated by renormalization of fields and coupling constants

$$g_B = \mu^{2-\frac{n}{2}} Z_g g, \quad (A_\mu^a)_B = \mu^{\frac{n}{2}-2} Z_A^{1/2} A_\mu^a , \quad Z_g Z_A^{1/2} = 1, \tag{4.27a}$$

$$\psi_B = \mu^{\frac{n}{2}-2} Z_\psi^{1/2} \psi, \quad \bar{\psi}_B = \mu^{\frac{n}{2}-2} Z_\psi^{1/2} \bar{\psi} , \quad M_B = Z_B M, \tag{4.27b}$$

$$\varphi_B^i = \mu^{\frac{n}{2}-2} \varphi^i Z_3^{1/2} , \quad m_B^2 = Z_m m^2 , \quad \lambda_B = \mu^{4-n} Z_\lambda \cdot \lambda, \tag{4.27c}$$

$$(\xi_i)_B = Z_{\xi_i} \xi_i , \quad i = 0, \ldots, 3, \quad h_B = \mu^{2-\frac{n}{2}} Z_h \cdot h, \tag{4.27d}$$

$$\alpha_B = Z_\alpha \cdot \alpha , \quad \beta_B = Z_\beta \cdot \beta , \tag{4.27e}$$

where $Z = 1 + \delta Z$ and

$$\delta Z_g = \frac{1}{\varepsilon}\left(7 - \frac{8P}{3}\right), \quad \delta Z_A = -2\delta Z_g, \tag{4.28a}$$

$$\delta Z_S = \frac{1}{\varepsilon}(-16g^2 + 8Ph^2), \tag{4.28b}$$

$$\delta Z_\psi = \frac{1}{\varepsilon}(4g^2 + 2h^2), \tag{4.28c}$$

$$m^2 \delta Z_m = \frac{1}{\varepsilon}(m^2\gamma + 48Ph^2M^2), \tag{4.28d}$$

$$\lambda \delta Z_\lambda = \frac{1}{\varepsilon}(2\lambda\gamma - 2\lambda^2 + 16Ph^4 - 12g^4) \tag{4.28e}$$

$$\delta Z_{\xi_1} = \delta Z_{\xi_3} = \frac{1}{\varepsilon}\gamma, \tag{4.28f}$$

$$\xi_0 \delta Z_{\xi_0} = \frac{1}{\varepsilon}\left(\xi_0 + \frac{1}{6}\right), \tag{4.28g}$$

$$\xi_2 \delta Z_{\xi_2} = \frac{1}{\varepsilon}(\xi_2\gamma - 32Ph^2\beta^2), \tag{4.28h}$$

$$\delta Z_M = \frac{1}{\varepsilon}(12g^2 - 16h^2), \tag{4.28i}$$

$$\delta Z_\alpha = 0, \quad \delta Z_\beta = -\frac{1}{\varepsilon}4h^2, \tag{4.28j}$$

$$\delta Z_h = \frac{1}{\varepsilon}(12g^2 - 4h^2(P+1)), \tag{4.28k}$$

$$\gamma \equiv 12g^2 - 8Ph^2 - 10\lambda. \tag{4.28l}$$

The Lagrangian of the second model is

$$L = -\frac{1}{4}F^a_{\mu\nu}F^{\mu\nu}_a + \frac{1}{2}|\overset{0}{D}{}^{AB}_\mu \varphi^B|^2 - \frac{1}{2}m^2|\varphi|^2 - \frac{\lambda}{4}(|\varphi|^2)^2 - \frac{1}{2}\Lambda(R,Q^2)|\varphi|^2 + \sum_{I=1}^{P}\Big\{ i\bar{\chi}^A_I \gamma^\mu \overset{1/2}{D}{}^{AB}_\mu \chi^B + i\bar{\eta}\gamma^\lambda \overset{1/2}{\nabla}_\lambda \eta - M_1\bar{\chi}^A_I\chi^A_I - M_2\bar{\eta}_I\eta_I - h(\bar{\chi}^A_I\eta_I\varphi^A + \varphi^{*A}\bar{\eta}_I\chi^A_I) + \alpha Q_\mu \bar{\chi}^A_I\gamma^\mu\chi^A_I + \alpha Q_\mu \bar{\eta}_I\gamma^\mu\eta_I + \beta_1 \check{Q}_\mu \bar{\chi}^A_I\gamma^\mu\gamma_5\chi^A_I + \beta_2\check{Q}_\mu\bar{\eta}_I\gamma^\mu\gamma_5\eta_I\Big\}. \tag{4.29}$$

Here

$$\overset{s}{D}{}^{AB}_\mu = \delta^{AB}\overset{s}{\nabla}_\mu - g\frac{i}{2}(\sigma_a)^{AB}A^a_\mu, \quad s = 0, 1/2. \tag{4.30}$$

The model includes the doublet of complex scalar fields $\{\varphi^A\}$ the set of spinor doublets $\{\chi^A_I\}$ and the set of spinor singlets $\{\eta_I\}$, A = I, 2 and I = I,2,....P. The representation of SU(2) is realized on the full set of spinors $\psi^l_I = \{\chi^1_I, \chi^2_I, \eta_I\}$.

It is natural to propose that the masses of fermions in the doublet and in the singlet are different, and also the constants of torsion coupling β_1 and β_2 are different. The last hypothesis call the universality of spin-torsion interaction in question, but we will see that this is true.

To apply the master formula to the model we replace the doublet of complex scalar fields on the quadruplet of real ones

$$\varphi_1 = Re\,\Phi^1,\quad \varphi_2 = Im\,\Phi^1,\quad \varphi_3 = Re\,\Phi^2,\quad \varphi_4 = Im\,\Phi^2. \tag{4.31}$$

Then the generators T will be the antisymmetrical (4 x 4) matrices and $(U_a)^{lk}_{I\gamma}$ and $(W)^{ke}_{I\gamma}$ will be the direct product of (3 x 3) matrices on the unit (P x P) matrix. The explicit form of them are

$$T_1 = \frac{1}{2}\begin{pmatrix} 0 & 0 & 0 & 1 \\ 0 & 0 & -1 & 0 \\ 0 & 1 & 0 & 0 \\ -1 & 0 & 0 & 0 \end{pmatrix},\ T_2 = \frac{1}{2}\begin{pmatrix} 0 & 0 & -1 & 0 \\ 0 & 0 & 0 & -1 \\ 1 & 0 & 0 & 0 \\ 0 & 1 & 0 & 0 \end{pmatrix},\ T_3 = \frac{1}{2}\begin{pmatrix} 0 & 1 & 0 & 0 \\ -1 & 0 & 0 & 0 \\ 0 & 0 & 0 & -1 \\ 0 & 0 & 1 & 0 \end{pmatrix}, \tag{4.32}$$

$$U_1 = -\frac{i}{2}\begin{pmatrix} 0 & 1 & 0 \\ -1 & 0 & 0 \\ 0 & 0 & 0 \end{pmatrix}\delta_{I\gamma},\ U_2 = -\frac{i}{2}\begin{pmatrix} 0 & -i & 0 \\ i & 0 & 0 \\ 0 & 0 & 0 \end{pmatrix}\delta_{I\gamma}\ \ U_3 = -\frac{i}{2}\begin{pmatrix} 1 & 0 & 0 \\ 0 & -1 & 0 \\ 0 & 0 & 0 \end{pmatrix}\delta_{I\gamma}, \tag{4.33}$$

$$W_1 = h\cdot\begin{pmatrix} 0 & 0 & 1 \\ 0 & 0 & 0 \\ 1 & 0 & 0 \end{pmatrix}\delta_{I\gamma},\quad W_2 = h\begin{pmatrix} 0 & 0 & i \\ 0 & 0 & 0 \\ -i & 0 & 0 \end{pmatrix}\delta_{I\gamma},$$

$$W_3 = h\begin{pmatrix} 0 & 0 & 0 \\ 0 & 0 & 1 \\ 0 & 1 & 0 \end{pmatrix}\delta_{I\gamma},\quad W_4 = h\cdot\begin{pmatrix} 0 & 0 & 0 \\ 0 & 0 & i \\ 0 & -i & 0 \end{pmatrix}\delta_{I\gamma} \tag{4.34}$$

The calculation of the all traces is not difficult problem. After some algebra we obtain

$$\Gamma^{\infty}_{(1)} = \frac{1}{\varepsilon}\int d^4x\sqrt{-g}\Big\{grav.\,terms - g^2\Big(\frac{43-4P}{12}\Big)F^a_{\mu\nu}F^{\mu\nu}_a +$$
$$+ (3g^2 - 4Ph^2)\,|D^{0AB}_{\mu}\Phi^B|^2 + |\Phi|^2\Big(12Ph^2(M_1^2+M_2^2) - 6m^2\lambda - \frac{3}{4}m^2g^2\Big) +$$
$$+ (|\Phi|^2)^2\Big(4Ph^4 - 6\lambda - \frac{9}{32}\lambda g^2\Big) + R\,|\Phi|^2\Big(-\frac{2}{3}Ph^2 - 6\lambda\xi_0 - \frac{3}{4}g^2\xi_0 +$$
$$+ \frac{3}{8}g^2\lambda\Big) - \Big(6\lambda + \frac{3}{4}g^2\Big)\Big(\xi_1 Q_\mu Q^\mu + \xi_3 \bar{Q}_{\alpha\mu\nu}\bar{Q}^{\alpha\mu\nu}\Big)|\Phi|^2 -$$

$$-\check{Q}_\alpha \check{Q}^\alpha |\Phi|^2 (4ph^2(\beta_1+\beta_2)^2 + 6\lambda\xi_2 + \frac{3}{4}g^2\xi_2) +$$

$$+ \sum_{I=1}^{p} [-i\bar{\chi}_I \gamma^\mu D_\mu^{1/2} \chi_I (\frac{3g^2}{2} + 2h^2) - 4h^2 i \bar{\eta}_I \gamma^\mu \nabla_\mu^{1/2} \eta_I -$$

$$-\alpha Q_\mu (\frac{3g^2}{2} + 2h^2) \bar{\chi}_I \gamma^\mu \chi_I + \alpha 4h^2 Q_\mu \bar{\eta}_I \gamma^\mu \eta_I +$$

$$+(2h^2\beta_2 - \frac{3}{2}\beta_1 g^2)\check{Q}_\mu \bar{\chi}_I \gamma^\mu \gamma_5 \chi_I + 4h^2\beta_1 \check{Q}_\mu \bar{\eta}_I \gamma^\mu \gamma_5 \eta_I +$$

$$+(6g^2M_1 - 4h^2M_2)\bar{\chi}_I \chi_I - 8h^2 M_1 \bar{\eta}_I \eta_I]\} . \qquad (4.35)$$

The renormalization of fields and coupling constants leads to the following expressions

$$\delta Z_g = \frac{1}{\varepsilon}\frac{2g^2}{3}(43-4p), \quad \delta Z_A = -2\delta Z_g, \qquad (4.36a)$$

$$\delta Z_\Phi = \frac{1}{\varepsilon}(-6g^2 + 8ph^2), \qquad (4.36b)$$

$$\delta Z_\chi = \frac{1}{\varepsilon}(\frac{3}{2}g^2 + 2h^2), \quad \delta Z_\eta = \frac{1}{\varepsilon}4h^2, \qquad (4.36c)$$

$$m^2 \delta Z_m = \frac{1}{\varepsilon}(m^2\gamma + 24ph^2M_1^2 + 24ph^2M_2^2), \qquad (4.36d)$$

$$M_1 \delta Z_{M_1} = \frac{1}{\varepsilon}[M_1(\frac{9}{2}g^2 - 2h^2) - 4h^2M_2], \qquad (4.36e)$$

$$M_2 \delta Z_{M_2} = \frac{1}{\varepsilon}(4h^2M_2 - 8h^2M_1), \qquad (4.36f)$$

$$\lambda \delta Z_\lambda = \frac{1}{\varepsilon}(2\lambda\gamma + 16ph^4 - \frac{9}{8}g^2), \qquad (4.36g)$$

$$\delta Z_h = \frac{1}{\varepsilon}(\frac{9}{2}g^2 - 3h^2 - 4h^2p), \qquad (4.36h)$$

$$\xi_0 \delta Z_{\xi_0} = \frac{1}{\varepsilon}(\xi_0 + \frac{1}{6}), \quad \delta Z_{\xi_1} = \delta Z_{\xi_3} = \frac{1}{\varepsilon}\gamma, \qquad (4.36i)$$

$$\xi_2 \delta Z_{\xi_2} = \frac{1}{\varepsilon}[\xi_2\gamma - 8ph^2(\beta_1+\beta_2)^2], \qquad (4.36j)$$

$$\beta_1 \delta Z_{\beta_1} = -\frac{1}{\varepsilon}2h^2(\beta_1+\beta_2), \qquad (4.36k)$$

$$\beta_2 \delta Z_{\beta_2} = -\frac{1}{\varepsilon}4h^2(\beta_1+\beta_2), \quad \delta Z_\alpha = 0, \qquad (4.36l)$$

$$\gamma = \frac{9}{2}g^2 - 8ph^2 - 12\lambda . \qquad (4.36m)$$

We can see that the constants β_1 and β_2 are renormalized in the different way. It is the confirm of our proposition on the non-universality of spin-torsion interaction.

Y. RENORMALIZATION OF NON-ABELIAN GAUGE FIELDS INTERACTING WITH TORSION

In the previous sections we have regarded the field models in that the vector fields interacted with gravity in the non-minimal manner and coupling with torsion was excluded into consideration. Here we will investigate the purely Yang-Mills fields interacting with torsion via the minimal copling recipe.

5.I Gauge invariance of non-Abelian fields

Let us consider the Yang-Mills theory in U_4. It will be convenient to use the following Lie algebra matrix valued connection $A_\mu = A_\mu^a T_a$, where $\{T_a\}$ are the set of generators of any Lie group G of internal symmetries.

In the flat spacetime the Yang-Mills action functional is invariant under the following gauge transformations

$$^{\omega}A_\mu = \omega A_\mu \omega^{-1} + \frac{1}{g}\,\omega \partial_\mu \omega^{-1}, \qquad (5.I)$$

here $\omega(x)$ belongs to G.

The field strength

$$F_{\mu\nu} = F^a_{\mu\nu} T_a = \partial_\mu A_\nu - \partial_\nu A_\mu + g[A_\mu \times A_\nu]$$

is transformed as

$$^{\omega}F_{\mu\nu} = \omega F_{\mu\nu} \omega^{-1}. \qquad (5.2)$$

To introduce the minimal interaction of gauge fields with gravity it is necessary to replace the ordinary derivatives in the strength tensor by covariant with respect to the connection form (I.2)

$$\partial_\mu A_\nu \to \tilde{\nabla}_\mu A_\nu = \partial_\mu A_\nu - \tilde{\Gamma}^\lambda{}_{\nu\mu} A_\lambda = \nabla_\mu A_\nu - K^\lambda{}_{\nu\mu} A_\lambda . \qquad (5.3)$$

This allows the following expression for strength tensor

$$\tilde{F}_{\mu\nu} = \nabla_\mu A_\nu - \nabla_\nu A_\mu + g[A_\mu \times A_\nu] + 2Q^\lambda{}_{\mu\nu} A_\lambda . \qquad (5.4)$$

It is easy to verify that gauge transformations (5.I) do not keep the invariance of action, because the strength tensor is transformed as

$$ {}^{\omega}\tilde{F}_{\mu\nu} = \omega \hat{F}_{\mu\nu} \omega^{-1} + \frac{1}{g} 2Q^{\lambda}{}_{\mu\nu} \omega \partial_{\lambda} \omega^{-1}, \tag{5.5} $$

since

$$ [\tilde{\nabla}_{\mu}, \hat{\nabla}_{\nu}] \omega^{-1} = 2Q^{\lambda}{}_{\mu\nu} \partial_{\lambda} \omega^{-1}. \tag{5.6} $$

To restore the gauge invariance of action

$$ S_{YM} = -\frac{1}{4} \int d^4x \sqrt{-g}\, \hat{F}^{a}_{\mu\nu} \hat{F}_{a}^{\mu\nu}, \tag{5.7} $$

admit that the constant of interaction is a function of coordinates. After replacing g by $g(x)$ in (5.I) and (5.4) and substituting this expression in (5.5) the following form of transformations of the strength tensor takes place

$$ {}^{\omega}\tilde{F}_{\mu\nu} = \omega \hat{F}_{\mu\nu} \omega^{-1} + $$

$$ + \frac{1}{g(x)} \omega [\delta^{\lambda}_{\mu} \nabla_{\nu} \ln g(x) - \delta^{\lambda}_{\nu} \nabla_{\mu} \ln g(x) + 2Q^{\lambda}_{\mu\nu}] \partial_{\lambda} \omega^{-1}. \tag{5.8} $$

If we require

$$ 2Q^{\lambda}{}_{\mu\nu} + \delta^{\lambda}_{\mu} \nabla_{\nu} \ln g(x) - \delta^{\lambda}_{\nu} \nabla_{\mu} \ln g(x) = 0, \tag{5.9} $$

then the transformed strength tensor will have the correct form

$$ {}^{\omega}\hat{F}_{\mu\nu} = \omega \hat{F}_{\mu\nu} \omega^{-1} $$

and the action (5.7) will be invariant with respect to these modified transformations.

Equation (5.9) imposes the linkage of torsion with the force of interaction through the influence on the effective constant $g(x)$. I think that the linkage ruled by (5.9) is deeper then it might seem from a first sight and its origin lies in the spin-isospin structures, which have been observed for SU(2) multiplet scalar fields (Jackiw, I977). The another physical treatment of such the linkage has been proposed in (Gasperini, I983; Nishioka, I983) via Lie-isotopic extension (Santilli, I978; Myung and Santilli, I982) of gauge theories. It was established the similarity of the Lie-isotopic extention of gauge theories and background geometry of gravitational theories.

5.2 One-loop divergences and renormalization

It is clear that in the background field method the gauge fields may be regarded as being split into the sum of a classical part A_μ and a quantum part a_μ, which gets integration over in the functional integration. The effective action is formed from one-particle irreducible vacuum averages. In order to obtain them, action (5.7) is expanded in powers of the quantum fields, and after some manipulations the bilinear terms can be written

$$S^{(bil)}_{YM} = -\frac{1}{2}\int d^4x\sqrt{-g}\Big\{a^c_\alpha\Big[-(D_\mu D^\mu)^{\alpha\beta}_{cb} - 4Q^{\alpha\beta}{}_\lambda(D^\lambda)_{cb} + \delta_{cb}(R^{\alpha\beta} + 2Q^\alpha{}_{\mu\nu}Q^{\beta\mu\nu}) + g(x)f^n{}_{cb}F_n{}^{\alpha\beta}\Big]a^b{}_\beta\Big\} + \frac{1}{2}\int d^4x\sqrt{-g}\,(D^\mu a^c_\mu)^2. \qquad (5.10)$$

To obtain (5.10) we took into account that

$$[D_\mu, D_\nu]^{\lambda\, a}_{\alpha\, b} = \delta^a_b R^\lambda{}_{\alpha\mu\nu} + g(x)f^a{}_{nb}\tilde{F}^n{}_{\mu\nu}\delta^\lambda_\alpha . \qquad (5.11)$$

The gauge for quantum fields is chosen in the form

$$D^\mu a^c{}_\mu = 0 \qquad (5.12)$$

With this choice of the gauge fixing condition the ghost part of the action will be

$$S_{FP} = -\int d^4x\sqrt{-g}\,\bar{C}_a\Delta^{ab}_{(FP)}C_b\,, \quad \Delta_{(FP)} = D_\mu D^\mu . \qquad (5.13)$$

Adding the gauge breaking term to $S^{(bil)}_{YM}$ we get

$$S^{(bil)}_{YM} + S_{g.f} = -\frac{1}{2}\int d^4x\sqrt{-g}\,a_c{}^\mu\,\Delta^{cb}_{\mu\nu}\,a_b{}^\nu, \qquad (5.14)$$

where operator Δ has the form of (2.10) and $\mathcal{D}_\mu$ and X are

$$(\mathcal{D}_\mu)^{a\alpha}_{b\beta} = \delta^a_b(\delta^\alpha_\beta\nabla_\mu + 2Q^\alpha{}_{\beta\mu}) + g(x)f^a{}_{nb}A^n_\mu\delta^\alpha_\beta, \qquad (5.15)$$

$$X^{a\alpha}_{b\beta} = \delta^a_b\mathcal{R}^\alpha_\beta + 2g(x)f^a{}_{nb}\tilde{F}^{n\alpha}{}_\beta, \qquad (5.16)$$

where $\mathcal{R}_{\alpha\beta} = R_{\alpha\beta} + 2\nabla_\nu Q_{\alpha\beta}{}^\nu + 4Q_{\alpha\mu\nu}Q^{\nu\mu}{}_\beta + 2Q_{\alpha\lambda\sigma}Q_\beta{}^{\lambda\sigma}$

Using the method of heat-kernel technique and dimensional regularization we can evaluate the infinite part of the effective action, which will be

$$\Gamma^{\infty}_{(1)} = \frac{1}{\varepsilon}\int d^4x\sqrt{-g}\Big\{ \frac{11}{6} g^2(x) n_A \tilde{F}^{a}_{\mu\nu}\tilde{F}^{\mu\nu}_{a} +$$
$$+ n_A \Big(-\frac{1}{3} R^{*}_{\alpha\beta\mu\nu} R^{*\alpha\beta\mu\nu} + \frac{1}{2}\mathcal{R}_{\alpha\beta}\mathcal{R}^{\alpha\beta} +$$
$$+ \frac{1}{3}\mathcal{R}R - \frac{1}{30} R_{\mu\nu}R^{\mu\nu} - \frac{7}{15}R^2 - \frac{1}{6}R_{\mu\nu\alpha\beta}R^{\mu\nu\alpha\beta}\Big)\Big\},$$
$$R^{*\alpha}{}_{\beta\mu\nu} = R^{\alpha}_{\beta\mu\nu}(\Gamma^{*}), \quad \Gamma^{*\lambda}{}_{\mu\nu} = \{^{\lambda}{}_{\mu\nu}\} + 2Q^{\lambda}{}_{\mu\nu}. \quad (5.17)$$

The detailes of calculation can be found in (Pronin, 1988).

We will deal with the problem of renormalization of a matter sector in the effective action, bearing in mind linkage (5.9). The ordinary procedure of renormalization consists in substitution

$$g_B = \mu^{2-\frac{n}{2}} Z_g g, \quad (A_\mu)_B = \mu^{\frac{n}{2}-2} Z_A^{1/2} A_\mu, \quad Z_g Z_A^{1/2} = 1, \quad (5.18)$$

where the index B marks as usually the bare values and due to (5.17)

$$Z_A = 1 + \delta Z_A, \quad Z_g = 1 + \delta Z_g,$$
$$\delta Z_A = -2\delta Z_g \quad (5.19)$$

where

$$\delta Z_g = \frac{1}{\varepsilon}\frac{11}{3} g^2(x) n_A \quad (5.20)$$

The naive use of these relations creates an extraordinary situation. Really $\delta Z_A(g)$ is the function of coordinates and

$$\nabla_\mu (A_\nu)_B \neq \mu^{\frac{n}{2}-2} Z_A^{1/2} \nabla_\mu A_\nu.$$

We have to use the following expression in

$$\nabla_\mu (A_\nu)_B = \mu^{\frac{n}{2}-2}\Big\{ Z_A^{1/2}\nabla_\mu A_\nu + A_\nu \nabla_\mu Z_A^{1/2}\Big\}. \quad (5.21)$$

It seems that the last term in (5.21) breaks the renormalizability as it was indeed belived by Trautman (1973) and Hehl (1974), because

$$(\tilde{F}_{\mu\nu})_B = \mu^{\frac{n}{2}-2} Z_A^{1/2} \{ \tilde{F}_{\mu\nu} +$$

$$+ A_\lambda (\delta^\lambda{}_\nu \nabla_\mu \ln Z_A^{1/2} - \delta^\lambda_\mu \nabla_\nu \ln Z_A^{1/2}),$$

$$\nabla_\mu \ln Z_A^{1/2} = \frac{1}{2} \frac{\nabla_\mu \delta Z_A}{1 + \delta Z_A}, \qquad (5.22)$$

has additional terms which are not contained in the original strength tensor. To overcame this difficulty it is necessary to use the link equation.

Since equation (5.9) has been introduce to keep the gauge invariance, it must hold for renormalized values as well as for bare ones. This implies that the background torsion due to renormalization of gauge fields and coupling constant should be shifted to

$$Q^\lambda{}_{\mu\nu} \to \overset{*}{Q}{}^\lambda{}_{\mu\nu} +$$

$$+ \delta^\lambda_\mu \nabla_\nu \ln Z_A^{1/2} - \delta^\lambda{}_\nu \nabla_\mu \ln Z_A^{1/2}, \qquad (5.23)$$

Inserting (5.23) in (5.22) we can get the correct expression

$$(\tilde{F}_{\mu\nu})_B = \mu^{\frac{n}{2}-2} Z_A^{1/2} \{ \nabla_\mu A_\nu - \nabla_\nu A_\mu + g(x)[A_\mu \times A_\nu] + + 2 \overset{*}{Q}{}^\lambda{}_{\mu\nu} A_\lambda \} \qquad (5.24)$$

This procedure may seem to be very artificial and incompatible with the usual ideology of quantum field theory in an arbitrary background geometry. But let us consider the self-consistent case when the torsion is defined only by induced structure by (5.9) then

$$\tilde{F}_{\mu\nu} = \nabla_\mu A_\nu - \nabla_\nu A_\mu + g(x)[A_\mu \times A_\nu] +$$

$$+ A_\lambda (\delta^\lambda_\mu \nabla_\nu \ln g(x) - \delta^\lambda{}_\nu \nabla_\mu \ln g(x)). \qquad (5.25)$$

The renormalization of this theory can be formed in the usual way. The unnecessary terms will be reduced and we get the true formula to renormalized the strength tensor without any additional propositions. It is then the remarkable fact that the theory with induced geometrical structure is the renormalizable theory.

Unfortunately the induced torsion does not cover all possible configurations in the semiconsistent approach. We have to assume that, in general, the torsion field should be split into background part which is not affected by quantum corrections and the quantum (or induced) part is renormalizable along with matter fields. Then the physical meaning of (5.23) becomes clear. This is the renormalization of quantum torsion.

YI. SUMMARY AND DISCUSSION

The investigation of the structure of divergences of the material fields effective action is an important aspect within the problem of the gravity covariant quantization. Great attention was payed to this aspect in Einstein's theory of gravity. Here we have considered the renormalization of the various field models in the Riemann-Cartan manifold. The minimal and non-minimal torsion couplings with matter were examined.

It was found that for some models the proposition of universality of the interaction of quantum fields with classical background torsion caused the many difficulties in the renormalization procedure.

We took into account that the Poincaré group (or the inhomogeneous Lorentz group) is the semidirect product of the translation group and the Lorentz group. So, it contains two invariant subgroups. In general it is possible to introduce in such the theory two independent coupling constants of interaction of gauge gravitational field with material ones. In our models these were the Einstein's constant and the constant of interaction of matter with torsion.

We can see that such the generalization of gravitational interaction structure was true but unfortunately it was not sufficient to solve the problem. In opposite to Einsteins theory the coupling interaction constants are depended on the particular isotopic structure of particles multiplets. This fact can be considered as an evident indication on the existence of different types of spin-torsion gravitational interactions. Besides, it not only returns us to the old question on the classical torsion nature, but leads to the need of a new revision of gauge treatment of gravity.

REFERENCES

Aragone C. and Deser S., I980, Nuovo Cim. B57:33.

Barth N.H., I987, J.Phys: Math. and Gen. A20:857.

Baulieu L., I986, Phys.Lett. BI75:I33.

Benn I.M., Dereli T. and Tucker P.M., I980, Phys.Lett. B96:I00.

Birrell N.D., I980, J.Phys: Math. and Gen. AI3:596.

Birrell N.D. and Davis P., I982, Quantum fields in curved space, Cambridge, University press.

Buchbinder I.L., Odintsov S.D. and Shapiro I.L., I985, Preprint of Tomsk Pedagog. Inst. N I0.

Bukhdal H.A., I962, Nuovo Cim. 25:486.

Bukhdal H.A., I982, J.Phys: Math. and Gen. AI5:I.

Bunch T.C., I98I, Gen. Rel. Grav. I3:7II.

Bunch T.S. and Parker L., I979, Phys.Rev. D20:2499.

Denardo G. and Spallucci E., I987, Class. Quan. Grav. 4:89.

De Sabbata V. and Gasperini M., I980, Phys.Lett. A77:300.

De Sabbata V. and Gasperini M., I98I, Lett.Nuovo Cim. 30:I93;

DeWitt B.S., I965, Dynamical theory groups and fields, Gordon and Breach, New York.

Drummond I.T. and Shore C.M., Ann.Phys. (NY), I979, 89:II7.

Freedan D.Z., Muzinich I.J. and Weinberg E., I974, Ann;Phys. (NY) 87:959.

Gasperini M., I983, Hadronic J. 6:I462.

Gass R., I986, Ann.Phys.(NY) I7I:I32.

Gilkey P.B., I975, J.Diff.Geom. I0:60I.

Goldthorpe W.H., I980, Nucl. Phys. B2I2:I237.

Gvozdev A.A. and Pronin P.I., I985, Izvestia Vusov (Fisika) N I:7.

Grib A.A., Mamaev S.G. and Mostepanenko V.M., I980, Quantum effects in strong external fields, Energoatomizdat, Moscow.

Hehl F.W., I974, Gen. Rel. Grav. 5:49I.

Hehl F.W., von der Heyde P, Kerlick G.D; and Nester J.M., I976, Rev. Mod. Phys. 48:393.

Hojman S., Rosenbaum M, Ryan M, and Shepley L., I978, Phys. Rev. DI7:3I4I.

Ivanenko D.D., Pronin P.I. and Sardanashvily G.A., I985, Gauge theory of gravity, University Publ. House, Moscow.

Ivanenko D.D., Budylin S.L. and Pronin P.I., I988, Ann. Phys. (Leipzig) B45:I9I.

Jack I., I984, Nucl. Phys. B234:365.
Jack I. and Osborn H., I985, Nucl. Phys. B249:472.
Jackiw R., I977, Rev. Mod. Phys. 49:68I.
Kibble T.W.B., I96I, J. Math. Phys. 2:2I2.
Kimura T., I98I, J.Phys; Math. and Gen. AI4:L329.
Kunfuss R. and Nitsch J., I986, Gen. Rel. Grav. I8:I207.
Lee C. and Rim C., I985, Nucl. Phys. B255:439.
Leen T.K., I983, Ann. Phys.(NY) I47:4I7.
Mukky C. and Sayed W., I979, Phys. Lett. B82:383.
Myung H.C. and Santilli R.M., I982, Hadronic J. 5:I272.
Nelson B.L. and Panangaden P., I982, Phys; Rev. D25:I0I9.
Neville D.E., I98I, Phys. Rev. D23:I244.
Nieh H.T. and Yan M.L., I982, Ann. Phys.(NY) I38:237.
Nishioka M., I983, Hadronic J. 6:I480.
Novello M., I976, Phys. Lett. A59:I05.
Obukhov Yu.N., I983a, Nucl. Phys. B2I2:237.
Obukhov YU.N., I983b, J.Phys; Math. and Gen. AI6:3795.
Obukhov Yu.N; and Nazarovsky E.A., I984, In: Relativistic astrophysics and cosmology - Proc. of the Sir A.Eddington Centen. Symp., Eds. De Sabbata V. and Karade T.M., World Scientific, Singapore, vol.I:I37.
Obukhov Yu.N. and Pronin P.I., I986, In: Problems of Gravity, Ed. Galtzov D.V., University, Publ. House, Moscow, I30.
Obukhov Yu.N; and Pronin P.I., I988, Acta Phys. Pol. BI9:34I.
Omote M. and Ishinose S., I983, Phys. Rev. D27:234I.
Panangaden P., I98I, Phys. Rev. D23:I735.
Ponomariev V.N. and Pronin P.I., I979, Teor. Mat. Fis. 39:425.
Ponomariev V.N. and Smetanin E.V., I978, Vestnik Mosk. Univ. N 5 :29.
Ponomariev V.N., Barvinsky A.O. and Obukhov Yu.N., I985, Geometrodynamical method and gauge approach in the theory of gravity, Energoatomizdat, Moscow.
Rumpf H., I979, Gen. Rel. Grav. I0:525.
Santilli R.M., I978, Hadronic J. I:228.
Sezgin E. and van Nieuwenhuizen P., I980, Phys. Rev. D2I: 3269.
Sezgin E., I98I, Phys. Rev. D24:I6I7.
Sciama D.W., I962, In: Recent developments in general relativity, Oxford, Pergamon, 4I5.

t'Hooft G., 1975, Lect. Notes Phys. 37:92.
Toms D.J., 1982, Phys. Rev. D26:2713.
Toms D.J., 1983a, Phys. Rev. D27:1803.
Toms D.J., 1983b, Phys. Lett. B126:37.
Trautman A., 1973, Symp. Math. 12:139.
Tseytlin A.A., 1982, Phys. Rev. D26:3327.
Yan M.L., 1983, Comm. Math. Phys. (China) 2:1281.

INDEX